T0140133

Lecture Notes in Networks and Systems   661

The series "Lecture Notes in Networks and Systems" publishes the latest developments in Networks and Systems—quickly, informally and with high quality. Original research reported in proceedings and post-proceedings represents the core of LNNS.

Volumes published in LNNS embrace all aspects and subfields of, as well as new challenges in, Networks and Systems.

The series contains proceedings and edited volumes in systems and networks, spanning the areas of Cyber-Physical Systems, Autonomous Systems, Sensor Networks, Control Systems, Energy Systems, Automotive Systems, Biological Systems, Vehicular Networking and Connected Vehicles, Aerospace Systems, Automation, Manufacturing, Smart Grids, Nonlinear Systems, Power Systems, Robotics, Social Systems, Economic Systems and other. Of particular value to both the contributors and the readership are the short publication timeframe and the world-wide distribution and exposure which enable both a wide and rapid dissemination of research output.

The series covers the theory, applications, and perspectives on the state of the art and future developments relevant to systems and networks, decision making, control, complex processes and related areas, as embedded in the fields of interdisciplinary and applied sciences, engineering, computer science, physics, economics, social, and life sciences, as well as the paradigms and methodologies behind them.

Indexed by SCOPUS, INSPEC, WTI Frankfurt eG, zbMATH, SCImago.

All books published in the series are submitted for consideration in Web of Science.

For proposals from Asia please contact Aninda Bose (aninda.bose@springer.com).

Leonard Barolli
Editor

Advanced Information Networking and Applications

Proceedings of the 37th International Conference on Advanced Information Networking and Applications (AINA-2023), Volume 1

 Springer

Editor
Leonard Barolli
Department of Information and Communication
Engineering, Faculty of Information Engineering
Fukuoka Institute of Technology
Fukuoka, Japan

ISSN 2367-3370 ISSN 2367-3389 (electronic)
Lecture Notes in Networks and Systems
ISBN 978-3-031-29055-8 ISBN 978-3-031-29056-5 (eBook)
https://doi.org/10.1007/978-3-031-29056-5

This Springer imprint is published by the registered company Springer Nature Switzerland AG
The registered company address is: Gewerbestrasse 11, 6330 Cham, Switzerland

Welcome Message from AINA-2023 Organizers

Welcome to the 37th International Conference on Advanced Information Networking and Applications (AINA-2023). On behalf of AINA-2023 Organizing Committee, we would like to express to all participants our cordial welcome and high respect.

AINA is an International Forum, where scientists and researchers from academia and industry working in various scientific and technical areas of networking and distributed computing systems can demonstrate new ideas and solutions in distributed computing systems. AINA was born in Asia, but it is now an international conference with high quality thanks to the great help and cooperation of many international-friendly volunteers. AINA is a very open society and is always welcoming international volunteers from any country and any area in the world.

AINA International Conference is a forum for sharing ideas and research work in the emerging areas of information networking and their applications. The area of advanced networking has grown very rapidly, and the applications have experienced an explosive growth especially in the areas of pervasive and mobile applications, wireless sensor and ad hoc networks, vehicular networks, multimedia computing, social networking, semantic collaborative systems, as well as IoT, Big Data and Cloud Computing. This advanced networking revolution is transforming the way people live, work, and interact with each other and is impacting the way business, education, entertainment, and health care are operating. The papers included in the proceedings cover theory, design, and application of computer networks, distributed computing, and information systems.

Each year AINA receives a lot of paper submissions from all around the world. It has maintained high-quality accepted papers and is aspiring to be one of the main international conferences on the information networking in the world.

We are very proud and honored to have two distinguished keynote talks by Dr. Leonardo Mostarda, Camerino University, Italy, and Prof. Flávio de Oliveira Silva, Federal University of Uberlândia, Brazil, who will present their recent work and will give new insights and ideas to the conference participants.

An international conference of this size requires the support and help of many people. A lot of people have helped and worked hard to produce a successful AINA-2023 technical program and conference proceedings. First, we would like to thank all the authors for submitting their papers, the session chairs, and the distinguished keynote speakers. We are indebted to Program Track Co-Chairs, Program Committee Members, and Reviewers, who carried out the most difficult work of carefully evaluating the submitted papers.

We would like to thank AINA-2023 General Co-Chairs, PC Co-Chairs, and Workshops Co-chairs for their great efforts to make AINA-2023 a very successful event. We have special thanks to Finance Chair and Web Administrator Co-Chairs.

We do hope that you will enjoy the conference proceedings and readings.

Organization

AINA-2023 Organizing Committee

Honorary Chair

Makoto Takizawa   Hosei University, Japan

General Co-chairs

Mario A. R. Dantas   Federal University of Juiz de Fora, Brazil
Tomoya Enokido   Rissho University, Japan
Isaac Woungang   Toronto Metropolitan University, Canada

Program Committee Co-chairs

Victor Ströele   Federal University of Juiz de Fora, Brazil
Flora Amato   University of Naples "Federico II", Italy
Marek Ogiela   AGH University of Science and Technology,
                                Poland

International Journals Special Issues Co-chairs

Fatos Xhafa   Technical University of Catalonia, Spain
David Taniar   Monash University, Australia
Farookh Hussain   University of Technology Sydney, Australia

Award Co-chairs

Arjan Durresi   Indiana University Purdue University in
                                Indianapolis (IUPUI), USA
Fang-Yie Leu   Tunghai University, Taiwan
Kin Fun Li   University of Victoria, Canada

Publicity Co-chairs

Markus Aleksy   ABB Corporate Research Center, Germany
Omar Hussain   University of New South Wales, Australia

| Lidia Ogiela | AGH University of Science and Technology, Poland |
| Hsing-Chung Chen | Asia University, Taiwan |

International Liaison Co-chairs

Nadeem Javaid	COMSATS University Islamabad, Pakistan
Wenny Rahayu	La Trobe University, Australia
Beniamino Di Martino	University of Campania "Luigi Vanvitelli", Italy

Local Arrangement Co-chairs

| Regina Vilela | Federal University of Juiz de Fora, Brazil |
| José Maria N. David | Federal University of Juiz de Fora, Brazil |

Finance Chair

| Makoto Ikeda | Fukuoka Institute of Technology, Japan |

Web Co-chairs

Phudit Ampririt	Fukuoka Institute of Technology, Japan
Kevin Bylykbashi	Fukuoka Institute of Technology, Japan
Ermioni Qafzezi	Fukuoka Institute of Technology, Japan

Steering Committee Chair

| Leonard Barolli | Fukuoka Institute of Technology, Japan |

Tracks and Program Committee Members

1. Network Protocols and Applications

Track Co-chairs

Makoto Ikeda	Fukuoka Institute of Technology, Japan
Sanjay Kumar Dhurandher	Netaji Subhas University of Technology, New Delhi, India
Bhed Bahadur Bista	Iwate Prefectural University, Japan

TPC Members

Admir Barolli	Aleksander Moisiu University of Durres, Albania
Elis Kulla	Fukuoka Institute of Technology, Japan
Keita Matsuo	Fukuoka Institute of Technology, Japan
Shinji Sakamoto	Kanazawa Institute of Technology, Japan
Akio Koyama	Yamagata University, Japan
Evjola Spaho	Polytechnic University of Tirana, Albania
Jiahong Wang	Iwate Prefectural University, Japan
Shigetomo Kimura	University of Tsukuba, Japan
Chotipat Pornavalai	King Mongkut's Institute of Technology Ladkrabang, Thailand
Danda B. Rawat	Howard University, USA
Amita Malik	Deenbandhu Chhotu Ram University of Science and Technology, India
R. K. Pateriya	Maulana Azad National Institute of Technology, India
Vinesh Kumar	University of Delhi, India
Petros Nicopolitidis	Aristotle University of Thessaloniki, Greece
Satya Jyoti Borah	North Eastern Regional Institute of Science and Technology, India

2. Next-Generation Wireless Networks

Track Co-chairs

Christos J. Bouras	University of Patras, Greece
Tales Heimfarth	Universidade Federal de Lavras, Brazil
Leonardo Mostarda	University of Camerino, Italy

TPC Members

Fadi Al-Turjman	Near East University, Cyprus
Alfredo Navarra	University of Perugia, Italy
Purav Shah	Middlesex University London, UK
Enver Ever	Middle East Technical University, Northern Cyprus Campus, Cyprus
Rosario Culmone	University of Camerino, Italy
Antonio Alfredo F. Loureiro	Federal University of Minas Gerais, Brazil
Holger Karl	University of Paderborn, Germany
Daniel Ludovico Guidoni	Federal University of São João Del-Rei, Brazil
João Paulo Carvalho Lustosa da Costa	Hamm-Lippstadt University of Applied Sciences, Germany
Jorge Sá Silva	University of Coimbra, Portugal

Apostolos Gkamas	University Ecclesiastical Academy of Vella, Greece
Zoubir Mammeri	University Paul Sabatier, France
Eirini Eleni Tsiropoulou	University of New Mexico, USA
Raouf Hamzaoui	De Montfort University, UK
Miroslav Voznak	University of Ostrava, Czech Republic
Kevin Bylykbashi	Fukuoka Institute of Technology, Japan

3. Multimedia Systems and Applications

Track Co-chairs

Markus Aleksy	ABB Corporate Research Center, Germany
Francesco Orciuoli	University of Salerno, Italy
Tomoyuki Ishida	Fukuoka Institute of Technology, Japan

TPC Members

Tetsuro Ogi	Keio University, Japan
Yasuo Ebara	Osaka Electro-Communication University, Japan
Hideo Miyachi	Tokyo City University, Japan
Kaoru Sugita	Fukuoka Institute of Technology, Japan
Akio Doi	Iwate Prefectural University, Japan
Hadil Abukwaik	ABB Corporate Research Center, Germany
Monique Duengen	Robert Bosch GmbH, Germany
Thomas Preuss	Brandenburg University of Applied Sciences, Germany
Peter M. Rost	NOKIA Bell Labs, Germany
Lukasz Wisniewski	inIT, Germany
Angelo Gaeta	University of Salerno, Italy
Graziano Fuccio	University of Salerno, Italy
Giuseppe Fenza	University of Salerno, Italy
Maria Cristina	University of Salerno, Italy
Alberto Volpe	University of Salerno, Italy

4. Pervasive and Ubiquitous Computing

Track Co-chairs

Chih-Lin Hu	National Central University, Taiwan
Vamsi Paruchuri	University of Central Arkansas, USA
Winston Seah	Victoria University of Wellington, New Zealand

TPC Members

Hong Va Leong	Hong Kong Polytechnic University, Hong Kong
Ling-Jyh Chen	Academia Sinica, Taiwan
Jiun-Yu Tu	Southern Taiwan University of Science and Technology, Taiwan
Jiun-Long Huang	National Chiao Tung University, Taiwan
Thitinan Tantidham	Mahidol University, Thailand
Tanapat Anusas-amornkul	King Mongkut's University of Technology North Bangkok, Thailand
Xin-Mao Huang	Aletheia University, Taiwan
Hui Lin	Tamkang University, Taiwan
Eugen Dedu	Universite de Franche-Comte, France
Peng Huang	Sichuan Agricultural University, China
Wuyungerile Li	Inner Mongolia University, China
Adrian Pekar	Budapest University of Technology and Economics, Hungary
Jyoti Sahni	Victoria University of Technology, New Zealand
Normalia Samian	Universiti Putra Malaysia, Malaysia
Sriram Chellappan	University of South Florida, USA
Yu Sun	University of Central Arkansas, USA
Qiang Duan	Penn State University, USA
Han-Chieh Wei	Dallas Baptist University, USA

5. Web-Based and E-learning Systems

Track Co-chairs

Santi Caballe	Open University of Catalonia, Spain
Kin Fun Li	University of Victoria, Canada
Nobuo Funabiki	Okayama University, Japan

TPC Members

Jordi Conesa	Open University of Catalonia, Spain
Joan Casas	Open University of Catalonia, Spain
David Gañán	Open University of Catalonia, Spain
Nicola Capuano	University of Basilicata, Italy
Antonio Sarasa	Complutense University of Madrid, Spain
Chih-Peng Fan	National Chung Hsing University, Taiwan
Nobuya Ishihara	Okayama University, Japan
Sho Yamamoto	Kindai University, Japan
Khin Khin Zaw	Yangon Technical University, Myanmar
Kaoru Fujioka	Fukuoka Women's University, Japan

Kosuke Takano	Kanagawa Institute of Technology, Japan
Shengrui Wang	University of Sherbrooke, Canada
Darshika Perera	University of Colorado at Colorado Spring, USA
Carson Leung	University of Manitoba, Canada

6. Distributed and Parallel Computing

Track Co-chairs

Naohiro Hayashibara	Kyoto Sangyo University, Japan
Minoru Uehara	Toyo University, Japan
Tomoya Enokido	Rissho University, Japan

TPC Members

Eric Pardede	La Trobe University, Australia
Lidia Ogiela	AGH University of Science and Technology, Poland
Evjola Spaho	Polytechnic University of Tirana, Albania
Akio Koyama	Yamagata University, Japan
Omar Hussain	University of New South Wales, Australia
Hideharu Amano	Keio University, Japan
Ryuji Shioya	Toyo University, Japan
Ji Zhang	The University of Southern Queensland, Australia
Lucian Prodan	Universitatea Politehnica Timisoara, Romania
Ragib Hasan	The University of Alabama at Birmingham, USA
Young-Hoon Park	Sookmyung Women's University, South Korea
Dilawaer Duolikun	Cognizant Technology Solutions, Hungary
Shigenari Nakamura	Tokyo Metropolitan Industrial Technology Research Institute, Japan

7. Data Mining, Big Data Analytics, and Social Networks

Track Co-chairs

Omid Ameri Sianaki	Victoria University, Australia
Alex Thomo	University of Victoria, Canada
Flora Amato	University of Naples "Frederico II", Italy

TPC Members

Eric Pardede	La Trobe University, Australia
Alireza Amrollahi	Macquarie University, Australia
Javad Rezazadeh	University of Technology Sydney, Australia

Farshid Hajati	Victoria University, Australia
Mehregan Mahdavi	Sydney International School of Technology and Commerce, Australia
Ji Zhang	University of Southern Queensland, Australia
Salimur Choudhury	Lakehead University, Canada
Xiaofeng Ding	Huazhong University of Science and Technology, China
Ronaldo dos Santos Mello	Universidade Federal de Santa Catarina, Brazil
Irena Holubova	Charles University, Czech Republic
Lucian Prodan	Universitatea Politehnica Timisoara, Romania
Alex Tomy	La Trobe University, Australia
Dhomas Hatta Fudholi	Universitas Islam Indonesia, Indonesia
Saqib Ali	Sultan Qaboos University, Oman
Ahmad Alqarni	Al Baha University, Saudi Arabia
Alessandra Amato	University of Naples "Frederico II", Italy
Luigi Coppolino	Parthenope University, Italy
Giovanni Cozzolino	University of Naples "Frederico II", Italy
Giovanni Mazzeo	Parthenope University, Italy
Francesco Mercaldo	Italian National Research Council, Italy
Francesco Moscato	University of Salerno, Italy
Vincenzo Moscato	University of Naples "Frederico II", Italy
Francesco Piccialli	University of Naples "Frederico II", Italy

8. Internet of Things and Cyber-Physical Systems

Track Co-chairs

Euripides G. M. Petrakis	Technical University of Crete (TUC), Greece
Tomoki Yoshihisa	Osaka University, Japan
Mario Dantas	Federal University of Juiz de Fora (UFJF), Brazil

TPC Members

Akihiro Fujimoto	Wakayama University, Japan
Akimitsu Kanzaki	Shimane University, Japan
Kawakami Tomoya	University of Fukui, Japan
Lei Shu	University of Lincoln, UK
Naoyuki Morimoto	Mie University, Japan
Yusuke Gotoh	Okayama University, Japan
Vasilis Samolada	Technical University of Crete (TUC), Greece
Konstantinos Tsakos	Technical University of Crete (TUC), Greece
Aimilios Tzavaras	Technical University of Crete (TUC), Greece

Spanakis Manolis	Foundation for Research and Technology Hellas (FORTH), Greece
Katerina Doka	National Technical University of Athens (NTUA), Greece
Giorgos Vasiliadis	Foundation for Research and Technology Hellas (FORTH), Greece
Stefan Covaci	Technicak University of Berlin (TUB), Germany
Stelios Sotiriadis	University of London, UK
Stefano Chessa	University of Pisa, Italy
Jean-Francois Méhaut	Université Grenoble Alpes, France
Michael Bauer	University of Western Ontario, Canada

9. Intelligent Computing and Machine Learning

Track Co-chairs

Takahiro Uchiya	Nagoya Institute of Technology, Japan
Omar Hussain	UNSW, Australia
Nadeem Javaid	COMSATS University Islamabad, Pakistan

TPC Members

Morteza Saberi	University of Technology Sydney, Australia
Abderrahmane Leshob	University of Quebec in Montreal, Canada
Adil Hammadi	Curtin University, Australia
Naeem Janjua	Edith Cowan University, Australia
Sazia Parvin	Melbourne Polytechnic, Australia
Kazuto Sasai	Ibaraki University, Japan
Shigeru Fujita	Chiba Institute of Technology, Japan
Yuki Kaeri	Mejiro University, Japan
Zahoor Ali Khan	HCT, UAE
Muhammad Imran	King Saud University, Saudi Arabia
Ashfaq Ahmad	The University of Newcastle, Australia
Syed Hassan Ahmad	JMA Wireless, USA
Safdar Hussain Bouk	Daegu Gyeongbuk Institute of Science and Technology, South Korea
Jolanta Mizera-Pietraszko	Military University of Land Forces, Poland
Shahzad Ashraf	NFC Institute of Engineering and Technology, Pakistan

10. Cloud and Services Computing

Track Co-chairs

Asm Kayes	La Trobe University, Australia
Salvatore Venticinque	University of Campania "Luigi Vamvitelli", Italy
Baojiang Cui	Beijing University of Posts and Telecommunications, China

TPC Members

Shahriar Badsha	University of Nevada, USA
Abdur Rahman Bin Shahid	Concord University, USA
Iqbal H. Sarker	Chittagong University of Engineering and Technology, Bangladesh
Jabed Morshed Chowdhury	La Trobe University, Australia
Alex Ng	La Trobe University, Australia
Indika Kumara	Jheronimus Academy of Data Science, The Netherlands
Tarique Anwar	Macquarie University and CSIRO's Data61, Australia
Giancarlo Fortino	University of Calabria, Italy
Massimiliano Rak	University of Campania "Luigi Vanvitelli", Italy
Jason J. Jung	Chung-Ang University, South Korea
Dimosthenis Kyriazis	University of Piraeus, Greece
Geir Horn	University of Oslo, Norway
Gang Wang	Nankai University, China
Shaozhang Niu	Beijing University of Posts and Telecommunications, China
Jianxin Wang	Beijing Forestry University, China
Jie Cheng	Shandong University, China
Shaoyin Cheng	University of Science and Technology of China, China

11. Security, Privacy, and Trust Computing

Track Co-chairs

Hiroaki Kikuchi	Meiji University, Japan
Xu An Wang	Engineering University of PAP, P.R. China
Lidia Ogiela	AGH University of Science and Technology, Poland

TPC Members

Takamichi Saito	Meiji University, Japan
Kouichi Sakurai	Kyushu University, Japan
Kazumasa Omote	University of Tsukuba, Japan
Shou-Hsuan Stephen Huang	University of Houston, USA
Masakatsu Nishigaki	Shizuoka University, Japan
Mingwu Zhang	Hubei University of Technology, China
Caiquan Xiong	Hubei University of Technology, China
Wei Ren	China University of Geosciences, China
Peng Li	Nanjing University of Posts and Telecommunications, China
Guangquan Xu	Tianjing University, China
Urszula Ogiela	AGH University of Science and Technology, Poland
Hoon Ko	Chosun University, Republic of Korea
Goreti Marreiros	Institute of Engineering of Polytechnic of Porto, Portugal
Chang Choi	Gachon University, Republic of Korea
Libor Měsíček	J. E. Purkyně University, Czech Republic

12. Software-Defined Networking and Network Virtualization

Track Co-chairs

Flavio de Oliveira Silva	Federal University of Uberlândia, Brazil
Ashutosh Bhatia	Birla Institute of Technology and Science, Pilani, India
Alaa Allakany	Kyushu University, Japan

TPC Members

Rui Luís Andrade Aguiar	Universidade de Aveiro (UA), Portugal
Ivan Vidal	Universidad Carlos III de Madrid, Spain
Eduardo Coelho Cerqueira	Federal University of Pará (UFPA), Brazil
Christos Tranoris	University of Patras (UoP), Greece
Juliano Araújo Wickboldt	Federal University of Rio Grande do Sul (UFRGS), Brazil
Yaokai Feng	Kyushu University, Japan
Chengming Li	Chinese Academy of Science (CAS), China
Othman Othman	An-Najah National University (ANNU), Palestine
Nor-masri Bin-sahri	University Technology of MARA, Malaysia
Sanouphab Phomkeona	National University of Laos, Laos
Haribabu K.	BITS Pilani, India

Shekhavat, Virendra   BITS Pilani, India
Makoto Ikeda   Fukuoka Institute of Technology, Japan
Farookh Hussain   University of Technology Sydney, Australia
Keita Matsuo   Fukuoka Institute of Technology, Japan

AINA-2023 Reviewers

Admir Barolli
Ahmed Bahlali
Aimilios Tzavaras
Akihiro Fujihara
Akimitsu Kanzaki
Alaa Allakany
Alba Amato
Alberto Volpe
Alex Ng
Alex Thomo
Alfredo Navarra
Anne Kayem
Antonio Esposito
Arcangelo Castiglione
Arjan Durresi
Ashutosh Bhatia
Asm Kayes
Bala Killi
Baojiang Cui
Beniamino Di Martino
Bhed Bista
Bruno Zarpelão
Carson Leung
Chang Choi
Changyu Dong
Chih-Peng Fan
Christos Bouras
Christos Tranoris
Chung-Ming Huang
Darshika Perera
David Taniar
Dilawaer Duolikun
Donald Elmazi
Elis Kulla
Eric Pardede
Euripides Petrakis

Evjola Spaho
Fabian Kurtz
Farookh Hussain
Fatos Xhafa
Feilong Tang
Feroz Zahid
Flavio Corradini
Flavio Silva
Flora Amato
Francesco Orciuoli
Gang Wang
Goreti Marreiros
Hadil Abukwaik
Hiroaki Kikuchi
Hiroshi Maeda
Hiroyoshi Miwa
Hiroyuki Fujioka
Hsing-Chung Chen
Hyunhee Park
Indika Kumara
Isaac Woungang
Jabed Chowdhury
Jana Nowaková
Ji Zhang
Jiahong Wang
Jianfei Zhang
Jolanta Mizera-Pietraszko
Jörg Domaschka
Jorge Sá Silva
Juliano Wickboldt
Julio Costella Vicenzi
Jun Iio
K. Haribabu
Kazunori Uchida
Keita Matsuo
Kensuke Baba

Kin Fun Li
Kiplimo Yego
Kiyotaka Fujisaki
Konstantinos Tsakos
Kouichi Sakurai
Lei Shu
Leonard Barolli
Leonardo Mostarda
Libor Mesicek
Lidia Ogiela
Lucian Prodan
Luciana Oliveira
Makoto Ikeda
Makoto Takizawa
Marek Ogiela
Marenglen Biba
Mario Dantas
Markus Aleksy
Masakatsu Nishigaki
Masaki Kohana
Masaru Kamada
Mingwu Zhang
Minoru Uehara
Miroslav Voznak
Mohammad Faiz Iqbal Faiz
Nadeem Javaid
Naohiro Hayashibara
Neder Karmous
Nobuo Funabiki
Omar Hussain
Omid Ameri Sianaki
Paresh Saxena
Pavel Kromer
Petros Nicopolitidis
Philip Moore Fatos Xhafa
Purav Shah
Rajdeep Niyogi

Rodrigo Miani
Ronald Petrlic
Ronaldo Mello
Rui Aguiar
Ryuji Shioya
Salimur Choudhury
Salvatore Venticinque
Sanjay Dhurandher
Santi Caballé
Satya Borah
Shahriar Badsha
Shengrui Wang
Shigenari Nakamura
Shigetomo Kimura
Somnath Mazumdar
Sriram Chellappan
Stelios Sotiriadis
Takahiro Uchiya
Takamichi Saito
Takayuki Kushida
Tetsuya Oda
Tetsuya Shigeyasu
Thomas Dreibholz
Tomoki Yoshihisa
Tomoya Enokido
Tomoyuki Ishida
Vamsi Paruchuri
Wang Xu An
Wei Lu
Wenny Rahayu
Winston Seah
Yoshihiro Okada
Yoshitaka Shibata
Yusuke Gotoh
Zahoor Khan
Zia Ullah

AINA-2023 Keynote Talks

Blockchain and IoT Integration: Challenges and Future Directions

Leonardo Mostarda

Camerino University, Camerino, Italy

Abstract. Massive overhead costs, concerns about centralized data control, and single point of vulnerabilities are significantly reduced by moving IoT from a centralized data server architecture to a trustless, distributed peer-to-peer network. Blockchain is one of the most promising and effective technologies for enabling a trusted, secure, and distributed IoT ecosystem. Blockchain technology can allow the implementation of decentralized applications that not only perform payments but also allow the execution of smart contracts. This talk will investigate the state of the art and open challenges that are related to IoT and blockchain integration. We review current approaches and future directions.

Toward Sustainable, Intelligent, Secure, Fully Programmable, and Multisensory (SENSUOUS) Networks

Flávio de Oliveira Silva

Federal University of Uberlândia, Uberlândia, Brazil

Abstract. In this talk, we will discuss and present the evolution of current networks toward sustainable, intelligent, secure, fully programmable, and multisensory (SENSUOUS) networks. The evolution of networks happens through these critical attributes that will drive the next-generation networks. Here networks consider data networks capable of transmitting audio and video in computer or telecommunication systems. While there is an established process for the evolution of telecommunication networks, regarding computer networks, this area is still open and has several challenges and opportunities. So far, networks can transmit audio and video data, which sensitize only part of our senses. Still, new senses must be considered in the evolution of networks, expanding the multisensory experience. SENSUOUS networks will shape and contribute to scaling our society's sustainable, smart, and secure digital transformation.

Contents

Energy-Efficient Two Phase Locking (2PL) Protocol by Not Performing Meaningless Methods in Virtual Machine Environments

Tomoya Enokido[1]([✉]), Dilawaer Duolikun[2], and Makoto Takizawa[3]

[1] Faculty of Business Administration, Rissho University,
4-2-16, Osaki, Shinagawa-ku, Tokyo 141-8602, Japan
eno@ris.ac.jp
[2] Department of Advanced Sciences, Faculty of Science and Engineering,
Hosei University, 3-7-2, Kajino-cho, Koganei-shi, Tokyo 184-8584, Japan
[3] Research Center for Computing and Multimedia Studies, Hosei University,
3-7-2, Kajino-cho, Koganei-shi, Tokyo 184-8584, Japan
makoto.takizawa@computer.org

Abstract. Distributed applications are becoming more scalable and high performance computing systems are essential to realize the applications. High performance computing systems like cloud computing systems are equipped with a huge number of servers and each server performs a large amount of methods issued by transactions to provide application services. Hence, servers in a system consume the large amount of electric energy. It is required to design not only high performance but also energy-efficient computing systems. In this paper, the EEL (Energy-Efficient Locking) protocol is newly proposed to energy-efficiently perform methods issued by transactions. We show servers in a system consumes smaller electric energy to perform methods issued by transactions by using the EEL protocol through the evaluation.

Keywords: Energy-efficient locking (EEL) protocol · Two-phase locking protocol · Energy-efficient systems · Green computing

1 Introduction

Various kinds of distributed application services like vehicle network services [1] are realized to support our life. These application services are becoming more scalable and a huge volume of data is required to realize the application services. Data required by each application service is gathered by various types of devices like IoT (Internet of Things) devices [2,3]. An object [4–6] is a computation resource like a database to realize an application service and is defined to be an encapsulation of data and methods to manipulate the data in the object. Data gathered from IoT devices is stored in an object and manipulated through only methods provided by the object. A transaction [7,8] is an atomic sequence of methods provided by objects. Application users initiate transactions on clients to utilize application services. A huge number of methods issued by transactions are concurrently performed on each object in information systems. Hence, not only scalable but also high performance computing systems like cloud computing systems [4,5,9–11] have to be needed to implement scalable application services. A cloud computing system is composed of a cluster of servers and multiple virtual machines [6,11] are performed on each server. Objects are distributed on virtual machines in a cluster of servers. Here, conflicting transactions have to be serialized [7,8] in order to keep all the objects mutually consistent. In addition, a cluster of servers consumes a large amount of electric energy since a huge number of methods are concurrently performed on virtual machines in each server.

The *EE2PL-VM* (*Energy-Efficient Two-Phase Locking in Virtual Machine environment*) protocol [6] is proposed to energy-efficiently perform methods issued by transactions on each object in our previous studies. In the EE2PL-VM protocol, *meaningless write methods* [6] are not performed on each object. As a result, the EE2PL-VM protocol can reduce not only the total processing electric energy consumption (PEE) of servers but also the average execution time (AET) of transactions. We newly introduce *meaningless read methods* in this paper. Then, the *EEL* (*Energy-Efficient Locking*) protocol is newly proposed to furthermore reduce the total PEE of servers and the AET of transactions by not performing meaningless read methods in addition to meaningless write methods. The EEL protocol is evaluated compared with the EE2PL-VM protocol. We show the EEL protocol can reduce the total PEE of servers and the AET of transactions than the EE2PL-VM protocol in the evaluation.

We explain the 2PL (Two-Phase Locking) protocol, data access model, and power consumption model in Sect. 2. The EEL protocol is newly proposed in Sect. 3. The evaluation results of the EEL protocol compared with the EE2PL-VM protocol are shown in Sect. 4.

2   System Model

2.1   2PL Protocol

Servers s_1, ..., s_n ($n \geq 1$) construct a cluster CL of servers. Each server s_t holds one multi-core CPU which is equipped with the total number nc_t (≥ 1) of cores. There is a set $C_t = \{c_{1t}, ..., c_{nc_t t}\}$ of cores in each server s_t. Each core c_{lt} supports the total number ct_t (≥ 1) of threads. Let TH_t be a set $\{th_{1t}, ..., th_{nt_t t}\}$ of threads in each server s_t where $nt_t = ct_t \cdot nc_t$. A server s_t supports a set $VM_t = \{vm_{1t}, ..., vm_{nt_t t}\}$ of virtual machines. Each thread th_{kt} is occupied to perform one virtual machine vm_{kt}. Let O be a set $\{o_1, ..., o_m\}$ ($m \geq 1$) of objects [4–6], which are distributed to virtual machines in a cluster CL. Each object o_h provides *read*, *full write*, and *partial write* methods. A full write method completely writes data while a read method completely reads data in an object o_h. A partial write method writes a part of data in an object o_h. A notation $op^i \circ op^j(o_h)$ means a state of an object o_h where a method op^j is performed after another method op^i on the object o_h. A pair of methods op^i and op^j are *compatible* on an object o_h if and only if (iff) $op^i \circ op^j(o_h) = op^j \circ op^i(o_h)$. Otherwise, a method op^i *conflicts* with another method op^j on an object o_h. The conflicting relations among methods are symmetric.

A *transaction* T^i [8] is defined as an atomic sequence of write and read methods supported by objects. A set $\mathbf{T} = \{T^1, ..., T^k\}$ ($k \geq 1$) of transactions initiated in a system. A transaction T^i *precedes* a transaction T^j ($T^i \rightarrow_{sch} T^j$) in a schedule *sch* of transactions in $\mathbf{T}$ iff for every pair of conflicting methods op^i and op^j issued by different transactions T^i and T^j, respectively, op^i is performed before op^j in *sch*. A schedule *sch* of transactions is serializable iff the precedent relation $\rightarrow_{sch}$ is acyclic [8]. Every pair of objects have to be mutually consistent. Hence, conflicting transactions have to be *serializable* [7,8]. Conflicting transactions can be serializable by using the *2PL* (*Two-Phase Locking*) protocol [7,8,12]. Let *mode(op)* indicate a *lock mode* of a method *op*. A transaction T^i manipulates an object o_h by a method *op* if the transaction T^i could lock the object o_h by a lock mode *mode(op)* in the 2PL protocol [7,8,12]. Before the transaction T^i commits or aborts, locks on each object o_h are released.

2.2 DAVM Model

In our previous studies, we measured data access rate [Bytes/sec] to concurrently perform write and read methods on objects supported by virtual machines in a server. Then, the *DAVM* (*Data Access in Virtual Machine environments*) model [11] is proposed based on the experiments. Notations $r_{kt}^i(o_h)$ and $w_{kt}^i(o_h)$ show read and write methods which are issued by a transaction T^i to read and write data of an object o_h stored in a virtual machine vm_{kt} in a server s_t. Let $Re_t(\tau)$ and $Wr_t(\tau)$ indicate sets of read and write methods being performed on virtual machines in a server s_t at time τ, respectively. Let $M_t(\tau)$ shows a set read and write methods on a server s_t at time τ. $M_t(\tau) = Re_t(\tau) \cup Wr_t(\tau)$. The maximum write rate $maxWR_t$ and the maximum read rate $maxRR_t$ of a server s_t depends on the specification of the server s_t. At time τ, each read method $r_{kt}^i(o_h)$ reads data in an object o_h stored in a virtual machine vm_{kt} at read rate $rr_{kt}^i(\tau)$ [B/sec]. At time τ, each write method $w_{kt}^i(o_h)$ writes data in an object o_h stored in a virtual machine vm_{kt} at write rate $wr_{kt}^i(\tau)$ [B/sec]. The read rate $rr_{kt}^i(\tau)$ ($\leq maxRR_t$) of a read method $r_{kt}^i(o_h)$ is $fr_t(\tau) \cdot maxRR_t$ at time τ where $0 \leq fr_t(\tau) \leq 1$. The write rate $wr_{kt}^i(\tau)$ ($\leq maxWR_t$) of a write method $w_{kt}^i(o_h)$ is $fw_t(\tau) \cdot maxWR_t$ at time τ where $0 \leq fw_t(\tau) \leq 1$. Here, $fr_t(\tau)$ and $fw_t(\tau)$ are degradation ratios of read and write rates, respectively, at time τ. Here, $fr_t(\tau) = 1 / (|Re_t(\tau)| + RW_t \cdot |Wr_t(\tau)|)$ where $0 \leq RW_t \leq 1$ and $fw_t(\tau) = 1 / (WR_t \cdot |Re_t(\tau)| + |Wr_t(\tau)|)$ where $0 \leq WR_t \leq 1$.

2.3 PCDAVM Model

In our previous experiments [11], we also measured the electric power of a real server to concurrently perform write and read methods on objects stored in virtual machines. Then, the *PCDAVM* (*Power Consumption for Data Access in*

Virtual Machine environment) model [11] is proposed based on the experimental results. Let $Base_t(\tau)$ be the base electric power of a server s_t at time τ. The base electric power $Base_t$ [W] of a sever s_t depends on both the number $actC_t(\tau)$ of active cores and the number $actT_t(\tau)$ of active threads in a server s_t at time τ. If $actC_t(\tau) \geq 1$, a server s_t consumes the electric power $minC_t$ [W]. A server s_t consumes the electric power cE_t and tE_t [W] to make one core and one thread active, respectively. The maximum electric power $maxE_t$ [W] and minimum electric power $minE_t$ [W] of a server s_t depends on the specification of the server s_t. The base electric power $Base_t(\tau) = minE_t + \gamma_t \cdot (minC_t + actC_t(\tau) \cdot cE_t + actT_t(\tau) \cdot tE_t)$ [W]. If no virtual machines are active in a server s_t, $\gamma_t = 0$. Otherwise, $\gamma_t = 1$.

A server s_t consumes the electric power $E_t(\tau)$ [W] to perform read and write methods on objects stored in virtual machines at time τ. In PCDAVM model, the electric power $E_t(\tau)$ is given as follows:

$$
E_t(\tau) = \begin{cases}
Base_t(\tau) & \text{if } |Wr_t(\tau)| = |Re_t(\tau)| = 0. \\
Base_t(\tau) + wE_t & \text{if } |Re_t(\tau)| = 0 \text{ and } |Wr_t(\tau)| \geq 1. \\
Base_t(\tau) + wrE_t(\alpha(\tau)) & \text{if } |Re_t(\tau)| \geq 1 \text{ and } |Wr_t(\tau)| \geq 1. \\
Base_t(\tau) + rE_t & \text{if } |Re_t(\tau)| = 1 \text{ and } |Wr_t(\tau)| \geq 0.
\end{cases} \tag{1}
$$

If no method is performed in every virtual machine in a server s_t, a server s_t consumes the minimum electric power $minE_t$ [W]. If only write methods are performed on virtual machines in a server s_t, the electric power $Base_t(\tau) + wE_t$ [W] is consumed in the server s_t where wE_t is the electric power to perform only write methods on the server s_t. If only read methods are performed on virtual machines in a server s_t, the electric power $Base_t(\tau) + rE_t$ [W] is consumed in the server s_t where rE_t is the electric power to perform only read methods on the server s_t. If both write and read methods are performed in a server s_t, the electric power $Base_t(\tau) + wrE_t(\alpha(\tau)) = \alpha(\tau) \cdot wE_t + (1 - \alpha(\tau)) \cdot rE_t$ [W] is consumed in the server s_t where $\alpha(\tau) = |Wr_t(\tau)|/|M_t(\tau)|$. Here, $rE_t \leq wrE_t(\alpha(\tau)) \leq wE_t$. A server s_t consumes the total processing electric energy (PEE) $TPEE_t(\tau_1, \tau_2) = \Sigma_{\tau=\tau1}^{\tau2}(E_t(\tau) - minE_t)$ [J] from time τ_1 to τ_2.

3   Energy-Efficient Locking (EEL) Protocol

Let *sch* and *sch'* be a pair of schedules in a system. *sch* and *sch'* are state-equivalent iff the same state is obtained from every state of the system by *sch* and *sch'*. A method *op* in a schedule *sch* is *meaningless* iff *sch* is state-equivalent to another schedule *sch'* obtained by just removing *op* from *sch*. Suppose a pair of methods op^1 and op^2 are performed on an object o_h. Here, the method op^1 *locally precedes* the method op^2 in a local schedule sch_h of the object o_h ($op^1 \Rightarrow_{sch_h} op^2$) iff $op^1 \rightarrow_H op^2$.

[Absorption of Methods]

1. A read method r^1 absorbs another read method r^2 in a local schedule sch_h of an object o_h iff (1) $r^1 \Rightarrow_{sch_h} r^2$ and there is no write method w such that $r^1 \Rightarrow_{sch_h} w \Rightarrow_{sch_h} r^2$, or (2) r^1 absorbs r^3 and r^3 absorbs r^2 for some read method r^3.
2. A full write method w^1 absorbs another full write or partial write method w^2 in a local schedule sch_h of an object o_h iff (1) $w^2 \rightarrow_{sch_h} w^1$ and there is no read method r such that $w^2 \Rightarrow_{sch_h} r \Rightarrow_{sch_h} w^1$, or (2) w^1 absorbs w^3 and w^3 absorbs w^2 for some write method w^3.

[Definition]. A method op is a *meaningless* method on an object o_h iff the method op is absorbed by another method op' in the local schedule sch_h of the object o_h.

The *EEL (Energy-Efficient Locking)* protocol is newly proposed to furthermore reduce the total PEE of servers and the AET of transactions by not performing meaningless read methods in addition to meaningless write methods on each object. Meaningless read and write methods are not performed on each object in the EEL protocol. A method op^i issued by a transaction T^i is performed on each object o_h supported by a virtual machine vm_{kt} in a server s_t by the EEL procedure shown in Algorithm 1. A variable $o_h.w_{wait}$ shows a write method $w^i_{kt}(o_h)$ issued by a transaction T^i to a object o_h. An object o_h waits for performing a write method $w^i_{kt}(o_h)$ shown by the variable $o_h.w_{wait}$ until the object o_h receives the next method op'. If the next method op' is a full write method and absorbs the write method $w^i_{kt}(o_h)$ shown by $o_h.w_{wait}$, the next method op' is performed on the object o_h without performing the write method $w^i_{kt}(o_h)$ shown by $o_h.w_{wait}$ since the write method $w^i_{kt}(o_h)$ is a meaningless write method on the object o_h. A variable $o_h.r_{current}$ shows a read method being performed on an object o_h. Suppose a read method r' is being performed on an object o_h, i.e. $o_h.r_{current} = r'$ when the object o_h receives a read method $r^i_{kt}(o_h)$ from a transaction T^i. Here, the read method $r^i_{kt}(o_h)$ is a meaningless read method since the read method r' absorbs the read method $r^i_{kt}(o_h)$. The read method $r^i_{kt}(o_h)$ is not performed on the object o_h and the result obtained performing the read method r' is also sent to the transaction T^i.

In the EEL protocol, meaningless read methods are not performed on each object in addition to meaningless write methods. Hence, the total PEE of servers and the AET of transactions can be furthermore reduced in the EEL protocol than the EE2PL-VM protocol.

4   Evaluation

Five homogeneous servers $s_1, ..., s_5$ constructs a cluster CL. We evaluate the EEL protocol in terms of the total PEE [J] of a cluster CL and the AET [sec] of transactions compared with the EE2PL-VM protocol [6]. Parameters of the PCDAVM model and DAVM model [11] for every server s_t in the cluster CL are the same. The parameters are obtained from the experiments in our previous studies [11]. The maximum electric power $maxE_t$ of each server s_t is 24.3 [W] and the minimum

---

Algorithm 1. EEL procedure

---

  procedure EEL($op^i(o_h)$)
   if $op^i(o_h)$ = a *read* method **then**   ▷ o_h receives a read method.
     if $o_h.w_{wait} = \phi$ **then**
       if $o_h.r_{current} = \phi$ **then**
         $o_h.r_{current} = op^i(o_h)$;
         perform($op^i(o_h)$);
         $o_h.r_{current} = \phi$;
       else   ▷ $op^i(o_h)$ is a meaningless read method.
         a result of $o_h.r_{current}$ is sent to a transaction T^i;
       end if
     else
       perform($o_h.w_{wait}$);
       $o_h.w_{wait} = \phi$;
       $o_h.r_{current} = op^i(o_h)$;
       perform($op^i(o_h)$);
       $o_h.r_{current} = \phi$;
     end if
   else   ▷ o_h receives a write method.
     if $o_h.w_{wait} = \phi$ **then**
       $o_h.w_{wait} = op^i(o_h)$;
     else
       if $op^i(o_h)$ absorbs $o_h.w_{wait}$ **then**
         perform($op^i(o_h)$);   ▷ $o_h.w_{wait}$ is a meaningless write method.
       else
         perform($o_h.w_{wait}$);
         $o_h.w_{wait} = op^i(o_h)$;
       end if
     end if
   end if
  end procedure

---

electric power $minE_t$ of each server s_t is 17 [W]. The electric power cE_t and tE_t to make one core and one thread active in each server s_t are 0.6 and 0.5 [W], respectively. $minC_t = 1.1$ [W], $WE_t = 4$ [W], $RE_t = 1$ [W], $maxRR_t = 98.5$ [MB/sec], $maxWR_t = 85.3$ [MB/sec], $RW_t = 0.077$, and $WR_t = 0.667$. Sixty objects o_1, ..., o_{60} are randomly distributed to virtual machines in the cluster CL. Each object o_h holds a randomly selected size of data between 10 to 100 [MB]. The number tn ($0 \leq tn \leq 500$) of transactions are initiated and each transaction issues five methods randomly selected on sixty objects to manipulate objects.

Figure 1 shows the total PEE [KJ] of the cluster CL in the EEL and EE2PL-VM protocols to perform the number tn of transactions. For $0 \leq tn \leq 500$, the EEL protocol can reduce the total PEE of the cluster CL than the EE2PL-VM protocol. In the EEL protocol, meaningless read methods are not performed on each object in addition to meaningless write methods. Hence, the total PEE of the cluster CL can be furthermore reduced in the EEL protocol than the EE2PL-VM protocol.

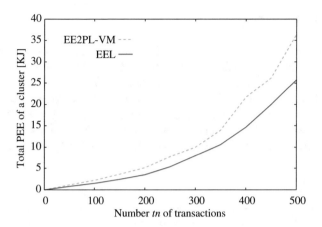

Fig. 1. Total PEE [KJ] of a cluster CL.

Figure 2 shows the AET [sec] of the tn transactions in the EEL and EE2PL-VM protocols. For $0 \le tn \le 500$, the AET of transactions in the EEL protocol is shorter than the EE2PL-VM protocol since both meaningless read and write methods are not performed in the EEL protocol and computation resources to perform meaningless read and write methods can use to perform other methods on each object.

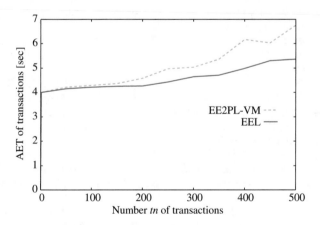

Fig. 2. AET [sec] of transactions.

From the evaluation results, we conclude the EEL protocol is more useful than the EE2PL-VM protocol.

5   Concluding Remarks

In this paper, we newly proposed the EEL protocol to reduce the total PEE of a server cluster and the AET of transactions. In the EEL protocol, meaningless read methods are not performed on each object in addition to meaningless write methods. Therefore, the EEL protocol can furthermore reduce the total PEE of a server cluster and the AET of transactions than the EE2PL-VM protocol. We show the total PEE of a server cluster and the AET of transactions can be furthermore reduced in the EEL protocol than the EE2PL-VM protocol.

References

1. Qafzezi, E., Bylykbashi, K., Ampririt, P., Ikeda, M., Matsuo, K., Barolli, L.: FSAQoS: a fuzzy-based system for assessment of QoS of V2V communication links in SDN-VANETs and its performance evaluation. Int. J. Distrib. Syst. Technol. **13**(1), 1–13 (2022)
2. Enokido, T., Takizawa, M.: The redundant energy consumption laxity based algorithm to perform computation processes for IoT services. Internet Things **9** (2020). https://doi.org/10.1016/j.iot.2020.100165
3. Nakamura, S., Enokido, T., Takizawa, M.: Information flow control based on capability token validity for secure IoT: implementation and evaluation. Internet Things **15**, 100423 (2021)
4. Enokido, T., Duolikun, D., Takizawa, M.: An energy-efficient quorum-based locking protocol by omitting meaningless methods on object replicas. J. High Speed Netw. **28**(3), 181–203 (2022)
5. Enokido, T., Duolikun, D., Takizawa, M.: Energy consumption laxity-based quorum selection for distributed object-based systems. Evol. Intel. **13**, 71–82 (2020)
6. Enokido, T., Duolikun, D., Takizawa, M.: Energy-efficient locking protocol in virtual machine environments. Accepted for publication in Proceedings of the 11-th International Conference on Emerging Internet, Data & Web Technologies (EIDWT-2023) (2023)
7. Gray, J.N.: Notes on data base operating systems. In: Bayer, R., Graham, R.M., Seegmüller, G. (eds.) Operating Systems. LNCS, vol. 60, pp. 393–481. Springer, Heidelberg (1978). https://doi.org/10.1007/3-540-08755-9_9
8. Bernstein, P.A., Hadzilacos, V., Goodman, N.: Concurrency Control and Recovery in Database Systems. Addison-Wesley (1987)
9. Enokido, T., Duolikun, D., Takizawa, M.: Energy-efficient concurrency control by omitting meaningless write methods in object-based systems. In: Proceedings of the 36th International Conference on Advanced Information Networking and Applications (AINA-2022), pp. 129–139 (2022)
10. Enokido, T., Duolikun, D., Takizawa, M.: The improved redundant active time-based (IRATB) algorithm for process replication. In: Proceedings of the 35th IEEE International Conference on Advanced Information Networking and Applications (AINA-2021), pp. 172–180 (2021)
11. Enokido, T., Takizawa, M.: The power consumption model of a server to perform data access application processes in virtual machine environments. In: Proceedings of the 34th International Conference on Advanced Information Networking and Applications (AINA-2020), pp. 184–192 (2020)
12. Garcia-Molina, H., Barbara, D.: How to assign votes in a distributed system. J. ACM **32**(4), 814–860 (1985)

System Design for DDoS Traffic Mitigation by a Collaboration of Mobile and IP Networks in 5G/6G

Kenichi Okonogi[✉], Masaki Suzuki, and Atsushi Tagami

Advanced Technology Laboratories, KDDI Research, Inc., Fujimino-shi, Japan
{ke-okonogi,mak-suzuki,at-tagami}@kddi.com

Abstract. The flooding distributed denial of service (DDoS) attacks are recognized as one of the most considerable threats to security. In the B5G/6G era, as devices connected to mobile networks increase, it is easily predicted that mobile data communication traffic will increase massively and rapidly. This suggests that DDoS attacks traffic via mobile networks can become a thread more than before. To defend against DDoS attacks, the best way is to collaborate with multiple domain networks, i.e., source network, core network, and victim-end network. An ideal approach is to detect and protect against DDoS attacks in a location close to the attack source. Therefore in this paper, we design a collaborative system working between IP and mobile networks to defend against DDoS attacks and introduce a prototype of the proposed scheme. By demonstrating the packet filtering function in UPF, we show its feasibility and effectiveness in 5G/6G networks.

1 Introduction

Mobile communication data quantities have continued to increase for many years, and mobile communications have generated one of the world's largest and most significant infrastructure deployments today. According to a study by the International Telecommunication Union, the global mobile traffic per month would then be estimated to reach 543 EB in 2,025 and 4,394 EB in 2030 [1]. The huge data traffic demands have been accompanied by increasing requirements for various services, full converge, ultra-high-speed wireless communications with high reliability, and ultra-low-latency. Besides, many devices, i.e., rich devices like smartphones and single-function devices like the internet of things (IoT) devices, will connect to a mobile network with data traffic. If devices with weak security are included, those devices may be infected with malicious software and will cause distributed denial of service (DDoS) attacks via mobile networks. Since the number of subscribers is significantly large and it is still increasing as well as the data volume, if the DDoS attack is triggered in the mobile network, it results in a network delay and failure in addition to the overload on the targeted application server. This quality of service (QoS) degradation may affect the other subscribers who share the same communication resources with the vulnerable devices due to the congestion in the control plane caused by the attack. Therefore, it is necessary to mitigate DDoS traffic in mobile networks.

© The Author(s), under exclusive license to Springer Nature Switzerland AG 2023
L. Barolli (Ed.): AINA 2023, LNNS 661, pp. 9–20, 2023.
https://doi.org/10.1007/978-3-031-29056-5_2

Considering the distributed nature of DDoS, a distributed mechanism is necessary for a successful defense. The critical DDoS defense functionalities, i.e., attack detection, rate limitation, and traffic differentiation, are best performed at different locations on the Internet [2]. Figure 1 shows the end-to-end communication consisting of victim-end, core, and source networks. A victim-end network quickly detects DDoS attacks with high accuracy since it can observe all aggregated traffic and recognize if the traffic includes attack traffic [3]. A core network limits the rate to avoid large floods that would overwhelm the victim's access links [4]. A source network can distinguish the attack traffic and legitimate user traffic because of a low address diversity and low traffic rate [5]. Hence to mitigate DDoS traffic in the mobile network, it is good to detect an attack in the victim-end network and to differentiate the attack traffic in the source network, i.e., the mobile network. Thus, collaborative defense is essential to mitigate DDoS traffic. The fundamental idea of this paper is to distinguish the attack traffic at the front of the victim side, then block the traffic at the source network.

For such collaboration in a mobile network, it is necessary to use an End to End (E2E) orchestrator, which manages and controls multiple domains, such as radio access network (RAN), transport network, mobile core network, and networks with different administrators. Additionally, the mobile network only supports specific tunnel protocols, e.g., GRPS tunnel protocol, mobility support, QoS guarantee, etc. A well-designed scheme is required to correctly distinguish the attack traffic detected out of the mobile network. In this paper, we investigate the system design to identify the attack traffic and block it in a mobile network with commercial considerations. The research contributions of this paper are listed as follows.

- System design to mitigate DDoS traffic with the collaboration of mobile and IP networks by using an E2E orchestrator.
- Interface design between mobile network functions according to standard specifications.
- Performance evaluations by the pilot implementations using open-source software on virtualized infrastructure.

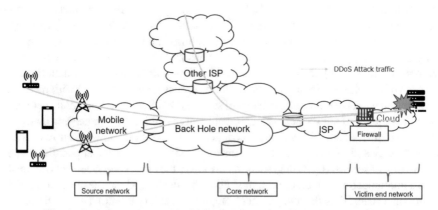

Fig. 1. Locations where critical DDoS defense functionalities are deployed.

The rest of this paper is organized as follows. Sections 2 and 3 introduce related works and the proposed system, respectively. In Sect. 4, we show the evaluation indicating the feasibility and efficiency of the proposed system. Finally, Sect. 5 describes the conclusions and future work.

2 Related Works

2.1 Traditional DDoS Protection Solution

The major open-source and commercial solutions mainly work in traditional IP networks. For example, Linux netfilter [6] and Remote Triggered Black Hole Filtering (RTBH) [7] are widely used and well-known as practical techniques for the mitigation of DDoS attacks. Linux netfilter can set a rule of packet hooks that work inside the Linux kernel. A hooked packet will be processed according to the forwarding rule. RTBH drops large traffic with specific IP prefixes. They use border gateway protocol (BGP) to trigger blackholing drops in DDoS traffic. However, these traditional solutions do not support overlay networks, including the GPRS Tunneling Protocol (GTP) adopted in the 5G networks. This is because the packet information to be essentially checked cannot be looked up due to the encapsulation in the case of overlay networks. Encapsulation may lead to wrong decisions, miss-dropping packets, and other misoperations.

2.2 Collaborative DDoS Mitigation System

Collaborative defense schemes have been suggested [2, 8–10]. The common idea is to eliminate the DDoS attack by having different centers or router capabilities working in collaboration across the other networks, targeting each attack. An internet engineering task force (IETF) specifies signaling information about DDoS attacks as IETF DDoS Open Threat Signaling (DOTS) standard [10]. The DOTS system would provide real-time signal information related to DDoS and handling requests to support collaborative detection, identification, and mitigation of DDoS attacks. To our knowledge, the collaborative defense can be implemented only when both are IP networks. The previously mentioned cooperative defense mechanism cannot handle the GTP capsuled packet. Therefore, it is not supported to defend in End-to-End networks, including mobile networks, collaboratively. This paper deals with the collaboration between the IP and mobile networks.

2.3 Interaction with Mobile Network Functions

Figure 2 shows the system architecture of the 5G system (5GS) [11]. The Control/User (C/U)-plane functions are clearly separated, and C-Plane is organized in a service-based architecture associated with network functions for flexible utilization and better scalability. The network functions in C-Plane communicate with each other with http/2 protocol [12, 13]. 5GC service-based interface APIs are based on the concept of REST according to Open API specification [14]. Using the REST API in this way, 5G defined the ability to disclose APIs to the outside of the mobile network and collaborate with external

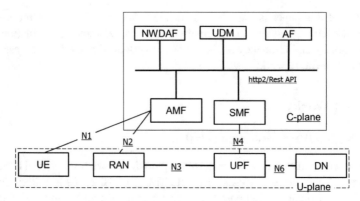

Fig. 2. 5GS architecture.

systems. Also, C-plane is responsible for session management, mobility, and authentication. For example, SMF allocates IP addresses to UEs and manages the user-plane via the N4 interface [15]. SMF also can send Packet Detection Rules (PDR) and forwarding rules (FDR) to UPF, and these rules can change the routing or drop the incoming packet on U-Plane. U-plane traffic generated from the UE is transmitted to the UPF via GPRS tunneling through the RAN. UPF performs de-encapsulation of the incoming packet before applying the defined rule one by one. Additionally, the UPF is flexible to locate near gNB for low latency and at the central site towards the Internet.

We believe UPF is suitable for distinguishing attack traffic and legitimate traffic and stopping the DDoS attack via the mobile network. Therefore, we propose a DDoS mitigation system at UPF. The scope of this paper includes the mitigation or elimination of the attack traffic at the user plane function in mobile networks.

2.4 E2E Orchestrator

In the 5G/6G era, with the advent of Network Function Virtualization (NFV), flexibility and agility were brought to the networks. E2E orchestrators have been defined to orchestrate virtualized network functions (VNFs) within various networks and to manage multi-network coordination. E2E orchestrators are defined in ETSI-NFV Management and Orchestration (MANO) [16], which is deployed as OSM, Open Baton, Open Network Automation Platform (ONAP) [17], etc.

According to [18], The E2E Orchestrator provides E2E services to customers, network operators, etc., and controls the networks. The E2E orchestrator needs the following functions.

- **E2E service orchestration**: adjust End-to-End network configuration or settings for customers/operators via network controllers.
- **E2E service intelligence**: Automated decision-making or decision support via technologies such as artificial intelligence, machine learning, and knowledge management.
- **E2E service analytics**: Analyze the network's performance under your control based on the collected data and detect abnormal traffic.

- **E2E service data collection**: Collect attack traffic and network performance data from own networks.

In addition, [18] says that each network under an E2E orchestrator should have the following capabilities.

- **Data collection**: Provide live performance and fault data for optimization and share information about the network.
- **Analytics**: Provide specific insights and generate specific predictions by using data collection.
- **Intelligence**: Decision support/making and Action planning via AI/ML.
- **Control**: Control each entity individually to change the configuration.

In addition, E2E orchestrators and interfaces for interoperation and coordination between each network are defined as **Integration fabric** and **cross-domain integration fabric**. These functions can be used to build a collaborative DDoS mitigation system.

3   Collaborative DDoS Mitigation System with E2E Orchestrator

3.1   System Design to Mitigate DDoS Traffic with the Collaboration of Mobile and IP Networks by Using an E2E Orchestrator

We propose an attack traffic mitigation system that works by collaborating with mobile and IP network controllers using the E2E orchestrator. Each network's controller shares traffic and other information within each network with the E2E orchestrator. They also execute attack mitigation requests to the E2E Orchestrator or receive attack mitigation instructions from the E2E Orchestrator.

Figure 3 depicts a high-level architecture of the proposed system. Since E2E orchestrators are equipped with the functions shown in Sect. 2.4, we believe they can perform end-to-end DDoS attack mitigation by linking data collection, data analysis, and control with various networks. The functions of the DDoS mitigation system using E2E orchestrators can be applied to the functions as follows.

- **E2E service data collection**: Collects the traffic information of each network under the coverage of the E2E orchestrator.
- **E2E service analytics**: Correlate observed traffic with information obtained from various networks (mobile, IP, cloud, etc.) to find attack traffic.
- **E2E service intelligence**: Use AI techniques, e.g., machine learning to make DDoS attack assessments and decisions from analytics results.
- **E2E service orchestration**: Collaborate with various networks to mitigate attacks. It will be linked to the mobile network via the E2E Orchestrator.

Using this system, the E2E orchestrator can determine the attack source network and block attack traffic based on information collected from the controller of the victim network.

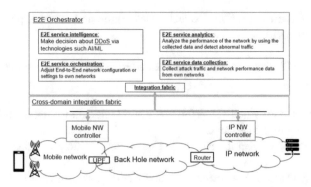

Fig. 3. A high-level architecture of the proposed system.

3.2 Interface Design Between Mobile Network Functions According to Standard Specifications

Each network to be linked to the E2E orchestrator must have the same functionality as described in Sect. 2.4, and we assume that similar functions are also needed for mobile networks. Applying these functions to the network functions defined in the 3GPP for mobile networks. The focus here is on the design for functionality within the mobile network since existing technology will be used in IP networks.

As **Data Collection** and **Analytics**, **Intelligence**, the Network Data Analytics Function (NWDAF) is considered to apply. Since NWDAF is a function that supports various analytics for optimizing the network configuration and resource allocation based on the collected traffic data. As **Control**, we need to make a function as an Application Function (AF). The controller needs another function to communicate with the E2E orchestrator, so this controller has the ability of **Integration fabric**.

The controller applies blocking rules based on DDoS mitigation requests from the E2E orchestrator to the UPF via C-Plane functions. UPF de-capsulated incoming packets from UE and performed the process for DDoS mitigation (Fig. 4).

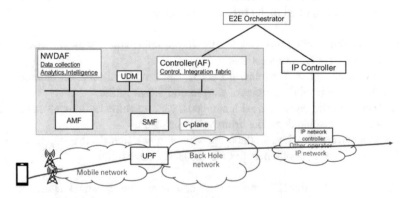

Fig. 4. The proposed functional-level architecture in a mobile network.

4   Experiment Results

4.1   Experimental Environment

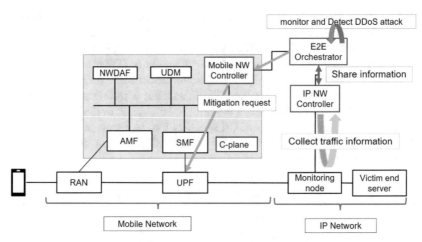

Fig. 5. Experimental environment.

Figure 5 shows the pilot implementation of the proposed system to mitigate DDoS traffic with the collaboration of mobile and IP networks. The IP controller in the victim-end network collects the incoming traffic information from the monitoring node. IP NW controller transfers collected information to the E2E orchestrator. E2E orchestrator analytics receives information and monitors it. If a detected DDoS attack occurs, it will request to mobile NW Controller. When the mobile NW Controller receives a defense request from the E2E orchestrator, it sends a request for dropping the target packets to the UPF via the C-plane bus. The experiment environment was deployed on a virtual machine using UERANSIM [19] and free5GC [20]. Free5GC organizes a 5G network consisting of various functions, i.e., C-Plane functions and UPF. UERANSIM performs the roles of gNBs and UEs. Using the implemented system, we simulate data traffic behavior in the mobile network. In the experiment, we prepared 20 UEs, 1 UPF, and 1 monitoring node for checking IP traffic and the victim-end server in this environment. The data traffic comes from UE to Victim-end via UPF. UPF can set a filter of mitigation based on the request from the source-side E2E orchestrator.

The experiment was conducted on a host server with 20 core processor running at 1.7 GHz using 64 GB of memory, 1 TB of SSD storage, and Ubuntu 22.04 TLS. Free5GC/UERANSIM, the router, and the victim end server were implemented with the virtual machine on the host server.

4.2 Performance Evaluations by the Pilot Implementations Using Open-Source Software on Virtualized Infrastructure

4.2.1 Feasibility of the Proposed Architecture and Interface Design to Mitigate DDoS Traffic

The pilot system collects traffic information at the victim-end monitoring node. The node can get the source IP/Port, Destination IP/Port, and protocol using the t-shark [21] command. The t-shark command makes CSV files of 5 tuples, which are sent to the IP NW controller by named pipe simulated collecting path. IP NW controller receives the CSV file and then shares it with the E2E orchestrator, and the E2E orchestrator visualizes data traffic at the IP network. When the E2E orchestrator detects that this threshold has been exceeded, the orchestrator sends the mitigation request to the mobile controller via a REST API following the Open API specification, as shown in Fig. 6. Open API specification E2E orchestrator transfers it to UPF, UPF receives a drop rule, and UPF shuts out the targeted traffic.

```
$curl -i -X POST http://{IP address }/controller/v1/rules -H "Content-Type:
application/json" ¥
         -H "accept: application/json" -d @- << EOF
         {
            "Action":0,(0: drop)
            "Protocol":17,
            "Filter_SrcIP":"60.60.0.1",
            "Filter_SrcPort":0,
            "Filter_DstIP":"192.168.56.201",
            "Filter_DstPort":5201
         }
         EOF
```

Fig. 6. Example API request about drop packet.

The pilot system monitored all traffic on the monitoring node and detected abnormal traffic from UE 1 at the closest victim end. After that, it set the packet drop rule by requesting from the E2E orchestrator. Figure 7 shows the number of a packet after setting packet drop for UE 1. We demonstrated the feasibility of our collaborative DDoS mitigation system.

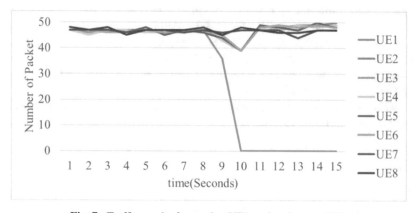

Fig. 7. Traffic graph after setting UE1 packet drop on UPF.

4.2.2 Performance Evaluation of UPF Filtering

We measured the following two metrics to analyze filter performance on UPF.

1. Number of filters that can be set on the UPF without decreasing throughput.
2. The throughput of the legitimate traffic indicates the effective use of resources after dropping the DDoS attack.

Metrics 1 and 2 are measured with iperf3 [22]/UDP protocol.

4.3 Open-Source Software for Filtering Function on UPF

The implemented system used the gtp5g [23] and libtp5gnl [24] modules. Using these modules, we set the filter information and forward information by using the gtp5g-tunnel command and processing PDR for looking up the packet coming in UPF. Also, the gtp5g module capsulizes and decasualizes the user plane packets at UPF. We focus on this module to use a DDoS mitigation system and modify these modules to enable understanding of five-tuple information (i.e., Source IP, Source Port, Destination IP, Destination Port, and Protocol) (Fig. 8).

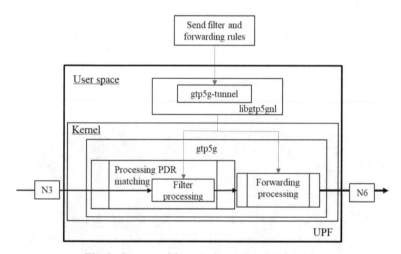

Fig. 8. Process of filter setting and packet parsing.

4.4 Evaluation by Experiments

First, we confirmed the number of filters that can be set in UPF without degrading the throughput. We installed a traffic filter that did not relate to UE's traffic, which performed 1 Mbit/sec from 20 UEs with iperf3. As the number of filters increased, the N6 interface throughput started degrading when 100 filters were activated. The throughput was approximately 3% less than that of no filter, as shown in Fig. 9. The throughput decreased to 90.3% when 500 filters were added. UPF performance decreased as the number of filters increased. Second, we demonstrated the effective use of resources after dropping the DDoS attack. We show the throughput percentage of each UE, as illustrated in Fig. 10. This experiment dropped traffic from UE1-5 in sequence every 5 s; however, the total amount of outgoing traffic from UPF was almost the same with no setting filter. If traffic from a UE is dropped, the other UE would use the room of the resource, so the sum of traffic volume is not changed. So, UPF reallocates the resources to the user's traffic fairly, which leads to effective use of the mobile network resource.

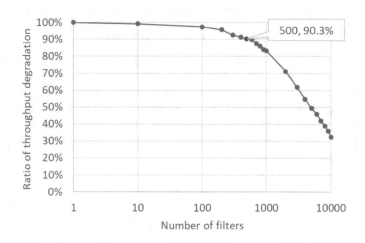

Fig. 9. The ratio of throughput degradation according to the number of filters set in UPF.

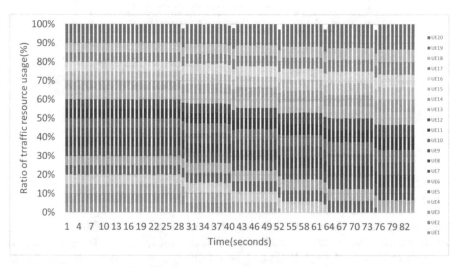

Fig. 10. The ratio of traffic resource usage.

5 Conclusions

We have developed a new approach to defend against DDoS attacks by collaborating with mobile and IP networks with the E2E orchestrator and showed the system's feasibility by constructing a pilot open-source-based environment. We confirmed the performance of UPF filtering according to 3GPP by using open-source software on a virtualized infrastructure.

Regarding the UPF performance, we found that setting a filter of 500 reduced UPF performance to 90%. In late 2016, The Mirai botnet overwhelmed several well-known targets with a massive distributed denial-of-service (DDoS) attack, significantly impacting the network. The number of devices with the malware is reaching a population of 200,000–300,000 infections, and the DDoS attack was performed from 906 ASes and from 85 countries [25]. Assuming an even distribution of devices per AS, the required filters in a single UPF would be less than 300 since it is common to install multiple UPF units connecting to other networks, such as for redundancy. The performance of UPF when the number of filters is 300 decreases by 8%, but since UPF is expected to be redundant, the performance degradation rate is not considered to be large. To investigate further scalability to deploy the commercial is necessary.

In addition, after the attack traffic was dropped at UPF, and legitimate user traffic increased. In this experiment, when traffic was dropped from five units, the traffic usage of the other 15 legitimate users increased by a factor of 1.3. Therefore, we believe that fair use of the mobile network is feasible.

We plan to use open-source E2E orchestrators called ONAP [17]. Using ONAP will further make the proposed system compliant with standard specifications. Also, studies will be conducted to improve the scalability of the UPF and performance of the UPF.

Acknowledgments. This work is partly supported by the commissioned research (02501) of the National Institute of Information and Communications Technology (NICT), Japan.

References

1. ITU-R, IMT Traffic Estimates for the Years 2020 to 2030, Report ITU-R M.2370-0 (2015)
2. Oikonomou, G., Mirkovic, J., Reiher, P., Robinson, M.: A framework for a collaborative DDoS defense. In: Proceedings of Annual Computer Security Applications Conference (ACSAC), pp. 33–42, December 2006
3. Shin, S., Kim, K., Jang, J.: D-SAT: detecting SYN flooding attack by two-stage statistical approach. In: Proceedings of Symposium on Applications and the Internet (SAINT), pp. 430–436, February 2005
4. Wang, H., Shin, K.G.: Transport-aware IP routers: a built-in protection mechanism to counter DDoS attacks. IEEE Trans. Parallel Distrib. Syst. **14**(9), 873–884 (2003)
5. Mirkovic, J., Prier, G., Reiher, P.L.: Attacking DDoS at the source. In: Proceedings of IEEE International Conference on Network Protocols (ICNP), pp. 312–321, November 2002
6. netfilter/iptables project. https://www.netfilter.org/
7. Nawrocki, M., Blendin, J., Christoph, D., Schmidt, T.C., Wählisch, M.: Down the block hole: dismantling operational practices of BGP blackholing at IXPs. In: Proceedings of Internet Measurement Conference (IMC), October 2019

8. Saad, R., Nait-Abdesselam, F., Serhrouchni, A.: A collaborative peer-to-peer architecture to defend against DDoS attacks. In: Proceedings of IEEE Conference on Local Computer Networks (LCN), pp. 427–434, October 2008
9. Rashidi, B., Fung, C., Bertino, E.: A collaborative DDoS defence framework using network function virtualization. IEEE Trans. Inf. Forensics Secur. **12**(10), 2483–2497 (2017)
10. Mortrnsen, A., Reddy, T., Moskowitz, R.: DDoS open threat signaling (DOTS) requirements. IETF, RFC 8612 (2019)
11. 3GPP, System Architecture for the 5G System, 3GPP TS 23.501 V17.5.0, June 2022
12. Thomson, M., Benfield, C.: HTTP/2, IETF, RFC 9113, June 2022
13. 3GPP, 5G System; Technical Realization of Service Based Architecture, 3GPP TS 29.500 V17.5.0, June 2022
14. OpenAPI, OpenAPI Specification Version 3.0.0. https://spec.openapis.org/oas/v3.0.0
15. 3GPP, Interface between the Control Plane and the User Plane nodes, 3GPP TS 29.244, V17.5.0, June 2022
16. ETSI, Network Functions Virtualisation (NFV); architectural framework, Technical report ETSI GS NFV 002 (2018)
17. ONAP. https://www.onap.org/
18. ETSI, Zero-touch network and service management, Technical report ETSI GS ZSM 002 (2019)
19. Güngör, A.: UERANSIM. https://github.com/aligungr/UERANSIM
20. The free5GC project, free5GC. https://www.free5gc.org/
21. tshark. https://tshark.dev/
22. iperf3. https://iperf.fr/
23. The free5GC project, gtp5g - 5G compatible GTP kernel module. https://github.com/free5gc/gtp5g
24. The free5GC project, libgtp5gnl - netlink library for Linux kernel module 5G GTP-U. https://github.com/free5gc/libgtp5gnl
25. Antonakakis, M., April, T., Bailey, M., et al.: C. Understanding the Mirai botnet. In: USENIX Security Symposium, pp. 1093–1110, August 2017

Distributed Algorithm for Localization of Localizable Wireless Sensor Networks

Saroja Kanchi[✉]

Department of Computer Science, Kettering University, Flint, MI 48504, USA
skanchi@kettering.edu

Abstract. In this paper, a distributed algorithm is presented that localizes large number of sensors in localizable wireless sensor networks. It is well known that a network is localizable if and only if the underlying graph is globally rigid, however, even for globally rigid graphs, localization problem is proven to be NP-Hard. We develop a polynomial time distributed algorithm for localization on localizable networks based on annexation of localizable nodes using trilateration, bilateration and edge reconciliation. We study the performance of the algorithm for networks that have varying node densities, sensor radii and connectivity of the underlying graph. We assume that there are three anchor nodes available for localization. Even for networks that whose underlying globally rigid graph is sparse, we demonstrate that the algorithm performs exceptionally well by localizing large percentage of nodes.

1 Introduction

Wireless sensor networks (WSN) are a set of sensor nodes deployed in a geographical area with a specific mission such as surveillance, monitoring and rescue operations. Sensors are equipped with trans-receivers, limited energy source, limited computing capability and small amount storage capability. Locations of sensors are not known due to aerial deployments, movement of sensors post deployment or movement of devices equipped with sensors. Locating sensors in a WSN is essential to success of these missions. Location of sensors is indicative of the location of sensed data and is therefore required in tasks such as providing imminent medical care, land mine detection and toxic agent detection etc. Finding the geo-locations of sensors is called the localization problem of wireless sensor networks.

Clearly, one of the techniques that could be used for locating sensors is to equip the sensors with a (Global Positioning System) GPS. However, this technique is not feasible due to the cost of equipping large number of sensors with GPS and large power consumption given the limited energy available at the sensor node. Moreover, GPS does not work in indoor, underwater and environments with obstructions. Therefore, various novel techniques have been developed for localization of wireless sensor networks.

Anchors are special sensors whose location is known either due to being equipped with GPS or due to being placed at a fixed location. Algorithms for localization use the availability of anchors as a means of localizing non-anchor nodes. In addition, signals are sent/received to/from other sensors that are within the communicating radius of the

© The Author(s), under exclusive license to Springer Nature Switzerland AG 2023
L. Barolli (Ed.): AINA 2023, LNNS 661, pp. 21–30, 2023.
https://doi.org/10.1007/978-3-031-29056-5_3

sensor. This signal exchange mechanism helps determine how far the nearby sensor is, by using received signal strength (RSS) or time difference of arrival between signals (TDOA). Localization algorithms use this signal information to estimate the (range) distance between sensors. Often angle of arrival (AoA) of signal from neighboring sensors might be available which helps determine the position of neighbor sensor relative to itself.

There have been numerous techniques for localization [5, 16]. The approaches vary depending on different assumptions of the WSN including, if range information is available (range-based) or not (range-free), angle is available (AoA) or not, whether any anchor is available (anchor-based) or not (anchor-free), and, if the algorithm propis centralized and distributed. Localization techniques based on MDS (multidimensional scaling) use a reduction technique that use dissimilarity matrix to find relative positions of nodes using the values of the dissimilarity. These values are then mapped to absolute positions by using available anchors [11]. Another technique is to formulate it as an optimization problem and then solve it by using semi-definite programming (SDP) [2]. One of the popular range-free approaches in centroid technique [4], in which all the anchors broadcast their positions and centroid of the locations of the anchors is used as an approximation to the location of the node. A variation of this called DV-hop where hop count to the anchor is used in determining the location.

In range-based localization, the problem of localization can be viewed as a mapping of nodes of the underlying graph of WSN on to a 2D plane. Rigidity properties are graph are used for localization in [10, 14, 17], by merging rigid patches. In [17], graph rigidity is used to guide robotic navigation in mobile sensor networks. In [10] the authors propose an algorithm for finding large rigid regions in a wireless sensor network. In [14], graph theoretic path planning is used localize in the presence of cases of radio irregularity, obstacle, asymmetrical distribution.

It is known that a WSN is localizable if and only if the underlying graph is globally rigid [1, 6]. However, even for globally rigid graphs, localization problem is NP-hard [7]. In this paper, we present a range-based distributed algorithm for localizing wireless sensor networks which are localizable, i.e., given that underlying graph is globally rigid. We assume the presence three anchors in the network.

The paper is organized as follows: In Sect. 2, we provide brief introduction to graph rigidity and approaches to localization. The distributed algorithm is presented in Sect. 3 with results of simulation discussed in Sect. 4. Conclusion and future research are presented in Sect. 5.

2   Graph Rigidity and Localization

Here we introduce graph rigidity and its relation to localization. For detailed discussion please see [1]. In this paper, we assume that the sensor radius is same for all the sensors in the network and is denoted by r. While nodes are aware of distance to other nodes that are within sensor radius, nodes do not their own location or the location of their neighbors. The three anchor nodes in the network have the same sensor radius as non-anchor nodes.

Given such a wireless sensor network, we can represent the underlying structure as an undirected graph $G = (V, E)$ where V represents the set of sensor nodes and E represents the set of connections or edges between sensor nodes that are within the sensor radius. The weight of each edge in E is the distance between two neighboring sensor nodes. The problem of 2D localization of wireless sensor network is the problem of determining the geolocations of vertices of V on the plane, such that, the edge weight for each edge (u, v) in E is equal to the Euclidean distance between geolocations of u and v.

Rigidity of graphs have studied as a graph-theoretic property much before wireless sensor networks existed [11]. A bar-and-joint framework was termed *generically rigid* if it has only trivial deformations, i.e., translations and rotations. The bars can be thought of as edges and joints as vertices of the underlying graph of a WSN. The concept of generic rigidity was characterized combinatorially by Laman [13] as given in Theorem 1.

Theorem 1 (Laman). The edges of a graph $G = (V, E)$ are independent in two dimensions if and only if no induced subgraph $G' = (V', E')$ has more than $2n' - 3$ edges, where n' is the number of nodes in G'.

Corollary 1. A graph with $2n - 3$ edges is generically rigid in two dimensions if and only if no induced subgraph G' of G has more than $2n' - 3$ edges, where n' is the number of nodes in G'.

Laman's theorem can be intuitively explained as follows. For a two-dimensional graph with n vertices, each having two degrees of freedom, the positions of its n vertices have $2n$ degrees of freedom. Of these possible degrees of freedom, three are rigid body motions of translation, rotation and reflection. Therefore, one can see that a graph is generically rigid if there are $2n - 3$ constraints. If each edge adds an independent constraint, then $2n - 3$ edges should be required to eliminate all nonrigid motions of the graph. A graph therefore needs $2n - 3$ independent edges to be generically rigid.

Clearly, if any induced subgraph with n' vertices has more than $2n' - 3$ edges then these edges cannot be independent which leads the Laman's theorem stated above. The corollary follows since the graph with $2n-3$ edges has no non-independent edges. Generically rigid graphs have flip ambiguity as seen in Fig. 1.

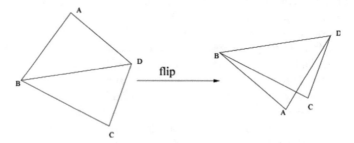

Fig. 1. Generically rigid graph.

Consider a graph $G = (V, E)$ is generically rigid but contains more than $2n - 3$ edges. An edge is called a *redundant edge* if graph remains generically rigid after its removal. G is called a *redundantly rigid* graph if $G - e$ is generically rigid for all $e \in E$. A graph $G = (V, E)$ is *globally rigid*, on R^2, if, there is mapping the set V onto R^2, such that length of edges (bar length) match with distance between pair of points corresponding to joints. It turns out that realization of globally rigid graph onto plane corresponds to localization of wireless sensor network represented by the graph. Therefore, determining if the underlying graph of a WSN is globally rigid and thus realizable on the plane, and, finding a realization of the globally rigid graph onto the plane are central to localization of wireless sensor networks.

The computational feasibility of determining if a graph is globally rigid therefore requires combinatorial characterization of global rigidity and this was provided by [13].

Theorem 2: A graph G globally rigid in R^2 if and only if G is *3*-connected and redundantly rigid.

Therefore, combining Laman's characterization of generically rigid graphs in combination with Theorem 2 gives a computationally feasible algorithm for determining if a graph is globally rigid. However, given Laman's theorem, checking every induced subgraph is computationally exponential, therefore, pebble game algorithm by [9] makes the determination of generic rigidity polynomially computable.

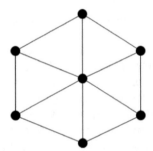

Fig. 2. A Wheel Graph with 6 rim vertices

Using pebble game algorithm, one can compute if a graph is generically rigid, and by removing each edge and checking for generic rigidity, one can determine if a graph is redundantly rigid. By Theorem 2, a graph that is redundantly rigid and 3-connected is globally rigid, thus, leading to polynomial time algorithm for checking global rigidity.

On the other hand, it is proved in [7], the problem of localization of a globally rigid graph is NP-Hard, thus making the localization of localizable graph intractable.

A special case of globally rigid graph, i.e. a trilateration is graph is one that admits ordering of its vertices in V as $\{v_1, v_2, \ldots v_n\}$ such that each v_i $(i \geq 3)$ has at least 3 edges to vertices in the set $\{v_1, v_2, \ldots v_{(i-1)}\}$. For trilateration graphs, the localization problem is polynomially solvable [1]. The wheel in Fig. 2 is an example of trilateration graph. It is known that when the connectivity of graph is 6 or above, the graphs can be proven to be trilateration graphs and are thus localizable. In this paper we focus on developing an algorithm for localizable WSN whose connectivity is less than 6.

3   Distributed Algorithm for Localization

In this section we describe the distributed algorithm for localization. The algorithm is based on trilateration, bilateration and edge reconciliation techniques. In each iteration, each node n that is not yet localized, processes each of its incident edges. It scans its neighbors along each edge to check if any of the neighbors were localized in the previous iteration. If a node finds its neighbor localized, the node looks for two other localized neighbors, and localizes itself using trilateration, as in Fig. 3. The node n localizes itself to position A, if it finds three localized neighbors positioned at P_1, P_2 and P_3 at distances d_1, d_2 and d_3 respectively by finding the intersection the three circles.

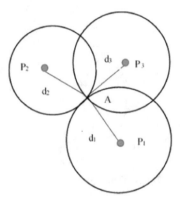

Fig. 3. Trilateration

However, if the node n finds two of its neighbors localized to positions P_1 and P_2 at distances d_1 and d_2, it can find two possible locations A_1 and A_2, using bilateration which at the intersection of the two circles as shown in Fig. 4. Node n adds the positions A_1 and A_2, to the list of possible positions for n.

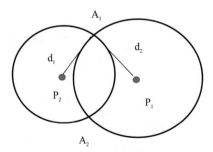

Fig. 4. Bilateration

If node n finds only one neighbor l localized at position P at distance d_1, it performs enhanced unilateration as follows: it looks for one other neighbor n_1 which is at distance

d_2 that have two (say $P^1{}_1$ and $P^2{}_1$) or more possible positions. Using the p possible locations of its neighbor and the one unique location of its localized neighbor l, the node determines at most $2p$ possible locations for itself which are obtained by intersection points of circles of radius d_1 with center as position of l and other as positions of n_1. Since n_1 has p possible locations, the node n adds at most $2p$ possible locations. Note that not all positions of n_1 may at distance d_2 intersect with circle centered at the position of l. See Fig. 5.

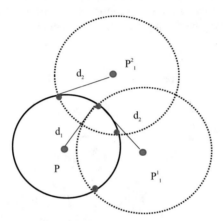

Fig. 5. Enhanced Unilateration

If a node n finds the neighbor n_1 at distance d that has q possible positions and the node n has p possible positions, algorithm jointly localizes n and n_1 to positions using edge reconciliation. In edge reconciliation, distance between possible position of n and possible position of n_1 are matched with d. Once such position pair is found, the algorithm jointly localizes node n and its neighbor n_1. See Fig. 6.

If the node n finds that all its neighbors have no possible positions then no update occurs at n.

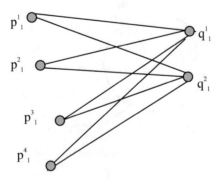

Fig. 6. Edge Reconciliation

The algorithm continues either until all nodes are localized or localization information of a node has reached along the diameter of the network.

Anchors are placed on 3 vertices that belong to triangle to facilitate the wheel around
it to be localized over the first few iterations.

```
program DistriButedLocalization

        While (there is an unlocalized node or number of
iterations is less than the diameter of the graph ) each
unlocalized node n do:

   For each neighbor of n that has reported change in
possible positions in previous step:

            a.1 If a neighbor becomes localized,
increase number of localized neighbors.

            a.1.1:   If number of localized
neighbors is now 3, localize node n using trilateration.

            a.1.2: If    number of localized
neighbors is now 2, determine two possible positions for
n using bilateration.

            a.1.3: If    number of localized
neighbors is now is 1, then if n has 2 or more possible
positions, resolve the positions by using the edge weight
to the newly localized neighbor.

            a.2 If a neighbor m goes from 0 to higher
number of positions and i  f n has   non-zero number of
possible positions resolve the edge weight between n and
m and jointly localize n and m.

            a.3 If a neighbor goes from 0 to higher
number of positions and n has zero possible positions,
and there is   only one other  localized neighbor, then
create positions for each pair of positions of m.

   endFor

endWhile
```

3.1 Time and Message Complexity

Since the algorithm has D iterations where D is the diameter of the network, and in each
iteration, each edge is processed once by each endpoint of the edge, the time complexity
is $O(D*e)$ leading to a polynomial algorithm.

In each iteration, each node exchanges information along each edge. The amount of
information is the possible positions at the neighbor. Since the graph is globally rigid and
therefore 3-connected, the number of possible positions is limited to a constant K. The
value of K that is reached for random geometric graphs is shown in the results section.

4   Results

The simulation was performed by creating random geometric graphs, by placing nodes in a 100 × 100 square. Given a specified radius, nodes were placed using uniform distribution with real value co-ordinates in the 100 × 100 square. The sensor radius was 16 or higher since it is shown in [1] that for a network with 100 × 100, as the radius approaches 16, the graphs becomes globally rigid graphs with probability close to 1. To ensure global rigidity, 3-connectivity is verified and redundant rigidity is checked by using pebble game algorithm.

Node density d is the number of nodes that is within the radio range of nodes. If there are n nodes, the node density of $d = (n*\pi*r^2)/10,000$, were r is the sensor radius of the network. Sparse graphs have node density below *10*.

The Figs. 7 and 8 and show globally rigid networks generated with radius *16* and *17* with *80* and *100* nodes respectively. The node density of the network in Figs. 7 and 8 are *6* and *9* respectively.

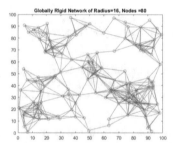   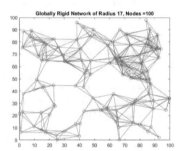

Fig. 7. Network of Radius 16   **Fig. 8.** Network of Radius 17

Figure 9 and Fig. 10 show the performance of the algorithm with increasing radius from *16* to *20* keeping the node density fixed at 9 and 10 respectively. Each data point is collected by averaging *50* randomly generated networks of the specific radius and node density. It can be seen that percentage of localized nodes is above *80%* for node density 9 for all radii between *16* and *20*, and, above *90%* for networks of node density 10 for all radii between *16* to *20*. In addition, when node density increases the percentage of localized nodes increases as well.

Next, we examine the performance on networks that 3-connected, 4-connected and 5-connected. Note that globally rigid graphs with connectivity 6 or above are trilateration graphs and thus are localizable in polynomial time by previously known algorithms.

To generate graphs that k-connected, ($k = 3, 4, 5$), graphs with minimum degree k are generated, since, for geometric graphs of minimum degree k are also k-connected with asymptotic probability of 1 [1]. It can be seen in Fig. 11 that when 3-connectivity is present, 80% of the nodes are localized and as connectivity increases to 4 or 5 then 100% of the nodes are localized. Figure 12 shows the maximum number of possible values at any node of the network for random networks of different number of nodes. It can be observed that as number of nodes increases the maximum length of messages reduces and it is in the order of number of nodes.

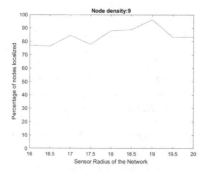

Fig. 9. Localization with d = 9

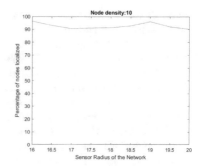

Fig. 10. Localization with d = 10

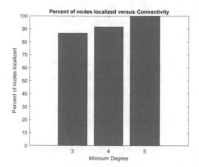

Fig. 11. Localization versus Connectivity

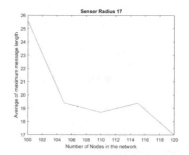

Fig. 12. Message Length versus nodes

5   Conclusion and Future Work

In this paper, we have presented a distributed algorithm for localization of localizable graph. The algorithm localizes 80% to 100% of the nodes even for sparse networks with low node density. The future work would be to examine how to identify specific structures within the globally rigid graph that help with localizing in polynomial time.

References

1. Aspnes, J., et al.: A theory of network localization. IEEE Trans. Mob. Comput. **5**(12), 1663–1678 (2006)
2. Biswas, P., Toh, K., Ye, Y.: A distributed SDP approach for large-scale noisy anchor-free graph realization with applications to molecular conformation. SIAM J. Sci. Comput. **30**(3), 1251–1277 (2008)
3. Bose, K., Kundu, M.K., Adhikary, R., Sau, B.: Distributed localization of wireless sensor network using communication wheel. Inf. Comput. **289**, 104962 (2022)
4. Bulusu, N., Heidemann, J.S., Estrin, D.: GPS-less low-cost outdoor localization for very small devices. IEEE Wirel. Commun. **7**(5), 28–34 (2000)
5. Chowdhury, T.J., Elkin, C., Devabhaktuni, V., Rawat, D.B., Oluoch, J.: Advances on localization techniques for wireless sensor networks: a survey. Comput. Netw. **110**, 284–305 (2016)

6. Eren, T.: Graph invariants for unique localizability in cooperative localization of wireless sensor networks: rigidity index and redundancy index. Ad Hoc Netw. **44**, 32–45 (2016)
7. Eren, T., et al.: Rigidity, computation, and randomization in network localization. In: IEEE INFOCOM 2004, vol. 4, pp. 2673–2684. IEEE, March 2004
8. Fang, J., Morse, A.S.: Network localization using graph decomposition and rigidity. In: 2008 47th IEEE Conference on Decision and Control, pp. 1091–1096. IEEE, December 2008
9. Jacobs, D.J., Hendrickson, B.: An algorithm for two-dimensional rigidity percolation: the pebble game. J. Comput. Phys. **137**(2), 346–365 (1997)
10. Kanchi, S., Welch, C.: An efficient algorithm for finding large localizable regions in wireless sensor networks. Procedia Comput. Sci. **19**, 1081–1087 (2013)
11. Jackson, B., Jordán, T.: Connected rigidity matroids and unique realizations of graphs. J. Comb. Theory Ser. B **94**(1), 1–29 (2005)
12. Jia, D., Li, W., Wang, P., Feng, X., Li, H., Jiao, Z.: An advanced distributed MDS-MAP localization algorithm with improved merging strategy. In: 2016 IEEE International Conference on Information and Automation (ICIA), pp. 1980–1985. IEEE, August 2016
13. Laman, G.: On graphs and rigidity of plane skeletal structures. J. Eng. Math. **4**, 331–340 (1970)
14. Ma, X., Yu, N., Zhou, T., Feng, R., Wu, Y.: Graph theory based localization of wireless sensor networks for radio irregularity cases. In: 2020 International Conference on Sensing, Measurement & Data Analytics in the Era of Artificial Intelligence (ICSMD), pp. 51–56. IEEE, October 2020
15. Rai, B.S., Varma, S. (2016). An algorithmic approach to wireless sensor networks localization using rigid graphs. J. Sens. **2016**, 1–11 (2016)
16. Saeed, N., Nam, H., Al-Naffouri, T.Y., Alouini, M.S.: A state-of-the-art survey on multidimensional scaling-based localization techniques. IEEE Commun. Surv. Tutorials **21**(4), 3565–3583 (2019)
17. Wu, C., Zhang, Y., Sheng, W., Kanchi, S.: Rigidity guided localisation for mobile robotic sensor networks. Int. J. Ad Hoc Ubiquitous Comput. **6**(2), 114–128 (2010)

Availability Model for Byzantine Fault-Tolerant Systems

Marco Marcozzi[1]([✉]), Orhan Gemikonakli[2], Eser Gemikonakli[3], Enver Ever[4], and Leonardo Mostarda[1]

[1] Computer Science Division, University of Camerino, Camerino, Italy
{marco.marcozzi,leonardo.mostarda}@unicam.it
[2] Faculty of Engineering, Final International University, Kyrenia, Cyprus
orhan.gemikonakli@final.edu.tr
[3] Department of Computer Engineering, Faculty of Engineering,
University of Kyrenia, Girne, Mersin 10, Kyrenia, Turkey
eser.gemikonakli@kyrenia.edu.tr
[4] Computer Engineering, Middle East Technical University Northern Cyprus
Campus, Mersin 10, 99738 Guzelyurt, Turkey
eever@metu.edu.tr

Abstract. The growth in the complexity and extensibility of computer systems have caused vulnerabilities such as exploitable software bugs and configuration flaws. In turn, confirming computer security is becoming an increasingly important task. Byzantine fault-tolerant algorithms are popularly used to allow systems automatically continue operating. In addition, Byzantine Fault-Tolerant Systems are used in blockchain networks, commonly in tandem with other consensus mechanisms. This study proposes an analytical availability model which is critical for the evaluation of fault-tolerant multi-server systems. A model is proposed based on continuous-time Markov chains to analyse the availability of Byzantine Fault-Tolerant systems. Numerical results are presented reporting availability as a function of the number of participants and the relative number of honest actors in the system. It can be concluded from the model that there is a non-linear relationship between the number of servers and availability inversely proportional to the number of nodes in the system. This relationship is further strengthened as the ratio of honest malicious nodes to the total number of nodes increases.

1 Introduction

Fault-Tolerance is a fundamental concept widely applied in industrial, information, and communication systems [11,15]. Tolerance to faults allows systems to continue working, even when some system components are not available. Undoubtedly, fault-tolerance found many applications in engineering (e.g. aerospace [28], avionics [8], automotive [5]) and computing (e.g. cloud computing [1,4,13,16], distributed systems [3,7]).

L. Barolli (Ed.): AINA 2023, LNNS 661, pp. 31–43, 2023.
https://doi.org/10.1007/978-3-031-29056-5_4

However, a vast interest on distributed computer systems came with the advent of blockchains and their applications in cryptocurrencies, decentralised and distributed computing (e.g. Smart Contracts), and in cyber-physical systems (e.g. IoT and Industry 4.0). Such distributed systems require a consensus protocol to handle the write/read operations on their memories, or to perform some actuation scheme. Moreover, these types of networks need a certain degree of fault-tolerance.

There are various possible implementations to achieve fault-tolerance in a computer system, as well as different levels of tolerance, but this work focuses on systems tolerant to Byzantine faults [17].

A Byzantine fault occurs when a node is acting maliciously in the network, e.g. sending contradictory messages to separate servers or being unresponsive. However, a node might also not act maliciously in the network, and yet be unresponsive due to crash or connection failures. In both cases, the system may fail to reach consensus.

In an implementation of Byzantine Fault-Tolerant (BFT) consensus protocol with $N \geq 4$ servers (if $N < 4$ the problem would not have a solution) exchanging unsigned messages, quorum (the minimum amount of committing messages to achieve consensus) is reached when the number of honest responsive nodes H is

$$H > 2N/3. \tag{1}$$

Therefore, such fault-tolerant systems, implemented by means of unsigned messages, can handle a number of Byzantine faults F, such that

$$F < N/3. \tag{2}$$

When considering developing a BFT protocol, availability and performance evaluation can be used effectively to ensure a successful working protocol. In fact, it is economically and technically challenging to implement a network based on a BFT consensus, and realise subsequently that the network is not achieving the expected level of availability. Indeed, to evaluate availability of complex systems, Continuous-Time-Markov-chains have been widely and successfully applied in the last few decades [6, 12].

In this work, a model based on a continuous-time Markov chain, describing systems in which participants achieve consensus through a Byzantine Fault-Tolerant process is proposed. The relationship between the number of participants, the breakdown and repair rates are investigated to find system configurations for the best availability.

The rest of the paper is organised as follows: a review of previous work on analytical availability evaluation is reported in Sect. 2; Sect. 3 presents the proposed availability model together with assumptions made; in Sect. 4 the mathematical requirements and conditions required to reach a solution are explored and presented together with analysis carried out; lastly, Sect. 5 gives a summary of the results obtained, with possible applications and future developments of the presented model.

2   Related Work

Availability can be defined as the ability of a considered system to be in a state to perform an operation at any instant time within a given time interval. In other words, it refers to failure-free operation at a given instant of time [25]. It is also possible to define instantaneous Availability or point availability $A(t)$ (the probability that the component is properly functioning at time t) as:

$$A(t) = R(t) + \int_0^t R(t-x)m(x)dx$$

where $R(t)$ is the probability of having no failure in interval $(0, t]$ and $m(x)$ is the repair density. Clearly considering this equation, the system can be available either if there were no failures in interval $(0, t]$, or in case there were failures, their repair is completed before time t [25].

Following this, it is possible to introduce the mean time to failure $MTTF$ and the mean time to repair $MTTR$. Then the mathematical definition of limiting availability A can be given as

$$A = \lim_{t \to \infty} A(t) = \frac{MTTF}{MTTF + MTTR} \tag{3}$$

The limiting availability depends only on the mean time to failure and mean time to repair, and not on the nature of the distributions of failure times and repair times.

Availability evaluation of multi-server systems has attracted considerable interest in the relevant literature. Research articles, books, and surveys provide a wealth of information on the subject [26]. Along with other examples, availability models relying on Markov chains are used in various studies.

In [23] an approach is presented for a healthcare IoT infrastructure. There, two-dimensional continuous-time Markov chains are presented for functional states of the healthcare IoT infrastructures and the end nodes. A case study with a Markov model considering attacks on vulnerabilities of the healthcare IoT system is also considered. In turn, a state diagram is also provided for attacks on healthcare IoT infrastructure. Availability is presented as a function of intensities of service requests flow. The main focus is on safety and security-related issues. The availability of healthcare IoT systems is studied in [24] as well. Two groups of structures which are the components of the IoT system considered are described by separate Markov state-space models. A two-dimensional state space representation is established and the system balance equations are solved similar to our approach. Availability related performance metrics of interest such as probabilities of full service, degraded service, and system unavailability, are presented.

The availability of IoT systems is considered in studies such as [10, 21], while facilitating infrastructures are modelled in presence of failures in studies such as [14]. In [21] analytical availability models are presented and the availability of physical edge and fog nodes running applications are evaluated. $MTTF$ and

$MTTR$ values are computed for the systems under study, and a two-dimensional Markov model is presented which includes failures and repairs. In [10] the availability, performance, and energy consumption-related measures are presented as results of the evaluation for clustered IoT systems. Two-dimensional models are solved for steady-state probabilities. In turn, these probabilities are used to compute critical availability-related measures such as the probability of being in a fully operative state, and performance measures such as mean energy consumption.

When cloud systems, particularly Infrastructure as a Service (IaaS) based ones are considered, one of the limiting factors for modelling attempts is the scale of the systems. In [2], the scalability-related problems for large cloud systems are tackled using approximate Stochastic Reward Net (SRN) models together with folding and fixed-point iteration techniques. The presented approach is able to capture the failure/repair behaviour of the physical machines. The percentage of available physical machines is analysed for different failure and repair rates. In [18], the main focus is on the high availability of IaaS cloud systems. This time, to reduce the complexity in terms of analysis and the solution time, an interacting Markov chain based approach is employed. SRNs are used in this study as well for the solution of the Markov chains. The presented availability models are used to perform trade-off analysis of longer MTTF vs faster MTTR on system availability as well as the effect of having multiple concurrent repair facilities. In [9], a novel approximate solution approach is introduced which allows consideration of large numbers of servers for cloud-based systems. The analytical models and solutions are monolithic unlike the other studies considered, but they are still capable of considering large numbers of facility nodes typically up to orders of hundreds or thousands. Quality of service for cloud centers is considered together with server availabilities, and performability measures are obtained in presence of server failures and repairs.

Finally, in [20], blockchain-based systems that can support service provisioning over cloud infrastructures are considered. In turn, models are presented for assessing the availability and capacity-oriented availability of cloud computing infrastructures running blockchain's distributed applications based on the Ethereum blockchain platform. The system's availability is also represented following the traditional approach as the ratio between the $MTTF$ and $MTTR$. Availability results are presented as functions of MTTF and MTTR for server, miner node and bootnodes.

In studies considered where various distributed systems are modelled for availability evaluation, the failure and repair times are assumed to follow an exponential distribution similar to our approach in this study. In this work, our main focus is on analyzing the availability of the system by means of a model based on Markov chains. When the existing literature is analysed, many of the systems are modeled using Markovian processes, with a common formalism and terminology. However, to the best of our knowledge, we are the first in using this formalism to model the availability of BFT systems.

3   System Model

We consider a system with N participants, entrusted to work on certain tasks by exchanging messages with each other, either in a point-to-point or in a broadcast fashion, as shown in Fig. 1.

Fig. 1. Representation of a BFT scheme with $N = 7$. The white and black nodes represent non-Byzantine and Byzantine nodes respectively.

Even though this scheme might be applied to a wide range of systems, as a working example, we use a network of N servers (also called replicas or nodes) committing messages, each to their own storage, as it happens in a blockchain or a distributed database.

The model is based on a continuous-time Markov chain in the form of a quasi-birth-death process. Continuous-time Markov chains are stochastic processes in which random variables are exponentially distributed and the system can change state at a rate defined in the stochastic transition matrix of the process. As its name suggests, a continuous-time Markov chain satisfies the Markov property, i.e. the probability distribution of future states of the process conditioned on both the past and present states depends only on the present state.

36   M. Marcozzi et al.

A quasi-birth-death process is a special case of continuous-time Markov chains, where the parameters ξ and η are the rates at which servers break-down ("death") and are repaired ("birth"), as represented in Fig. 2.

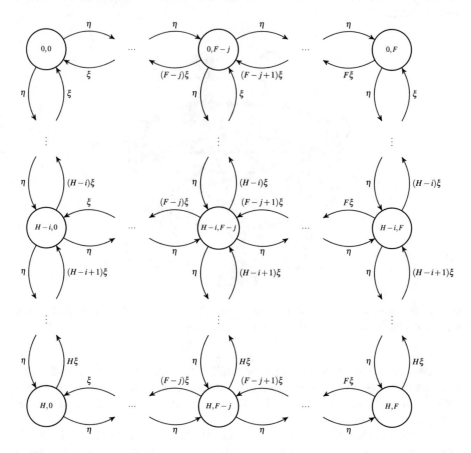

Fig. 2. Availability model for a BFT consensus protocol.

In this model, it is assumed that servers can break-down independently, but they are repaired sequentially, one at the time. Thus, the break-down rate ξ is multiplied by a number reflecting the current number of available nodes, i.e. if there are F nodes, the break-down rate is $F\xi$, while if there is only one node available ξ is the corresponding break-down rate. This is not the case for the repair process, since we are assuming that repairs can occur only one at a time, with repair rate η.

The model is proposed with the following assumptions[1]: in the system there are N servers, of which $H \leq N$ are nodes participating honestly to the network and $F \leq N$ are nodes acting maliciously (F does not take into account breakdowns).

The state diagram, as depicted in Fig. 2, is composed of $(H+1)(F+1)$ states. All the states can be eventually visited from any starting point, thus the chain is irreducible and ergodic. These two conditions are sufficient for the chain to admit a stationary distribution.

Associated with the continuous-time Markov chain, there is the stochastic transition matrix $\mathbf{A}$, in which the transition rates from a state to another are shown. For this model, the transition matrix $\mathbf{A}$ is written by means of the simultaneous equations in Eq. 4. In these equations, $P_{i,j}$ is the probability that the system is in state (i,j), where $i \in [0, H]$ and $j \in [0, F]$, while the coefficients of $P_{i,j}$s are the entries in the matrix $\mathbf{A}$. The goal is to determine the vector of unknowns, $\mathbf{P}$, which is the limiting distribution for the process.

$$
\begin{cases}
2\eta P_{0,0} - \xi(P_{1,0} + P_{0,1}) = 0 \\
(\eta + H\xi)P_{H,0} - \eta P_{H-1,0} - \xi P_{H,1} = 0 \\
(F + H)\xi P_{H,F} - \eta(P_{H-1,F} + P_{H,F-1}) = 0 \\
(\eta + F\xi)P_{0,F} - \eta P_{0,F-1} - \xi P_{1,F} = 0 \\
[2\eta + (H - i)\xi]\,P_{H-i,0} - \eta P_{H-i-1,0} - (H - i + 1)\xi P_{H-i+1,0} - \xi P_{H-i,1} = 0 & \forall i \in [1, H - 1] \\
[2\eta + (F - j)\xi]\,P_{0,F-j} - \eta P_{0,F-j-1} - (F - j + 1)\xi P_{0,F-j+1} - \xi P_{1,F-j} = 0 & \forall j \in [1, F - 1] \\
[\eta + (H - i + F)\xi]\,P_{H-i,F} - \eta(P_{H-i,F-1} + P_{H-i-1,F}) - (H - i + 1)\xi P_{H-i+1,F} & \forall i \in [1, H - 1] \\
[\eta + (H + F - j)\xi]\,P_{H,F-j} - \eta(P_{H-1,F-j} + P_{H,F-j-1}) - (F - j + 1)\xi P_{H,F-j+1} & \forall j \in [1, F - 1] \\
[2\eta + (H + F - i - j)\xi]\,P_{H-i,F-j} - \eta(P_{H-i,F-i-1} + P_{H-i-1,F-j}) + \\
\quad -(H - i + 1)\xi P_{H-i+1,F-j} - (F - j + 1)\xi P_{H-i,F-j+1} & elsewhere \\
\sum_{i=0}^{H} \sum_{j=0}^{F} P_{i,j} = 1
\end{cases}
$$

$$(4)$$

4   Availability Analysis

Availability can be computed from the limiting distribution of the continuous-time Markov process. However, in order to compute the stationary distribution of the chain, it is needed to solve the system of simultaneous linear equations in Eq. 4. This system can be rewritten in the form of $\mathbf{Ax} = \mathbf{b}$, where $\mathbf{A}$ is the coefficient matrix, $\mathbf{x}$ the vector of unknowns, and $\mathbf{b}$ the vector of constants.

Nevertheless, it is important to notice that we are solving a set of linear simultaneous equations, $\mathbf{AP} = 0$, where $\mathbf{A}$ is the transition matrix associated with the continuous-time Markov process, $\mathbf{P}$ the vector of unknowns (probabilities $P_{i,j}$), and 0 the vector of constants (zeros).

There are several techniques that can be used to solve systems of linear equations. However, since it is assumed that a non-trivial solution to the system

---

[1] For simplicity of exposition, it is assumed that N, H, and $F \in \mathbb{N}_0$. Therefore, when dealing with divisions, we are implicitly applying the *ceiling* $\lceil \cdot \rceil$ and *floor* $\lfloor \cdot \rfloor$ functions to H and F, respectively.

of simultaneous linear equations exists, it means that $\mathbf{A}$ is a singular matrix, or at least a matrix with a large condition number, i.e. $\mathbf{P}$ belongs to the null space of $\mathbf{A}$. Since common methods for linear algebra, e.g. LU decomposition, can not be applied effectively (or at all) for nearly-singular matrices, we use the Singular Value Decomposition (SVD) method. Indeed, matrix $\mathbf{A}$ has at least a singular value close to zero, therefore $\mathbf{A}$ admits a non-trivial solution to the linear simultaneous equations.

The linear simultaneous equations are solved for various values of N, H, F, η, and ξ. As computations show, the individual values of η and ξ do not mean much in calculating availability, what matters is the ratio of these two parameters.

The pseudo-code at Algorithm 1 illustrates the procedure used to compute solutions for the system of linear equations.

Algorithm 1. Pseudo-code to obtain the probability vectors $\mathbf{P}$

Require: $N_{max} \geq N_0 \geq 4$ and $0 < \xi/\eta \ll 1$
   $N \leftarrow N_0$
   while $N \leq N_{max}$ **do**
     for $H \in [N, 5N/6, 2N/3 + 1]$ **do**
       $F \leftarrow N - H$
       $\mathbf{A} \leftarrow \mathbf{A}(N, H, F, \xi/\eta)$
       $\mathbf{P} \leftarrow SVD(A, 0)$ ▷ solution through SVD
     end for
     $N \leftarrow N + 1$
   end while

The computational process starts with setting the ratio $\frac{\xi}{\eta}$ to a constant and taking N as an integer constant. After forming the matrix $\mathbf{A}$, $\mathbf{P}$ can be computed through SVD. The process is iterated choosing at each step three different risk levels, i.e. $H = N$ (low), $H = 5N/6$ (medium), and $H = 2N/3 + 1$ (high), computing F accordingly. For each pair of (H, F) we compute the probabilities $P_{i,j}$ for the vector $\mathbf{P}$ (vector $\mathbf{P}$ is technically reshaped to reflect the 2-dimensional structure in Fig. 2). The algorithm runs until a predefined value of N, N_{max}, is reached. At that point, a different ratio ξ/η is selected and the computation process is repeated.

Finally, once all the $\mathbf{P}$ values are computed, availability can be determined. Availability is the cumulative probability that the system is working and it can commit messages. Thus, all the states with $H > 2N/3$ show system availability and the corresponding state probabilities are summed up to calculate the availability for the given values of N, H, and F as shown in Eq. 5.

$$A_H^{(N)} = \sum_{i=\frac{2N}{3}+1}^{H} \sum_{j=0}^{F} P_{i,j}. \tag{5}$$

Results shown in Fig. 3, Fig. 4, and Fig. 5 (where ratios $\xi/\eta = 0.01, 0.015, 0.02$ have been used respectively) are reporting the trend of the availability respect

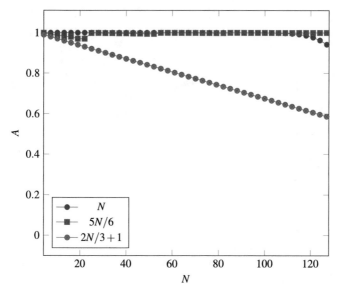

Fig. 3. Availability as a function of N and for different values of H at ratio $\xi/\eta = 0.01$.

to three malicious nodes ratio levels: low, medium, and high. From these results, it appears that when there are no malicious nodes ($H = N$), the availability is non-linearly inversely related to N and availability is degrading faster than when there are malicious nodes for large N. Furthermore, it is remarkable that for $H = 5N/6$ the availability tends to 1 and when $H = 2N/3+1$ the availability decreases linearly with N. Even though results might look counter-intuitive, actually, when compared with the exact solution from Eq. 4, the trend for $H = N$ is legitimate. As a matter of fact, from Eq. 4, the probability for the system to be in state i is

$$P_i = \frac{\eta^i}{i!\xi^i}, \tag{6}$$

therefore the availability tend to zero for large values of N, i.e.

$$A = \sum_{i=2N/3+1}^{N} P_i = \sum_{i=2N/3+1}^{N} \frac{\eta^i}{i!\xi^i} \xrightarrow{N\to\infty} 0. \tag{7}$$

Moreover, the comparison between different ratios ξ/η shows that indeed increasing the ratio ξ/η lowers the value of availability for large N.

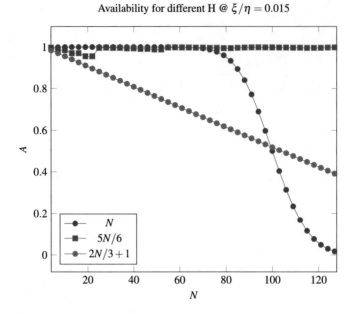

Fig. 4. Availability as a function of N and for different values of H at ratio $\xi/\eta = 0.015$.

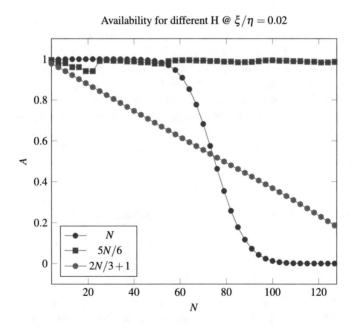

Fig. 5. Availability as a function of N and for different values of H at ratio $\xi/\eta = 0.02$.

5   Conclusion and Future Work

Many distributed systems implement mission-critical applications and hence require high availability. This study proposes an analytical availability model which is critical for the evaluation of fault-tolerant multi-server systems. A model is proposed based on continuous-time Markov chains to analyse the availability of Byzantine Fault-Tolerant systems in the presence of break-downs and repairs and malicious nodes. A system is assumed available when the ratio of honest nodes to malicious nodes is one or higher. The total number of nodes considered changes between 4 and 128. The proportion of malicious nodes has been taken as small, medium and large to represent different risk levels. Numerical results are presented reporting availability as a function of the number of participants and a relative number of honest actors in the system. The contribution of this work is to extend the availability calculation to take into account the presence of malicious nodes. It can be concluded from the model that there is a non-linear relationship between the number of servers and availability inversely proportional to the number of nodes in the system. This relationship is further strengthened as the ratio of honest malicious nodes to the total number of nodes increases. In addition to these findings, the model can be the first step in the performability modelling of blockchain systems.

Further work: In the analyses, the proportion of malicious nodes have been chosen as low, medium, and high. It is of interest to further develop the model to consider the ratio of malicious nodes as a random variable and generalise the findings beyond these three classes. Furthermore, when sufficient data will be available (e.g. from [19, 22, 27]), the model can be enriched with a probabilistic distribution. In the absence of such data, it is possible to consider different probabilistic distributions such as uniform, exponential and Gaussian and obtain results. Finally, the available model can be a good step in modelling blockchain systems for performability evaluation.

References

1. Amin, Z., Singh, H., Sethi, N.: Review on fault tolerance techniques in cloud computing. Int. J. Comput. Appl. **116**(18) (2015)
2. Ataie, E., Entezari-Maleki, R., Rashidi, L., Trivedi, K.S., Ardagna, D., Movaghar, A.: Hierarchical stochastic models for performance, availability, and power consumption analysis of IaaS clouds. IEEE Trans. Cloud Comput. **7**(4), 1039–1056 (2017)
3. Avizienis, A., Laprie, J.-C., Randell, B., Landwehr, C.: Basic concepts and taxonomy of dependable and secure computing. IEEE Trans. Dependable Secure Comput. **1**(1), 11–33 (2004)
4. Bala, A., Chana, I.: Fault tolerance-challenges, techniques and implementation in cloud computing. Int. J. Comput. Sci. Issues (IJCSI) **9**(1), 288 (2012)
5. Baleani, M., Ferrari, A., Mangeruca, L., Sangiovanni-Vincentelli, A., Peri, M., Pezzini, S.: Fault-tolerant platforms for automotive safety-critical applications. In: Proceedings of the 2003 International Conference on Compilers, Architecture and Synthesis for Embedded Systems, pp. 170–177 (2003)

6. Bolch, G., Greiner, S., De Meer, H., Trivedi, K.S.: Queueing Networks and Markov Chains: Modeling and Performance Evaluation with Computer Science Applications. Wiley (2006)
7. Cristian, F.: Understanding fault-tolerant distributed systems. Commun. ACM **34**(2), 56–78 (1991)
8. Edwards, C., Lombaerts, T., Smaili, H., et al.: Fault tolerant flight control. Lecture Notes Control Inform. Sci. **399**, 1–560 (2010)
9. Ever, E.: Performability analysis of cloud computing centers with large numbers of servers. J. Supercomput. **73**(5), 2130–2156 (2017)
10. Ever, E., Shah, P., Mostarda, L., Omondi, F., Gemikonakli, O.: On the performance, availability and energy consumption modelling of clustered IoT systems. Computing **101**(12), 1935–1970 (2019). https://doi.org/10.1007/s00607-019-00720-9
11. Gao, Z., Cecati, C., Ding, S.X.: A survey of fault diagnosis and fault-tolerant techniques-part I: fault diagnosis with model-based and signal-based approaches. IEEE Trans. Industr. Electron. **62**(6), 3757–3767 (2015)
12. Goyal, A., Lavenberg, S.S.: Modeling and analysis of computer system availability. IBM J. Res. Dev. **31**(6), 651–664 (1987)
13. Jhawar, R., Piuri, V.: Fault tolerance and resilience in cloud computing environments. In: Computer and Information Security Handbook, pp. 165–181. Elsevier (2017)
14. Kirsal, Y., Ever, E., Kocyigit, A., Gemikonakli, O., Mapp, G.: Modelling and analysis of vertical handover in highly mobile environments. J. Supercomput. **71**(12), 4352–4380 (2015). https://doi.org/10.1007/s11227-015-1528-3
15. Koren, I., Krishna, C.M.: Fault-Tolerant Systems. Morgan Kaufmann (2020)
16. Kumari, P., Kaur, P.: A survey of fault tolerance in cloud computing. J. King Saud Univ. Comput. Inf. Sci. **33**(10), 1159–1176 (2021)
17. Lamport, L., Shostak, R., Pease, M.: The Byzantine generals problem. ACM Trans. Program. Lang. Syst. **4**(3), 382–401 (1982)
18. Longo, F., Ghosh, R., Naik, V.K., Trivedi, K.S.: A scalable availability model for infrastructure-as-a-service cloud. In: 2011 IEEE/IFIP 41st International Conference on Dependable Systems & Networks (DSN), pp. 335–346. IEEE (2011)
19. Mehmood, N.Q., Culmone, R., Mostarda, L.: Modeling temporal aspects of sensor data for MongoDB NoSQL database. J. Big Data **4**(1) (2017)
20. Melo, C., Dantas, J., Pereira, P., Maciel, P.: Distributed application provisioning over ethereum-based private and permissioned blockchain: availability modeling, capacity, and costs planning. J. Supercomput. **77**(9), 9615–9641 (2021)
21. Pereira, P., Araujo, J., Melo, C., Santos, V., Maciel, P.: Analytical models for availability evaluation of edge and fog computing nodes. J. Supercomput. **77**(9), 9905–9933 (2021). https://doi.org/10.1007/s11227-021-03672-0
22. Russello, G., Mostarda, L., Dulay, N.: A policy-based publish/subscribe middleware for sense-and-react applications. J. Syst. Softw. **84**(4), 638–654 (2011)
23. Strielkina, A., Kharchenko, V., Uzun, D.: Availability models for healthcare IoT systems: classification and research considering attacks on vulnerabilities. In: 2018 IEEE 9th International Conference on Dependable Systems, Services and Technologies (DESSERT), pp. 58–62. IEEE (2018)
24. Tang, S., Xie, Y.: Availability modeling and performance improving of a healthcare internet of things (IoT) system. IoT **2**(2), 310–325 (2021)
25. Trivedi, K.S.: Probability & Statistics with Reliability, Queuing and Computer Science Applications. Wiley (2008)

26. Trivedi, K.S., Bobbio, A.: Reliability and Availability Engineering: Modeling, Analysis, and Applications. Cambridge University Press (2017)
27. Vannucchi, C., et al.: Symbolic verification of event–condition–action rules in intelligent environments. J. Reliable Intell. Environ. **3**(2), 117–130 (2017)
28. Yin, S., Xiao, B., Ding, S.X., Zhou, D.: A review on recent development of spacecraft attitude fault tolerant control system. IEEE Trans. Industr. Electron. **63**(5), 3311–3320 (2016)

A Fuzzy-Based Scheme for Selection of Radio Access Technologies in 5G Wireless Networks: QoE Assessment and Its Performance Evaluation

Phudit Ampririt[1]([⊠]), Makoto Ikeda[2], Keita Matsuo[2], and Leonard Barolli[2]

[1] Graduate School of Engineering, Fukuoka Institute of Technology,
3-30-1 Wajiro-Higashi, Higashi-Ku, Fukuoka 811-0295, Japan
`bd21201@bene.fit.ac.jp`
[2] Department of Information and Communication Engineering,
Fukuoka Institute of Technology,
3-30-1 Wajiro-Higashi, Higashi-Ku, Fukuoka 811-0295, Japan
`makoto.ikd@acm.org`, {`kt-matsuo,barolli`}`@fit.ac.jp`

Abstract. The 5-th Generation (5G) heterogeneous networks are expected to provide dense network services and a plethora of different networks for fulfilling the user requirements. They are supposed to give User Equipment (UE) the ability to connect with the appropriate Radio Access Technology (RAT). However, many parameters should be considered for the selection of RAT, which makes the problem HP-hard. Especially, Quality of Experience (QoE) is one of the important parameters for the selection of RAT in 5G wireless networks. For this reason, in this paper, we propose a fuzzy-based scheme for evaluating QoE considering three parameters: Network Capacity (NC), Experienced End-User Throughput (EEUT) and Connectivity (Cn). From simulation results, we found that when NC, EEUT and Cn are increased, the QoE parameter value is increased.

1 Introduction

In the 5-th Generation (5G) wireless networks, the unprecedently explosive growth of user devices with erratic traffic patterns will cause a huge volume of data, congesting the Internet and affecting Quality of Service (QoS) [1]. In order to meet the QoS requirements in many application situations, the 5G Wireless Networks will offer enhanced reliability, throughput, latency, and mobility in order to overcome these problems.

For improving the performance of 5G Radio Access Technologies (RATs), multiple base stations (BSs) use heterogeneous RATs (such as GSM, HSPA, LTE, LTE-A, Wi-Fi, and so on) which provide different radio coverages (such as macrocell, microcell, picocell, femtocell, Wi-Fi, etc.) with different transmission power levels in order to provide the mobile users with the best Quality of Experience (QoE), Energy Efficiency (EE), redundancy and reliability [2,3].

© The Author(s), under exclusive license to Springer Nature Switzerland AG 2023
L. Barolli (Ed.): AINA 2023, LNNS 661, pp. 44–53, 2023.
https://doi.org/10.1007/978-3-031-29056-5_5

In the future, 5G will need to provide a diverse range of services for activities, including lifestyle, working, entertainment and transportation. The 5G will solve the challenges by considering three main different application scenarios: enhanced Mobile Broadband (eMBB), massive Machine Type Communication (mMTC) and Ultra-Reliable & Low Latency Communications (URLLC). The eMBB enhances seamless QoE and has good accessibility to services and multimedia information related to the human-essential. The mMTC provides extended battery life despite supporting many connected devices. Finally, by effectively decreasing the latency and enhancing reliability, the URLLC can enable operating in real-time, such as those associated with transport security, remote surgeries and the automation of industrial processes [4–6].

Recently, many research works deal with design of systems appropriate for 5G wireless networks. One example is the utilization of Network Function Virtualization (NFV) and Software-Defined Networking (SDN) for several administrative and technological networks, including massive computing resources [7,8]. Also, the mobile handover approach and SDN are used to minimize processing delays [9–11]. In addition, QoS is improved by implementing Fuzzy Logic (FL) to SDN controllers.

In our previous work [12–16], we presented some Fuzzy-based systems for Call Admission Control (CAC), Handover in 5G wireless networks. In this paper, we propose a Fuzzy-based scheme for evaluating Quality of Experience (QoE) considering three parameters: Network Capacity (NC), Experienced End-User Throughput (EEUT) and Connectivity (Cn).

The rest of the paper is organized as follows. In Sect. 2 is presented an overview of SDN. In Sect. 3, we present Fuzzy Logic. In Sect. 4, we describe the proposed Fuzzy-based system and its implementation. In Sect. 5, we discuss the simulation results. Finally, conclusions and future work are presented in Sect. 6.

2 Software-Defined Networks (SDNs)

The SDN is one of the most promising methods to make networks programmable and virtualizable by separating the network's data plane from control plane. The SDN structure is shown in Fig. 1.

The Application Layer gathers data from the controller to create an integrated overview of the network for the purpose of making decision. The Northbound Interfaces create a plethora of possibilities for network programming by enabling communication between the Application Layer and the Control Layer. According to the requirements of the application, it will send instructions and data to the control layer, where the controller will establish the best software network feasible with the necessary service quality and security. The Control Layer manages the data plane and transmits various sorts of rules and policies to the Infrastructure Layer via Southbound Interfaces after receiving requests or orders from the Application Layer. The Southbound Interfaces are protocols that allow the controller to set rules for the forwarding plane and offer connectivity

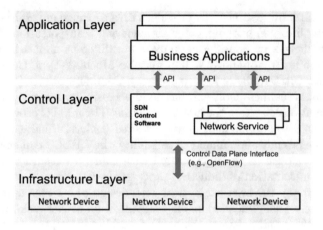

Fig. 1. Structure of SDN.

and interaction between the Control Plane and the Data Plane. The Infrastructure Layer represents the network's forwarding devices, such as load balancers, switches, and routers, and receives instructions from the SDN controller.

These components of SDN can be efficiently controlled and used by a centralized control plane. The SDN can regulate and modify resources on the control plane effectively in situations of traffic congestion. Forwarding data over many wireless technologies is faster and simpler for mobility management [17,18].

3   Fuzzy Logic

A Fuzzy Logic (FL) system is an extension of multivalued logic that uses a non-linear mapping of an input data vector into an output scalar in effort to accomplish approximate reasoning rather than an absolute solutions. Additionally, the FL can instantly handle both linguistic and numerical data. Traditional Crisp Logic, also known as Binary Logic, only considers truth values that are true and false, denoted by the numbers 1 and 0. On the other hand, FL has a truth value that is between 0 and 1. So, FL is flexible and can simulate non-linear functions of any complexity [19,20].

Fuzzifier, Inference Engine, Fuzzy Rule Base, and Defuzzifier are the four main parts of the fuzzy logic controller (FLC) structure, as shown in Fig. 2. The use of fuzzy sets with rules, which are linguistic variables, requires the use of a Fuzzifier to combine the crisp values with the fuzzy sets. The Rules can be given by a specialist or can be obtained from numerical data. The rules are presented in engineering cases as a series of IF-THEN expressions. For example, "IF t_1 is very hot and t_2 is cold, THEN change v somewhat to the left." By utilizing fuzzy rules and fuzzified input values, the Inference Engine derives fuzzy output. Mamdani-type and Sugeno-type fuzzy inference systems are the two basic types that may be used. The Defuzzifier converts the fuzzy values into crisp values received from fuzzy inference engine. There are various defuzzification techniques, such as The Centroid Method, The Center of Are (COA) Method and Tsukamoto's Defuzzification Method [21].

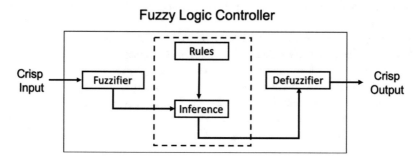

Fig. 2. FLC structure.

4 Proposed Fuzzy-Based System

We use FL to implement the proposed system. In Fig. 3, we present the overview of our proposed system. The SDN controller will provide commands to each evolve Base Station (eBS), allowing them to communicate and transfer data to User Equipment (UE). Additionally, each eBS contains a variety of slices for multiple purposes.

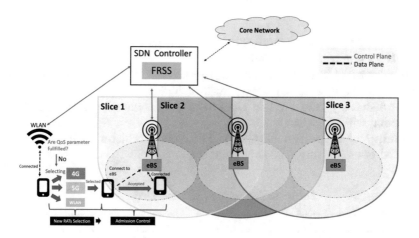

Fig. 3. Proposed system overview.

The proposed Fuzzy-based system for selecting a new RAT will be implemented in SDN controller, which will control eBS and other RAT's base stations and collect all the data regarding network traffic situation. The SDN controller will act as a transmission medium between the RAT's base station and the core network. For example, when the UE is connected to Wireless LAN (WLAN) but its QoS is not good, the SDN controller will collect other RAT networks data and decides whether the UE will still be connected with WLAN or connect to other RATs.

The proposed system is called Fuzzy-based RATs Selection System (FRSS) in 5G Wireless Networks. The structure of FRSS is shown in Fig. 4. For the implementation of our system, we consider four input parameters: Coverage (CV), User Priority (UP), Spectral Efficiency (SE) and Quality of Experience (QoE) and the output parameter is Radio Access Technology Decision Value (RDV). In this paper, we applied FL to evaluate QoE. For QoE, we are considering three parameters: Network Capacity (NC), Experienced End-User Throughput (EEUT) and Connectivity (Cn).

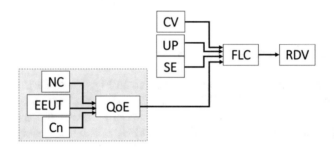

Fig. 4. Proposed system structure.

Network Capacity (NC): The network capacity is traffic volume density. When NC value is high, the QoE is high.

Experienced End-User Throughput (EEUT): When users experience a high-throughput, their satisfaction will be high.

Connectivity (Cn): When users have good connection with a RAT, their satisfaction will be higher.

Quality of Experience (QoE): The QoE parameter is the output value evaluated based on three input parameters.

Table 1. Parameter and their term sets.

Parameters	Term Sets
Network Capacity (CV)	Small (Sm), Intermediate (In), Big (Bi)
Experienced End-User Throughput (EEUT)	Slow (Sl), Medium (Mi), Fast (Fs)
Connectivity (Cn)	Low (Lo), Medium (Mu), High (Hg)
Quality of Experience (QoE)	QoE1, QoE2, QoE3, QoE4, QoE5, QoE6, QoE7

Table 2. FRB.

Rule	NC	EEUT	Cn	QoE
1	Sm	Sl	Lo	QoE1
2	Sm	Sl	Mu	QoE2
3	Sm	Sl	Hg	QoE3
4	Sm	Mi	Lo	QoE2
5	Sm	Mi	Mu	QoE3
6	Sm	Mi	Hg	QoE4
7	Sm	Fs	Lo	QoE3
8	Sm	Fs	Mu	QoE4
9	Sm	Fs	Hg	QoE5
10	In	Sl	Lo	QoE2
11	In	Sl	Mu	QoE3
12	In	Sl	Hi	QoE4
13	In	Mi	Lo	QoE3
14	In	Mi	Mu	QoE4
15	In	Mi	Hg	QoE5
16	In	Fs	Lo	QoE4
17	In	Fs	Mu	QoE5
18	In	Fs	Hg	QoE6
19	Bi	Sl	Lo	QoE3
20	Bi	Sl	Mu	QoE4
21	Bi	Sl	Hg	QoE5
22	Bi	Mi	Lo	QoE4
23	Bi	Mi	Mu	QoE5
24	Bi	Mi	Hg	QoE6
25	Bi	Fs	Lo	QoE5
26	Bi	Fs	Mu	QoE6
27	Bi	Fs	Hg	QoE7

The membership functions are shown in Fig. 5. We use triangular and trapezoidal membership functions because they are more suitable for real-time operations [22–25]. We show parameters and their term sets in Table 1. The Fuzzy Rule Base (FRB) is shown in Table 2 and has 27 rules. The control rules have the form: IF "condition" THEN "control action". For example, for Rule 1: "IF NC is Sm, EEUT is Sl and Cn is Lo, THEN QoE is QoE1".

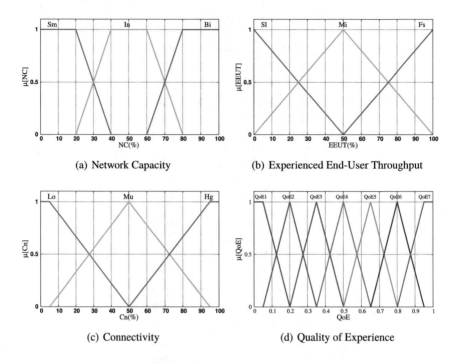

Fig. 5. Membership functions.

5 Simulation Results

In this section, we present the simulation result of our proposed system. The simulation results are shown in Fig. 6, Fig. 7 and Fig. 8. They show the relation of QoE with Cn for different EEUT values considering NC as a constant parameter.

In Fig. 6, we consider the NC value 10%. When Cn is increased from 10% to 50% and 50% to 90% for EEUT 90%, we see that QoE is increased by 15% and 12%, respectively. This is because when users have good network connection, the QoE value will be higher. When we increased the EEUT value from 10% to 50% and 50% to 90%, both QoE values are increased by 11% when Cn is 50%. This indicates that when users experience higher throughput, the user satisfaction will be higher.

We compare Fig. 6 with Fig. 7 to see how NC has affected QoE. We change the NC value from 10% to 50%. The QoE is increased by 15% when the EEUT value is 90% and Cn is 90%. When the RAT has better network capacity, the user satisfaction value also is higher.

We increase the value of NC to 90% in Fig. 8. Comparing the results with Fig. 6 and Fig. 7, we can see that the QoE values are increased significantly.

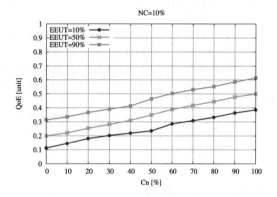

Fig. 6. Simulation results for NC = 10%.

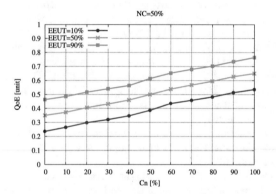

Fig. 7. Simulation results for NC = 50%.

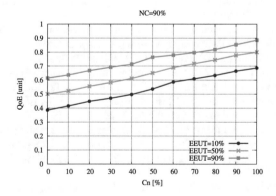

Fig. 8. Simulation results for NC = 90%.

6   Conclusions and Future Work

In this paper, we proposed and implemented a Fuzzy-based scheme for evaluating QoE. The QoE parameter will be considered as an input parameter for deciding RAT decision value. We evaluated the proposed scheme by simulations. From the simulation results, we found that when NC, EEUT and Cn are increasing, the QoE parameter is increased.

In the future work, we will consider different parameters and perform extensive simulations to evaluate the proposed system.

References

1. Navarro-Ortiz, J., Romero-Diaz, P., Sendra, S., Ameigeiras, P., Ramos-Munoz, J.J., Lopez-Soler, J.M.: A survey on 5G usage scenarios and traffic models. IEEE Commun. Surv. Tutorials **22**(2), 905–929 (2020). https://doi.org/10.1109/COMST.2020.2971781
2. Pham, Q.V., et al.: A survey of multi-access edge computing in 5G and beyond: fundamentals, technology integration, and state-of-the-art. IEEE Access **8**, 116,974–117,017 (2020). https://doi.org/10.1109/ACCESS.2020.3001277
3. Orsino, A., Araniti, G., Molinaro, A., Iera, A.: Effective rat selection approach for 5G dense wireless networks. In: 2015 IEEE 81st Vehicular Technology Conference (VTC Spring), pp. 1–5 (2015). https://doi.org/10.1109/VTCSpring.2015.7145798
4. Akpakwu, G.A., Silva, B.J., Hancke, G.P., Abu-Mahfouz, A.M.: A survey on 5G networks for the internet of things: communication technologies and challenges. IEEE Access **6**, 3619–3647 (2018)
5. Palmieri, F.: A reliability and latency-aware routing framework for 5G transport infrastructures. Comput. Netw. **179**(9) (2020). Article 107365. https://doi.org/10.1016/j.comnet.2020.107365
6. Kamil, I.A., Ogundoyin, S.O.: Lightweight privacy-preserving power injection and communication over vehicular networks and 5G smart grid slice with provable security. Internet Things **8**(100116), 100–116 (2019). https://doi.org/10.1016/j.iot.2019.100116
7. Hossain, E., Hasan, M.: 5G cellular: key enabling technologies and research challenges. IEEE Instrum. Meas. Mag. **18**(3(3)), 11–21 (2015). https://doi.org/10.1109/MIM.2015.7108393
8. Vagionas, C., et al.: End-to-end real-time service provisioning over a SDN-controllable analog mmWave fiber-wireless 5G X-haul network. J. Lightwave Technol., 1–10 (2023). https://doi.org/10.1109/JLT.2023.3234365
9. Yao, D., Su, X., Liu, B., Zeng, J.: A mobile handover mechanism based on fuzzy logic and MPTCP protocol under SDN architecture*. In: 18th International Symposium on Communications and Information Technologies (ISCIT-2018), pp. 141–146 (2018). https://doi.org/10.1109/ISCIT.2018.8587956
10. Lee, J., Yoo, Y.: Handover cell selection using user mobility information in a 5G SDN-based network. In: 2017 Ninth International Conference on Ubiquitous and Future Networks (ICUFN-2017), pp. 697–702 (2017). https://doi.org/10.1109/ICUFN.2017.7993880
11. Moravejosharieh, A., Ahmadi, K., Ahmad, S.: A fuzzy logic approach to increase quality of service in software defined networking. In: 2018 International Conference on Advances in Computing, Communication Control and Networking (ICACCCN-2018), pp. 68–73 (2018). https://doi.org/10.1109/ICACCCN.2018.8748678

12. Ampririt, P., Qafzezi, E., Bylykbashi, K., Ikeda, M., Matsuo, K., Barolli, L.: IFACS-Q3S-a new admission control system for 5G wireless networks based on fuzzy logic and its performance evaluation. Int. J. Distrib. Syst. Technol. (IJDST) **13**(1), 1–25 (2022)

13. Ampririt, P., Qafzezi, E., Bylykbashi, K., Ikeda, M., Matsuo, K., Barolli, L.: A fuzzy-based system for handover in 5G wireless networks considering network slicing constraints. In: Barolli, L. (ed.) Computational Intelligence in Security for Information Systems Conference, pp. 180–189. Springer, Cham (2022). https://doi.org/10.1007/978-3-031-08812-4_18

14. Ampririt, P., Qafzezi, E., Bylykbashi, K., Ikeda, M., Matsuo, K., Barolli, L.: A fuzzy-based system for handover in 5G wireless networks considering different network slicing constraints: effects of slice reliability parameter on handover decision. In: Barolli, L. (ed.) International Conference on Broadband and Wireless Computing, Communication and Applications, pp. 27–37. Springer, Cham (2022). https://doi.org/10.1007/978-3-031-20029-8_3

15. Ampririt, P., Ohara, S., Qafzezi, E., Ikeda, M., Matsuo, K., Barolli, L.: An integrated fuzzy-based admission control system (IFACS) for 5G wireless networks: its implementation and performance evaluation. Internet Things **13**, 100,351 (2021). https://doi.org/10.1016/j.iot.2020.100351

16. Ampririt, P., Qafzezi, E., Bylykbashi, K., Ikeda, M., Matsuo, K., Barolli, L.: Application of fuzzy logic for slice QoS in 5G networks: a comparison study of two fuzzy-based schemes for admission control. Int. J. Mob. Comput. Multimedia Commun. (IJMCMC) **12**(2), 18–35 (2021)

17. Li, L.E., Mao, Z.M., Rexford, J.: Toward software-defined cellular networks. In: 2012 European Workshop on Software Defined Networking, pp. 7–12 (2012). https://doi.org/10.1109/EWSDN.2012.28

18. Mousa, M., Bahaa-Eldin, A.M., Sobh, M.: Software defined networking concepts and challenges. In: 2016 11th International Conference on Computer Engineering & Systems (ICCES-2016), pp. 79–90. IEEE (2016)

19. Lee, C.: Fuzzy logic in control systems: fuzzy logic controller I. IEEE Trans. Syst. Man Cybern. **20**(2), 404–418 (1990). https://doi.org/10.1109/21.52551

20. Jantzen, J.: Tutorial on fuzzy logic. Technical University of Denmark, Department of Automation, Technical Report (1998)

21. Mendel, J.: Fuzzy logic systems for engineering: a tutorial. Proc. IEEE **83**(3), 345–377 (1995). https://doi.org/10.1109/5.364485

22. Norp, T.: 5G requirements and key performance indicators. J. ICT Stand. **6**(1), 15–30 (2018)

23. Parvez, I., Rahmati, A., Guvenc, I., Sarwat, A.I., Dai, H.: A survey on low latency towards 5G: ran, core network and caching solutions. IEEE Commun. Surv. Tutorials **20**(4), 3098–3130 (2018)

24. Kim, Y., Park, J., Kwon, D., Lim, H.: Buffer management of virtualized network slices for quality-of-service satisfaction. In: 2018 IEEE Conference on Network Function Virtualization and Software Defined Networks (NFV-SDN-2018), pp. 1–4 (2018)

25. Barolli, L., Koyama, A., Yamada, T., Yokoyama, S.: An integrated CAC and routing strategy for high-speed large-scale networks using cooperative agents. IPSJ J. **42**(2), 222–233 (2001)

Lotka-Volterra Applied to Misinformation Extinction in Opportunistic Networks

Victor Messner[1]([⊠]), Anderson Zudio[2], Diego Dutra[1], and Claudio Amorim[1]

[1] Universidade Federal do Rio de Janeiro, Av. Pedro Calmon,
550 - Cidade Universitária, Rio de Janeiro-RJ 21941-901, Brazil
{vcmessner,ddutra,amorim}@cos.ufrj.br
[2] Universidade Federal Fluminense, Av. Gal. Milton Tavares de Souza,
s/n - São Domingos, Niterói - RJ 24210-240, Brazil
azudio@id.uff.br

Abstract. This work addresses a distributed misinformation neutralization model that extinguishes misinformation without requiring centralized authority or acknowledging the falsehood of the content by statistical analysis. The proposed approach creates a competitive relationship between misinformation and truthful information toward the susceptible nodes. We model this behavior by an ordinary differential equations (ODEs) system to analyze the competitive relationship between messages in the network and simulate the proposed approach using The ONE Simulator. The simulation results corroborates that our ordinary differential equations system solutions are a close approximation of the model, and that our distributed approach suppresses misinformation in the network faster than the other evaluated approaches in the verified scenarios.

1 Introduction

Mobile Wireless Sensor Networks is a popular theme in research as the number of network nodes grows faster than the expected worldwide population [1], thus making the amount of information in the network to grow even faster. Unfortunately, a consequence of the information growth is the increase in misinformation dissemination which has negative consequences for national security and public health [2]. Misinformation identification and removal is a relevant issue in social networks with multiple solutions [3–5], yet this issue is an open problem in an opportunistic network scenario.

Opportunistic networks are hardware-restrained networks without a granted end-to-end connectivity, and solutions to the wireless technologies issues focus on the node mobility and the diffusion algorithm adopted [6]. Solutions in this type of network should overcome intermittency, high node mobility, low node hardware availability, and scale with the number of nodes in the network, making it capable of being applied from scenarios with a low density, such as rangers applications, to high-density scenarios, such as crowded events and smart-cities. The misinformation suppression in opportunistic networks considers the amount

L. Barolli (Ed.): AINA 2023, LNNS 661, pp. 54–65, 2023.
https://doi.org/10.1007/978-3-031-29056-5_6

of information, the network node mobility, and the consumption of limited resources, thus making a cost-effective evaluation necessary. For this reason, we propose a low-cost misinformation suppression and study message dissemination behavior under a competitive view.

This paper proposes a novel approach for opportunistic networks based on the Lotka-Volterra competition models. Nodes in this kind of network transmit information during opportunistic encounters, referred to as contacts, that define the system's growth rate of both types of conflicting information in our system. The message transmission range, the number of nodes in the network, the area of the network, and the node mobility impact the number of contacts between nodes in the network in a given time, known as the contact rate. Similarly to an epidemic approach [7], our proposed approach uses a story-carry-forward characteristic [12] to guarantee the persistence of the information in the network during a possible intermittence. Finally, since our proposed model is a distributed approach that benefits from the inherent competition between the truthful messages and the misinformation by the nodes that desire the information, it does not require a centralized entity capable of contacting every node in the network or analyzing the messages in the network.

Our system uses two predator groups, represented by the nodes that disseminate truthful information and those that disseminate misinformation, and one prey group, represented by the nodes interested in the data. As other works in this area [7,8], we assume that at least one source only provides truthful information and that its integrity and authenticity are verifiable. We develop an ordinary differential equation (ODE) model to analyze the competitive relationship between conflicting information and the impact these pieces of information exert on the network nodes. The distributed approach presented in this paper suppress misinformation independently of the chosen communication protocol.

The remainder of this paper is organized as follows: Section 2 presents the related work. Section 3 discusses our model and presents the results of the proposed ordinary differential equations. Section 4 presents the computational results through simulations, and, Sect. 5 concludes this paper.

2   Related Work

When applied to a communication protocol, epidemic models guarantee information delivery in a partially connected ad-hoc network [9,10]. Other Epidemic model applications include centralized authorities that vaccinate and treat the network nodes to neutralize false information in an area, thus creating a safe area free from misinformation [7]. The former approach uses the Double-pulse vaccination [11] technique to vaccinate nodes in the network, and it extinguishes misinformation in the network, i.e., the impact of misinformation dissemination in nodes that want the information is null, in 2×10^3 steps for speed and direction variations while using a 3.5×10^3 pause duration and movement variation [7]. The authors verify the variation when other vaccination techniques are applied to conclude that the double pulse vaccination approach presents better results under Mobile Opportunistic Networks.

In the context of misinformation removal in a network, the Lotka-Volterra is an alternative solution. Zhang et al. [8] proposed a Lotka-Volterra competition model for social networks based on a modified interactive system with Holling type for social networks where the different pieces of information may influence each other while competing for the same readership and theme. Their work considers that messages may reach larger audiences depending on their publisher and how the reliability of the publishers impact message diffusion. The authors focus on the coexistence of both messages in the network through a qualitative analysis of the system, showing that controlling a strategic node or a publisher with the highest reliability influences the system's stability and resource availability for covering a larger audience in the targeted community. Their work also shows that maturation parameters and assimilation coefficient significantly impact the steady-state equilibrium dynamics.

3   Proposed Model

This work uses a Lotka-Volterra model to represent the network's misinformation and truthful information competition, resulting in a distributed approach capable of neutralizing misinformation in an intermittent hardware-restrained network where there is no granted end-to-end connectivity. Similar to biological models [13], our competition model considers the existence of two types of predators nodes and one type of prey node, where one of the predators consumes nodes that contain misinformation and nodes that do not have any information. This approach is promising in an opportunistic network since the primary variable is the contact rate which increases with the node mobility on the network and lowers the cost of identifying misinformation by creating natural segregation of this type of information instead of requiring a centralized entity to verify the messages in every node.

Our approach considers that nodes accept any information received until receiving truthful information, and that a node, upon receiving a message that is truthful and its integrity is attested, discards conflicting misinformation and accepts the truthful one. We use these premises to create a ODEs system that reveal the conditions where a piece of truthful information extinguishes conflicting misinformation and the ones where there is a coexistence between the two types of information. The presented system represents only one type of message, but its behavior is analogous to any message in the network.

Latent nodes are defined as a set of nodes that transmit information, whether true or false, without accepting it. In our work, for the modeling simplicity we will consider that latent nodes only increase the probabilistic range of nodes that transmit information, so in our model, there are three types of nodes:

- **Malicious:** nodes that propagate misinformation, but upon receiving truthful information that came from a secure source, they discard the false conflicting message and accept the truthful one;
- **Benigns:** nodes who receive truthful information and transmits it;
- **Susceptible:** nodes that have an interest in a piece of information, and upon receiving it, they become a malicious or benign node according to the veracity of the information received.

In ODE systems we highlight six types of parameters that impact the nodes:

- **Intra-specific competition:** In our model this parameter represent the competition to access an overloaded router;
- **Consumption rate:** In our model this parameter represent the node accept rate;
- **Contribution of prey to predator increase:** In our model we set this parameter as one since the nodes change from their set, and no new nodes enter the network;
- **Substrate Availability:** In our model the substrate availability is the load capacity of the service provider from an area;
- **Mortality rate:** In our model the mortality rate is the rate that the nodes in a system variable group leave the network.

Ordinary Differential equations (ODEs) models the relations between the malicious, truthful, and susceptible nodes, and its solutions are steady-states. In our modeling, we consider the intra-specific competition, the consumption rate, the contribution of prey to predator increase, the growth of Susceptible nodes, the number of nodes in each group (susceptible, malicious, and benign), and the mortality rate of the predators. We argue that the impact caused by the misinformation transmission by a set of nodes that transmits misinformation even upon receiving a truthful information is null if we prove the existence of a misinformation extinction steady-state within the network. We achieve this by analyzing the stability of the solution to demonstrate how the systems behave around the expected solution.

System 1 shows this system modeling over time.

$$
1 \begin{cases} dS/d\tau = S(K - a_{11}S - a_{12}M - a_{13}B) & \text{(1a)} \\ dM/d\tau = M(-D_1 + a_{21}S - a_{23}B) & \text{(1b)} \\ dB/d\tau = B(-D_2 + a_{31}S + a_{32}M) & \text{(1c)} \end{cases}
$$

Table 1 shows the meaning of all the variables presented in the system 1.

Table 1. The variables of the unmodified system.

(a_{ij} where i=j)	Intra-specific competition
(a_{ij} where i< j)	Consumption rate
(a_{ij} where i> j)	Contribution of prey to predator increase
K	Substrate Availability for S growth
M	Malicious nodes
B	Benign nodes
S	Susceptible nodes
D1	Mortality rate of Malicious nodes;
D2	Mortality rate of Benign nodes;
τ	Unit of time

We transform these ODEs into a dimensionless system with fewer variables due to the lack of measurement units [14] and thus facilitating the analysis of the System 1. We consider the following new variables values: $t = K\tau$, $x = a_{11}S/K$, $y = a_{12}M/K$, $z = B/K$, d1 = D1/K, d2 = D2/K, $\alpha = a_{21}/a_{11}$, $\beta = a_{23}$, $\gamma = a_{31}/a_{11}$, $\bar{\gamma} = a_{13}$, $\delta = a_{32}/a_{12}$. System 2 exhibits the new system.

$$2\begin{cases} dx/dt = x(1 - x - y - \bar{\gamma}z) & (2a) \\ dy/dt = y(-d1 + \alpha x - \beta z) & (2b) \\ dz/dt = z(-d2 + \gamma x + \delta y) & (2c) \end{cases}$$

4 Experiments

This section details the study in this work through the results attained with computational simulation using the ONE Simulator 1.6.0 [15] executing over Microsoft Windows 10 operational system. For this task, we consider the existence of 100 susceptible nodes that use the Random Direction model [16], a well-known mobility model with close analytical and simulation results [17]. The random direction model utilizes all the possible directions to the movement by default. Still, since the direction is limited in some experiments, we need to change the adequate direction pool to be the same as the double pulse vaccination experiment [7].

We consider the number of nodes with misinformation and the number of nodes that had contact with truthful information during each simulation test. This data shows the impact of the misinformation in the system and the number of nodes that had prior contact with the truthful information. Table 2 summarizes our default input variables for the simulation.

Each input value shown in Table 2 has to follow some guidelines to ensure that the simulation conforms to the double pulse-vaccination specified standards.

Table 2. Simulation parameters and their values.

Parameter	Value
Total number of nodes	105
Number of susceptible nodes that contains misinformation	100
Number of stationary misinformation spreader	2
Number of stationary truthful information spreader	2
Number of stationary blackholes	2
Buffer size	5 MB
Message Time To Leave (TTL)	1500 s (25 min)
Mobility Model	Random Direction
Message generation interval	1500 s (25 min)
Simulation time	4500 s (1 h 15 min)
Warmup time	1500 s (25 min)
Message Size	75 KB
Message transmission radius	4 m (Close contact)
V_μ	Average speed the range
V_ρ	Std deviation the range
H_μ	Average pause duration the range
H_ρ	Std deviation the range
Bandwidth	250 KB/s
Message Retention	40%
Probability of failing on transmitting the message upon contact	0.2

The ratio between the available buffer size and the message size guarantees that the nodes receive information properly. Likewise, the chosen TTL and time to generate a new message has to guarantee a period where the truthful information is nonexistent in the network, making it possible to evaluate the impact of another truthful information on the network.

The number of secure sources is the same as the number of nodes that disseminate misinformation for all tests. Therefore, we use two static nodes that only provide truthful information and two that spread misinformation. We also want to consider the existence of selfish nodes in the network, so we utilize one selfish node that behaves like a black hole, discarding any received message.

We must guarantee that misinformation persists in the network, so the contact rate between malicious and susceptible nodes must be greater than the number of malicious nodes leaving the network. Namely, we want values that guarantee that $\alpha > D1$, and for that, we consider that nodes in the opportunistic network have a 60% chance of discarding their messages, so the following message retention is 40%. We use weighted averages to determine the flux through the remaining nodes considering $\tau = 2$ units of time. Unlike the double pulse approach, our model does not assume vaccination and treatment of a fixed population, so a low message retention value degrades our misinformation extinction performance when the truthful secure source is unavailable and the misinformation spreader is available. We also set a small transmission radius since it directly affects each node's contact rate, decreasing the information spread in our simulation. These chosen values are essential to validate the system performance under non-ideal networks and in near worst-case scenarios.

In the proposed model, the nodes maintain their behavior independently of the message's source during the simulation. Since we vary the activity of the secure source while retaining the misinformation spreader always active, the values of δ and γ vary.

4.1   Results

This section provides the results of 100 independent simulation executions where the availability of the truthful message spreader as 360 s, and all susceptible nodes contain misinformation at before the system generate a truthful message. It is essential to notice that the cases where the standard deviation of the results presents minor variations indicates a small dispersion over time.

The Figs. 1, 2, 3, and 4 show the time step of the simulation on the x-axis. On the y-axis, we display two types of data: the number of nodes that contain misinformation in the network ($\mu_I(t)$) and the number of nodes that accepted the truthful information at least once ($\mu_T(t)$). In the following section we analyze the impact of the speed, and the pause duration in our solution.

4.1.1   Case 1

In this case the nodes speed ranges are $[0, 10]$, $[10, 20]$, $[20, 30]$, and $[30, 40]$, so their average speeds are $v_\mu = 5$, $v_\mu = 15$, $v_\mu = 25$, $v_\mu = 35$ respectively. Since each speed value in the range has the same probability of being chosen, we have the speeds standard deviation, $\rho = 2.89$, for each range. The direction range is $[0, 2\pi]$, the pause duration range is $[0, 10]$, the number of nodes that desire the information is 100, and the experiment area is 200×200 meters. The number of nodes that have a misinformation in a given time t is $\mu_I(t)$ and the number of nodes that had contact with the truthful information at least one time at a given this is $\mu_T(t)$.

Figure 1 shows that the nodes converge faster to misinformation extinction in the network at higher speed rates, and Fig. 1a shows that after 3000 steps, the number of misinformation in the network increases since the first message reached its TTL. However, similarly to the first message behavior, the system reaches a steady state where the system extinguishes the misinformation on average.

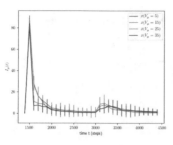

(a) The misinformation extinction data.

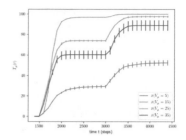

(b) The truthful message spread data.

Fig. 1. The impact of average speeds on the system simulation when the availability of the truthful message spreader is 360 s.

Figure 1b shows that higher speeds do not guarantee an increase in the nodes that had contact at least once with the truthful information on the network. The increase in the number of nodes that had contact at least once with a truthful message after the second message dissemination indicates that we would have better results in intermittent systems with high availability. In this kind of system, the message will have time to converge to a steady state during its intermittence period.

4.1.2   Case 2

In this case the nodes speed ranges are $[0, 70]$, $[20, 50]$, $[30, 40]$, and $[34, 36]$, so their standard deviations are $H_\rho = 20.2$, $H_\rho = 8.66$, $H_\rho = 2.89$, $H_\rho = 0.58$, respectively. The direction range is $[0, 2\pi]$, the pause duration range is $[0, 10]$, the number of nodes that desire the information is 100, and the experiment area is 200×200 meters. The number of nodes that have a misinformation in a given time t is $\mu_I(t)$ and the number of nodes that had contact with the truthful information at least one time ate a given this is $\mu_T(t)$. In each of the chosen speed ranges, we have the same average value of 35.

As in previous case Fig. 2 shows that the nodes converge to misinformation extinction shows that almost all nodes were capable of acquiring truthful information during the first message dissemination. Also as previously stated after 3000 steps, the number of misinformation in the network increases since the first message reached its TTL, as shown in Fig. 2a.

Figure 2a shows that, on average, a negligible number of nodes on the network contain the misinformation sustaining that the system is capable, on average, of extinguishing the misinformation on the system at the same time as the results presented in Fig. 1a. The results indicate that increases in message retention will increase the number of nodes that contain the truthful information at a given time, making the number making the number of nodes that accepted a truthful message at least one time go near 100%, as shown in Fig. 2b, and bringing this small number of nodes that contains the misinformation down to 0 after the second truthful information dissemination, as shown in Fig. 1a. Note that at

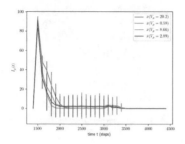

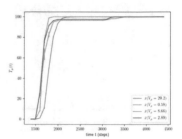

(a) The misinformation extinction data.   (b) The truthful message spread data.

Fig. 2. The impact of the standard deviation of speed range on the system simulation when the availability of the truthful message spreader is 360 s

higher speed deviations, the system also presents higher deviation values during the first truthful information dissemination and that the impact of the second truthful message in the simulation is negligible.

4.1.3 Case 3

In this case the pause duration ranges of nodes are $[0, 2]$, $[1, 3]$, $[3, 5]$, and $[10, 12]$, and their average duration are $H_\mu = 1$, $H_\mu = 2$, $H_\mu = 4$, $H_\mu = 11$ respectively. The direction range is $[-\pi/4, \pi/4]$, the speed range is $[30, 40]$, the number of nodes that desire the information is 100, and the experiment area is 200×200 meters. The number of nodes that have a misinformation in a given time t is $\mu_I(t)$ and the number of nodes that had contact with the truthful information at least one time at a given this is $\mu_T(t)$. Since each value in the range has the same probability of being chosen, we have $\rho = 0.58$ for each range.

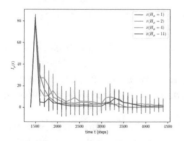

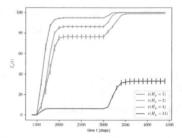

(a) The misinformation extinction data.   (b) The truthful message spread data.

Fig. 3. The impact of average pause duration rate on the system simulation when the availability of the truthful message spreader is 360 s.

Figure 3a reveals that, on average, the pause duration decreases the misinformation extinction performance of our system. Differently from the other cases, at a pause duration of 4 s, our system establishes a steady-state where approximately 10% of the nodes that contains misinformation on the network. This figure also shows that at the 11 s pause duration the misinformation was extinct in the network even faster than the lowest pause duration, which can be explained by the negative impact of the low communication radius and the low message retention to misinformation and truthful information dissemination, that when combined with a high pause duration, the system becomes inconsistent since the lack of mobility of the nodes causes a diminishing contact rate value. The results in Fig. 3b corroborates the inconsistent behavior by presenting a negligible number of nodes that had contact with the truthful message at least once during the first truthful message dissemination.

4.1.4 Case 4

In this case the pause duration ranges of nodes are $[0, 10]$, $[4, 6]$, $[4.6, 5.4]$, and $[4.9, 5.1]$, and their standard deviations are $H_\rho = 2.89$, $H_\rho = 0.58$, $H_\rho = 0.23$, $H_\rho = 0.06$, respectively. The direction range is $[-\pi/4, \pi/4]$, the speed range is $[30, 40]$, the number of nodes that desire the information is 100, and the experiment area is 200×200 meters. In each of the chosen pause duration ranges, we have the same average value of 5.

Figure 4a shows that, on average, the time needed to extinguish the misinformation on the network remains nearly the same as in the first two cases, but the deviation in the number of nodes that contains the misinformation on the network increase in this case when compared to Fig. 3a. Figure 4b shows that all nodes were able to acquire the first truthful information, and all nodes were able to acquire the truthful information at least once at the end of the simulation, indicating that lower values on the standard deviation of the pause duration range favor our system, increasing the number of nodes that had contact with the truthful information at least once.

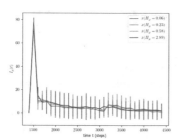

(a) The misinformation extinction data.

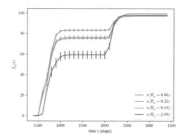

(b) The truthful message spread data.

Fig. 4. The impact of the standard deviation of pause duration range on the system simulation when the availability of the truthful message spreader is 360 s.

5   Concluding Remarks and Future Work

This paper proposes a competitive relationship between nodes that disseminate truthful and false messages inspired a Lotka-Volterra system to model the competitive relationship between misinformation and truthful messages in the network. We verify the cases where the misinformation and the truthful information persist in the networks and cases where the misinformation is extinct by using an ODE model when the nodes discard the misinformation upon receiving a guaranteed truthful message.

We simulate the ODE system behavior in an opportunistic network using the previously mentioned parameters to find a lower bound result. Upon analyzing the results we conclude that in most cases our system reaches a misinformation extinction steady-state. In the cases that the misinformation persisted in the network, the maximum number of nodes that contained this type of information on average remained under 10%. Our results show that with a 360s availability, our system reaches its steady state in less than 1000 steps, taking half of the time of an epidemic approach while having 12.5 times less availability to achieve nearly the same result in most cases.

In future work, we intend to increase the communication range and, consequently, our system performance. We also aim to analyze the availability trade-off and verify the effect of other variables that affect our system, such as the area, the number of nodes, buffer size, and message retention.

References

1. Mattisson, S.: Overview of 5G requirements and future wireless networks. In: 43rd IEEE European Solid State Circuits Conference (ESSCIRC), pp. 1–6 (2017)
2. Sługocki, W., Sowa, B.: Disinformation as a threat to national security on the example of the COVID-19 pandemic. In: Security and Defence Quarterly, vol. 35 (2021)
3. Guo, F., Blundell, C., Wallach, H., Heller, K.: The Bayesian echo chamber: modeling social influence via linguistic accommodation. arXiv preprint (2014). https://arxiv.org/abs/1411.2674
4. Murayama, T., Wakamiya, S., Aramaki, E., Kobayashi, R.: Modeling the spread of fake news on Twitter. Plos One **16**, e0250419 (2021)
5. Xu, H., Farajtabar, M., Zha, H.: Learning granger causality for hawkes processes. arXiv preprint (2016). https://arxiv.org/abs/1602.04511
6. Chancay-García, L., Hernández-Orallo, E., Manzoni, P., Calafate, C., Cano, J.: Evaluating and enhancing information dissemination in urban areas of interest using opportunistic networks. IEEE Access **6**, 32514–32531 (2018)
7. Wang, X., Lin, Y., Zhao, Y., Zhang, L., Liang, J., Cai, Z.: A novel approach for inhibiting misinformation propagation in human mobile opportunistic networks. Peer-to-Peer Network. Appl. **10**, 377–394 (2017)
8. Zhang, Y., Liu, F., Koura, Y., Wang, H.: Dynamic of interactive model for information propagation across social networks media. Adv. Differ. Equ. **318** (2020)
9. Garg, P., Kumar, H., Johari, R., Gupta, P., Bhatia, R.: Enhanced epidemic routing protocol in delay tolerant networks. In: 2018 5th International Conference on Signal Processing And Integrated Networks (SPIN), pp. 396–401 (2018)

10. Zhang, X., Neglia, G., Kurose, J., Towsley, D.: Performance modeling of epidemic routing. Comput. Netw. **51**, 2867–2891 (2007). https://doi.org/10.1016/j.comnet.2006.11.028
11. Yang, L., Yang, X.: The pulse treatment of computer viruses: a modeling study. Nonlinear Dyn. **76** (2014)
12. Dias, G., Rezende, J., Salles, R.: Semi-Markov model for delivery delay prediction in multi-copy opportunistic networks with heterogeneous pairwise encounter rates. Ad Hoc Netw. **123** (2021). https://doi.org/10.1016/j.adhoc.2021.102655
13. Hsu, S., Ruan, S., Yang, T.: Analysis of three species Lotka-Volterra food web models with omnivory. J. Math. Anal. Appl. **426**, 659–687 (2015). https://www.sciencedirect.com/science/article/pii/S0022247X15000542
14. Holmes, M.H.: Dimensional analysis. In: Introduction to the Foundations of Applied Mathematics. TAM, vol. 56, pp. 1–47. Springer, Cham (2019). https://doi.org/10.1007/978-3-030-24261-9_1
15. Keränen, A., Ott, J., Kärkkäinen, T.: The ONE simulator for DTN protocol evaluation. In: 2nd International Conference on Simulation Tools and Techniques (2009). https://doi.org/10.4108/ICST.SIMUTOOLS2009.5674
16. Nain, P., Towsley, D., Liu, B., Liu, Z.: Properties of random direction models. In: 24th Annual Joint Conference of the IEEE Computer and Communications Societies, vol. 3, pp. 1897–1907. IEEE (2005)
17. Groenevelt, R., Nain, P., Koole, G.: The message delay in mobile ad hoc networks. In: Perform. Eval. **62**, 210–228 (2005). https://doi.org/10.1016/j.peva.2005.07.018

A Comparison Study of FC-RDVM and RIWM Router Placement Methods for WMNs: Performance Evaluation Results by WMN-PSOHC Simulation System Considering Chi-Square Distribution and Different Instances

Shinji Sakamoto[1]([✉]), Admir Barolli[2], Yi Liu[3], Elis Kulla[4], Leonard Barolli[5], and Makoto Takizawa[6]

[1] Department of Information and Computer Science,
Kanazawa Institute of Technology, 7-1 Ohgigaoka,
Nonoichi, Ishikawa 921-8501, Japan
shinji.sakamoto@ieee.org

[2] Department of Information Technology, Aleksander Moisiu University of Durres,
L.1, Rruga e Currilave, Durres, Albania
admirbarolli@uamd.edu.al

[3] Department of Computer Science, National Institute of Technology, Oita College,
1666, Maki, Oita 870-0152, Japan
y-liu@oita-ct.ac.jp

[4] Department of System Management, Fukuoka Institute of Technology,
3-30-1 Wajiro-Higashi, Fukuoka, Higashi-Ku 811-0295, Japan
kulla@fit.ac.jp

[5] Department of Information and Communication Engineering,
Fukuoka Institute of Technology, 3-30-1 Wajiro-Higashi, Fukuoka,
Higashi-Ku 811-0295, Japan
barolli@fit.ac.jp

[6] Department of Advanced Sciences, Faculty of Science and Engineering,
Hosei University, Kajino-Machi, Koganei-Shi, Tokyo 184-8584, Japan
makoto.takizawa@computer.org, fatos@lsi.upc.edu

Abstract. In this work, we deal with the node placement problem in Wireless Mesh Networks (WMNs). We present a hybrid intelligent simulation system called WMN-PSOHC, which combines Particle Swarm Optimization (PSO) and Hill Climbing (HC). We implement in WMN-PSOHC system the Fast Convergence Rational Decrement of Vmax Method (FC-RDVM) and Random Inertia Weight Method (RIWM) router replacement methods. By using WMN-PSOHC system, we carry out simulations the access the performance of these two methods considering Chi-Square distribution of mesh clients and two different instances. The simulation results show that FC-RDVM performs better then RIWM in the considered scenario.

© The Author(s), under exclusive license to Springer Nature Switzerland AG 2023
L. Barolli (Ed.): AINA 2023, LNNS 661, pp. 66–73, 2023.
https://doi.org/10.1007/978-3-031-29056-5_7

1   Introduction

The advantages of Wireless Mesh Networks (WMNs) compared with the conventional Wireless Local Area Networks (WLANs) are that they have low upfront cost, ease of deployment and high robustness [1,6]. Also, WMNs can provide more useful services than traditional WLANs. WMNs have a wide range of potential applications, including medical, transportation, and surveillance in urban, metropolitan, municipal, and community areas [2,5,7].

However, several issues must be considered in WMNs. One of the most critical challenges is ensuring network connectivity and coverage, which is closely related to the class of node placement problems known to be NP-hard.

We have previously developed a Particle Swarm Optimization (PSO) based simulation system, called WMN-PSO [10], as well as a Hill Climbing (HC) based simulation system, called WMN-HC [9], to solve the node placement problem in WMNs. Then, we implemented a hybrid intelligent simulation system by integrating PSO and HC, called WMN-PSOHC [11]. We implemented the Random Inertia Weight Method (RIWM) and Fast Convergence Rational Decrement of Vmax Method (FC-RDVM) router replacement methods in WMN-PSOHC [12].

In this paper, by using WMN-PSOHC system, we carry out simulation to access the performnce of these two methods considering Chi-Square distribution of mesh clients and two different instances. The simulation results show that FC-RDVM performs better then RIWM in the considered scenario.

The rest of this paper is organized as follows. In Sect. 2, we present relevant, intelligent algorithms. The proposed hybrid simulation system and FC-RDVM are discussed in Sect. 3. The simulation results are presented in Sect. 4, followed by our conclusions and suggestions for future work in Sect. 5.

2   Intelligent Algorithms

2.1   Particle Swarm Optimization

In the Particle Swarm Optimization (PSO) algorithm, a group of simple entities, known as particles, are placed in a search space to evaluate an objective function by using their current location. The objective function is typically minimized and the search space is explored through the movement of particles rather than through evolution [8].

In this work, we consider a bi-objective optimization considering network connectivity and mesh client coverage. Each particle determines its movement through the search space by combining its own current and best-fitness locations with those of other particles in the swarm. After all particle have moved, the next iteration begins. Eventually, the movement of the swarm as a whole is likely to converge to an optimal solution the same as a flock of birds collectively searching for food.

Each particle in the swarm is composed of three $\mathcal{D}$-dimensional vectors, where $\mathcal{D}$ is the dimensionality of the search space, which are the current position, previous best position and particle velocity. The swarm is more than just a collection of individual particles, as the progress made in solving a problem emerges from the interactions and behavior of the particles as a whole.

The population topology or how the particles are connected, often resembles a social network, with bidirectional edges connecting pairs of particles. Each particle communicates with other particles and is affected by the best point found by members of its neighborhood, represented by the vector $\vec{p}_g$. During the PSO process, the velocity of each particle is iteratively adjusted in order that the particle oscillates stochastically around the $\vec{p}_i$ and $\vec{p}_g$ locations.

2.2   Hill Climbing

The Hill climbing (HC) is a heuristic algorithm that relies on a simple concept. The current solution is accepted as the new solution if $\delta \leq 0$, where $\delta = f(s') - f(s)$ and f is the fitness function, which evaluates the current solution s and the next solution s'.

The key factor in the success of HC is the effective definition of a neighbor solution, as this directly impacts the performance of the algorithm. In our WMN-PSOHC system, we use the next step of particle-pattern positions as the neighbor solutions for the HC part.

3   WMN-PSOHC Hybrid Intelligent Simulation System

The flowchart of WMN-PSOHC system is shown in Fig 1. We combine PSO and HC intelligent algorithms in order to make a better optimization of mesh routers in order to cover many mesh clients. Thus, the network connectivity and client coverage will be improved.

We present in following some processes, functions and methods used in WMN-PSOHC system such as initialization, particle-pattern, fitness function and router replacement methods.

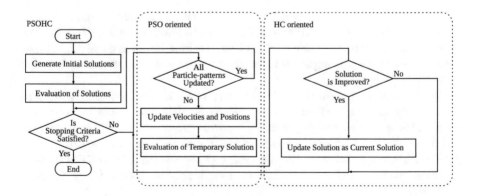

Fig. 1. WMN-PSOHC flowchart.

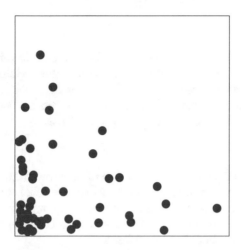

Fig. 2. Chi-Square distribution of mesh clients.

Initialization

The WMN-PSOHC system generates an initial solution randomly by using *ad hoc* methods [16]. The velocity of particles is decided randomly by considering the area size. For example, when the area size is $W \times H$, the velocity is decided randomly from $-\sqrt{W^2 + H^2}$ to $\sqrt{W^2 + H^2}$.

The WMN-PSOHC system can generate many mesh client distributions such as Normal, Uniform, Weibull, Chi-Square, Stadium, Two Islands, Subway and Boulevard distributions. In this paper, as shown in Fig. 2, we consider Chi-Square distribution of mesh clients.

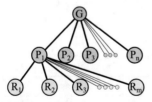

G: Global Solution
P: Particle-pattern
R: Mesh Router
n: Number of Particle-patterns
m: Number of Mesh Routers

Fig. 3. Relationship among global solution, particle-patterns and mesh routers.

Table 1. Instances parameters.

Parameters	Instance 1	Instance 2
Area Size	32×32	64×64
Number of Mesh Routers	16	32
Number of Mesh Clients	48	96

Particle-pattern

We consider a particle as a mesh router and the fitness value of a particle-pattern is computed by combination of mesh routers and mesh clients positions. Each particle-pattern is considered as a solution. Therefore, the number of particle-patterns is the number of solutions. The relationship between global solution, particle-patterns and mesh routers is shown in Fig. 3.

Fitness function

In optimization problems the determination of an appropriate objective function and its encoding is a very important issue. In our case, the fitness function follows a hierarchical approach, with the main objective to maximize the SGC and then improving the NCMC parameter. We use weight coefficients α and β in the fitness function, which is defined as follows.

$$\text{Fitness} = \alpha \times \text{SGC}(\boldsymbol{x}_{ij}, \boldsymbol{y}_{ij}) + \beta \times \text{NCMC}(\boldsymbol{x}_{ij}, \boldsymbol{y}_{ij})$$

Router replacement methods

The mesh routers movement is based on their velocities. Their placement is done by different router replacement methods [4,13–15]. In this paper, we compare two replacement methods RIWM and FC-RDVM.

In RIWM, the ω parameter changes randomly from 0.5 to 1.0. The C_1 and C_2 are kept 2.0. The ω can be estimated by the week stable region. The average of ω is 0.75 [3,15].

In FC-RDVM, the V_{max} decreases with increasing the number of iterations as shown in Eq. (1).

$$V_{max}(k) = \sqrt{W^2 + H^2} \times \frac{T - k}{T + \gamma k} \tag{1}$$

Table 2. Parameter settings.

Parameters	Values
Clients Distribution	Chi-Square Distribution
Instances	Instance 1, Instance 2
Total Iterations	800
Iteration per Phase	4
Number of Particle-patterns	9
Radius of a Mesh Router	From 2.0 to 3.0
Fitness Function Weight-coefficients (α, β)	0.7, 0.3
Curvature parameter (γ)	10.0
Replacement Methods	RIWM, FC-RDVM

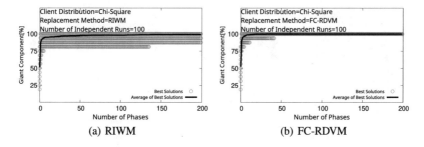

Fig. 4. Simulation results of WMN-PSOHC for SGC considering Instance 1.

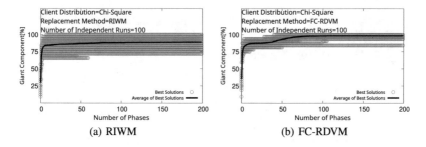

Fig. 5. Simulation results of WMN-PSOHC for SGC considering Instance 2.

The γ parameter is curvature parameter.

4   Simulation Results

In this section, we show simulation results. We consider Chi-Square distribution of mesh clients and the number of particle-patterns 9. We consider two instances: Instance 1 and Instance 2 as shown in Table 1. We show the parameter settings for WMN-PSOHC in Table 2.

We show the simulation results from Fig. 4 to Fig. 7. Considering SGC parameter, both replacement methods can reach 100% for both instances as shown in Fig. 4 and Fig. 5. This shows that all mesh routers are connected and the WMN has a good connectivity. However, the RIWM convergence is slower than FC-RDVM and has more oscillations. For NCMC, the FC-RDVM covers more mesh clients than RIWM for both instance as shown in Fig. 6 and Fig. 7. This shows that FC-RDVM has better coverage than RIWM.

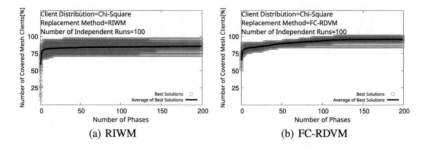

Fig. 6. Simulation results of WMN-PSOHC for NCMC considering Instance 1.

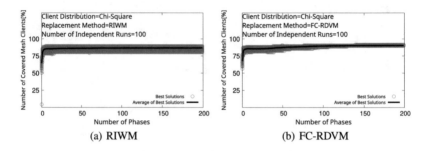

Fig. 7. Simulation results of WMN-PSOHC for NCMC considering Instance 2.

5   Conclusions

In this work, we presented WMN-PSOHC hybrid intelligent simulation system. We implemented in WMN-PSOHC system FC-RDVM and RIWM router replacement methods and carried out simulations considering Chi-Square distribution of mesh clients and two different instances. Simulation results show that FC-RDVM performs better than RIWM. In our future work, we would like to evaluate the performance of the proposed system for different parameters and scenarios.

References

1. Akyildiz, I.F., Wang, X., Wang, W.: Wireless mesh networks: a survey. Comput. Netw. **47**(4), 445–487 (2005)
2. Amaldi, E., Capone, A., Cesana, M., Filippini, I., Malucelli, F.: Optimization models and methods for planning wireless mesh networks. Comput. Netw. **52**(11), 2159–2171 (2008)
3. Barolli, A., Sakamoto, S., Ohara, S., Barolli, L., Takizawa, M.: Performance analysis of WMNs by WMN-PSOHC-DGA simulation system considering linearly decreasing inertia weight and linearly decreasing Vmax replacement methods. In: Barolli, L., Nishino, H., Miwa, H. (eds.) INCoS 2019. AISC, vol. 1035, pp. 14–23. Springer, Cham (2020). https://doi.org/10.1007/978-3-030-29035-1_2

4. Clerc, M., Kennedy, J.: The particle swarm-explosion, stability, and convergence in a multidimensional complex space. IEEE Trans. Evol. Comput. **6**(1), 58–73 (2002)
5. Franklin, A.A., Murthy, C.S.R.: Node placement algorithm for deployment of two-tier wireless mesh networks. In: Proceedings of Global Telecommunications Conference, pp. 4823–4827 (2007)
6. Islam, M.M., Funabiki, N., Sudibyo, R.W., Munene, K.I., Kao, W.C.: A dynamic access-point transmission power minimization method using PI feedback control in elastic WLAN system for IoT applications. Internet Things **8**(100), 089 (2019)
7. Muthaiah, S.N., Rosenberg, C.P.: Single gateway placement in wireless mesh networks. In: Proceedings of 8th International IEEE Symposium on Computer Networks, pp. 4754–4759 (2008)
8. Poli, R., Kennedy, J., Blackwell, T.: Particle Swarm Optimization. Swarm Intell. **1**(1), 33–57 (2007)
9. Sakamoto, S., Lala, A., Oda, T., Kolici, V., Barolli, L., Xhafa, F.: Analysis of WMN-HC simulation system data using friedman test. In: The Ninth International Conference on Complex, Intelligent, and Software Intensive Systems (CISIS-2015), pp. 254–259. IEEE (2015)
10. Sakamoto, S., Oda, T., Ikeda, M., Barolli, L., Xhafa, F.: Implementation and evaluation of a simulation system based on particle swarm optimisation for node placement problem in wireless mesh networks. Int. J. Commun. Netw. Distrib. Syst. **17**(1), 1–13 (2016)
11. Sakamoto, S., Ozera, K., Ikeda, M., Barolli, L.: Implementation of intelligent hybrid systems for node placement problem in WMNs considering particle swarm optimization, hill climbing and simulated annealing. Mob. Netw. Appl. **23**(1), 27–33 (2018)
12. Sakamoto, S., Barolli, L., Okamoto, S.: A comparison study of linearly decreasing inertia weight method and rational decrement of Vmax method for WMNs using WMN-PSOHC intelligent system considering normal distribution of mesh clients. In: Barolli, L., Natwichai, J., Enokido, T. (eds.) EIDWT 2021. LNDECT, vol. 65, pp. 104–113. Springer, Cham (2021). https://doi.org/10.1007/978-3-030-70639-5_10
13. Schutte, J.F., Groenwold, A.A.: A study of global optimization using particle swarms. J. Glob. Optim. **31**(1), 93–108 (2005)
14. Shi, Y.: Particle swarm optimization. IEEE Connections **2**(1), 8–13 (2004)
15. Shi, Y., Eberhart, R.C.: Parameter selection in particle swarm optimization. Evolutionary Programming VII, pp. 591–600 (1998)
16. Xhafa, F., Sanchez, C., Barolli, L.: Ad hoc and neighborhood search methods for placement of mesh routers in wireless mesh networks. In: Proceedings of 29th IEEE International Conference on Distributed Computing Systems Workshops (ICDCS-2009), pp. 400–405 (2009)

Performance Analysis of Cooperative LoRa with Multihop Relaying for Rayleigh and Nakagami Channels

Takoua Mahjoub[1,2(✉)], Adel Ben Mnaouer[2], Hatem Boujemaa[1],
and Maymouna Ben Said[1]

[1] UCAR COSIM Lab, Higher School of Communications of Tunis, Aryanah, Tunisia
{takoua.mahjoub,boujemaa.hatem,maymouna.bensaid}@supcom.tn
[2] Faculty of Engineering and Architecture, Canadian University Dubai,
Dubai, United Arab Emirates
adel@cud.ac.ae

Abstract. In this paper, we derive the Bit Error Probability (BEP) of cooperative LoRa using multihop Decode and Forward (DF) relaying. We have fixed the total transmitted energy per bit for any number of hops with uniform power allocation to the source and relay nodes. The analysis is valid for Rayleigh and Nakagami channels with arbitrary fading figure $m = 1, 2, 3, 4$, etc. The theoretical results have been confirmed with computer simulations and show that the BEP decreases as the number of hops increases and the fading figure m increases.

1 Introduction

Long Range (LoRa) technology has been studied in many papers [1–8] as it is a new trend in the development of wireless communications. LoRa uses Chirp Spread Spectrum (CSS) transmission technique. CSS has been used because it offers a high processing gain, a low power consumption and a good resistance to multipath fading in the context of Rayleigh, Nakagami and Rice fading channels. The performance of LoRa has been studied first for Additive White Gaussian Noise (AWGN) channels without fading [1,2]. The packet error rate has been derived in [3] and experimental results of LoRa in urban environment has been provided in [4]. Non cooperative LoRa performance in the presence of Rayleigh, Nakagami and Rice fading channels were derived in [5,6]. Cooperative LoRa has been studied in [7,8]. However, the Bit Error Probability (BEP) of LoRa using multihop relaying for Nakagami and Rayleigh channels was not derived in the literature which is the aim of this paper. The main contributions of the paper are:

- We derive the BEP of non cooperative and cooperative LoRa using multihop relaying for Rayleigh and Nakagami channels.
- The obtained results of BEP are compared to the simulated Bit Error Rate (BER). We verify the exactness of our derivations by comparing to computer simulations.
- The derived results are valid for any number of hops using decode and forward relaying and any fading figure of Nakagami channel.

© The Author(s), under exclusive license to Springer Nature Switzerland AG 2023
L. Barolli (Ed.): AINA 2023, LNNS 661, pp. 74–81, 2023.
https://doi.org/10.1007/978-3-031-29056-5_8

The paper contains five section. Section 2 derives the BEP of non cooperative LoRa for Nakagami and Rayleigh channels. Section 3 computes the BEP of LoRa using multihop decode and forward relaying. Section 4 comments on the obtained results while Sect. 5 concludes the article.

2 BEP Analysis for Nakagami and Rayleigh Channels

2.1 Nakagami Channels

LoRa uses a Chirp Spread Spectrum, CSS, modulation to encode data in a symbol. Each symbol is formed by SF bits, $SF = \{7,8,9,10,11,12\}$ where SF is the spreading factor. To encode a symbol s_k, $k = \{0,1,...,2^{SF}-1\}$, a specific frequency varying chirp signal is produced. The chirp signal, with a bandwidth $B = f_{max} - f_{min}$, is spread out over 2^{SF} frequency samples during the symbol transmission time T_S. The symbol k defines the starting frequency of the corresponding chirp signal, which is equal to $f_k = f_{min} + k\frac{B}{2^{SF}}$. Starting from f_k, the chirp increases (or decreases) to reach f_{max} (f_{min}) and then shrinks and starts over from f_{min} (f_{max}) to f_k.

In LoRa systems, the encoder converts SF bits to a symbol s_k with a sampling period $T = \frac{T_S}{2^{SF}}$. The transmitted signal waveform is given by

$$s_k(nT) = \sqrt{E_s}a_k(nT) = \sqrt{\frac{E_s}{2^{SF}}} e^{\frac{2\pi jn}{2^{SF}}[(k+n)mod2^{SF}]} \tag{1}$$

T is the sampling period, $n = \{0,1,2,...,2^{SF}-1\}$ is the sample index, E_s is the signal energy and $a_k(nT)$ are orthogonal functions forming the multi-dimensional LoRa space.

The receiver uses the orthogonality property of $a_i * (nT)$, $i = \{0,1,2,...,2^{SF}-1\}$, to extract the sent symbol k. The received signal $r_k(nT)$ is then multiplied by the conjugates of waveforms $a_i(nT)$ as follows

$$\sum_{n=0}^{2^{SF}-1} r_k(nT)a_i^*(nT) \tag{2}$$

The transmitted symbol k is identified as the index of LoRa waveform that yields the highest correlation value.

For Nakagami channel, the demodulation process gives

$$\sum_{n=0}^{2^{SF}-1} r_k(nT)a_i^*(nT) = \begin{cases} \sqrt{\alpha E_s} + n_i, & \text{if } i = k \\ n_i, & \text{if } i \neq k \end{cases} \tag{3}$$

where n_i is a complex Gaussian noise and $\sqrt{\alpha}$ is a channel coefficient that has a Nakagami distribution. The selected waveform should have the highest correlation

$$\hat{k} = \underset{i}{argmax} |\delta_{k,i}\sqrt{\alpha E_s} + n_i| \tag{4}$$

where $\delta_{k,i} = 1$ if $k = i$ and $\delta_{k,i} = 0$ if $k \neq i$ is the Kronecker index.

Let $\rho_i = |n_i|$ for $i \neq k$, $i = 1, 2, ..., 2^{SF}$ and $\rho = \underset{i \neq k}{max} \rho_i$. The Cumulative Distribution Function (CDF) of ρ_i is

$$F_{\rho_i}(x) = 1 - e^{-\frac{x^2}{2\sigma^2}} \tag{5}$$

where $\sigma^2 = 0.5N_0$ and N_0 is noise power spectral density.

We deduce the CDF of ρ as the maximum of independent and identically distributed (i.i.d.) $(2^{SF} - 1)$ random variables (r.v.)

$$F_\rho(x) = [1 - e^{-\frac{x^2}{2\sigma^2}}]^{2^{SF}-1} = \sum_{q=0}^{2^{SF}-1} \frac{(2^{SF}-1)!}{q!(2^{SF}-1-q)!}(-1)^q e^{-\frac{qx^2}{2\sigma^2}} \tag{6}$$

The BEP is computed as

$$P_e = P(\rho > \beta_k = |\sqrt{\alpha E_s} + n_k|) \tag{7}$$

We deduce

$$P_e = \int_0^{+\infty} [1 - F_\rho(x)]f_{\beta_k}(x)dx. \tag{8}$$

where $f_{\beta_k}(x)$ is the Probability Density Function, *PDF*, of β_k computed as

$$f_{\beta_k}(x) = \int_0^{+\infty} f_{\beta_k|\alpha}(x)f_\alpha(\alpha)d\alpha \tag{9}$$

$f_\alpha(\alpha)$ is the PDF of α that has a Nakagami distribution [9]

$$f_\alpha(\alpha) = (\frac{m}{\Omega})^m \frac{\alpha^{m-1}}{\Gamma(m)} e^{-\frac{\alpha m}{\Omega}} \tag{10}$$

where $\Omega = \frac{1}{d^{ple}}$ is the average power of channel coefficient, *ple* is the path loss exponent, $d = \frac{d^{effective}}{d_0}$ is the normalized distance between the transmitter and receiver and d_0 is a reference distance.

For a given α, β_k has a Rician distribution given by

$$f_{\beta_k|\alpha}(x) = \frac{x}{\sigma^2} e^{-\frac{(x^2+\alpha E_s)}{2\sigma^2}} I_0(\frac{x\sqrt{\alpha E_s}}{\sigma^2}) \tag{11}$$

We deduce

$$P_e = \sum_{q=1}^{2^{SF}-1} \frac{(2^{SF}-1)!(-1)^{q+1}}{q!(2^{SF}-1-q)!} \int_0^{+\infty} \int_0^{+\infty} e^{-\frac{qx^2}{2\sigma^2}} f_{\beta_k|\alpha}(x)f_\alpha(\alpha)$$

$$dxd\alpha \tag{12}$$

The inner integral that depends on x is written as

$$\int_0^{+\infty} \frac{x}{\sigma^2} e^{-\frac{qx^2}{2\sigma^2}} e^{-\frac{(x^2+\alpha E_s)}{2\sigma^2}} I_0\left(\frac{x\sqrt{\alpha E_s}}{\sigma^2}\right)$$

$$= e^{-\frac{\alpha E_s q}{(q+1)2\sigma^2}} \int_0^{+\infty} \frac{x}{\sigma^2} e^{-\frac{(q+1)x^2+\frac{\alpha E_s}{q+1}}{2\sigma^2}} I_0\left(\frac{x\sqrt{\alpha E_s}}{\sigma^2}\right) dx$$

$$= e^{-\frac{\alpha E_s q}{(q+1)2\sigma^2}} \frac{1}{q+1} \tag{13}$$

Using (12)–(13), we obtain

$$P_e = \sum_{q=1}^{2^{SF}-1} \frac{(2^{SF}-1)!(-1)^{q+1}}{q!(2^{SF}-1-q)!} \frac{1}{q+1} \left(\frac{m}{\Omega}\right)^m \frac{1}{\Gamma(m)}$$

$$\times \int_0^{+\infty} e^{-\frac{\alpha E_s q}{(q+1)2\sigma^2}} \alpha^{m-1} e^{-\frac{\alpha m}{\Omega}} d\alpha \tag{14}$$

We finally get the close-form expression of P_e as follows

$$P_e = \sum_{q=1}^{2^{SF}-1} \frac{(2^{SF}-1)!(-1)^{q+1}}{q!(2^{SF}-1-q)!} \frac{1}{q+1} \left(\frac{m}{\Omega}\right)^m$$

$$\times \frac{1}{[\frac{m}{\Omega} + \frac{E_s q}{2(q+1)\sigma^2}]^m} \tag{15}$$

2.2 Rayleigh Channels

For Rayleigh channels, we replace m by one to get the following expression

$$P_e = \sum_{q=1}^{2^{SF}-1} \frac{(2^{SF}-1)!(-1)^{q+1}}{q!(2^{SF}-1-q)!} \frac{1}{q+1} \left(\frac{1}{\Omega}\right)$$

$$\times \frac{1}{[\frac{1}{\Omega} + \frac{E_s q}{2(q+1)\sigma^2}]} \tag{16}$$

3   Multihop DF Relaying

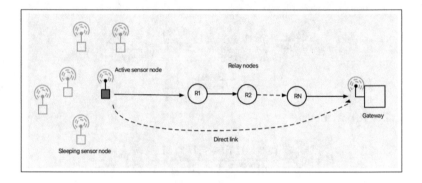

Fig. 1. Multihop system model

We consider a wireless network with n LoRa sensing nodes and L relay nodes as illustrated in Fig. 1. We assume all sensor nodes are battery powered and class A LoRa devices. Relay nodes are uniformly placed between sensor nodes and the gateway and are all operating in class C LoRa devices. Relay nodes are powered locally to support message receiving at nearly any time except when they are transmitting. We also assume all sensor nodes are synchronized in a manner that in each transmit window, only one node is active. When a sensor node transmits its sensed data, relays overhear the transmission but operate over orthogonal time slots. So only the first relay decodes the received packet from the sensor node before forwarding it to the subsequent relay and so on until the packet reaches the gateway. All relay nodes are operating in Decode and Forward (DF) protocol. For each transmitted packet, the gateway receives two copies, one from the direct link (single hop) and the other from the relaying link.

The bit is correctly received at D if it has been correctly transmitted over $(L+1)$ hops. The corresponding BEP is written as

$$BEP_{multihop} = 1 - \prod_{i=1}^{L+1} [1 - BEP_i] \qquad (17)$$

where BEP_i is the BEP over the i-hop written as (15) and (16) for Nakagami and Rayleigh channels

4   Theoretical and Simulation Results

We did some simulations for $SF = 4$ and distance between consecutive relay nodes $d_i = \frac{1}{L+1}$ so that the total normalized distance between the sensor node and the gateway is one. The power allocated to the source and relay nodes is the same so that the total transmitted energy per symbol is E_s.

Figure 2 shows the BEP of single hop relaying without relay nodes for $m = 1, 2, 3, 4$. The BEP decreases as the fading figure m increases. Figure 3-4 show the BEP of multi-hop DF relaying for $m = 1, 4$. We observe that the BEP improves as the number of hops is increased since relays will be closer to each others.

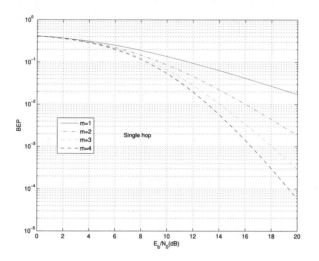

Fig. 2. BEP of single hop relaying for $m = 1, 2, 3, 4$

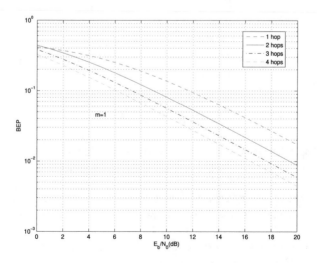

Fig. 3. BEP of multihop relaying for $m = 1$

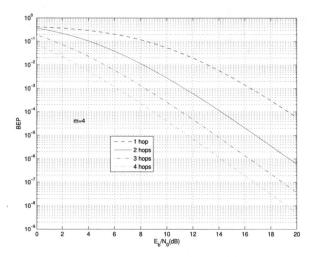

Fig. 4. BEP of multihop relaying for $m = 4$

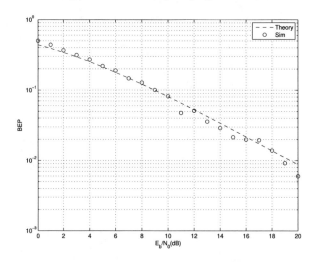

Fig. 5. Theoretical and simulation results for $m = 1$ and $L = 2$ hops

Figure 5 compares the BEP and BER for $m = 1$ and $L = 2$ hops. There is a perfect match between theoretical and simulation results which confirms the validity of our derivations.

Figure 6 compares the obtained results for Nakagami and AWGN channels. When m increases, the BEP becomes closed to that of AWGN [3].

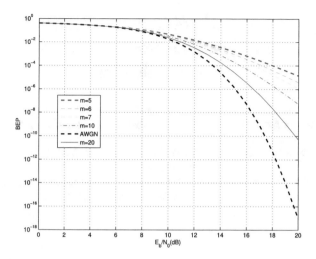

Fig. 6. BEP for Nakagami and AWGN channels

5 Conclusion and Perspectives

In this paper, we derived the bit error probability of LoRa systems using multihop decode and forward relaying. The analysis is valid for Nakagami fading channels with arbitrary fading figure m as well as Rayleigh channels. We have shown that the BEP decreases as the number of hops and fading figure m increases. As a perspective, we can derive the performance of cooperative LoRa for Rician fading channels.

References

1. Elshabrawy, T., Robert, J.: Closed form approximation of LoRa modulation BER performance. IEEE Commun. Lett. **22**(9), 1–3 (2018)
2. Mroue, H., Nasser, A., Perrein, B., Hamriou, S., Motta-Cruz, E., Rouyer, G.: Analytical and simulation study for LoRa Modulation. In: 2018 IEEE International Conference on Telecommuniation (IEEE ICT) (2018)
3. Mahjoub, T., Said, M.B., Boujemaa, H.: On the performance of Packet error rate for LoRa networks. In: 2022 International Conference on Wireless Communications and Mobile Computing (2022)
4. Mahjoub, T., Said, M.B., Boujemaa, H.: Experimental analysis of LoRa signal in urban environment. In: 2022 International Conference on Wireless Communications and Mobile Computing (2022)
5. Dias, C.F., de Lima, E.R., Fraidenraich, G.: Bit error closed form expressions for LoRa systems under nakagami and rice fading channels. Sensors **4412**(19), 1–11 (2019)
6. Peppas, K., Chronopoulos, S.K., LoukatosLi, D., Arvanitis, K.: New results for the error rate performance of LoRa systems over fading channels. Sensors **3350**(22), 20 (2022)
7. de Oliveira Alvesn, L.H., Rebellato, J.L., Souza, R.D., Brante, G.: Network coded cooperative LoRa network with D2D communications. IEEE Internet Things J. **9**(7), 4997–5008 (2022)
8. Borkotoky, S.S., Schilcher, U., Bettsletter, C.: Cooperative relaying in LoRa sensor networks. In: 2019 IEEE GLOBECOM Global Communication Conference (2019)
9. Proakis, J.G.: Digital Communications. Third edn. Mc Graw Hill (1995)

Rigorous Analysis of Induced Electrical Field in Human Tissues for Body Centric Wireless Communication

Intissar Krimi[1(✉)], Sofiane Ben Mbarek[2], and Fethi Choubani[1]

[1] Innov'Com Laboratory, SUPCOM, University of Carthage, Ariana, Tunisia
intissar.krim@supcom.tn
[2] Physical Science and Engineering Division, King Abdullah University of Science and Technology, Thuwal, Saudi Arabia
sofien.benmabrek@isimg.tn

Abstract. The increasing use of wireless technology in the Body Area Network (WBAN) system has created a demand for accessible data showing the depth of penetration of electromagnetic waves (EMW) into human tissue and how this varies with the wavelength. This information would be especially useful in medical studies and would aid in the development of WBAN for specific diseases based on a better understanding of the electromagnetic penetration effects. In this paper, we estimate the magnitude of the electric field inside muscle tissue at various depths and with the most important frequencies in medical applications. Moreover, this investigation provides evidence that the penetration of EMW in biological tissue strongly depends on the frequency and thickness of the tissue involved. The method used here is based on an analytical model, and it focuses on the impact of incident plane waves incoming in a lossy medium on muscle tissue using the Method of Moments (MoM). All results are finally validated and compared through Finite Element Method (FEM) computations. The presented study is interesting in view of a practical application of the effects of deep wave penetration.

1 Introduction

Over the last decade, the interaction between incident EMW and human tissue has been studied by using a several methods. As we know, the human body is a lossy channel that significantly attenuates the propagation of EMW [1,2]. Hence, the investigation of the EM characteristics of biological tissues and the properties of the body channel are the keys to establishing wireless communication [3,4]. The electric properties (permittivity and conductivity) of human tissue change with the frequency that varies [5]. Several bands are employed in medical application. In this investigation, we focused on ISM (Industrial, Scientific, and Medical) bands [6,7], which are commonly used in therapeutic and diagnostic applications [8]. It includes the frequencies : 434 MHz, 915 MHz, 2.45 MHz and 5.8 GHz. The main purpose of the recent research is to estimate the path loss (PL) [9] when determining the required implanted transceiver properties, such as the required input power or receiver sensitivity, or when evaluating the communication

© The Author(s), under exclusive license to Springer Nature Switzerland AG 2023
L. Barolli (Ed.): AINA 2023, LNNS 661, pp. 82–91, 2023.
https://doi.org/10.1007/978-3-031-29056-5_9

range. We need to identify the electric field at a specific frequency and apply Eq. (1) to evaluate the PL at a particular depth [5].

$$PL(dB) = 10log10(P_i/P_r) = 10log10(s_i/s_{av}), (dB) \tag{1}$$

where, P_i is the transmitted signal, P_r is the received power which equals to the time-average power density s_{av}.

There are numerous applications for assessing the E-field [10]. The most important applications are:

- Diagnostic application, the electromagnetic field (EMF) at microwave frequency can provide a convenient approach to detect and monitor physiological responses;
- Therapeutic application, like hyperthermia, by focusing the EMF at a certain location;
- Study the effects of electromagnetic waves that penetrate tissues, for example by using ultra-wide frequency bands.

The analytical method will provide an accurate solution only in simplified cases. Therefore, the numerical method, although more computationally demanding than an analytical technique [11], can provide physical insight into the propagation behaviour and precise solutions. Several numerical techniques are available and each has its own advantages and limitations. The Finite Difference Time Domain (FDTD) technique is based on a direct time-domain solution. It is calculated by an iterative algorithm. The actual human condition is extremely complicated. So, FDTD must be used to mesh the model with high precision, which increases the time required [12, 13]. Subsequently, the method of moments (MoM) is an extremely efficient computational technique that can present in the frequency domain an efficient solution of EMW for handling large, complex structures with short calculation times and sufficient computing resources [14, 15]. Another method is using Comsol Multiphysics software for investigating the E-field inside human tissue. The Comsol Multiphysics software applies the Finite element method (FEM) [16]. This paper provides interesting information to evaluate the magnitude of the electric field inside the muscle tissues at different points and frequencies using two alternative approaches. The main reason for the selection of muscle tissue is its high thickness [17]. Therefore, the investigation will be conducted more clearly. The magnitude of the E-field inside another tissue can be evaluated using the techniques described in this contribution.

The rest of the paper is set as follows: Sect. 2 presents the parameters and characteristics of human tissue. Section 3 describes the two proposed methods. The results and comparison are explained in Sect. 4. In Sect. 5, the paper is finally summarized and concluded.

2 Characteristics of Human Tissue

According to previous studies, the electrical properties of human tissues change with frequency. There is a database available for tissue parameters [5]. Table 1 includes information for different tissues and frequencies. The penetration depth (Pd), as defined by the Eq. (2), is the distance over which the field decreases to (1/e) of its value just inside the layer boundary. Due to reflection, the E-field inside the boundary may already be much smaller than the incident external field.

$$\delta = \frac{1}{w}\left[\frac{\mu\varepsilon}{2}\left(\sqrt{1+\left(\frac{\sigma}{w\varepsilon}\right)^2}-1\right)\right]^{-\frac{1}{2}},\qquad(2)$$

where δ is the penetration depth, w is the angular frequency, μ is the electromagnetic permeability, ε is the electromagnetic permittivity, and σ is the conductivity.

Table 1. Electric Properties of Human Tissues at ISM band [5].

Frequency	Tissue parameters	Skin	Fat	Muscle
915 MHz	ε_r	41.33	5.45	54.99
	σ (s/m)	0.87	0.05	0.94
	Pd (mm)	39.95	242.3	42.1
2.45 GHz	ε_r	38	5.28	52.73
	σ (s/m)	1.46	0.1	1.73
	Pd (mm)	22.57	117.02	22.33
5.8 GHz	ε_r	35.11	4.95	48.48
	σ (s/m)	3.71	0.29	4.96
	Pd (mm)	8.57	40.48	7.54

Figure 1 depicts the penetration depth of human tissue as a function of frequency. To penetrate the tissue, the Pd must be greater than the tissue thickness.

According to [18], we can state the decrease of the field strength by the Eq. (3):

$$E = E_i e^{-\frac{z}{\delta}},\qquad(3)$$

where E is the E-field at a certain point, E_i is the reference E-field, and z is the distance from E_i. As mentioned in Eq. (4), the E-field inside the human tissue will combine the wave that arrived and the reflected wave from the next tissue.

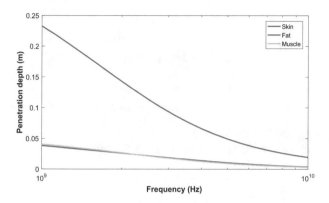

Fig. 1. The relation between frequency (1000MHz-10GHz) and the penetration depth for different tissues [5].

$$E = E_i(e^{-\frac{z}{\delta}} - \tau e^{\frac{z}{\delta}}) \tag{4}$$

τ represents the reflection coefficient of the next tissue, because there are multiple reflections within each tissue. Therefore, the question will be more complicated.

3   Calculation and Numerical Simulation

3.1   MoM Method

This method considers the human tissues as non-uniform transmission lines. The MoM consists essentially in solving numerically the differential or integral forms of Maxwell's equations. The equations are then transformed into matrix systems of linear equations, where the parameters to be solved are the electromagnetic wave. In the first step, these unknown coefficients are developed as a set of known linearly independent basis functions multiplied by unknown weighting parameters.

To apply this method, we will consider the layers of human as one-dimensional layers, as shown in Fig. 2-a ,where B is the thickness of bone (20 mm), M is the thickness of muscle (20 mm), F is the thickness of fat (4 mm), and S is the thickness of skin (2 mm), then we calculate the electromagnetic wave using the MoM method. $E_i = 10(V/m)$ is the incident signal that is perpendicular to the skin layer.

In this study, the mathematical formulation of the MoM method is based on the numerical solution of the differential and integral Maxwell's equations:

$$\frac{dE_t}{dz} = Z(z)H_t, \tag{5}$$

$$\frac{dH_t}{dz} = Y(z)E_t, \tag{6}$$

where,

$$Z(z) = \begin{cases} jw\mu(z), & TE \\ jw\mu(z) + k_x^2(\sigma(z) + jw\varepsilon(z)), & TM \end{cases} \tag{7}$$

$$Y(z') = \begin{cases} \sigma(z) + jw\varepsilon(z) + k_x^2 jw\mu(z), & TE \\ \sigma(z) + jw\varepsilon(z), & TM \end{cases} \tag{8}$$

$k_x = \frac{w}{c}sin(\theta_i)$, and c is the velocity.

In addition, there are two boundary conditions, as follows:

$$E_t(0) + \eta_s H_t(0) = 2Ei, \tag{9}$$

$$E_t(d) + \eta_l H_t(d) = 0, \tag{10}$$

where, d is thickness of the layer, and

$$\eta_s = \begin{cases} \sqrt{\frac{\mu_0}{\varepsilon_0}} \frac{1}{cos(\theta_i)}, & TE, \\ \sqrt{\frac{\mu_0}{\varepsilon_0}} cos(\theta_i), & TM, \end{cases} \tag{11}$$

and,

$$\eta_l = \begin{cases} \sqrt{\frac{jw\mu_0}{\sigma + jw\varepsilon}} \frac{1}{cos(\theta_i)}, & TE, \\ \sqrt{\frac{jw\mu_0}{\sigma + jw\varepsilon}} cos(\theta_i), & TM. \end{cases} \tag{12}$$

The integral form will be determined by using the Eq. (5) and Eq. (6), respectively, as follows:

$$E_t(z) = \frac{\eta_l}{\eta_s + \eta_l} E_i + \frac{\eta_s}{\eta_s + \eta_l} \int_0^d Z(z')H_t(z')\,dz' - \eta_l Y(z')E_t(z')\,dz' - \int_0^z Z(z')H_t(z')\,dz', \tag{13}$$

$$H_t(z) = \frac{E_i}{\eta_s + \eta_l} - \frac{1}{\eta_s + \eta_l} \int_0^d Z(z')H_t(z')\,dz'$$
$$- \eta_l Y(z')E_t(z')\,dz' - \int_0^z Y(z')E_t(z')\,dz'. \tag{14}$$

Finally, we can apply the MoM method to Eq. (13) and Eq. (14), we expand E and H in a set of pulse functions:

$$E_t(z) = \sum_{n=1}^{N} a_n E_n, \tag{15}$$

$$H_t(z) = \sum_{n=1}^{N} a_n H_n, \tag{16}$$

where a_n is a rectangular pulse function:

$$a_n = \begin{cases} 1, & (n-1)\Delta z < z < n\Delta z \\ 0, & elsewhere \end{cases} \tag{17}$$

with $\Delta z = d/N$. Substituting Eq. (15), Eq. (16) into Eq. (13), Eq. (14), and matching the equation at the points $z = (m-1/2)\Delta z$, for m=1, 2,... N, we obtained:

$$E_n \delta_{n-m} + \frac{\eta_s \eta_l}{\eta_l + \eta_s} \int_{(n-1)\Delta_z}^{n\Delta_z} Y(z')E_n \,dz'$$

$$+ \int_{(n-1)\Delta_z}^{(n-1+U_{mn})\Delta_z} Z(z')H_n \,dz'$$

$$- \frac{\eta_s \eta_s}{\eta_l + \eta_s} \int_{(n-1)\Delta_z}^{n\Delta_z} Z(z')H_n \,dz' = \frac{\eta_l}{\eta_l + \eta_s} E_i, \tag{18}$$

$$\eta H_n \delta_{n-m} + \frac{1}{\eta_l + \eta_s} \int_{(n-1)\Delta_z}^{n\Delta_z} Z(z')H_n \,dz') +$$

$$\int_{(n-1)\Delta_z}^{(n-1+U_{mn})\Delta_z} Y(z')E_n \,dz'$$

$$- \frac{\eta_l}{\eta_l + \eta_s} \int_{(n-1)\Delta_z}^{n\Delta_z} Y(z') \,dz' E_n = \frac{\eta_l}{\eta_l + \eta_s} E_i, \tag{19}$$

and δ_{n-m} is the Kronecker delta function, U_{mn} takes as input m and n, it returns 1/2 if they are identical, 1 if $m > n$ and 0 if $m < n$. We only extract the electric field, as result:

$$[M_{nm}] [E_n] = \left[\frac{\eta_l}{\eta_l+\eta_s}E_i\right], \tag{20}$$

where, M_{nm} is the matrix system that is found by combining in the Eq. (18) and Eq. (19):

$$M_{nm} = \begin{pmatrix} n_1...n_N \\ m_1...m_N \end{pmatrix}, \tag{21}$$

with M_{nm} where determined, respectively, by the following integral element:

$$n_N = \delta_{n-m} + \frac{\eta_s \eta_l}{\eta_s + \eta_l} \int_{(n-1)\Delta_z}^{n\Delta_z} Y(z') \,dz', \tag{22}$$

$$m_N = \int_{(n-1)\Delta_z}^{(n-1+U_{mn})\Delta_z} Y(z') \,dz' - \frac{\eta_l}{\eta_s + \eta_l} \int_{(n-1)\Delta_z}^{n\Delta_z} Y(z') \,dz'. \tag{23}$$

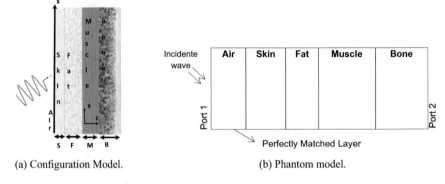

(a) Configuration Model.

(b) Phantom model.

Fig. 2. Proposed Model

3.2   FEM Method

This method uses Comsol Multiphysics, which applies the Finite Element Method. The dimensions of the proposed phantom model are $40mm \times 40mm$. The structure of human tissue consists of four layers (skin, fat, muscle, and bone) with the same thickness as those used in the first method. Figure 2-b includes the configuration of the phantom model that was used.

4   Results and Comparison

To compare the results obtained by the two methods, the dimensions and parameters of the tissues must be the same in both cases. The depths in the muscle tissue chosen for comparison were 3, 6, 9, 11, 14, and 17 mm. The magnitude of the E-field at these points has been calculated for different frequencies. For plotting the results we should take more points. Thus, the magnitude of E- field will calculate at each 3 mm into muscle tissue. When using a plane wave with frequencies of 915 MHz and 2.45 GHz, the

Table 2. Comparison between two methods at different frequencies.

Depth(mm)	3	6	9	11	14	17
915 MHz						
MoM method	2.30	1.875	2.087	2.25	2.83	2.883
FEM method	2.39	1.83	1.85	1.99	2.5	2.44
2.45 GHz						
MoM method	2.728	2.015	1.628	2.098	1.564	0.8924
FEM method	2.336	2.064	1.562	1.998	1.37	1.012
5.8 GHz						
MoM method	0.642	0.433	0.2863	0.22	0.162	0.117
FEM method	0.587	0.477	0.33	0.22	0.155	0.12

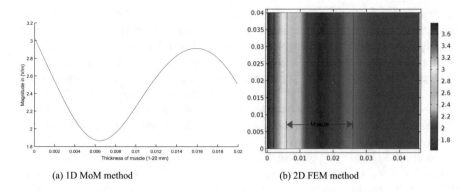

(a) 1D MoM method   (b) 2D FEM method

Fig. 3. Distribution of E-Field inside muscle tissue for both methods at 915 MHz.

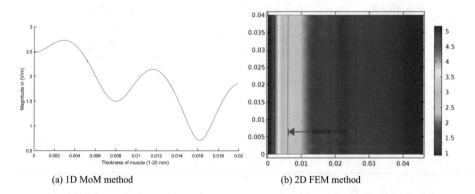

(a) 1D MoM method   (b) 2D FEM method

Fig. 4. Distribution of E-Field inside muscle tissue for both methods at 2.45 GHz.

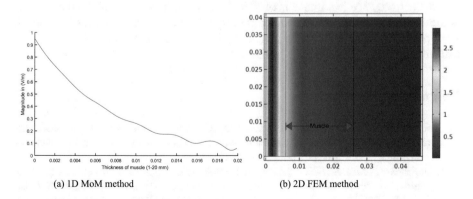

(a) 1D MoM method   (b) 2D FEM method

Fig. 5. Distribution of E-Field inside muscle tissue for both methods at 5.8 GHz.

penetration depth of each tissue is greater than its thickness. Therefore, the wave will penetrate the tissue and reflect from the next tissue, and producing fluctuation at different points and forming a standing wave. Figures 3, 4 and 5 show the results. The comparison of the two methods at different depths and frequencies is illustrated in Table 2.

According to Eq. (2), The penetration depth of a plane wave at a frequency of 5.8 GHz is 7.5413 mm mm, which is less than the thickness of the muscle. This means that the wave reflected from the bone will be minimal and result in exponential attenuation of the wave, particularly in the 12.45 mm to 20 mm region of the muscle tissue, as the wave will behave according to Eq. (3).

As shown in Figs. 5, the transmitted wave is significantly stronger than the reflected wave in most cases. However, in the region between 12.4587 mm mm and 20 mm, the reflected wave becomes more prominent as a result of the wave's behavior outlined in Eq. (4). This leads to fluctuations in the E-field and the formation of standing waves. Accordingly, If the tissue thickness is less than the penetration depth, the wave will reflect and mix with the transmitted wave, resulting in a slow fluctuation of the wave. If the penetration depth is less than the tissue thickness, the reflected wave will be very weak, causing exponential fading of the wave. At a certain depth, the penetration depth will exceed the tissue thickness, resulting in wave fluctuations caused by reflected waves.

5   Conclusion

A comparison of both proposed methods gives, approximately, the same results. The difference in these results can be overlooked when high values of electric field are used. Accordingly, the MoM and FEM are useful to investigate the E-field at any point in the muscle or other tissues. The impact of different frequencies and tissue thickness on the wave penetration were studied. This result has important implications for practical applications of WBAN, particularly for inter-/intra-body communication links, to evaluate the performance requirements of WBANs and beyond. For example, the analysis of the waveform that is emitted by a smart sensor may be necessary to investigate which wave produces the greatest penetration, in which case further investigations involving realistic structures will need to be performed.

References

1. Zeinelabedeen, W., Uyguroglu, R.: A study on health care monitoring of femoral shaft fracture healing by using implanted antenna for wireless in-to-out body channel communication. J. Electromagn. Waves Appl. **36**(5), 722–742 (2022)
2. Ben Saada, A., Mbarek, S.B., Choubani, F.: Whole-body exposure to far-field using infinite cylindrical model for 5G FR1 frequencies. In: Barolli, L., Hussain, F., Enokido, T. (eds.) AINA 2022. LNNS, vol. 449, pp. 471–478. Springer, Cham (2022). https://doi.org/10.1007/978-3-030-99584-3_41
3. Ying, C.Z., Gao, Y.M., Du, M.: Propagation characteristics of electromagnetic wave on multiple tissue interfaces in wireless deep implant communication. IET Microwaves Antennas Propag. **12**(13), 2034–2040 (2018)

4. Saada, A.B., Mbarek, S.B., Choubani, F.: Antenna polarization impact on electromagnetic power density for an off-body to in-body communication scenario. In: 2019 15th International Wireless Communications Mobile Computing Conference (IWCMC), pp. 1430–1433. IEEE (2019) https://doi.org/10.1109/IWCMC.2019.8766713
5. Carrara, N.: Dielectric Properties of Body Tissues. IFAC, Institute for Applied Physics, Italy (2007). http://niremf.ifac.cnr.it/tissprop/
6. Radiocommunications Agency, ?UK radio interface requirement 2030 short range devices,? ver. 1.2, Oct. 2002
7. Ung, J., Karacolak, T.: A wideband implantable antenna for continuous health monitoring in the MedRadio and ISM bands. IEEE Antennas Wirel. Propag. Lett. **11**, 1642–1645 (2012)
8. Leelatien, P.: Time-domain analysis of wireless telemetry for liver-implanted medical monitoring applications. In: 2020 17th International Conference on Electrical Engineering/Electronics, Computer, Telecommunications and Information Technology (ECTI-CON). IEEE (2020)
9. EbrahimiZadeh, J., et al.: Pathloss calculation for fat-intra body communication using poynting vector theory. In: 2020 14th European Conference on Antennas and Propagation (EuCAP). IEEE (2020)
10. Silue, D., Choubani, F., Labidi, M.: Enhanced meander antenna for in-body telemetry applications. In: 2022 18th International Conference on Wireless and Mobile Computing, Networking and Communications (WiMob). IEEE (2022)
11. Khaleda, A.: Numerical Techniques at Millimeter Wave Frequencies for Wireless Body Area Networks-A
12. Krimi, I., Mbarek, S.B., Hattab, H., et al.: Electromagnetic near-field study of electric probes for EMC applications. In: Innovative and Intelligent Technology-Based Services for Smart Environments Smart Sensing and Artificial Intelligence. CRC Press, pp. 45–50 (2021). https://doi.org/10.1201/9781003181545-8
13. Ben Mbarek, S., Choubani, F.: FDTD modeling and experiments of microfabricated coplanar waveguide probes for electromagnetic compatibility applications. J. Electromagn. Waves Appl. **35**(5), 634–646 (2021). https://doi.org/10.1080/09205071.2020.1851776
14. Khalaj-Amirhosseini, M.: Analysis of lossy inhomogeneous planar layers using the method of moments. J. Electromagn. Waves Appl. **21**(14), 1925–1937 (2007)
15. Rothwell, E.J.: Natural-mode representation for the field reflected by an inhomogeneous conductor-backed material layer-TE case. Prog. Electromagn. Res. **63**, 1–20 (2006)
16. COMSOL Multiphysics v. 5.2. www.comsol.com
17. Varotto, G., E.M. Staderini.: A 2D simple attenuation model for EM waves in human tissues: comparison with a FDTD 3D simulator for UWB medical radar. In: 2008 IEEE International Conference on Ultra-Wideband. vol. 3. IEEE (2008)
18. Kang, G., Gandhi, O.P.: Effect of dielectric properties on the peak 1-and 10-g SAR for 802.11 a/b/g frequencies 2.45 and 5.15 to 5.85 GHz. IEEE Trans. Electromagn. Compat. **46**(2), 268–274 (2004)

Investigation Model of Electromagnetic Propagation for Wireless Body Communication

Intissar Krimi[1(✉)], Sofiane Ben Mbarek[2], and Fethi Choubani[1]

[1] Innov'Com Laboratory, SUPCOM, University of Carthage, Ariana, Tunisia
intissar.krim@supcom.tn
[2] Physical Science and Engineering Division, King Abdullah University of Science and Technology, Thuwal, Saudi Arabia
sofien.benmabrek@isimg.tn

Abstract. This paper investigates the electromagnetic (EM) distribution in a human model irradiated by an incident EM plane wave. A planar inhomogeneous structure is used for modeling human tissue. Moreover, the steady-state EM distribution is calculated by solving the Differential and Integral Equations (DIE) by using the Method of Moments (MoM). The obtained results demonstrate the great of performing a theoretical analysis for Path Loss (PL) estimation. For the different examined conditions, an excellent agreement with the recent results, and the Finite Element (FEM), and MoM methods are verified to be valid in this investigation. It is found that the distribution of the field and the PL for different communication scenarios are very useful for estimating the quality of communication for implant communication.

1 Introduction

A wireless body area network (WBAN) connects nodes that are in/on the human body. WBANs Applications include health care, multimedia, and sports. As they facilitate movement among patients, WBANs have brought about a revolutionary change in healthcare facilities and patient monitoring [2,3]. Medical implants placed within the human body lead to a faster and better diagnosis, which improves the patient's quality of life. For example, the emergence of wireless capsule endoscopy (CE) enables the examination of the small intestine that was previously inaccessible by other types of endoscopies [4]. The advantages of CE in terms of better diagnosis are obvious. Additionally, it improves the comfort of the patient, as there are no tubes or wires involved in the treatment.

The modelling of the physical layer is an essential step in the development of human body communication [5,6]. Industrial, Scientific, and Medical (ISM,2.4 GHz) and Ultra Wideband (UWB,3.1–10.6 GHz) technologies are increasing in popularity because of their ability to transmit data at high data rates while consuming low power, which could improve the standard for wireless communications [6,7]. As we know, the human body is a lossy channel that

L. Barolli (Ed.): AINA 2023, LNNS 661, pp. 92–103, 2023.
https://doi.org/10.1007/978-3-031-29056-5_10

significantly attenuate the propagation of the EM wave. As a result, the main challenge is predicting the radio channel behaviour, which is critical for investigating propagation loss scenarios.

Analytically solving the equations of heterogeneous human tissues with different geometrical complexity is a very hard problem. For this reason, several approaches have been reported in the literature to solve the EM characteristics of a lossy medium numerically in either differential or integral form by solving Maxwell's equations [8,9]. These techniques are divided into two groups: frequency domain and time domain. The most successful method in the time-domain is the FDTD [10], it is an effective tool for modelling a miniature structure for analysing near-field microscopy for electromagnetic compatibility applications [11,12]. However, FDTD requires a large computation time and memory, as it uses a uniform mesh to model the entire computational domain, which does not conform to curved surfaces (as a spherical or cylindrical boundary). The frequency-domain approaches include the Method of Moments (MoM) [13]. As compared to FDTD, MoM is an interesting solution for such applications in terms of accuracy. It provides a direct numerical solution of an exact integral equation. Moreover, it is a very powerful method when studying inhomogeneous complex shapes [14].

This work presented an evaluation of the communication inside the human body at the ISM and UWB ranges. This understanding is important to confirm the potential of the modelling analysis for wireless propagation loss. It can be used as a reference to facilitate the development process of different implanted telehealth applications, such as organ monitoring in the abdomen. For these purposes, comparative results were computed to validate the body communication case.

This paper presents an investigation of a planar model of human tissue that will be used to calculate the EM fields (E and H) and the Path Loss (PL) within human tissue when exposed to an incident plane wave. Most of the research is focused on simulation rather than mathematical calculation, where the accuracy of the calculation is higher than the simulation. Therefore, this study demonstrates the importance and potential of MoM analysis to make results with very high precision. A suggestion for wireless in-body communication applications will be discussed.

The remaining part of this paper is divided into the following sections: Sect. 2 describes the proposed model and method while Sect. 3 contains the results and verification. The simulation results and discussion are described in Sect. 4. Lastly, Sect. 5 gives the conclusion.

2   The Proposed Model and Method

2.1   Configuration Setup

Two distinct scenarios were taken into consideration regarding the positioning of the antenna: the on-body antenna is in direct contact with the surface of the skin, whereas the implanted antenna is situated in the muscle, which looks like Fig. 1.

- In-Body to On-Body (IB2OB) communication.
- On-Body to In-Body (OB2IB) communication.

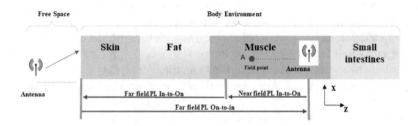

Fig. 1. Specific configuration of layered biological tissue.

The model was developed to study the effect of electromagnetic propagation loss on the communication channel of the body. It can be assumed to be sufficiently reliable for the intended calculations as the transmission waves are generally calculated considering the properties of each region of the tissue layer, which is always the case even if different model forms (circular or spherical) are considered.

The human tissues are inhomogeneous, and their dielectric properties depend on frequency. These properties have been measured and are available in a database [14]. Table 1 provides information about the electric properties of several tissues at three frequencies.

Table 1. Electric Properties of Human Tissues at Different Frequencies

Frequency	Tissue parameters	Skin	Fat	Muscle
2.45 GHz	ϵ	38	5.2	52.7
	σ (s/m)	1.46	0.1	1.73
6 GHz	ϵ	34.94	4.93	48.2
	σ (s/m)	8.	0.5	11.42
	Thickness (mm)	1.5	8.5	27.5

2.2   Mathematical Formulation

2.2.1   Selection of Computational Methods

The electromagnetic problem should be analysed with the configuration behaviour, and based on the modelling methods, they can be applied to solve the problem:

- MoM is most useful for unbounded problems related to radiation elements. It is efficient in terms of performance, precision, and execution time.
- FEM is based on the large-volume configuration for analysis. It is useful for large problems with complex, inhomogeneous configurations. Nevertheless, it does not support unbounded conditions.
- FDTD is a time-domain technique and is the most efficient for transient analysis problems. It is effective for complex, unbounded, and inhomogeneous problems. The disadvantage of FDTD is the complexity of the analysis, which requires more execution time.

In this study, we need a method that can be implemented for efficient communication inside a human body. MoM is the quintessential requirement for exploration.

The formulation has been validated by using the plane wave propagation formula and the concept of transmission line analogy. MoM is one of the most well-known approaches to electromagnetic problems [13]. This technique is based on reducing the operator equations to a system of linear equations that can be written in matrix form. One of the advantages of employing this approach is that the result is extremely precise since the equations are almost exact and MoM offers a direct numerical solution to these equations.

The incident wave may collide with the tissue, forming certain angles (other than 90°) with it. This is an instance of oblique incidence. There are two main varieties of oblique incidence, depending on whether the electric field is normal to it (perpendicular polarization) or is parallel to the plane of incidence (parallel polarization). In the next section, both kinds are explained.

2.2.2   TE Case

In perpendicular polarization, the y-component is the electric field E_y that is normal to the plane of incidence. We obtain from Faraday's law and Ampere's law the differential equations that describe the problem:

$$\frac{dE_y}{dz} = jw\mu(z)H_x, \tag{1}$$

$$\frac{dH_x}{dz} = [\sigma(z) + jw\epsilon(z) + \frac{k_x^2}{jw\mu(z)}]E_y, \tag{2}$$

$$H_z = \frac{k_x}{\mu(z)}E_y. \tag{3}$$

where $k_x = \frac{w}{c}sin(\theta_i)$, μ [H/m] is the magnetic permeability which is equal to μ_0 for the nonmagnetic human body tissues, ϵ [F/m] is the electric permittivity, and σ [S/m] is the electric conductivity.

2.2.3   TM Case

In this case, the plane of incidence can be determined by the axes of the interface, and normal to the interface which is the xz-plane,

$$\frac{dH_y}{dz} = -[\sigma(z) + jw\epsilon(z)]E_x, \tag{4}$$

$$\frac{dE_x}{dz} = -[jw\mu_r(z) + \frac{k_x^2}{\sigma(z) + jw\epsilon_r(z)}]H_y, \tag{5}$$

$$E_z = -\frac{jk_x}{\sigma(z) + jw\epsilon_0\epsilon_r(z)}H_y. \tag{6}$$

Furthermore, there are boundary conditions that enforce the continuity of tangential EM fields at the boundaries (at $z=0$ and $z=d$). According to Fig. 1, we can write:

$$E_t(0) + \eta H_t(0) = 2E_i, \tag{7}$$

$$E_t(d) + \eta H_t(d) = 0. \tag{8}$$

where

$$E_i = \begin{cases} 2E_0, & TE \\ 2E_0 cos(\theta_i), & TM \end{cases} \tag{9}$$

$$\eta = \begin{cases} \sqrt{\frac{\mu_0}{\epsilon_0}} \frac{1}{cos(\theta_i)}, & TE \\ \sqrt{\frac{\mu_0}{\epsilon_0}} cos(\theta_i), & TM \end{cases} \tag{10}$$

Differential equations (DE) do not have an analytical solution in most cases. So, DE must be solved numerically [8]. Alternatively, the differential equations can also be converted into integral equations. The efficiency with which the resulting integral equation may be solved makes this technique an interesting way to investigate the effects of the incident wave inside the tissue. To convert Eq. (1) to Eq. (6), into integral equations, as an example, we integrate the Eq. (1):

$$\frac{dE_y}{dz} = jw \int_0^z \mu(z)H_x \, dz + C. \tag{11}$$

Where C is a constant, which can be determined according to the boundary conditions Eq. (8). Thus, the integral equations obtained may be stated in the following general form:

$$E_t(z) = \frac{E_i}{2} + \frac{1}{2} \int_0^d Z(z')H_t(z') \, dz'$$

$$- \eta Y(z')E_t(z') \, dz' - \int_0^z Z(z')H_t(z') \, dz', \tag{12}$$

$$H_t(z) = \frac{E_i}{2\eta} - \frac{1}{2\eta} \int_0^d Z(z')H_t(z')\,\mathrm{d}z'$$

$$- \eta Y(z')E_t(z')\,\mathrm{d}z' - \int_0^z Y(z')E_t(z')\,\mathrm{d}z'.$$

(13)

where

$$Z(z') = \begin{cases} jw\mu(z), & TE \\ jw\mu(z) + k_x^2(\sigma(z) + jw\epsilon(z)), & TM \end{cases}$$

(14)

$$Y(z') = \begin{cases} \sigma(z) + jw\epsilon(z) + k_x^2 jw\mu(z), & TE \\ \sigma(z) + jw\epsilon(z), & TM \end{cases}$$

(15)

To solve the equations Eq. (12) and Eq. (13), the accuracy of MoM for a time-harmonic incident EM plane wave on a human tissue medium is required. Collocation yields excellent results according to [9]. We expand E and H in a set of pulse functions:

$$E_t(z) = \sum_{n=1}^{N} a_n E_n,$$

(16)

$$H_t(z) = \sum_{n=1}^{N} a_n H_n.$$

(17)

where a_n is a rectangular pulse function:

$$a_n = \begin{cases} 1, & (n-1)\Delta z < z < n\Delta z \\ 0, & elsewhere \end{cases}$$

(18)

With $\Delta z = d/N$. This partitions the structure into N regions. Substituting Eq. (16) and Eq. (17) into Eq. (12) and Eq. (13) and matching the equation at the points $z = (m - 1/2)\Delta z$, for m=1, 2,... N, we obtain:

$$2E_n\delta_{n-m} + \eta \int_{(n-1)\Delta_z}^{n\Delta_z} Y(z')E_n \,, \mathrm{d}z'$$

$$+ 2 \int_{(n-1)\Delta_z}^{(n-1+U_{mn})\Delta_z} Z(z')H_n \,\mathrm{d}z'$$

$$- \int_{(n-1)\Delta_z}^{n\Delta_z} Z(z')H_n \,\mathrm{d}z' = E_i,$$

(19)

$$2\eta H_n \delta_{n-m} + \int_{(n-1)\Delta_z}^{n\Delta_z} Z(z^{'}) H_n \, dz^{'}) +$$

$$2\eta \int_{(n-1)\Delta_z}^{(n-1+U_{mn})\Delta_z} Y(z^{'}) E_n \, dz^{'}$$

$$- \eta \int_{(n-1)\Delta_z}^{n\Delta_z} Y(z^{'}) \, dz^{'} E_n = E_i. \quad (20)$$

Finally, we can write the equations Eq. (19) and Eq. (20) as a matrix form:

$$\begin{bmatrix} P_{mn} \end{bmatrix} \begin{bmatrix} E_n \\ H_n \end{bmatrix} = \begin{bmatrix} E_i \end{bmatrix}. \quad (21)$$

where δ_{n-m} is the Kronecker function, U_{mn} takes as input m and n, it returns $1/2$ if they are identical, 1 if $m > n$ and 0 if $m < n$, and P_{mn} is defined by the following matrix:

$$P_{mn} = \begin{pmatrix} p_{11} & p_{12} \\ p_{13} & p_{14} \end{pmatrix}, \quad (22)$$

with P_{11}, P_{12}, P_{13} and P_{14} where determined, respectively, by the following equations:

$$p_{11} = 2\delta_{n-m} + \eta \int_{(n-1)\Delta_z}^{n\Delta_z} Y(z^{'}) \, dz^{'}, \quad (23)$$

$$p_{12} = 2 \int_{(n-1)\Delta_z}^{(n-1+U_{mn})\Delta_z} Z(z^{'}) \, dz^{'} - \int_{(n-1)\Delta_z}^{n\Delta_z} Z(z^{'}) \, dz^{'}, \quad (24)$$

$$p_{13} = 2\eta \int_{(n-1)\Delta_z}^{(n-1+U_{mn})\Delta_z} Y(z^{'}) \, dz^{'} - \eta \int_{(n-1)\Delta_z}^{n\Delta_z} Y(z^{'}) \, dz^{'}, \quad (25)$$

$$p_{14} = 2\eta\delta_{n-m} + \int_{(n-1)\Delta_z}^{n\Delta_z} Z(z^{'}) \, dz^{'}). \quad (26)$$

3   Results and Discussion

The proposed model of tissue is composed of consecutive layers: skin, fat, and muscle. We consider the electrical properties [1] in Table 1. The MoM calculates the EM fields, where $E_0 = 1\,\mathrm{V/m}$. Figure 2a–Fig. 3b reveal the impact of the gradual alteration of medium characteristics between the layers of the human body. This gradual change in medium between layers causes minor discrimination in both electric and magnetic fields. That will increase the propagation loss in the transmission channel. Furthermore, for the scenario OB2IB, the variation of the fields in the two frequencies peaks at the skin and is lowest in the muscle due to their higher electrical properties. Similarly, in the scenario IB2OB, the maximum amount of EM field is produced in the muscle and degrades slowly depending on the muscle thickness. This indicates the overall lower transmission energy at the boundaries and such attenuation is highly influenced by the dielectric properties (permittivity, conductivity) of the different biological tissues, which are frequency-dependent. As a result, we can conclude that human tissue is extremely absorbent for both OB2IB and IB2OB.

Table 2. Comparative study between calculation and simulation results on different communication scenario at Frequency of 2.45 GHz.

Communication scenario	Physics quantity	Calculation result	Simulation result	Reference
OB2IB	E (V/m)	0.6144	0.5033	Our work
		0.59746	0.59853	Reference [1]
	H (A/m)	3.3 e–3	3.8e–3	Our work
		4 e–3	4.02e–3	Reference [1]
	PL (dB)	47.63	46.29	Our work
IB2OB	E (V/m)	0.2495	0.2105	Our work
	H (A/m)	4.1e–3	4.6e–3	Our work
	PL (dB)	33.0103	33.1502	Our work

Table 2 includes the comparative results of IB2OB and OB2IB. The result of the EM field is obtained by solving Eq. (21). We notice that the variation is almost similar when comparing the MoM, FEM, and [1], which strongly points out that our results are correct and valid. Currently, all research is centred on various numerical techniques for analysing the communication body setup. Despite its limitations, the FDTD technique is the most widely used [15]. However, we conclude in this study that MOM in the frequency domain gives good results for EM propagation inside human issues in a relatively short amount of time and with greater accuracy.

To accommodate the wireless communication link, it is necessary to investigate the propagation loss for OB2IB and IB2OB channels. For that matter, we derived the path loss based on the average power density. In this way, the path loss in decibels for each link can be obtained from:

$$PL(dB) = 10log10(P_i/P_r) = 10log10(s_i/s_{av}), \qquad (27)$$

where, P_i is the transmitted signal, P_r is the received power which equals to the time-average power density s_{av}.

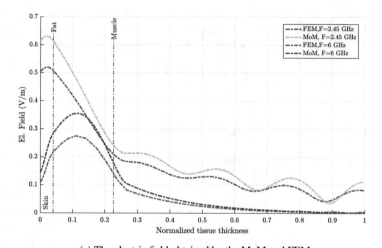

(a) The electric field obtained by the MoM and FEM

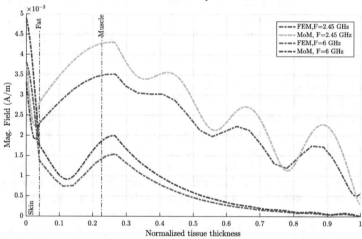

(b) The magnetic field obtained by the MoM and FEM.

Fig. 2. On-Body to In-Body (OB2IB) communication.

To develop a comprehensive channel model for typical implant applications, we considered IB2OB and OB2IB channel links based on different Tx locations. At each Tx location, a small antenna is inside or outside the body to transmit information. Meanwhile, the Rx was assumed on the body surface for IB2OB, and at X locations, the Rx was placed in the body. Table 3 presents an evaluation of the path loss level at different frequencies for two communication links. We can observe that for OB2IB, the PL increases as position depth increases. Likewise, a significant attenuation is visible due to the wide frequency, mainly in IB2OB. in Fig. 4, the EM wave has difficulty transmitting when the frequency is higher than 10 GHz. The minimum PL has occurred, and thus the smallest power budget is

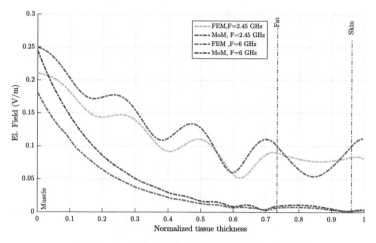

(a) The electric field obtained by the MoM and FEM.

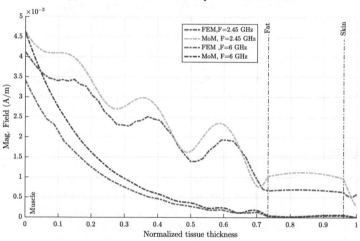

(b) The magnetic field obtained by the MoM and FEM.

Fig. 3. In-Body to On-Body (IB2OB) communication.

required at a frequency of about 1.5 GHz. This can be explained by the fact that the wide frequency influences on the propagation channel. It should also be noted that, after passing through the muscle, the EM wave decreases significantly. The main reason for this phenomenon is that the thickness and conductivity of muscle tissue are much larger than those of other tissues, and therefore, much more loss is generated in this region.

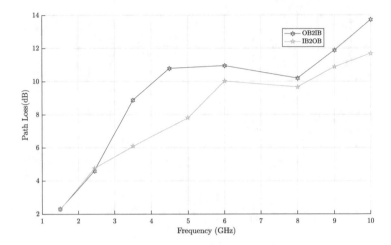

Fig. 4. Path Loss for different communication scenarios.

Table 3. Propagation loss scenarios.

Propagation loss scenarios (dB)	2.45 GHz	6 GHz
OB2IB in a different muscle position 5 mm; 20 mm	35.29; 41.35	45.75; 63.87
IB2OB on the interface of the skin	47.67	76.24

4   Conclusion

Understanding the characteristics inside the human body remains a challenge for massive use in the era of communication technologies. This work demonstrated the characteristics of inhomogeneous planar human tissue exposed to incident plane waves by using the MoM method. The difference in outcomes is approximately 10^2, which confirms the potential of the MOM method for analyzing complex media. Moreover, this model enabled us to make simulations with very high precision, thus, figuring out the EM wave's effect on planar thin layers in terms of PL, which will provide useful information for future wireless communication systems and IoT applications. In future work, more complex simulations will be done to demonstrate the ability of this method based on the MoM method to predict the PL levels, which offer useful insight and simplify physical layer modelling for implant communication systems.

References

1. Chen, Z.Y., Gao, Y.M., Du, M.: Propagation characteristics of electromagnetic wave on multiple tissue interfaces in wireless deep implant communication. IET Microw. Ant. Propag. **12**(13), 2034–2040 (2018)
2. Taleb, H., et al.: Wireless technologies, medical applications and future challenges in WBAN: a survey. Wirel. Netw. **27**(8), 5271–5295 (2021)

3. Teshome, A., Kibret, B., Lai, D.T.H.: A review of implant communication technology in WBAN, progresses and challenges. IEEE Rev. Biomed. Eng. **12**, 88–99 (2018)
4. Nikolayev, D., et al.: Reconfigurable dual-band capsule-conformal antenna array for in-body bioelectronics. IEEE Trans. Ant. Propag. **70**(5), 3749–3761 (2021)
5. Saada, A.B., Mbarek, S.B., Choubani, F.: Antenna polarization impact on electromagnetic power density for an off-body to in-body communication scenario. In: 2019 15th International Wireless Communications Mobile Computing Conference (IWCMC), pp. 1430-1433. IEEE, 2019. https://doi.org/10.1109/IWCMC.2019.8766713
6. Ben Saada, A., Mbarek, S.B., Choubani, F.: Whole-body exposure to far-field using infinite cylindrical model for 5G FR1 frequencies. In: Barolli, L., Hussain, F., Enokido, T. (eds.) AINA 2022. LNNS, vol. 449, pp. 471–478. Springer, Cham (2022). https://doi.org/10.1007/978-3-030-99584-3_41
7. Saada, A.B., Ben Mbarek, S., Choubani, F.: Towards a New Model of Human Tissues for 5G and Beyond. CRC Press, Boca Raton (2021). https://doi.org/10.1201/9781003181545-9
8. Rothwell, E.J.: Natural-mode representation for the field reflected by an inhomogeneous conductor-backed material layer-TE case. Prog. Electromagn. Res. **63**, 1–20 (2006)
9. Khalaj-Amirhosseini, M.: Analysis of lossy inhomogeneous planar layers using the method of moments. J. Electromagn. Waves Appl. **21**, 1925–1937 (2007)
10. Sullivan, D.M.: Electromagnetic Simulation Using the FDTD Method. John Wiley Sons, Hoboken (2013)
11. Ben Mbarek, S., Choubani, F.: FDTD modeling and experiments of microfabricated coplanar waveguide probes for electromagnetic compatibility applications. J. Electromagn. Waves Appl. **35**(5), 634–646 (2021). https://doi.org/10.1080/09205071.2020.1851776
12. Krimi, I., Mbarek, S.B., Hattab, H., et al.: Electromagnetic near-field study of electric probes for EMC applications. In: Innovative and Intelligent Technology-Based Services for Smart Environments Smart Sensing and Artificial Intelligence, pp. 45–50. CRC Press (2021). https://doi.org/10.1201/9781003181545-8
13. Gibson, W.C.: The Method of Moments in Electromagnetics. Chapman and Hall/CRC, Boca Raton (2021)
14. Happ, F., Schroder, A., Bruns, H.-D., et al.: A method for the calculation of electromagnetic fields in the presence of thin anisotropic conductive layers using the method of moments. In: 2013 International Symposium on Electromagnetic Compatibility. IEEE (2013)
15. Carrara, N.: Dielectric properties of body tissues. IFAC, Institute for applied physics, Italy (2007). http://niremf.ifac.cnr.it/tissprop/
16. Ali, K.: Numerical Techniques at Millimeter Wave Frequencies for Wireless Body Area Networks-A (2019)

Implementation of a Fuzzy-Based Testbed for Assessment of Neighbor Vehicle Processing Capability in SDN-VANETs

Ermioni Qafzezi[1]([✉]), Kevin Bylykbashi[2], Elis Kulla[3], Makoto Ikeda[2],
Keita Matsuo[2], and Leonard Barolli[2]

[1] Graduate School of Engineering, Fukuoka Institute of Technology (FIT),
3-30-1 Wajiro-Higashi, Higashi-Ku, Fukuoka 811–0295, Japan
`bd20101@bene.fit.ac.jp`
[2] Department of Information and Communication Engineering,
Fukuoka Institute of Technology (FIT), 3-30-1 Wajiro-Higashi,
Higashi-Ku, Fukuoka 811-0295, Japan
`kevin@bene.fit.ac.jp`, `makoto.ikd@acm.org`, {`kt-matsuo,barolli`}`@fit.ac.jp`
[3] Department of System Management, Fukuoka Institute of Technology,
3-30-1 Wajiro-Higashi, Higashi-Ku, Fukuoka 811-0295, Japan
`kulla@fit.ac.jp`

Abstract. In SDN-VANETs, the processing capabilities of new generation vehicles offer not only processing and data repository resources for their own applications, but also they are able to share their available resources with other neighbors within their range of communication. In this work, we implement a testbed to assess the processing and storage capability of vehicles for helping other vehicles in need for additional resources. We used our testbed and carried out some experiments. The results demonstrate the feasibility of the proposed approach in coordinating and managing the available SDN-VANETs resources.

1 Introduction

According to World Health Organization, around 1.3 million people die every year because of road traffic crashes [18]. The key risk factors come from human error (speeding, wrong decisions, etc.), irresponsible behavior (drinking, distracted driving, fatigue, etc.), unsafe road infrastructure, and bad weather conditions (e.g., inadequate visibility and slippery roads) [3,5,15]. Vehicular Ad hoc Networks (VANETs) have emerged as a solution to alleviate all these factors by means of different applications [9,14,16]. For example, implementing an accident prevention system in VANETs that considers velocity, weather condition,

© The Author(s), under exclusive license to Springer Nature Switzerland AG 2023
L. Barolli (Ed.): AINA 2023, LNNS 661, pp. 104–112, 2023.
https://doi.org/10.1007/978-3-031-29056-5_11

risk location, nearby vehicles density, and driver fatigue can reduce the number of road crashes and consequently the number of deaths [2]. Other applications, on the other hand, can improve traffic management and the driving experience [7, 10, 17]. VANETs goal is to benefit all road users, drivers, passengers and walkers, without any exception.

Vehicular networks applications rely on Vehicle-to-Vehicle (V2V) communications, Vehicle-to-Infrastructure (V2I), Vehicle-to-Network (V2N) and Vehicle-to-Pedestrian (V2P) communications. Through these communication links they exchange important information that come from other sources such as data about traffic lights, public safety information, weather conditions, the state of other vehicle and its surrounding environment, and so on. Such data improves the accuracy and decisions taken in the network.

However, there are still many challenges that are yet to be addressed. One of these challenges is the management of the abundant information and resources available in these networks [1, 4, 14, 17]. The volume generated data from VANETs keeps increasing as the number of vehicles and the sensors incorporated keeps increasing. On the other hand, many new applications have strict requirements for more resources, leading to increased complexity in network management. Dealing with resource management problems while still satisfying application requirements is the focus on our proposed approach.

In a previous work [13], we considered the resource management problem and proposed an integrated intelligent architecture based on Fuzzy Logic (FL) and Software Defined Networking (SDN) approach that could efficiently manage cloud-fog-edge storage, computing, and networking resources in VANETs, from a bottom-up perspective by exploiting the resources of edge layer first and then the fog and cloud resources. The integrated system is composed of three subsystems, namely Fuzzy-based System for Assessment of QoS (FS-AQoS), Fuzzy-based System for Assessment of Neighbor Vehicle Processing Capability (FS-ANVPC), and Fuzzy-based System for Cloud-Fog-Edge Layer Selection (FS-CFELS), each having a key role in the proposed approach.

In order to evaluate the simulation results of the aforementioned integrated system, in this work we implement a testbed and carry out experiments. We consider FS-ANVPC system and analyze it experimental results. The experimental results show the feasibility of FS-ANVPC system in assessing the processing capability of vehicles to help other vehicles in need for additional resources.

The rest of this paper is organized as follows. In Sect. 2, we present an overview of Cloud-Fog-Edge SDN-VANETs. The proposed approach is presented in Sect. 3. The details of testbed implementation are given in Sect. 4. Section 5 discusses the evaluation results. The last section, Sect. 6, gives some concluding remarks and ideas for future work.

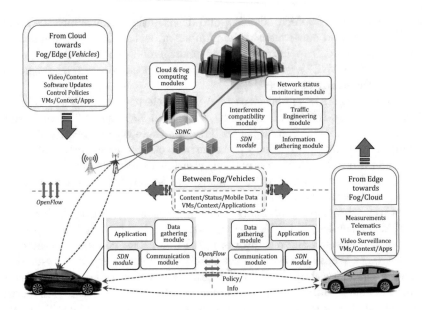

Fig. 1. Logical architecture of Cloud-Fog-Edge SDN-VANET with content distribution.

2   Cloud-Fog-Edge SDN-VANETs

The integration of Cloud-Fog-Edge computing in VANETs is the solution to handle complex computation, provide mobility support, low latency and high bandwidth. While they offer scalable access to storage, networking and computing resources, SDN provides higher flexibility, programmability, scalability and global knowledge. In Fig. 1, we give a detailed structure of SDN-VANET architecture. It includes the topology structure, its logical structure and the content distribution in the network. Specifically, the architecture consists of cloud computing data centers, fog servers with SDN Controllers (SDNCs), Road-Side Units (RSUs), RSU Controllers (RSUCs), Base Stations (BSs) and vehicles.

The implementation of this architecture can enable and improve the VANET applications such as road and vehicle safety services, traffic optimization, video surveillance, telematics, commercial and entertainment applications.

3   Description of FS-ANVPC

This section presents the architecture of our proposed approach for assessment of neighbor vehicle processing capability in a SDN-VANETs environment. Vehicles are capable to share their available resources with the vehicles that need additional resources (hereinafter will be referred to as *the vehicle*). So, in the case when more storing and computing resources are needed, the *vehicle* can request

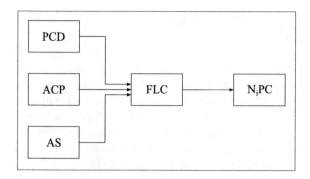

Fig. 2. Structure of FS-ANVPC system.

to use those of the adjacent vehicles, assuming a connection can be established and maintained between them for a while.

The proposed approach, named FS-ANVPC, assesses the processing capability of all neighboring vehicles within the *vehicle*'s communication range. It represents the processing and storage capability of the edge layer, which is comprised by the total number on vehicles that are able to communicate and share their storage and processing capabilities with each other. Using the edge layer and V2V communication links exploits the high capacity of neighbors which at this point is being unused and it avoids unnecessary traffic being sent in the core network. They offer real-time computing without the need of infrastructure. However, V2V communication are constrained by vehicles limited resources, their high mobility and inter-vehicle communication duration.

FS-ANVPC is implemented in the SDNC and the vehicles which are equipped with an SDN module. SDNC manages the resource management between vehicles and determines the most suitable vehicle for storing and processing data application.

In the following, we give details of the composition of FS-ANVPC proposed system, describe the input and output parameters and present the design and implementation of FS-ANVPC testbed.

The input parameters of FS-ANVPC system do not correlate to one another, leading to an NP-hard problem. FL can deal with these problems. Moreover, we want our systems to make decisions in real time and fuzzy systems can give very good results in decision making and control problems [6,8,11,12,19,20]. The parameters of FS-ANVPC are described in following. The structure of FS-ANVPC system is given in Fig. 2.

Available Storage (AS): In SDN-VANETs vehicles have limited storage resources, however they increase when the total amount of storage resources of the edge layes is able to be shared among them. There should be also free storage to handle the virtual machine created for V2V communication and sharing resources.

Available Computing Power (ACP): New generation vehicles are equipped with additional computing resources to process and analyze their sensed data and run their own application. But, they are also able to share their computing resources with other vehicles in need via V2V communication links. Distributed computing augments the processing capability of the edge layer, makes them able to handle complex processes and helps *the vehicles* which have shortage of these resources and could not process the required data otherwise.

Predicted Contact Duration (PCD): The longer their communication time, the higher the number of exchanged information and amount of processed data. However, due to their high mobility, vehicles in SDN-VANETs have a very short communication time. In our work we make a prediction of V2V communication period, based on vehicles speed, direction, communication range and the initial distance between two vehicles.

Neighbor i Processing Capability (NiPC): The output parameter values range between 0 and 1, with the value 1 indicating a potential neighbor vehicle in the best conditions to share its available resources to a *vehicle*, and the value 0 indicating a neighbor which is not capable at all to help a *vehicle* in need of additional resources.

4   Testbed Design

To evaluate the feasibility of FS-ANVPC, we have designed and implemented a small-scale testbed using Raspberry Pis (RPis). We use five RPis with 50 m communication range that represent the vehicles moving for about 25 min in a 200m x 200m urban area. One of the vehicles is *the vehicle* in need of resources, whereas the other four are the vehicles that could turn into potential neighbors for sharing their resources with the vehicle in need. The setup of the testbed is summarized in Table 1. The movement of vehicles are generated using the *sumo* simulator and the layout is given in Fig. 3.

Table 1. Testbed setup.

Vehicles	5 RPi Model 3B+
Mobility trace generator	Sumo
Area size	200 m × 200 m
Communication range	50 m
TS	[0.1, 0.5, 0.9]
DC	[0.1, 0.5, 0.9]
Acceptable PCD	2 s
Experimental time	1500 s

The position of vehicles changes every second. Information about vehicles current position, together with vehicle previous position, vehicle id, speed, direction and timestep are gathered via sensors implemented in vehicles. Then, this

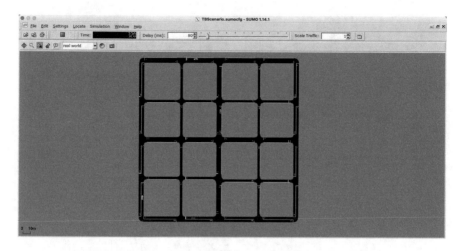

Fig. 3. A screenshot of the vehicles moving around the considered area in sumo simulator.

information is shared with the neighbors though beacon messages which are broadcasted every second. Neighbor vehicles receive these beacon messages, for the vehicles which are inside their communication range, and extract the necessary information. In our system, these data are used to inform about vehicles geographic position, their speed, direction, transmission power, available storage, available computing power. Based on this data, FS-ANVPC calculates the current condition of all input parameters of the system, specifically for updating PCD. Then, the fuzzy system FS-ANVPC implemented in neighbor vehicles determines NiPC, which shows whether this neighbor is capable of helping *the vehicle* in need for additional resources. This process is shown in Fig. 4.

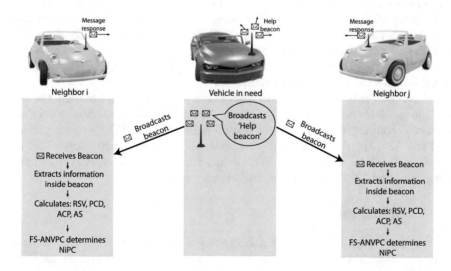

Fig. 4. A scheme of the communication between the *vehicle* and its neighbors.

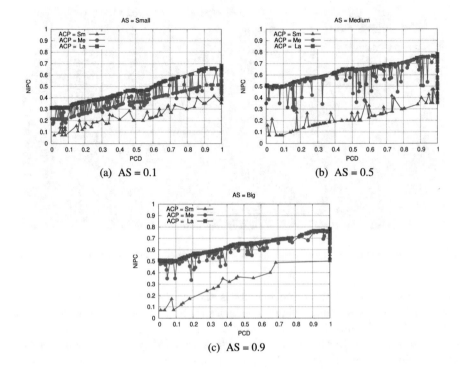

Fig. 5. Testbed results for FS-ANVPC.

5   Experimental Results

The experimental results of FS-ANVPC are given in Fig. 5. We show the results for three scenarios: for small, medium and big AS. We show the relation between NiPC and PCD for different ACP values. We see some oscillations occurring, especially for ACP small and medium. This is due to the fluctuations of the ACP of RPi while running many other applications simultaneously, thus imitating a similar scenario of a vehicle operating its own applications and the applications of *vehicle* in need for additional resources.

For small AS (see Fig. 5(a)), vehicles that are considered helpful are the ones with large ACP and long PCD. Despite the ACP values, vehicles with short predicted contact duration are not considered as helpful. Whereas, for medium and big AS (Fig. 5(b) and Fig. 5(c) respectively) vehicles with large ACP are considered helpful, despite the PCD value. However, vehicles with small ACP are considered helpful only when PCD is very long.

6   Conclusions

In this paper, we implemented a fuzzy-based testbed to assess the available edge computing resources in a layered Cloud-Fog-Edge architecture for SDN-VANETs. The proposed FS-ANVPC system decides if a neighboring vehicle is capable to help a vehicle that lacks the appropriate resources to accomplish certain tasks based on ACP, AS and PCD values. We have shown through experimental results by using the implemented testbed the effect of considered parameters on our system and its feasibility. From the experimental results we conclude as follows.

- The neighboring vehicles which have small ACP and small AS are not capable to help other vehicles in need.
- For medium and big AS, vehicles offering large ACP amount are considered helpful, despite the PCD value.
- The highest value of NiPC is achieved when the neighboring vehicle has a large ACP, big AS and long PCD.

In the future, we would like to improve FS-ANVPC by implementing a testbed with mobile RPis moving randomly and not using sumo simulator for extracting location of vehicles.

References

1. Al-Heety, O.S., Zakaria, Z., Ismail, M., Shakir, M.M., Alani, S., Alsariera, H.: A comprehensive survey: benefits, services, recent works, challenges, security, and use cases for sdn-vanet. IEEE Access **8**, 91028–91047 (2020). https://doi.org/10.1109/ACCESS.2020.2992580
2. Aung, N., Zhang, W., Dhelim, S., Ai, Y.: Accident prediction system based on hidden markov model for vehicular ad-hoc network in urban environments. Information **9**(12), 311 (2018). https://doi.org/10.3390/info9120311
3. Colagrande, S.: A methodology for the characterization of urban road safety through accident data analysis. Transp. Res. Procedia **60**, 504–511 (2022). https://doi.org/10.1016/j.trpro.2021.12.065
4. Fadhil, J.A., Sarhan, Q.I.: Internet of vehicles (iov): a survey of challenges and solutions. In: 2020 21st International Arab Conference on Information Technology (ACIT), pp. 1–10 (2020). https://doi.org/10.1109/ACIT50332.2020.9300095
5. Wu, G.F., Liu, F.J., Dong, G.L.: Analysis of the influencing factors of road environment in road traffic accidents. In: 2020 4th Annual International Conference on Data Science and Business Analytics (ICDSBA), pp. 83–85 (2020). https://doi.org/10.1109/ICDSBA51020.2020.00028
6. Kandel, A.: Fuzzy Expert Systems. CRC Press Inc., Boca Raton (1992)
7. Karagiannis, G., et al.: Vehicular networking: a survey and tutorial on requirements, architectures, challenges, standards and solutions. IEEE Commun. Surv. Tutor. **13**(4), 584–616 (2011)
8. Klir, G.J., Folger, T.A.: Fuzzy Sets, Uncertainty, and Information. Prentice Hall, Upper Saddle River (1988)
9. Lu, N., Cheng, N., Zhang, N., Shen, X., Mark, J.W.: Connected vehicles: solutions and challenges. IEEE Internet Things J. **1**(4), 289–299 (2014). https://doi.org/10.1109/JIOT.2014.2327587
10. Luan, T.H., Cai, L.X., Chen, J., Shen, X.S., Bai, F.: Engineering a distributed infrastructure for large-scale cost-effective content dissemination over urban vehicular networks. IEEE Trans. Veh. Technol. **63**(3), 1419–1435 (2014). https://doi.org/10.1109/TVT.2013.2251924

11. McNeill, F.M., Thro, E.: Fuzzy Logic: A Practical Approach. Academic Press Professional Inc., San Diego (1994)
12. Munakata, T., Jani, Y.: Fuzzy systems: an overview. Commun. ACM **37**(3), 69–77 (1994)
13. Qafzezi, E., Bylykbashi, K., Ampririt, P., Ikeda, M., Matsuo, K., Barolli, L.: An intelligent approach for cloud-fog-edge computing sdn-vanets based on fuzzy logic: effect of different parameters on coordination and management of resources. Sensors **22**(3) (2022). https://doi.org/10.3390/s22030878
14. Raza, S., Wang, S., Ahmed, M.: A survey on vehicular edge computing: architecture, applications, technical issues, and future directions. Wirel. Commun. Mob. Comput. **3159**, 762 (2019). https://doi.org/10.1155/2019/3159762
15. Rolison, J.J., Regev, S., Moutari, S., Feeney, A.: What are the factors that contribute to road accidents? an assessment of law enforcement views, ordinary drivers' opinions, and road accident records. Accid. Anal. Prev. **115**, 11–24 (2018). https://doi.org/10.1016/j.aap.2018.02.025
16. Seo, H., Lee, K.D., Yasukawa, S., Peng, Y., Sartori, P.: LTE evolution for vehicle-to-everything services. IEEE Commun. Maga. **54**(6), 22–28 (2016). https://doi.org/10.1109/MCOM.2016.7497762
17. Shrestha, R., Bajracharya, R., Nam, S.Y.: Challenges of future vanet and cloud-based approaches. Wirel. Commun. Mob. Comput. **2018** (2018)
18. World Health Organization. Global status report on road safety 2018: summary. World Health Organization, Geneva, Switzerland, (WHO/NMH/NVI/18.20). Licence: CC BY-NC-SA 3.0 IGO) (2018)
19. Zadeh, L.A., Kacprzyk, J.: Fuzzy Logic for the Management of Uncertainty. John Wiley & Sons Inc., New York (1992)
20. Zimmermann, H.J.: Fuzzy control. In: Fuzzy Set Theory and its Applications, pp. 203–240. Springer, Heidelberg (1996). https://doi.org/10.1007/978-94-010-0646-0_11

Frobenius Norm-Based Radio Propagation Path Selection Policy and Performance Analysis

Sainath Bitragunta$^{(\boxtimes)}$

Department of Electrical and Electronics Engineering, BITS Pilani,
Pilani 333031, Rajasthan, India
sainath.bitragunta@pilani.bits-pilani.ac.in

Abstract. A common problem in several wireless systems is multipath fading, which limits the performance of these systems. Different forms of diversity and combining techniques are proposed and investigated in the literature to deal with the fading problem and enhance the performance of wireless systems. This work proposes a novel best path selection policy based on the Frobenius norm of the multi-input multi-output (MIMO) fading channel matrix. I derive various insightful analytical expressions and numerically evaluate the performance of the proposed policy in terms of outage probability and spectral efficiency. When mean channel power gain is 0 dB, the proposed selection policy shows 75.78% and 71.74% improvement in average spectral efficiency for Rayleigh fading and Nakagami fading scenarios, respectively. I compare the proposed policy performance with the benchmark policy to validate its usefulness in advanced and intelligent wireless systems.

1 Background and Motivation

The most predominant channel impairment in wireless communication networks, long-range mobile cellular or short-range Wi-Fi, is multipath fading and shadowing [1,2]. Various forms of diversity and combining techniques are employed to improve the performance of fixed and mobile wireless systems and networks. The fundamental diversity techniques include time diversity, frequency diversity, and transmit-receive antenna diversity [3]. Popular and well-investigated combining techniques include maximal ratio combining (MRC), a linear combining technique, and selection combining [1]. Researchers widely investigated lower- to higher-order (massive) multi-input and multi-output (MIMO) antenna diversity performance for various cellular communication generations. Furthermore, additional performance improvements achievable with analog and digital beamforming (hybrid) for the fifth-generation (5G) millimeter wave (mmWave) systems have been studied qualitatively and quantitatively [4].

Multi-antenna-equipped communication systems can enhance reliability and spectral efficiency. However, the MIMO radio frequency (RF) chains are expensive due to size, power consumption, and hardware. Antenna selection is a low-cost, low-power, low hardware-complexity alternative to realize advantages of MIMO systems [5].

© The Author(s), under exclusive license to Springer Nature Switzerland AG 2023
L. Barolli (Ed.): AINA 2023, LNNS 661, pp. 113–124, 2023.
https://doi.org/10.1007/978-3-031-29056-5_12

Literature on Antenna Selection:

For the past decade or more, researchers investigated the performance of MIMO systems and massive MIMO systems with the antenna selection [5–7]. In [7], the authors presented an overview of MIMO wireless systems with antenna selection. In it, the authors discussed antenna selection algorithms and investigated the effect of channel estimation errors. Practical antenna selection algorithms and performance limits of massive MIMO wireless systems are studied in [6]. Other antenna selection problems investigated in the literature include transmit-antenna selection [8], optimal joint antenna selection [9], energy-efficient antenna selection [10] for various advanced wireless systems, for example, MIMO cognitive radio, massive MIMO, cooperative MIMO and more.

Unlike several antenna selection techniques in the literature, I propose a novel radio propagation path (RPP) selection policy (SP). For the best path selection, it is important to consider full channel state information (CSI) and determine the best path based on a performance measure that includes the full CSI. Based on the measure, an intelligent source node could decide on the best propagation path for the MIMO source transmitter to radiate energy in the desired direction toward the destination node. My goal is design a novel selection policy to choose the best RPP based on a novel measure, the Frobenius norm (FN). Therefore, the novelty in the presented article lies in the measure I use for the best path selection.

A MIMO channel is characterized by the fading channel matrix whose elements are complex fading coefficients. I consider the FN square of each MIMO matrix and use order statistics to determine the best RPP. However, practical implementation of this policy requires full CSI knowledge. With the recent developments of machine learning and deep learning algorithms for channel estimation [11], accurately determining the FN of the MIMO fading channel matrices is feasible. In this work, specific contributions are as follows.

1.1 Novelty and Key Contributions

Novelty: The novelty in the presented work on path selection in wireless systems over Rayleigh and Nakagami fading channels lies in the best path selection based on Frobenius norms of the MIMO channel matrices. This selection policy implementation requires CSI overhead. However, the rapid advancements in applying ML and DL in intelligent wireless communication systems, particularly in accurate CSI prediction, are useful and address the problem of CSI complexity involved in the proposed FN-based RPPSP. I present specific contributions below.

Contributions: i) *Novel selection policy for wireless systems:* I propose a novel FN-based RPPSP for MIMO wireless systems comprising a single source and multiple paths to destination nodes, which can also be intermediate nodes in multi-hop communication. I assume two different fading scenarios, namely, Rayleigh fading and Nakagami fading, and knowledge of perfect CSI in the system model.

ii) *Performance analysis of RPPSP and benchmarking:* I investigate the performance of the proposed RPPSP in terms of PHY measures outage probability and average spectral efficiency. Specifically, I derive four theorems associated with the two performance measures for Rayleigh and Nakagami fading scenarios. I compare the performance of the proposed policy with the benchmark policy, which is CSI-independent and inferior. Considering the advances in ML and DL-aided CSI prediction approaches, acquiring accurate CSI knowledge is practical and useful for next-generation intelligent wireless communication systems and networks.

1.2 Paper Outline and Notation

Outline: The paper's organization is as follows. Section 2 describes MIMO wireless system model that has a single intelligent source and multiple destination nodes. Section 3 presents problem formulations. Section 4 derives various theorems on outage probability and spectral efficiency for Rayleigh fading and Nakagami fading scenarios. My numerical results consist of various performance plots and conclusions with future research directions following in Sect. 5 and 6.

Notation: The following describes the key notation that I use in the paper. Let $\Gamma(.)$ denotes the gamma function defined as $\Gamma(N) = \int_0^\infty t^{N-1} e^{-t} dt$. For positive N, $\Gamma(N) = (N-1)!$. Further, the incomplete gamma function is defined as $\gamma_{\text{inc}}(N,z) = \frac{1}{N} \int_0^z t^{N-1} e^{-t} dt$. The probability of an outage event is defined as by $\mathscr{P}(\gamma_D < \gamma_{\text{Th}})$. For a random variable (RV) Ψ, its probability density function (pdf), expectation value, and variance are denoted by $p_\Psi(\psi)$, $\mathbf{E}[\Psi]$, and $\mathbf{var}[\Psi]$, respectively. Furthermore, $\mathscr{CN}(0, \sigma^2)$ represents a circularly symmetric complex Gaussian distribution with zero mean, variance σ^2.

2 Three Path, Four-Node MIMO System Model

Figure 1 depicts an intelligent, beamforming-enabled source transmitter and three destination nodes. In a general model, there could be L propagation paths. I assume half-duplex, multi-antenna nodes whose transmissions occur within the same RF or mmWave bandwidth. Furthermore, I assume that the source node and destination nodes are equipped with the same number of antennas. For ease of analysis, I assume that the MIMO source transmitter power P_s becomes $P_{s,\text{eff}}$ after accounting pathloss and beamforming gain. Without loss of generality, this value is set to 0 dB.

MIMO Channel Matrix and FN: Let N be a perfect square ≥ 4. Consider $\sqrt{N} \times \sqrt{N}$ MIMO. Therefore, the channel matrix has N channel fading coefficients. For $N = 4$, let

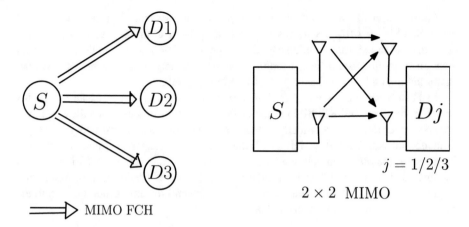

Fig. 1. A three path, four-node MIMO system model. Source selects the best radio propagation path. In general, the system could have L paths and N nodes.

the fading coefficients are h_{11}, h_{21}, h_{12} and h_{22}. The FN and its square, are defined as

$$\mathscr{F}_{\mathrm{N}} = \sqrt{\sum_{i=1}^{\sqrt{N}} \sum_{j=1}^{\sqrt{N}} |h_{ij}|^2}, \ \mathscr{F}_{\mathrm{N}}^2 = \sum_{i=1}^{\sqrt{N}} \sum_{j=1}^{\sqrt{N}} |h_{ij}|^2. \tag{1}$$

Rayleigh Fading Model and Order Statistics: Let $h_{ij} \sim CN\left(0, \sigma_{ij}^2\right)$. The complex fading coefficient's amplitude is Rayleigh distributed. I assume that all fading channels are statistically independent. Further, the fading channel power gain is exponentially distributed. Since $\mathscr{F}_{\mathrm{N}}^2$ is the sum of exponential random variables, I have gamma distributed random variable (GDRV). For the number of radio propagation paths $L = 3$, I have three statistically independent and identically distributed (i.i.d.) GDRVs, denoted by $\mathscr{Y}_{sd_1}, \mathscr{Y}_{sd_2}, \mathscr{Y}_{sd_3}$. I formulate a novel problem using order statistics. The problem formulation details are provided in the following section.

Nakagami Fading Model and Order Statistics:

Let the magnitude of $h_{ij}, 1 \le i \le \sqrt{N}, 1 \le j \le \sqrt{N}$ is Nakagami distributed with parameter m. I assume that all Nakagami fading channels are statistically independent. Therefore, the fading channel power gains are gamma distributed. Further, $\mathscr{F}_{\mathrm{N}}^2$ is the sum of GDRVs. For the number of radio propagation paths $L = 3$, I have three statistically distributed random variables, denoted by $\mathscr{Z}_{sd_1}, \mathscr{Z}_{sd_2}, \mathscr{Z}_{sd_3}$. I formulate a novel problem using order statistics. The problem formulation based on order statistics is provided in the following section.

3   Problem Formulations

I focus on two important PHY performance measures: outage probability and spectral efficiency. While the former is useful to predict coverage mapping, the latter is useful to measure the efficient use of the available spectrum.

Outage Probability (OP): This performance measure is defined as

$$p_{OP}^{RPPSP} = \mathscr{P}\left(\max\left(\mathscr{Z}_1, \ldots, \mathscr{Z}_L\right) < \gamma_{Th}\right). \tag{2}$$

Channel-Fading-Averaged Spectral Efficiency (CFASE):
The average spectral efficiency performance measure is defined as

$$\text{CFASE}^{RPPSP} = \mathbf{E}\left[\log_2\left(1 + \max\left(\mathscr{Z}_1, \ldots, \mathscr{Z}_L\right)\right)\right]. \tag{3}$$

Outage Probability and Average Spectral Efficiency: I aim to derive outage probability and channel-fading-averaged spectral efficiency for a general scenario of L possible paths from source to destination nodes with each MIMO channel matrix of N fading coefficients. For example, consider 2×2 MIMO. Each MIMO matrix has four complex fading coefficients.

I present four theorems on the two performance measures for the general scenario and for a specific scenario of $L = 3$ paths and $N = 4$ fading coefficients associated with 2×2 MIMO channel matrix. As mentioned before, I assume Rayleigh fading and Nakagami fading scenarios.

4 Analytical Results and Remarks

Below, I state theorems on the 1) outage probability and 2) spectral efficiency of the FN-based RPPSP for the MIMO wireless system over Rayleigh fading channels.

Theorem 1. *The outage probability is given by*

$$p_{OP}^{RPPSP} = \frac{L}{\lambda^N} \frac{1}{\Gamma(N)} \int_0^{\gamma_{Th}} \left(\gamma_{inc}\left(N, \frac{z}{\lambda}\right)\right)^{L-1} z^{N-1} \exp\left(-\frac{z}{\lambda}\right) dz, \tag{4}$$

where

$$\gamma_{inc}\left(N, \frac{z}{\lambda}\right) = \frac{1}{\Gamma(N)} \int_0^{\frac{z}{\lambda}} t^{N-1} \exp\left(-t\right) dt, \tag{5}$$

denote the lower incomplete Gamma function.
For the three propagation path-2×2 MIMO wireless system, I have

$$p_{OP}^{RPPSP} = \frac{1}{2\lambda^4} \int_0^{\gamma_{Th}} \left(\gamma_{inc}\left(4, \frac{z}{\lambda}\right)\right)^2 z^3 \exp\left(-\frac{z}{\lambda}\right) dz, \tag{6}$$

where

$$\gamma_{inc}\left(4, \frac{z}{\lambda}\right) = \frac{1}{6} \int_0^{\frac{z}{\lambda}} t^3 \exp\left(-t\right) dt, \tag{7}$$

Proof. Let $\mathscr{Z} = \max\left(\mathscr{Y}_{sd_1}, \mathscr{Y}_{sd_2}, \ldots, \mathscr{Y}_{sd_L}\right)$. I need to determine the pdf to obtain the analytical expression for outage probability. For statistically independent and identically distributed GDRVs, I have

$$\mathscr{P}\left(\mathscr{Z} = \max\left(\mathscr{Y}_{sd_1}, \mathscr{Y}_{sd_2}, \ldots, \mathscr{Y}_{sd_L}\right) \leq z\right) = \left(\mathscr{P}\left(\mathscr{Y} \leq z\right)\right)^L \triangleq P_{\mathscr{Z}}(z), \tag{8}$$

where $P_{\mathscr{Z}}(z)$ is the CDF of the $\mathscr{Z}$. To obtain the pdf $p_{\mathscr{Z}}(z)$, I differentiate the CDF with respect to z. The pdf is given by

$$p_{\mathscr{Z}}(z) = L(P_{\mathscr{Y}}(z))^{L-1} p_{\mathscr{Y}}(z) = \frac{L}{\lambda^N} \frac{1}{\Gamma(N)} \left(\gamma_{inc}\left(N, \frac{z}{\lambda}\right)\right)^{L-1} z^{N-1} e^{-\frac{z}{\lambda}}, z \geq 0, \quad (9)$$

where λ denotes fading-averaged channel power gain.

The outage probability is given by $\int_0^{\gamma_{Th}} p_{\mathscr{Z}}(z)\,dz$. Therefore, substituting the pdf in the definition of the outage probability yields the single integral expression. Further simplification after substituting $L = 3$ and $N = 4$ and using $\Gamma(N) = (N-1)!$ yields the result on outage probability for the four-node, three-path MIMO wireless system model. $\square$

For the benchmark policy, the outage probability is given by

$$p_{OP}^{BMP} = \frac{1}{\lambda^N} \frac{1}{\Gamma(N)} \int_0^{\gamma_{Th}} z^{N-1} \exp\left(-\frac{z}{\lambda}\right) dz. \quad (10)$$

Theorem 2. *The channel-fading-averaged exact spectral efficiency (CFASE in bps/Hz) for the proposed policy is given by*

$$CFASE^{RPPSP} = \frac{L(\log_2 e)}{\Gamma(N)\lambda^N} \int_0^\infty (\log(1+z)) z^{N-1} \left(\gamma_{inc}\left(\frac{z}{\lambda}, N\right)\right)^{L-1} \exp\left(-\frac{z}{\lambda}\right) dz. \quad (11)$$

For the three propagation path-2 $\times$ 2 MIMO system, I have

$$CFASE^{RPPSP} = \frac{\log_2 e}{2\lambda^3} \int_0^\infty (\log(1+z)) z^3 \left(\gamma_{inc}\left(\frac{z}{\lambda}, N\right)\right)^2 \exp\left(-\frac{z}{\lambda}\right) dz. \quad (12)$$

Further, the upper bound for the exact CFASE is given by

$$CFASE^{RPPSP,UB} = \log_2 \left(1 + \frac{L\lambda}{\Gamma(N)} \int_0^\infty t^N (\gamma_{inc}(t,N))^{L-1} \exp(-t)\,dt \right). \quad (13)$$

For the three propagation path-2 $\times$ 2 MIMO system, the upper bound is given by

$$CFASE^{RPPSP,\,UB} = \log_2 \left(1 + \frac{\lambda}{2} \int_0^\infty t^4 (\gamma_{inc}(t,4))^2 \exp(-t)\,dt \right). \quad (14)$$

Proof. By definition, I have

$$CFASE^{RPPSP} = (\log_2 e) \int_0^\infty \ln(1+z) p_{\mathscr{Z}}(z)\,dz, \quad (15)$$

where $p_{\mathscr{Z}}(z)$ is given in (9). Substituting the pdf in the above equation yields the desired analytical result on spectral efficiency. It is difficult to simplify further due to the incomplete gamma function. Hence, I evaluate the CFASE numerically. Further, it is easy to verify the single integral CFASE expression when $N = 4$, $L = 3$.

Upper Bound of CFASE: To obtain the expression for the upper bound, I use the simple Jensen's inequality [1] for the concave function of a random variable:

$$(\log_2 e)\mathbf{E}\left[\ln(1+\mathscr{Z})\right] \leq (\log_2 e)\ln\left(1+\mathbf{E}[\mathscr{Z}]\right) = (\log_2 e)\ln\left(1+\int_0^{\infty} z p_{\mathscr{Z}}(z)\,dz\right). \tag{16}$$

Substituting the expression for the pdf yields the desired expression for the upper bound. Finally, finding the expression for the specific scenario of 3 path and four-node 2 MIMO system is trivial.   □

For the benchmark policy, the CFASE is given by

$$\text{CFASE}^{\text{BMP}} = \frac{1}{6\lambda^4}\int_0^{\infty} z^3 \log(1+z)e^{-\frac{z}{\lambda}}\,dz. \tag{17}$$

I now state theorems on the outage probability and spectral efficiency of the FN-based RPPSP for the MIMO wireless system over Nakagami fading channels.

Theorem 3. *The outage probability is given by*

$$p_{OP}^{RPPSP} = L(\Gamma(Nm))^{L-2}\left(\frac{Nm}{\lambda}\right)^{Nm}\int_0^{\gamma_{Th}}\left(\frac{\gamma_{inc}\left(\frac{Nmz}{\lambda},mN\right)}{\Gamma(Nm)}\right)^{L-1} z^{Nm-1}e^{-\frac{Nmz}{\lambda}}\,dz, \tag{18}$$

where m is the Nakagami fading parameter.
    For the proposed three propagation path, 2×2 MIMO system, outage probability is given by

$$p_{OP}^{RPPSP} = \frac{3}{\Gamma(4m)}\left(\frac{4m}{\lambda}\right)^{4m}\int_0^{\gamma_{Th}}\left(\gamma_{inc}\left(\frac{4mz}{\lambda},4m\right)\right)^2 z^{4m-1}e^{-\frac{4mz}{\lambda}}\,dz, \tag{19}$$

Proof. Let $\Gamma = \max\left(\mathscr{Z}_{sd_1},\mathscr{Z}_{sd_2},\ldots,\mathscr{Z}_{sd_L}\right)$. As mentioned in the previous theorem, I need to determine the pdf to obtain the analytical expression for the outage probability. For statistically independent and identically distributed RVs, I have

$$\mathscr{P}(\Gamma \leq z) = \max\left(\mathscr{Z}_{sd_1},\mathscr{Z}_{sd_2},\ldots,\mathscr{Z}_{sd_L} \leq z\right) = (\mathscr{P}(\Gamma \leq z))^L \triangleq P_{\Gamma}(z), \tag{20}$$

where $P_{\Gamma}(z)$ is the CDF of the Γ. To obtain the pdf $p_{\Gamma}(z)$, I differentiate the CDF with respect to z. The pdf is given by

$$p_{\Gamma}(z) = L(\Gamma(Nm))^{L-2}\left(\frac{Nm}{\lambda}\right)^{Nm}\left(\frac{\gamma_{inc}\left(\frac{Nmz}{\lambda},mN\right)}{\Gamma(Nm)}\right)^{L-1} z^{Nm-1}e^{-\frac{Nmz}{\lambda}}, \quad z \geq 0, \tag{21}$$

where λ is fading-averaged channel power gain.
    As mentioned before, substituting the pdf in the definition of the outage probability yields the single integral expression. Simplifying after the substitution of $L = 3$ and $N = 4$ and using $\Gamma(N) = (N-1)!$ yields the desired result.   □

For the benchmark policy, the outage probability is given by

$$p_{\text{OP}}^{\text{BMP}} = \gamma_{\text{inc}}\left(\frac{Nm\gamma_{\text{Th}}}{\lambda}, Nm\right).$$

(22)

Theorem 4. *The channel-fading-averaged exact spectral efficiency (CFASE in bps/Hz) for the proposed policy is given by*

$$CFASE^{RPPSP} = L(\log_2 e)\left(\Gamma(Nm)\right)^{L-2}\left(\frac{Nm}{\lambda}\right)^{Nm}\int_0^\infty \log(1+z)$$
$$\times\left(\gamma_{\text{inc}}\left(\frac{Nmz}{\lambda}, \frac{Nm}{\Gamma(Nm)}\right)\right)^{L-1} z^{Nm-1}\exp\left(-\frac{Nmz}{\lambda}\right) dz.$$

(23)

For the three propagation path-2 × 2 MIMO system, the CFASE is given by

$$CFASE^{RPPSP} = 3(\log_2 e)\left(\Gamma(4m)\right)\left(\frac{4m}{\lambda}\right)^{4m}\int_0^\infty \log(1+z)$$
$$\times\left(\gamma_{\text{inc}}\left(\frac{4mz}{\lambda}, \frac{4m}{\Gamma(4m)}\right)\right)^2 z^{4m-1}\exp\left(-\frac{4mz}{\lambda}\right) dz.$$

(24)

Further, the upper bound for the exact CFASE is given by

$$CFASE^{RPPSP,UB} = \log_2\left[1 + (L-1)(\Gamma(Nm))\left(\frac{Nm}{\lambda}\right)^{Nm}\right.$$
$$\left.\times\int_0^\infty\left(\gamma_{\text{inc}}\left(\frac{Nmt}{\lambda}, \frac{Nm}{\Gamma(Nm)}\right)\right)^2 t^{Nm}e^{-\frac{Nmt}{\lambda}} dt\right].$$

(25)

For the three propagation path-2 × 2 MIMO system, the upper bound is given by

$$CFASE^{RPPSP,UB} = \log_2\left[1 + 2(\Gamma(4m))\left(\frac{4m}{\lambda}\right)^{4m}\right.$$
$$\left.\times\int_0^\infty\left(\gamma_{\text{inc}}\left(\frac{4mt}{\lambda}, \frac{4m}{\Gamma(4m)}\right)\right)^2 t^{4m}e^{-\frac{4mt}{\lambda}} dt\right].$$

(26)

Proof. By definition, we have

$$CFASE^{RPPSP} = (\log_2 e)\int_0^\infty \ln(1+z)\, p_{\Gamma}(z)\, dz,$$

(27)

where $p_\Gamma(z)$ is given in (21). Substituting the pdf in the above equation yields the desired analytical result on spectral efficiency of the MIMO wireless system with Nakagami fading channels. It is difficult to simplify further due to the complicated integrand consisting of the incomplete gamma function with an exponent over it. However, I can evaluate the CFASE numerically. Further, I can easily verify the single integral CFASE expression when $N = 4, L = 3$.

Upper Bound of CFASE: Substituting the expression for the pdf in (16) yields the desired expression for the upper bound. Finally, finding the expression for the specific scenario of 3 path and 4 node-2×2 MIMO system is trivial.   □

For the benchmark policy, the CFASE is given by

$$\text{CFASE}^{\text{BMP}} = \left(\frac{4m}{\lambda} \right)^{4m} \frac{1}{\Gamma(4m)} \int_0^\infty (\log(1+z)) z^{4m-1} e^{-\frac{4mz}{\lambda}} \, dz. \qquad (28)$$

5   Numerical Results and Interpretations

I numerically evaluate the performance of the proposed FN-based policy and compare it with the benchmark policy. I present various performance plots and remarks on the observations of variations in the performance measure as a function of channel model parameters. First, I present outage probability and spectral efficiency plots for the Rayleigh fading scenario. Later, I present outage probability and spectral efficiency plots for Nakagami fading scenario. I mention the plot-specific simulation parameters in figure captions.

5.1   Performance Plots of RPPSP: Rayleigh Fading

Outage Probability: Impact of Channel-Fading-Averaged Power Gain
Figure 2 evaluates the outage probability of the proposed policy and benchmark policy as a function of mean fading-channel power gain (λ). I evaluate the outage performance for two different threshold SNRs. I observe that as average channel power gain increases, outage performance improves. Further, for both thresholds, the FN-based RPPSP significantly outperforms the benchmark policy.

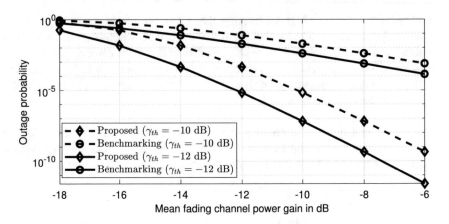

Fig. 2. Outage probability as a function of λ ($N = 4$, $L = 3$, $P_{s,\text{eff}} = 0$ dB, and $\sigma_n^2 = 1$).

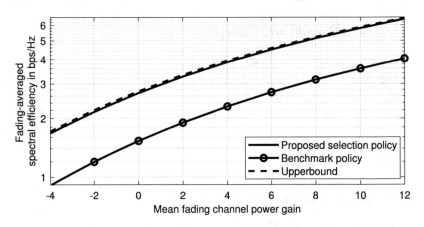

Fig. 3. CFASE as a function of λ in dB ($N = 4$, $L = 3$, $P_{s,\text{eff}} = 0$ dB, and $\sigma_n^2 = 1$).

CFASE and it's Upper Bound: Impact of Channel-Fading-Averaged Power Gain
Figure 3 evaluates the CFASE of the proposed policy and benchmark policy as a function of mean fading-channel power gain. It also plots the upper bound of CFASE. I find the following observations: i) As average channel gain increases, CFASE performance improves, as I expected. ii) The proposed FN-based RPPSP significantly outperforms the benchmark policy. iii) The upper bound tracks the exact CFASE well and is a tighter upper bound.

5.2   Performance Plots of RPPSP: Nakagami Fading

Outage Probability: Impact of Channel-Fading-Averaged Power Gain
Figure 4 evaluates the outage probability of the proposed policy and benchmark policy as a function of mean fading-channel power gain for the Nakagami fading scenario.

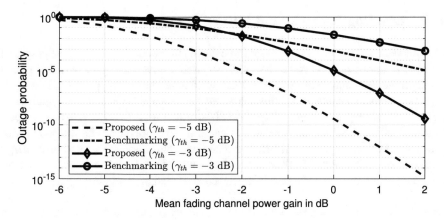

Fig. 4. Outage probability as a function of λ ($N = 4$, $L = 3$, $P_{s,\text{eff}} = 0$ dB, $\sigma_n^2 = 1$, and $m = 2.9$).

I evaluate the outage performance for two different threshold SNRs. From the plots, I observe that as mean channel gain increases, outage probability decreases. Further, for both thresholds, the FN-based RPPSP significantly outperforms the benchmark policy in terms of outage probability.

CFASE and it's Upper Bound: Impact of Channel-Fading-Averaged Power Gain

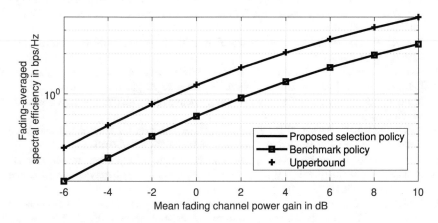

Fig. 5. CFASE as a function of λ ($N = 4$, $L = 3$, $P_{s,\text{eff}} = 0$ dB, $\sigma_n^2 = 1$, and $m = 2.9$).

Figure 5 evaluates the CFASE of the proposed policy and benchmark policy as a function of mean fading-channel power gain in the Nakagami fading scenario. It also plots the upper bound of CFASE. I find the following observations: i) As average channel gain increases, CFASE performance improves, as I expected. ii) The proposed FN-based RPPSP significantly outperforms the benchmark policy. iii) The upper bound tracks the exact CFASE well and is a much tighter upper bound.

6 Concluding Remarks and Future Research Directions

I considered a four-node (a source and three destination nodes) MIMO wireless system with multiple propagation paths. For it, I proposed a novel best path selection policy based on the Frobenius norm of MIMO fading channel matrix comprised of CSI coefficients. I derived various insightful analytical expressions for key performance PHY measures: outage probability and spectral efficiency. I numerically evaluated the performance of the proposed policy and compared the performance with the benchmark policy. Further, I also present an outage-constrained transmission policy to save energy consumption. I quantitatively showed that the proposed policy delivers superior performance that validates its usefulness in advanced and intelligent wireless systems.

Acknowledgement. The author would like to express his gratitude to the EEE department of BITS Pilani, Pilani campus, for providing the required software and infrastructure facilities.

References

1. Goldsmith, A.J.: Wireless Communications. Cambridge University Press, Cambridge (2005)
2. Biglieri, E., et al.: MIMO Wireless Communications. Cambridge University Press, Cambridge (2007)
3. Tse, D., Viswanath, P.: Fundamentals of Wireless Communication. Cambridge University Press, Cambridge (2005)
4. Zhang, R., Zhou, J., Lan, J., Yang, B., Yu, Z.: A high-precision hybrid analog and digital beamforming transceiver system for 5G millimeter-wave communication. IEEE Access **7**, 83012–83023 (2019)
5. Sanayei, S., Nosratinia, A.: Antenna selection in MIMO systems. IEEE Commun. Maga. **42**(10), 68–73 (2004)
6. Asaad, S., Rabiei, A.M., Müller, R.R.: Massive MIMO with antenna selection: fundamental limits and applications. IEEE Trans. Wirel. Commun. **17**(12), 8502–8516 (2018)
7. Molisch, A.F., Win, M.Z.: MIMO systems with antenna selection. IEEE Microw. Maga. **5**(1), 46–56 (2004)
8. Sarvendranath, R., Mehta, N.B.: Transmit antenna selection for interference-outage constrained underlay CR. IEEE Trans. Commun. **66**(9), 3772–3783 (2018)
9. Sarvendranath, R., Mehta, N.B.: Optimal joint antenna selection and power adaptation in underlay cognitive radios. In: Wireless Communications and Networking Conference (WCNC), pp. 3265–3270 (2013)
10. Naduvilpattu, S., Mehta, N.B.: Optimal energy-efficient antenna selection and power adaptation for interference-outage constrained underlay spectrum sharing. IEEE Trans. Commun. **70**(9), 6341–6354 (2022)
11. Han, S., Oh, Y., Song, C.: A deep learning based channel estimation scheme for IEEE 802.11p systems. In: Proceedings of ICC (2019)

5G Virtual Function Infrastructure Management in Adverse Scenarios Using LPWA

Rafael Soares Amaral, Priscila Solis Barreto, and Marcos F. Caetano[✉]

Department of Computer Science, University of Brasilia (UnB),
Brasilia, DF 70910-900, Brazil
{pris,mfcaetano}@unb.br

Abstract. 5G may provide support for applications that need to be deployed in adverse scenarios or dynamic conditions, for example, reduced wireless coverage or emergency situations. The advantage in energy consumption of LPWAN (*Low Power Wide Area Networks*) has the drawback of low data rates, which is unsuitable for 5G data traffic, nevertheless, LPWA devices may be an alternative to maintain and manage a NFVI (*Network Function Virtualization Infrastructure*) in 5G when there are not good 5G coverage conditions. Considering the need of 5G applications in adverse scenarios, this work proposes the use of LPWAN and compression techniques based on NFVI traffic redundancies to extend the coverage of NFV management traffic in a 5G network. The experimental results show the viability of applying the proposed solution to enable 5G/6G applications in rural or adverse scenarios.

1 Introduction

The use of NFV (*Network Function Virtualization*) in 5G [24] provides flexible and reconfigurable conditions for applications and operators, enabling the implementation of network slices and reducing operational and capital costs related to hardware. Network slices in 5G supply performance requirements to scenarios such as URLLC (*Ultra Reliable Low Latency Communications*), mMTC (*Massive Machine Type Communications*), and eMBB (*Enhanced Mobile Broadband*). These requirements also impose a significantly higher level of automation in which the efficient management and orchestration of all network resources are key elements [3].

Orchestration and management tasks depend on control traffic to manage the network infrastructure, which in some applications, whether the scenario is urban or rural, may be limited by wireless coverage and bandwidth constraints. Especially in the case of adverse conditions (natural disasters, massive events, etc.) it is important to at least maintain the NFVI (*Network Function Virtualization Infrastructure*) functioning, to provide conditions for the use of specific applications. One possible path to provide this feature, mainly in rural or remote areas, where there may be limitations in the 5G physical network infrastructure,

L. Barolli (Ed.): AINA 2023, LNNS 661, pp. 125–136, 2023.
https://doi.org/10.1007/978-3-031-29056-5_13

is the use of low-cost communication technologies that dynamically may provide 5G expected functionalities, for example, LPWAN (*Lower Power Wide Area Network*).

With the motivation to provide 5G/6G functionalities in adverse scenarios, where the 5G/6G coverage is at its limits, this article presents a study of the feasibility of using LPWAN technologies for NFVI traffic, to extend functionalities of a 5G network. The focus is on evaluating the capacity of LPWANs for transmitting NFVI traffic, which is the minimum requirement to keep a 5G application online. The study evaluates scenarios where this expansion may be necessary, investigates 5G NFVI traffic behavior and its redundancies, and proposes an architecture to implement compression algorithms in the 5G infrastructure, that may reduce the necessary bandwidth required by this traffic.

This paper is structured as follows: Section 2 presents theoretical concepts, exemplifies a 5G adverse scenario and discusses a compilation of related work. Section 3 details the proposed solution based on ETSI (*European Telecommunications Standards Institute*) specifications. Section 4 describes the implemented testbed and presents the evaluation of two experimental scenarios that exemplify 5G adverse coverage and evaluates the proposed solution and its results. The article ends with the conclusions and future work, in Sect. 5.

2   Theoretical Concepts and Related Work

2.1   5G and LPWA Combination

The ETSI reference architecture [13] for 5G is synthesized in Fig. 1, which shows a high-level view of the components, where the NFVI block is responsible for the management and provision of resources for VNFs (*Virtual Network Function*). The EM (*Element Management*) is for VNF control and the virtualized infrastructure manager (VIM) manages communication with the virtualized infrastructure. The VNFM manages the lifecycle of VNFs and the NFV Orchestrator (NFVO) is in charge of grouping the VNFs and ensuring their intrinsic communication. Also, Fig. 1 shows in its right side the NFV MANO orchestrator, which encompasses all components.

In remote, rural areas or adverse scenarios, many applications are mainly for environment control, remote monitoring, and automation. LPWA components have strong support for these applications, which has motivated advances by groups such as the LoRa Alliance[1], with the aim of improving and building protocols that reduce the amount of energy required for their operation. LPWANs have raised interest in remote applications for allowing communication over long distances with the least possible complexity [21]. The characterization of traffic in these scenarios shows that, despite the small size, thousands of messages are produced, typically in the *uplink* direction. LoRa, Sigfox and NB-IoT are examples of LPWA protocols. LoRa is one of the most promising due to its simplicity

---

[1] https://lora-alliance.org/.

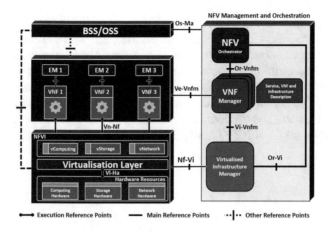

Fig. 1. ETSI NFV Architecture

and efficiency. Patented by the Semtech Group[2] and promoted by the LoRa Alliance group, this technology uses CSS (*Chirp Spread Spectrum*) modulation and is capable to adapt the baud rate according to two main parameters: SF (*Spreading Factor*), numbered 7 to 12, and bandwidth (125KHz, 250KHz, and 500kHz). The SF value defines how long a symbol will take for transmission, where each value is twice as long as the next. Therefore, the higher the value of SF, the lower the baud rate, on the other hand, the greater the value of SF, the greater the range of the signal. Furthermore, this type of modulation is resilient to the Doppler effect [4] making it suitable for mobile applications that do not need a high baud rate [9].

5G and LPWAN can be assembled in scenarios such as the one described in Fig. 2, which illustrates a use case where 5G and LPWA can be combined for the provision of voice and data connectivity in an emergency situation. If the 5G rural cell is not available to cover an area where a sinister event is occurring (fire, flood, etc.), then the rescue team may be unable to communicate and correctly coordinate rescue actions in the area. In such a situation, UAVs (*Unmanned Aerial Vehicles*) can be launched in the area, carrying low-cost devices that may include wireless transceivers (e.g., Wi-Fi or LPWA) to enable network communications towards ground units/devices and other aerial vehicles. Some UAVs may work as access points and mobile devices can communicate in the emergency region through VoIP clients, for example. Technology such as LoRa can be used to control the VNFs in this scenario. Unlike NB-IoT, LoRa does not use licensed spectrum and has a higher throughput than Sigfox [12]. In Fig. 2, the green line represents the communication between the UAVs and the base station using a LoRa interface. The blue line represents the communication between the UAVs using another wireless interface for shorter distances, for example, Wi-Fi. In such a scenario, a subset of UAVs may deploy a Wi-Fi access point,

---

[2] https://www.semtech.com/.

to provide network access connectivity to ground users and other UAVs can be positioned at specific locations and offer routing and forwarding functionalities for data traffic, creating a FANET (*Flying Adhoc Network*) that supports data communications over larger and dynamic areas. The 5G devices can access external networks and services through the 5G core (5GC) of the operator where the support of NFVI traffic is the minimum requirement for extending the 5G VNF infrastructure, then, the available bandwidth between UAVs should comply with the bandwidth needs of NFVI traffic for management and orchestration.

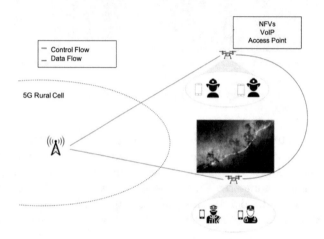

Fig. 2. Example of a 5G Adverse Scenario with LPWAN

2.2   Related Work

The combination of LPWA with 5G in rural areas is studied in [21], where the authors evaluate the communication in vehicular communication using LoRa in field tests evaluating signal strength, reception ratio, and signal-to-noise ratio. In [14] the authors evaluate a 5G and VoIP scenario. Few works in 5G present deeper studies in NFVI control traffic. In [5] is proposed a telemetry and monitoring tool for NFVI traffic and network slices. The work of [8] designs a tool to collect and classify NFVI traffic in the 5G network core. In [11], the authors use a testbed based on the 5G ETSI architecture to better analyze this traffic. The results propose future evaluations of compression algorithms to decrease the volume of NFVI control traffic to achieve rates compatible with LPWA networks.

Several compression techniques known in literature [6,25] basically exploit the number of byte sequences repeated within byte streams. On the other hand, Difference (delta) compression [18] removes redundancies in sets of different byte streams. A good example of these techniques is the Git[3] utility. Delta compression can be divided into two main steps: finding repeated byte sequences and

---

[3] https://git-scm.com/.

encoding the similar and dissimilar sequences in copy and insert instructions, respectively. Finding similar sequences in different byte streams involves using checksums [20] and the use of faster techniques for its calculation [22].

The use of compression techniques has already been applied in 5G traffic. In [19] is proposed a compression scheme based on header fields and SDN characteristics to improve performance in bandwidth utilization and delay. In [17] the authors propose a discrete cosine transform-based data compression scheme to enhance the effective data rate of the fronthaul link. All the cited works do not specifically evaluate the use of compression algorithms given the 5G control traffic redundancies, to optimize bandwidth for NFVI control flows in the 5G architecture.

3   Proposal for NFVI Traffic Compression

The proposal considers the use of the ETSI architecture for 5G, where the VIM/NFVI and the orchestrator are the vital components. The proposal considers OpenStack[4] as the centralized control of physical resources while OpenMano[5] as the assembler of the virtual infrastructure. The control of the virtual infrastructure made by OpenStack is based on two main protocols: HTTP (*Hypertext Transfer Protocol*) and AMQP (*Advanced Message Queuing Protocol*) [15]. Both allow services running on compute nodes and controllers to communicate for basic operations like authentication, image retrieval, and status reporting, among others. AMQP is a queue-oriented protocol, employing the concept of publisher-consumer where a middleware known as the broker is normally used. There are three main components in this protocol: the *exchange* which receives messages from the producer applications and transfers them to the appropriate message queue based on some arbitrary criterion; the message queue that stores them until they can be processed by a consumer application and the *biding* that defines the relationship between the queues and the *exchange* component providing the criteria for forwarding. These components provide a simple and lightweight form of communication and are predominantly used between OpenStack components. Further details about these methods can be found in the documentation [15].

The messages produced by OpenStack have specific characteristics depending on their type. A set of evaluations show that control traffic has peaks of 180 kbps, with mean and standard deviation relatively low, around 11.5 kbps and 26 kbps, respectively. A relevant characteristic observed in the traffic is the frequency of spikes every 60 s. The messages contain fields such as _msg_id, _reply_q and _unique_id, which are fields that have a *hash* value to identify the message, among others. The method field, placed as *method_name* defines the class of message being transmitted. This characteristic led to group the messages by the method field to verify similarities existing in the same group. Figure 3 shows some messages that were grouped by the method field. Each bar represents the average percentage of

---

[4] https://www.openstack.org/.

[5] https://osm.etsi.org/.

similarity to the first message of the class to which it belongs. To calculate the percentage of similarity, the instructions *COPY* and *ADD* are used to assemble a message using delta encoding. Assuming two-byte streams A and B, in which A is used to encode B, the instructions of type *COPY* contain the size and address of the byte stream that must be copied from A to decode B, which allows calculating the ratio between the sum of the sizes of the copied parts and the original message. Note that most of the values are above 50%, which means that half of the transmitted content was also present in previous messages.

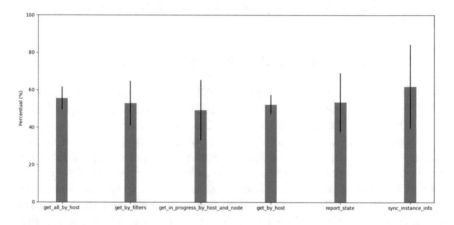

Fig. 3. Control Traffic Similarities.

The previous analysis confirms that there is considerable redundancy in the set of messages transmitted by OpenStack, which may allow the application of

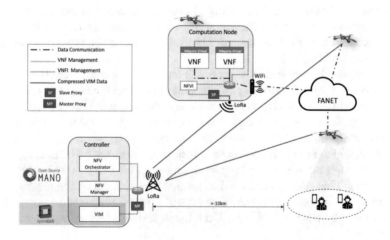

Fig. 4. Proposed Architecture and Components

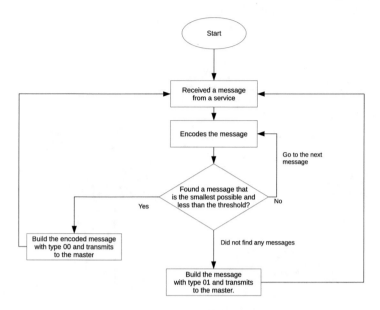

Fig. 5. Logic flow of slave proxy

compressing techniques to produce a sequence of NFVI messages that contain only differences from what has already been sent. Figure 4 shows the proposed architecture for NFVI traffic compressing, based on a master and a slave proxy. A message dictionary is assembled in both proxies. For each message redirected to/from the slave proxy, and when its counterpart is found in the dictionary, delta compression is applied to reach a pre-established threshold that depends on the maximum available data rate. The threshold value defines an optimal compression ratio and allows the best use of the dictionary by maximizing the probability of finding a message with more redundancy in the long run. To better understand these procedures in the master and slave proxies, Figs. 5 and 6 describe the logic flows for slave and master proxies respectively. The slave proxy connects with the master proxy and starts to listen to redirect connections after receiving a dictionary identifier. Upon receiving a message, it checks its similarity rate. When the ratio between the original message size and its compressed version is greater than a defined threshold, the compressed message is transmitted. Otherwise, the original message will be transmitted after being added to the dictionary. The master proxy upon connections generates a unique identifier for the slave's dictionary and thereafter for each message transmitted, it takes several actions such as encoding, decoding, dictionary adding, and forwarding the message to the service. For each message produced by the slave, there is a corresponding action on the master and its dictionaries are kept synchronized. The defined threshold plays an important role in the final size of the dictionary and its balance. Based on the literature survey, were chosen three encoders to be implemented in the proxies: XDelta [7], DDelta [22] and EDelta [23].

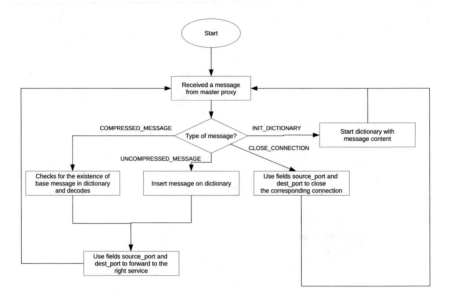

Fig. 6. Logic flow of master proxy

4 Experimental Results

The proposed architecture in Fig. 4 was implemented in a testbed based on low-budget equipments. Both OpenStack and OpenMano were configured using the standard installation guides. The computation nodes were implemented using Raspberries PI (RPI) v4 running *Nova* and *Neutron* components, where *Nova* manages the VMs (*Virtual Machines*) and *Neutron* provides virtual machine connectivity. Each device used two VxLAN-type interfaces, one carrying VNFI traffic and the other data traffic. Figure 4 also shows the blocks for traffic interception, using master and slave proxies, which were used to implement the compression algorithms.

LoRa, in the best case, achieves a transmission rate of 21 kbps. In the scenario described in Fig. 2, the average rate of NFVI traffic could not be higher than this peak transmission rate. As presented in Sect. 2.1, LoRa CSS modulation scheme defines six different SF values. The SF values and different options of bandwidth are combined by thirteen different data rate (DR) options defined in the LoRaWAN MAC layer protocol. Data rates options from DR0 to DR4 are reserved for the uplink channel, while from DR8 to DR13 are for the downlink channel, according to US 902–928 standard [1]. The DR4 and DR13 combinations provide the highest physical baud rate and are the chosen ones in our evaluation. To check the throughput achieved in LoRaWAN devices, we made three tests to reach the maximum uplink and downlink rates simultaneously. For the uplink, 10 rounds with 33 messages of type CONFIRMED_DATA_UP [2] were transmitted in their maximum size available (240 bytes). For downlinks, the same procedure was done using messages of the type CONFIRMED_DATA_DOWN. For the third test, the maximum size was transmitted in both ways. The mean values obtained in the tests were 6.825 kbps for the uplink, 8.231 kbps for the downlink, and 4.794 kbps for the uplink and downlink simultaneously. As can be seen, the use of the LoRaWAN MAC protocol drastically reduces the transmission rate since the MAC protocol includes encryption mechanisms and a data frame header, which decreases the effective throughput.

Considering the previous results and the limit of 4.794 kbps, to evaluate the proposed architecture described in Fig. 4, were defined the following scenarios:

1. Instance creation scenario: limiting the control interface throughput to 4.794 kbps within an interval of 500 seconds, which evaluates the viability of building a virtual infrastructure using the LoRa channel.
2. VoIP communication scenario: the infrastructure described in Fig. 4 simulates a VoIP communication between two firefighters, each one within an area covered by different UAVs using WiFi. The two UAVs are emulated by RPI devices, called Rasp1 and Rasp2. The VoIP workload in the simulations was generated using a 5G synthetic traffic generator developed in previous works, called EROS-5 [16]. The payload follows the G.729 coding according to a model presented in [10] in which each message has 20 bytes and the time between consecutive messages follows a normal distribution with a mean of 0.02 and a standard deviation of 0.0038. Thus, the average rate of this application is around 8 kbps over 170 seconds. This scenario aims to evaluate the LoRa channel for control communication with the VIM/NFVI components.

In both scenarios were evaluated 3 main combinations: non-proxy and non-bandwidth limitation (NonP-NonB), non-proxy and bandwidth limitation (NonP-CB), and proxy and bandwidth limitation (CP-CB), for three compression algorithms: XDelta, DDelta and EDelta. The set of simulations was made to achieve a 95% confidence interval.

Table 1 shows the results for Scenario 1, showing the type of messages exchanged, time to launch and tear down instances, and throughput between components. Among the compression algorithms, XDelta shows better performance than the others. Also, considering the 4.794 kbps limit, the creation of instances was only possible when using XDelta and EDelta, which shows that the use of compression is fundamental to guarantee NFVI maintenance. Table 2 shows the results for Scenario 2. Without the bandwidth limitation, the average throughput reaches a value of around 8 kbps while using the proxy with XDelta decreases this rate to 1.5 kbps. XDelta really shows remarkable superiority over other compression schemes. Also, only XDelta and EDelta allowed the creation of instances, suggesting that the high compression ratio of XDelta relative to DDelta shows room for improvement in finding similar *bytestreams* between two messages of the same type.

The experimental results in Table 1 and Table 2 show that the proposed solution using compression proxies achieves a relevant reduction in NFVI traffic in the evaluated scenarios, which suggest the viability of using LoRA as an alternative communication channel to manage 5G NFV services in scenarios with adverse or emergency conditions.

Table 1. Results for First Scenario

| | NoNP-NoNB | NonP-CB | CP-CB | | |
			XDelta	DDelta	EDelta
Retransmission (Fast/Ret/Spurious)	0	444	0	0	0
RESET	9	302	1	2	2
DUPACK	0	232	1	12	19
Lost Segments	0	6	1	0	1
Keep-alive	2	1	1	12	20
Time (launch/tear down) instance(sec)	7.05/6.21	–	46.52/12.36	–	259.75/78.96
Throughput (Controller-Rasp) bps	7579.89	1654.94	1319.77	1495.02	2237.81
Throughput (Rasp-Controller) bps	9370.40	3183.65	1335.14	3258.91	2029.52

Table 2. Results for Second Scenario

| | NonP-NonB | NonP-CB | CP-CB | | |
			XDelta	DDelta	EDelta
Retransmission (Fast/Ret/Spurious)	0	689	114	124	118
RESET	17	471	7	20	16
DUPACK	0	367	114	113	136
Lost Segments	0	4	0	9	0
Keep-alive	1	4	0	0	0
Time (launch/tear down) client (sec)	12.35/10.73	–	46.67/12.89	–	233.87/56.51
Time (launch/tear down) server (sec)	10.49/7.26	–	53.61/16.15	–	246.13/42.92
Throughput (Controller-Rasp1) bps	5105.61	1664.94	1666.65	1831.04	2012.09
Throughput (Rasp1-Controller) bps	7928.43	2941.34	1540.83	3350.23	2850.14
Throughput (Controller-Rasp2) bps	6425.55	1756.30	1355.21	1644.02	2156.71
Throughput (Rasp2-Controller) bps	8322.90	2660.30	1305.49	2884.17	2582.10

5 Conclusions and Future Work

In this work were exploited the redundancies of control traffic from OpenStack in a 5G architecture to evaluate the use of compressing algorithms for the case of 5G adverse scenarios and their extension with LPWANs. The proposed solution implements an architecture with proxies to intercept control traffic and evaluates three compression algorithms. The experimental results show a reduction of 80% in the average rate (11.5 kbps to 2.357 kbps) of control traffic, which is acceptable for virtual infrastructure management in 5G and shows that LoRA may be relevant to extend 5G applications in critical situations. The future work of this research aims to validate the results with real life scenarios implemented using LPWANs extensions in 5G.

References

1. Alliance, L.: Lorawan specification (2016). https://lora-alliance.org/wp-content/uploads/2020/11/lorawan1_0_2-20161012_1398_1.pdf
2. Alliance, L.: Lorawan 1.0.2 regional parameters (2017). https://lora-alliance.org/wp-content/uploads/2020/11/RP_2-1.0.2.pdf
3. Brown, G.: Service-based architecture for 5g core networks. https://tinyurl.com/ServiceBased-Architecture (2017)
4. Doroshkin, A.A., Zadorozhny, A.M., Kus, O.N., Prokopyev, V.Y., Prokopyev, Y.M.: Experimental study of lora modulation immunity to doppler effect in cubesat radio communications. IEEE Access **7**, 75721–75731 (2019). https://doi.org/10.1109/ACCESS.2019.2919274
5. Giannopoulos, D., Papaioannou, P., Tranoris, C., Denazis, S.: Monitoring as a service over a 5g network slice. In: 2021 Joint European Conference on Networks and Communications & 6G Summit (EuCNC/6G Summit), pp. 329–334 (2021). https://doi.org/10.1109/EuCNC/6GSummit51104.2021.9482534
6. Huffman, D.A.: A method for the construction of minimum-redundancy codes. Proceedings of the IRE **40**(9), 1098–1101 (1952). https://doi.org/10.1109/JRPROC.1952.273898
7. Korn, D., MacDonald, J.P., Mogul, J., Vo, K.P.: The VCDIFF Generic Differencing and Compression Data Format. RFC 3284 (2002). 10.17487/RFC3284. https://rfc-editor.org/rfc/rfc3284.txt
8. Leite, C., Barreto, P.S., Caetano, M.F., Amaral, R.: A framework for performance evaluation of network function virtualisation in 5g networks. In: 2020 XLVI Latin American Computing Conference (CLEI), pp. 314–321 (2020). https://doi.org/10.1109/CLEI52000.2020.00043
9. Li, Y., Han, S., Yang, L., Wang, F.Y., Zhang, H.: Lora on the move: Performance evaluation of lora in v2x communications. In: 2018 IEEE Intelligent Vehicles Symposium (IV), pp. 1107–1111 (2018). https://doi.org/10.1109/IVS.2018.8500655
10. Mattos, C.I., Ribeiro, E.P., Fernandez, E.M., Pedroso, C.M.: An unified voip model for workload generation. Multimedia Tools Appl. **70**(3), 2309–2329 (2014)
11. Medeiros, E., Barreto, P.S.: Analysis of vnf control traffic in 5g scenarios extended with lpwa networks. In: 2021 IEEE Latin-American Conference on Communications (LATINCOM), pp. 1–6 (2021).https://doi.org/10.1109/LATINCOM53176.2021.9647822
12. Mekki, K., Bajic, E., Chaxel, F., Meyer, F.: A comparative study of lpwan technologies for large-scale iot deployment. ICT Express **5**(1), 1–7 (2019)
13. NFV, E.: Network functions virtualisation: An introduction, benefits, enablers, challenges & call for action. issue 1 (2012)
14. Nogales, B., Sanchez-Aguero, V., Vidal, I., Valera, F.: Adaptable and automated small uav deployments via virtualization. Sensors 18(12) (2018). https://doi.org/10.3390/s18124116.https://www.mdpi.com/1424-8220/18/12/4116
15. OASIS: Amqp: Advanced message queuing protocol; protocol specification (2008). https://www.amqp.org/sites/amqp.org/files/amqp0-9-1.zip
16. Soares, R., Ferreira, G., Solis, P., Caetano, M.: Eros-5: Gerador de tráfego sintético para redes 5g. In: Anais Estendidos do XXXVIII Simpósio Brasileiro de Redes de Computadores e Sistemas Distribuídos, pp. 41–48. SBC, Porto Alegre, RS, Brasil (2020). https://doi.org/10.5753/sbrc_estendido.2020.12400.https://sol.sbc.org.br/index.php/sbrc_estendido/article/view/12400

17. Su, Y., Lu, X., Huang, L., Du, X., Guizani, M.: A novel dct-based compression scheme for 5g vehicular networks. IEEE Transactions on Vehicular Technology PP, 1–1 (2019). https://doi.org/10.1109/TVT.2019.2939619
18. Suh, Y.S.: Send-on-delta sensor data transmission with a linear predictor. Sensors **7**(4), 537–547 (2007). https://doi.org/10.3390/s7040437
19. Sun, J., Dong, P., Qin, Y., Zheng, T., Yan, X., Zhang, Y.: Improving bandwidth utilization by compressing small-payload traffic for vehicular networks. International Journal of Distributed Sensor Networks 15(4), 1550147719843,050 (2019). https://doi.org/10.1177/1550147719843050.https://doi.org/10.1177/1550147719843050
20. Tan, H., Zhang, Z., Zou, X., Liao, Q., Xia, W.: Exploring the potential of fast delta encoding: Marching to a higher compression ratio. In: 2020 IEEE International Conference on Cluster Computing (CLUSTER), pp. 198–208 (2020). https://doi.org/10.1109/CLUSTER49012.2020.00030
21. Torres, A., Silva, C., Filho, H.: An experimental study on the use of lora technology in vehicle communication. IEEE Access PP, 1–1 (2021). https://doi.org/10.1109/ACCESS.2021.3057602
22. Xia, W., Jiang, H., Feng, D., Tian, L., Fu, M., Zhou, Y.: Ddelta: A deduplication-inspired fast delta compression approach. Perform. Evaluation **79**, 258–272 (2014)
23. Xia, W., et al.: Edelta: a word-enlarging based fast delta compression approach. In: Proceedings of the 7th USENIX Conference on Hot Topics in Storage and File Systems, HotStorage'15, p. 7. USENIX Association (2015)
24. Yi, B., Wang, X., Li, K., k. Das, S., Huang, M.: A comprehensive survey of network function virtualization. Computer Networks 133, 212–262 (2018). https://doi.org/10.1016/j.comnet.2018.01.021.https://www.sciencedirect.com/science/article/pii/S1389128618300306
25. Ziv, J., Lempel, A.: A universal algorithm for sequential data compression. IEEE Transactions on Information Theory **23**(3), 337–343 (1977). https://doi.org/10.1109/TIT.1977.1055714

Healthcare Conversational Agents: Chatbot for Improving Patient-Reported Outcomes

Giuseppe Fenza, Francesco Orciuoli, Angela Peduto[✉],
and Alberto Postiglione

Dipartimento di Scienze Aziendali-Management and Innovation Systems,
Universitá di Salerno, Fisciano, Italy
{gfenza,forciuoli,anpeduto,ap}@unisa.it

Abstract. The patient's lack of adherence to the doctor's prescriptions limits the favorable effects that the optimal prescribed treatments have on the disease. Adherence to therapy can be measured by various methods, including self-assessment questionnaires (structured interviews) and electronic devices. The Patient Report Outcomes are derived from the self-assessment questionnaires. When implementing an effective patient-centered care strategy, clinicians must keep track of Patient Report Outcomes over time. The patient-reported outcomes are effective tools to better understand a patient's health conditions, goals, and specific factors related to his care. Conversational Agents are receiving increasing attention in healthcare and academia, but they are still little used to collect data. This paper describes how a conversational agent can intervene in the collection of data carried out directly by the patient and not by the doctor, using patient-reported outcomes and their electronic version. In the case study, the chatbot created with Dialogflow encourages patients to follow their therapy and report all the treatment's effects.

1 Introduction and Related Works

Patient Reported Outcomes (PROs) are reports of a patient's health status that are provided by the patient (or, in certain situations, a caregiver), without being interpreted by a professional or anyone else. [8]. PROs data can be gathered in several ways, including interviews, questionnaires (completed on paper, online), and diaries. PROs are an invaluable source of data that may inform a broad range of domains in health care, including clinical investigations, clinical practice, healthcare management, and decision-making on regulatory, coverage, and reimbursement aspects. They are also a tool that the patient can use to evaluate the healthcare system and therefore an important element for patient-centered healthcare [7]. Poor adherence to prescribed therapies negatively affects treatment results. The causes of non-adherence are numerous and understandable. They consist of forgetfulness, lack of coordination with a caregiver, and a perceived discrepancy or scarcity of resonance between the patient and the physician

L. Barolli (Ed.): AINA 2023, LNNS 661, pp. 137–148, 2023.
https://doi.org/10.1007/978-3-031-29056-5_14

concerning clinical realities or treatment priorities [11]. Remote patient monitoring and the use of PROs can play a significant role in correcting this discrepancy in the mental model and thereby closing a substantial gap in currently existing clinical care. It is possible to collect patient data using chatbots to extract patient information using simple questions. These Chatbots then store information in the medical facility system to simplify and speed up patient communication. This positively affects treatment results and medical record keeping. Various chatbot applications have been developed to address health management issues. ELIZA was the first chatbot used in the healthcare industry in 1966, mimicking a psychotherapist using role models and response selection. However, he had limited knowledge and skills. Today chatbots offer a diagnosis of symptoms, mental health, nutritional information, and follow-up. For example, in 2020, WhatsApp partnered with the World Health Organization to create a chatbot service that answers users' questions about COVID-19. There are chatbot technologies that motivate users to improve their standard of living, encouraging them towards healthy eating habits to prevent weight gain [9]. Some chatbots like HealthTap[1] and YourMD[2], provide online health care to the patient, while others, like MANDY, support doctors by automating routine operations [16]. As a multi-functional personal assistant, kBot stands out by providing a personalized approach to track patient health, warn patients about asthma triggers, and help them collectively manage their asthma to achieve an overall improved health outcome [12]. Users trust chatbots for rational reasons and doctors for emotional reasons. Transparency and the ability to naturalness interact are perceived fundamental elements for the acceptance of Conversational Agents (CA) in health care [17].

This paper first describes, in Sect. 2, how a conversational agent can be inserted into a healthcare system that involves the use of ePROs. The methodology for designing a conversational agent to collect ePROs is then described in Sect. 3 Finally, in Sect. 4, a case study is presented in which a chatbot has been developed that, conversing with the patient, can compile ePROs in oncology.

2 Overall Architecture

There are many approaches for using technology for PRO data collection and application, and there is often no "one size fits all" solution [2]. Figure 1 illustrates the data flow (DF) options and actors involved in the ePROs Life Cycle system. In the hypothesis depicted, both the Electronic Health Record (EHR) and a third-party tool are used with some degree of integration. This approach can capitalize on the flexibility of third-party tools while still maintaining a footprint in the EHR workflow and documentation. The EHR contained a designated section, where all assessment outcomes, including the date, finished, severity interpretation (normal, mild, moderate, or severe), questions, and patient replies, were recorded. Messages delivered to the EHR's messaging area notify clinicians

[1] https://www.healthtap.com/.
[2] https://www.your.md/.

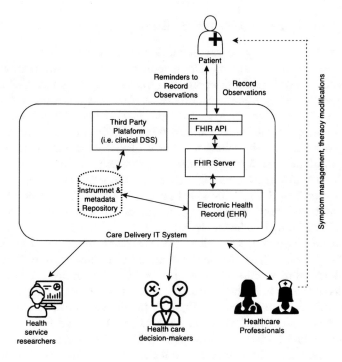

Fig. 1. ePROs life cycle system.

of severe symptoms and patient concerns. In response to the growth of ePROs tools, agencies such as Patient-Centered Outcomes Research Institute (PCORI)[3] have partnered on several initiatives to facilitate the implementation of ePROs in the clinical workflow like the Health Level Seven (HL7), PROs, Fast Healthcare Interoperability Resource (FHIR), Implementation Guide [3] that provides direction and technical specifications for the capture and exchange of PRO data using FHIR standards. FHIR aims to make electronic health records (EHRs) available, discoverable, and easily understandable for stakeholders as patients move within the healthcare ecosystem. The use of third-party platforms may allow each local practice within a healthcare organization to easily tailor the PRO content and cadence to their local workflows. The use of third-party platforms may allow each local practice within a healthcare organization to easily tailor the PROs content and cadence to their local workflows. This it can be an advantage as it minimizes the burden on the centralized IT resources needed to support, adjust, and maintain PRO tools. In the proposed approach, PROs collection is carried out using a chatbot. Conversational agents have emerged as a good choice to create human-machine interfaces that offer a more engaging and human-like contact between patients and physicians.

---

[3] PCORI is the leading founder of patient-centered comparative clinical effectiveness research in the United States.

2.1   Chatbot Roles and Functionalities

Chatbots might be a way to save time for medical staff and follow up with patients throughout treatments. They replicate human conversation through text or speech via cellphones or computers, provide a dynamic engagement, and are simple to use. People might feel more at ease sharing intimate information with a chatbot than with a person because chatbots don't have minds of their [14]. A well-designed healthcare process must take into account both the experiences of patients and physicians while still placing a strong focus on patients' activities [13]. The need for an architecture that can accommodate the requirements of both types of users is highlighted by the requirement for a Conversational User Interface (CUI) to be able to switch between these views (healthcare personnel and patients). A chatbot's design and development involve several techniques [1]. There are a few main components in the conversational platform: ASR (Automatic Speech Recognition), NLU (Natural Language Understanding), Dialog Management, NLG (Natural Language Generation), and TTS (Text to Speech). A Chatbot architecture requires a candidate reply generator and an answer selector to answer the user's questions through text, images, and voice. Chatbot architecture is shown in Fig. 2. A possible scenario of CA integration in the EPROs system is that based on a calendar defined by the doctors for verifying compliance with the drugs, the patient receives an alerting them to activate the chatbot. The CA first greets the patient, then verifies his identity. If successfully authenticated check which questionnaires the patient needs to complete and collect responses for each questionnaire accordingly. The idea that patient-identified (person-level) PROs data may be used for reasons other than direct patient care, such as patient-centered outcomes research, is implicit in this scenario [18].

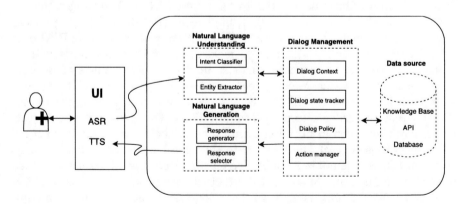

Fig. 2. Chatbot architecture.

3   Design and Implementation Methodologies

To understand how to develop the flow of dialogue between the patient and the healthcare system integrated with ePROs, it is necessary to define intent, entity, response generator, and response selector. Intents are goals or objectives expressed in a client's dialogue input. By recognizing the intent expressed in a client's input, the assistant can select the applicable next action. Entities are specific terms that allow giving context to the received message. There are three types of entities: system entity, developer entity, and session entity. An entity in the Chatbot is used to modify an intent. The Response Generator performs calculations using several algorithms to process the user's request. So, the result of these calculations is the candidate's answer. The Response Selector is used to select the word or text based on the user's questions giving the users the answer that should work better. There are several approaches to identifying Intents and Entities. The simplest one is to search for specific words within the received messages. This strategy can be quickly set up when creating a small prototype with limited functionality. The most complex and comprehensive approach is to use natural language processing (NLP) services.

3.1   Chatbot Framework Selection

To provide the best user experience, chatbot frameworks make it easier to create, connect, publish, and maintain intelligent, interactive CA. A comparison was made between the different frameworks, to then use the most compliant one for the case study. The chatbot frameworks analyzed are: Amazon Lex, Dialog Flow, IBM Watson Microsoft Bot, Rasa, and Wit.ai. All these platforms are powered by machine learning algorithms.

Amazon Lex is an open-source chatbot offered by Amazon Web Services (AWS) and it uses the Amazon AI suite. It can be said that the core features of this framework are: High-quality speech recognition and natural language understanding, 8 kHz telephony audio support, one-click deployment to multiple platforms, ease to add a new language to a bot at any time, support context management to understand user intent. Not multilingual, only English available. https://aws.amazon.com/it/lex/

DialogFlow is a Google-owned human-computer interaction technology based on NLP. It is utilized to create conversational apps for users across a variety of platforms and languages. This Google bot platform is adaptable and user-friendly. Core features: support rich intuitive customer conversations, interact naturally and accurately, reduce development time with interactive flow visualizations, build once, deploy everywhere on a required platform, BERT-based natural language processing (NLP) models to improve call/chat. The only flaw found is no live customer support. https://dialogflow.com/

IBM Watson is a Natural Language Classifier API that lets you use a custom classifier to decipher the natural language. It organizes text from many languages into unique or different groups. Core features are: automatically identifies topics from pre-existing chat logs, and uses artificial intelligence search to pull answers from your existing content, enabling a secure channel through authenticated SSL certificates. The use of this framework is not intuitive. https://www.ibm.com/watson

The **Microsoft Bot** A group of libraries, tools, and services known as the Microsoft Bot Framework and Azure Bot Service make it possible to create, test, deploy, and manage intelligent bots. A modular and expandable SDK for creating bots and establishing connections to AI services is included in the Bot Framework. The weak points: the NLU engine cannot be installed on-premise. https://dev.botframework.com

Rasa Stack is an open-source conversational AI platform. This chatbot framework is composed of two main parts: Rasa NLU (natural language understanding) and Rasa Core. The main features are: customizable NLU for any domain, industry, or use case, and handles complex conversations easily. Provides interactive learning, connects to the commonly used messaging channel, CMS, and CRM, and deploys anywhere. Contextual chatbot development needs knowledge of Python, NLP, and deep learning, making it unsuitable for beginners. https://rasa.com/

Wit.ai is an open-source NLP API. Important advantages include ease of application development and learning human language from every encounter, ready-made templates for easy integration, and SDK in multiple languages, including Python, Ruby, and Node.js. However, learning to use it is laborious. https://wit.ai/

- Speech Processing determines whether the framework is capable of processing speech.
- Pre-build Entities indicate if pre-build entities are offered (date, name, location, etc.).
- Pre-build Intents indicate if pre-build groups of intents are offered.
- Default Fallback Intent indicates availability to automatically configure with a variety of static text responses, such as "I don't understand. Can you say it again?" when the agent does not recognize an end user expression.
- Online Integration that is the possibility or not of third-party integration.
- Program Language.
- Multichannel indicates the possibility of integrating the chatbot in different applications
- Languages indicate how many languages are supported by the framework.
- Graphical interface tells if a visual editor to build a CA is available.
- Co-creation indicates the possibility to sharing portions of the dialog and co-creating is paramount
- License terms and conditions to use the software.

Table 1 summarizes the comparison between the main characteristics of the frameworks analyzed. Dialogflow CX was chosen to develop the chatbot of the

Table 1. Comparison of analyzed chatbot frameworks.

Framework	Speech proc.	Pre-build entities	Pre-build intents	Default fallback intent	Online integration	Prog. lan.	Multi channel	Languages	G.I.	Co-creation	License
Amazon Lex	X	X	X	X	X	Java, CLI, .Net, Python, Android, Go, ABAP, Ruby, CPP, IOS SDK, PHP	X	English	X	X	Paid
Dialogflow	X	X	X	X	X	C++, C#, Go, Java, Node.js, PHP, Python, Ruby	X	Multilingual	X	X	Free Trial
IBM Watson	X	X	X	X	X	Java, Node.js, .Net, Python, Android, Go, ABAP, Ruby, Salesforce, Swift, Unity	X	Multilingual	X	X	Free Trial
Microsoft Bot	X	X	X	X	X	C#, JavaScript, Typescript, Python	X	Multilingual	X	X	Free Trial
RASA	Ø	Ø	Ø	Ø	X	Python	X	English, German	Ø	Ø	Open source
Wit.ai	X	X	X	Ø	X	Node.js, Python, Go, Unity, Ruby	X	Multilingual	X	Ø	Free Trial

case study. Compared to other platforms that use predetermined questions, Dialogflow offers an excellent user experience thanks to Natural Language Processing [19]. Other reasons for choosing Dialogflow are sentiment analysis on queries and IoT integration allowed. Dialogflow is built in layers as a decision tree like Fig. 3. An important feature of Dialogflow is the ability to save contexts during the conversation.

3.2 Dialogue Design

Designing Voice User Interfaces (VUIs) requires a very different approach compared to building graphical user interfaces (GUIs). Directly translating a survey into a voice interface by reading through each option and asking the table yes or no would surely wear out the patience of any potential users quickly. Good practice in building natural and user-centric VUIs is allowing users to speak in their own words. It is possible thanks to NLU capabilities. Some good requirements for a conversational health agent that empowers users to understand health information, make informed decisions, and interact better with healthcare professionals are:

- to provide tailored information easily understood and relevant to patients.
- is objective and does not communicate any kind of bias or opinion.
- to explain the Tables of how information is selected in terms that a person/patient can understand, thus increasing his/her trust and acceptance.

The first step in the Dialog Management system design is to choose an interaction strategy, which determines who is leading the conversation: the user, the system, or both. In this study case, the user starts the conversation, but then the chatbot leads it by asking the questions of the questionnaire (PRO). To build PROs-chatbot, have been defined distinct intents that each represents a meaning the patient tries to convey. To develop a virtual assistant decision trees (conversational trees) are employed. Each of the nodes that make them up contains the response to a specific query. The developer adds a condition to each node that is activated when the user's query matches the node's response (with some degree of accuracy). Natural language processing and keywords are used for this phase. Have been provided "sample utterances" for each intent, which

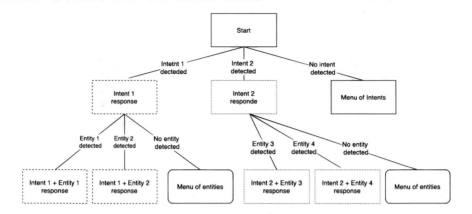

Fig. 3. Structure of proposed decision tree.

were used as training data to build machine-learning models to recognize the correct intent. It's important to provide a wide range of sample utterances when defining each intent as there may be a variety of ways users could respond to a given question. Testing with users with diverse backgrounds can help identify ways of expression you might not consider otherwise.

However, even with a thorough set of sample utterances in place, the bot could still fail to interpret the user from time to time. To handle errors using the built-in fallback intent (Fig. 3). Including disfluencies in dialogue makes bots appear human.

4   Case Study: PROsBot

The prototype for the PROs collection was made in the oncology field. There is growing evidence that PROs can be used as communication tools to enhance symptom control and treatment response tracking in the oncology environment [6]. Recent studies have revealed advantages of PRO monitoring including longer overall survival and less usage of emergency departments, in addition to improving patient care at the clinician level [5]. The information that the chatbot required to gather was created based on an analysis of the implementation sciences literature that already existed addressing the use of PRO in cancer care [15]. The survey was comprised of questions like: How do you feel today? How many times a day do you go to the toilet? What is the consistency of the stool? Have you taken any diarrhea medications? And so on.

4.1   Architectural Design

PROsBot is a client-server architecture where the client is a lightweight frontend chat interface, and the backend server is a standalone web application hosted in the cloud. The frontend interface communicates (now text but in next future voice) with patients. The data of interest (described in the previous paragraph) are extracted from conversations and stored in the PostgreSQL database. PostgreSQL, also known as Postgres, is a free and open-source relational database management system that emphasizes SQL extensibility and compliance. It used Psycopg to perform heavily multi-threaded applications and make many simultaneous inserts or updates. A table named "User" has been created in the database, this table has one row for each user, and the primary key consists of the user's email and date of compilation. Each macro argument is represented by a column in the table. Thanks to the help of the server, the chatbot checks the database to verify the presence of data and asks can only the questions necessary to complete the missing fields. Each argument has its specific function, and each function creates a connection to the database, if the connection fails the chatbot ends the conversation and gives the error of no connection. The other functions, on the other hand, once connected to the database, obtain a JSON object from the chatbot containing the collected data, and update the column. Clients are Google Android applications. Patients' primary interface for communicating with PROsBot is the chatroom activity in the client application. PROsBot has a standalone Python web application that instantiates and serves several clients on the server side. The web application is created using Flask, a micro web framework.

4.2   Implementation

DialogFlow's knowledge base comes preloaded with a list of PRO cancer concepts and vocabulary, including descriptions of symptoms, classes of medications, and categories of activity restrictions. This enhances DialogFlow's ability to find entities that are pertinent to PROs. In the "Path" function, which represents the initial page of the flow, once the connection has been created, the server checks the data in the database and directs the user to the Page useful for completing the missing column, sending a JSON object to the chatbot containing the id of the Page. The flow has been organized so that the chatbot can propose a "humanlike" conversation. In the prototype several webhooks[4] have been created, one for each page of the flow that collects data, which are activated when the conversation moves to the next topic of conversation and sends a JSON object containing the information collected. To read the database, first of all, a connection to the database is made through the create_connection () function. Class connection handles the connection to a PostgreSQL database instance. Class Cursor allows Python code to execute PostgreSQL. The fetchone() method retrieves rows using the previously-created cursor, reads the SQL response to the command, and obtained the list of the elements of the DB (row by row). To allow the chatbot to converse asking only for data not yet present in the PRO database, page leaps have been made. These jumps are implemented by creating a webhook response that returns an element of type JSON, in which the page to jump to is specified. In PROsBot the main element of the JSON file is the "fullfillment_response" and

---

[4] Webhooks are services that host your business logic.

its fundamental fields are "Messages" which specifies the reply message to the user and "MergeBehavior" which indicates whether the message should simply be returned to the user, or if it should "replace" the existing one. In the study case page, "id" are recovered by carrying out some conversation tests and reading the log of the same. To increase adherence to medicines PROsBot will look up the patient's medical history, for instance when the patient reports feeling queasy. When there is poor adherence, PROs Bot alerts the patient that their symptoms might be brought on by a failure to take the controller medication as prescribed. In another scenario, PROsBot might inquire as to whether nausea has been accompanied by episodes of vomiting or diarrhea, and if so, it might advise visiting a physician.

4.3   Experimentation and Evaluation

Chatbot quality, technology adoption, and system usability make up the evaluation criteria. The plan is to design and carry out a laboratory test in which three field operators with various levels of experience (very low, low, and medium) utilize the PROsBot (loaded on a smartphone with a 6.7-inch display) to be assisted in conducting PRO surveys. The PROsBot client application is given to each evaluator, who is then instructed to use it independently while interacting with fictitious patient scenarios. The scenarios are as illustrated in Sect. 2.1. Patients with a system account received an alerting them to activate the chatbot. If successfully authenticated, the PROsBot then checks which questionnaires the patient needs to complete and collects responses for each questionnaire accordingly. The chatbot's quality, according to the evaluation, was about 80%. More specifically, the percentage of operators with minimal and average experience was roughly 83%. The operator with less expertise was able to classify objects correctly in 9 out of 12 instances, or 75% of the time. Regarding the assessment of technological acceptability, the field operators' responses highlighted the chatbot's value in promoting medication adherence. Additionally, the system usability of the chatbot is rated as good by all three operators (possible values: excellent, good, fair, terrible, bad).

5   Conclusion

The result of this study confirmed, as claimed by Bibault et al. [4], that using chatbots might free up time for medical staff to focus on more complex work while providing a way to monitor patients throughout treatment. The ability to perform PRO surveys electronically eliminates some difficulties (such as automatic reminders and providing ePROs to patients before visits) and opens up new potential for enhancing care delivery [10]. PROsBot is one such solution that enhances patient participation and empowerment. Preliminary results suggest further examining the influence of the ePROs solution on common clinical practice. The action field could dramatically alter the workflows for various ePROs data even if it constantly reflected the necessity for thorough and effective PRO data gathering. For ePROs to be successful on a large scale, governance, workflow and IT must all be aligned to ensure that tools like PROsBot are usable. In this way, health systems can integrate the patient's voice into care delivery and further advance patient-centered personalized care. In the future, he plans to add more experimentation in the health and voice fields PROsBot.

Acknowledgement. This research was partially supported by MIUR (Ministero dell'Istruzione dell'Universitá e della Ricerca) under the national program PON 2014–2020, I.CARE.ME (ID ARS01_00707) research project in the area of digital healthcare.

References

1. Ahmad, N.K., Che, M.H., Zainal, A., Rauf, M.F.A., Adnan, Z.: Review of chatbots design techniques. Int. J. Comput. Appl. **181**(8), 7–10 (2018)
2. Austin, E., LeRouge, C., Hartzler, A.L., Chung, A.E., Segal, C., Lavallee, D.C.: Opportunities and challenges to advance the use of electronic patient-reported outcomes in clinical care: a report from amia workshop proceedings. JAMIA Open **2**(4), 407–410 (2019)
3. Benson, T., Grieve, G.: Principles of Health Interoperability: SNOMED CT, HL7 and FHIR. Springer, Heidelberg (2016). https://doi.org/10.1007/978-3-319-30370-3
4. Bibault, J.-E., Chaix, B., Nectoux, P., Pienkowski, A., Guillemasé, A., Brouard, B.: Healthcare ex machina: are conversational agents ready for prime time in oncology? Clin. Transl. Radiat. Oncology **16**, 55–59 (2019)
5. Boyce, M.B., Browne, J.P., Greenhalgh, J.: The experiences of professionals with using information from patient-reported outcome measures to improve the quality of healthcare: a systematic review of qualitative research. BMJ Qual. Saf. **23**(6), 508–518 (2014)
6. Cheung, Y.T., et al.: The use of patient-reported outcomes in routine cancer care: preliminary insights from a multinational scoping survey of oncology practitioners. Support. Care Cancer **30**(2), 1427–1439 (2022)
7. Ciani, O., Federici, C.B.: Value lies in the eye of the patients: the why, what, and how of patient-reported outcomes measures. Clin. Therap. **42**(1), 25–33 (2020)
8. Eton, D.T., et al.: Harmonizing and consolidating the measurement of patient-reported information at health care institutions: a position statement of the mayo clinic. Pat. Related Outcome Meas. **5**, 7 (2014)
9. Fadhil, A., Gabrielli, S.: Addressing challenges in promoting healthy lifestyles: the al-chatbot approach. In: Proceedings of the 11th EAI International Conference on Pervasive Computing Technologies for Healthcare, pp. 261–265 (2017)
10. Jensen, R.E., Gummerson, S.P., Chung, A.E.: Overview of patient-facing systems in patient-reported outcomes collection: focus and design in cancer care. J. Oncol. Pract. **12**(10), 873 (2016)
11. Jin, J., Sklar, G.E., Oh, V.M.S., Li, S.C.: Factors affecting therapeutic compliance: a review from the patient's perspective. Therap. Clin. Risk Manag. **4**(1), 269 (2008)
12. Kadariya, D., Venkataramanan, R., Yip, H.Y., Kalra, M., Thirunarayanan, K., Sheth, A.: KBOT: knowledge-enabled personalized chatbot for asthma self-management. In: 2019 IEEE International Conference on Smart Computing (SMARTCOMP), pp. 138–143. IEEE (2019)

13. Lee, D.H.: A model for designing healthcare service based on the patient experience. Int. J. Healthcare Manag. **12**(3), 180–188 (2019)
14. Lucas, G.M., Gratch, J., King, A., Morency, L.P.: It's only a computer: virtual humans increase willingness to disclose. Comput. Human Behav. **37**, 94–100 (2014)
15. Nguyen, H., Butow, P., Dhillon, H., Sundaresan, P.: A review of the barriers to using patient-reported outcomes (pros) and patient-reported outcome measures (proms) in routine cancer care. J. Med. Radiat. Sci. **68**(2), 186–195 (2021)
16. Ni, L., Lu, C., Liu, N., Liu, J.: MANDY: towards a smart primary care chatbot application. In: Chen, J., Theeramunkong, T., Supnithi, T., Tang, X. (eds.) KSS 2017. CCIS, vol. 780, pp. 38–52. Springer, Singapore (2017). https://doi.org/10.1007/978-981-10-6989-5_4
17. Seitz, L., Bekmeier-Feuerhahn, S., Gohil, K.: Can we trust a chatbot like a physician? a qualitative study on understanding the emergence of trust toward diagnostic chatbots. Int. J. Human-Comput. Stud. **165**, 102848 (2022)
18. Snyder, C., Wu, A.W.: Users' Guide to Integrating Patient-Reported Outcomes in Electronic Health Records. John Hopkins University, Baltimore (2017)
19. Zahour, O., El Habib Benlahmar, A.E., Ouchra, H., Hourrane, O.: Towards a chatbot for educational and vocational guidance in morocco: chatbot e-orientation. Int. J. **9**(2) (2020)

Realtime Visualization System of Various Road State Sensing Data in Winter Season

Yoshitaka Shibata[1]([⊠]), Akira Sakuraba[2], and Yasushi Bansho[3]

[1] Iwate Prefectual University, 152-89 Sugo, Takizawa, Iwate, Japan
shibata@iwate-pu.ac.jp
[2] Tokyo Metropolitan Industrial Technology Research Institute, Tokyo, Japan
sakuraba.akira@iri-tokyo.jp
[3] Holonic Systems, Ltd, Kamihirasawa 7-3, Shiwa 028-3411, Iwate, Japan
bansho@holonic-systems.com

Abstract. In order to maintain safe and secure driving, it is very important to correctly and quickly identify the road state on any time any locations. In this paper, we introduce a new road state viewing system using multiple on-board environmental sensors in winter season. The sensor data are processed by on-board edge computer to decide and show the road state in realtime. We build a prototype system and test the actual winter road to evaluate the accuracy and performance. As results, higher road state decision accuracy and faster viewing can be realized.

1 Introduction

According to the Unite Nation, the world population exceeded more than 8 billion in 2022 and will be expected 12 billion within 2030 year. For this reason, foods and energy problems are becoming serious around the word as the carbon emission problem is progressed. On the other hand, the aging and declining birthrate problems are serious in developed countries. Japan is not exception [1, 2] so that the scale of social and economic activities are shrinking where the most of younger people move and live in the larger cities, while the old people have to stay in the rural areas. Eventually public transportation services and road infrastructure in rural areas getting worse, and those people have to drive by themselves to continue their daily lives. However, the traffic accidents by the old people increase on bad road condition year and year and become serious problem. Particularly, as shown in Fig. 1, in northern part of countries, most of the road surfaces are occupied with heavy snow and iced surface in winter and many slip accidents occurred even though the vehicles attach snow specific tires. Therefore, safer and more liable mobile system which is effective to drive on worse and dangerous road infrastructure is required.

On the other hand, recently autonomous driving and connected car technologies cars have progressed and practically been running on highways and the dedicated roads in the world. For those operations of autonomous driving on ordinal or general road, the traffic rule is being revised year and year. In fact, the autonomous driving service by

L. Barolli (Ed.): AINA 2023, LNNS 661, pp. 149–156, 2023.
https://doi.org/10.1007/978-3-031-29056-5_15

revel 3 and 4 are realized in various countries. However, the infrastructures environment on general roads are not well maintained compared with highways and the dedicated roads for such cases where falling objects from other vehicles and overloaded trucks are frequently occurred. Particularly, the road states in rural areas and snow areas are even worse. Therefore, in order to maintain safe and reliable autonomous driving, the vehicles have to detect and avoid those obstacles in advance when they pass through.

Fig. 1. Road condition in snow country

So far, for those road infrastructure problems, we have designed and implemented a wide area road state platform based on crowd sensing V2X communication technologies [3, 4]. Many data from various on-board environmental sensors including 3-axis accelerator, road surface temperature sensor, humid sensor, RGB camera and GPS data are sampled and the road states are identified [5–7]. Those sensor data and the road states in time and location are shared with other vehicles and road side edge server by V2X communication network which is organized by N wavelength wireless cognitive network [8–10]. However, though the shared data were reasonable qualified results, those are not visualized. Therefore, it is required to visualize and understand the road surface states on the GIS at every moment according to the weather, e.g. dry, wet, snow, freezing in realtime.

In this research, in order to improve the previous road state platform, particularly for winter season, we introduce a new realtime visualization system of road state decision method using various new and high accuracy and sampled environmental sensors. As environmental sensors, six-axis dynamic sensors, road surface temperature sensor, near infrared laser sensors and GPS are used to determine the correct road state on every time and locations. As results, the sampled data and identified road states are visualized on time axis line and GIS in two dimensional GIS on Web based browser.

In the following, general system and architecture of Road Surface State Information Platform are explained in Sect. 2. Next, the sensing system and its functions with various sensors are precisely shown in Sect. 3. The prototype system and visualization system to evaluate function and performance of the proposed sensing system is explained in Sect. 4. Finally, conclusion and future works are summarized in Sect. 5.

2   Road Surface State Information Platform

In this section, we introduce a new wide area road surface state information platform based on crowd sensing and V2X technologies as shown in Fig. 2 [3, 4]. The wide area road surface state information platform mainly is organized by multiple road side wireless nodes so called Smart Relay Shelters (SRS) and Gateways located along the roads, and mobile nodes as vehicles, so called Smart Mobile Box (SMB). Each SMB is furthermore consisted of a sensor information part and communication network part. The sensor information part includes various sensor devices and edge computer as sensor server. Using those sensor devices, various road surface states can be quantitatively identified.

On the other hand, the communication network part is organized by multiple wireless network devices with different N-wavelength wireless networks such as 0.92 GHz, 2.4 GHz, 5.6 GHz in addition to sub 6 GHz, 28 GHz by 5G network. Those wireless network devices are integrated as a cognitive wireless node. 0.92 GHz wireless devices is used as control link to establish a connection while the other wireless devices are used to deliver the actual sensor data as data links. The 0.92 GHz control node selects the best link of cognitive wireless network depending on the possible data links which observe the network quality by Software Defined Network (SDN). If none of data link connection is existed, those sensing data are locally stored at mobile node server until the mobile node approaches to another mobile node or road side node. If the mobile node approaches to another mobile node or road side node, the mobile node starts to transmit sensor data by DTN Protocol. The collected sensor data and road state information at road side node are transmitted to the cloud system through the Gateway. Thus, data communication between nodes can be attained even though the network infrastructure is not existed in challenged network environment such as mountain areas or just after large scale disaster areas and all of the sensor data are widely collected in cloud system.

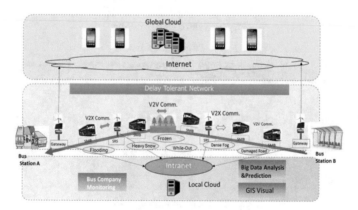

Fig. 2. Road surface state information platform

3   Sensing System with Various Sensors

Sensing system as SMB is installed on various mobile nodes and to sense quantitative road surface states, such as roughness coefficient and grip coefficient to identify and the qualitative road surface states, such dry, wet, dumpy, showy, frozen roads. The sensing system is organized by a sensor server and various sensing devices including accelerator, gyro sensor, infrared temperature sensor, humidity sensor, quasi electrical static sensor, camera and GPS as shown in Fig. 3. The sensor server periodically samples those sensor signals, performs signal filtering and analog/digital conversion and in Receiver module, and analyzes the sensor data in Analyzer module to quantitatively determine the road surface state and learning from the sensor data in AI module to classify the road surface state as shown in Fig. 4. Through those process, the correct road surface state can be quantitatively and qualitatively decided. The decision data with road surface condition in SMB are temporally stored in Regional Road Condition Data module and mutually exchanged when the SMB on one vehicle approaches to other SMB. Thus, by V2V and V2R communication, the both SMBs can obtain the most recent road surface state data with just forward road. By the same way, the SMB can also exchange and obtain the forward road surface data from road side SRS.

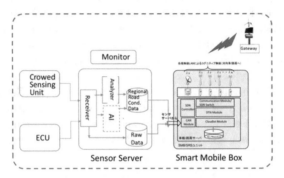

Fig. 3. Sensor server system

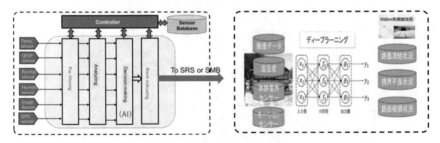

Fig. 4. Analyzer module by AI

4   Prototype System and Evaluation

In order to verify the effectiveness of proposed realtime visual system by various road state sensing, we build a prototype system and evaluate those functional and performance as shown in Fig. 5. The prototype system of consisted of 6-axis EQUAUS, Li-DAR by Hokuyo UXM-30LXH-EHA, VISARA MD-30 and GPS-M2 as environmental sensor, C-CR1000X as data logger and Lenovo PC as Sensor Processing server. The 6-axis EQUAUS is used observe roughness of road surface by changing the sampling rate from 50–2000 Hz. LiDAR is used to identify the obstacles ahead of vehicle. VISARA MD-30 is furthermore organized by a near infrared laser sensor, infrared temperature and humidity sensors to decide the road states. GPS-M2 GPS is used to precisely identify the locations and time of the running vehicle. The data logger C-CR1000X is used to temporally store the sensor data from the near infrared laser sensor. The processing server, Lenovo PC is used to process and analyze the sensor data by AI based road state decision and display the output of analyzed results. As field experiment, this road state sensing system is installed on the car which runs on the road while sensing the road state and visualizing the results in Morioka city which is located in northern snow part in Japan.

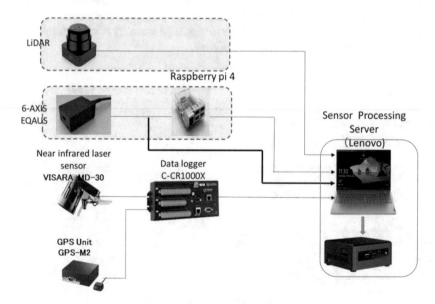

Fig. 5. Prototype of road state sensing system

In order to verify the effects of our proposed sensing system, an experimental prototype is constructed and its performance is evaluated. The six-axis acceleration sensor, the EQUAS is fixed on the front dashboard while the VISARA MD-30 is fixed to the front of car as shown in Fig. 6 and connected to the sensor processing server which is set to the inside car. The experiment is executed on the snow road in Morioka city which is located in northern part of Japan. The car is running on the road at average 40 km. The

Fig. 6. Road state sensor installation

output from the sensor server with 6 axis accelerator and roughness and road surface state are calculated on every 0.1 s and averaged in 1 s. and visualized on the monitor in realtime. Figure 7 shows the road surface state on every 1 s. on GIS map on the right side while the road surface temperature, air temperature, humidity, water, ice and snow thicknesses in time are indicated on the left side. On the other hand, Fig. 8 shows the roughness on GIS map on the right side while 6 axis acceleration and roughness rate in time are indicated on the left side. In this case the sampling rate of accelerator is 1 kHz. Thus, our proposed system can calculate and visualize not only road surface state but roughness rate with higher resolution in realtime.

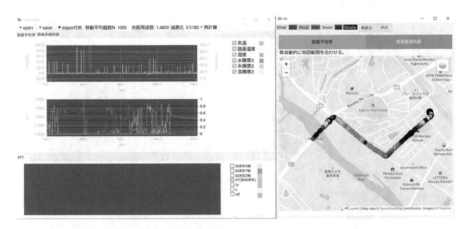

Fig. 7. Realtime road surface state and sensor data

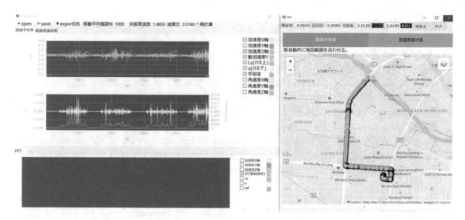

Fig. 8. Acceleration and roughness at sampling rate 1,000 Hz

5   Conclusion and Future Works

In this paper, we proposed a new road state visualizing system for road surface conditions using various sensor data. Particularly, we designed and implemented on-board visualizing system using relatively compact and cost-effective sensor devices, in which acceleration and environmental sensor are performed to monitor road surface in realtime. We actually constructed the experimental prototype to evaluate our system. It is clear that this six-axis sensor device and road state system can provide realtime and effective visual ability with high quality on real running road surface.

Now, we are analyzing the accuracy of the friction rate and roughness by comparing the results from our decision method and the actual measurement of road surface. As future works, we are going to continue the field experiment by considering surface temperature, humidity and reflection/absorption rates for more general environmental conditions such as dry, rainy, snowy and icy conditions. In addition, we will investigate a study of better road decision algorithm, e.g. the rule-based fuzzy algorithm, nonlinear time series analysis, and deep learning, to analyze and decide road surface conditions more precisely.

Acknowledgments. The research was supported by Japan Keiba Association Grant Numbers 2021M-198, JSPS KAKENHI Grant Numbers JP 20K11773, Strategic Information and Communications R&D Promotion Program Grant Number 181502003 by Ministry of Affairs and Communication.

References

1. A 2016 Aging Society White Paper. http://www8.cao.go.jp/kourei/whitepaper/w-014/zenbun/s1_1_1.html
2. A 2016 Decling Birthrate White Paper. http://www8.cao.go.jp/shoushi/shoushika/whitepaper/measures/english/w-2016/index.html

3. Shibata, Y., Sakuraba, A., Sato, Y., Arai, Y., Hakura, J.: Basic consideration of video applications system for tourists based on autonomous driving road information platform in snow country. In: The 15th International Conference on P2P, Parallel, Grid, Cloud and Internet Computing, (3PGCIC-2020), Tottori, Japan, pp. 348–355 (2020)
4. Shibata, Y., Arai, Y., Saito, Y., Hakura, J.: Development and evaluation of road state information platform based on various environmental sensors in snow countries. In: The 8-th International Conference on Emerging Internet, Data & Web Technologies (EIDWT-2020), Kitakyusyu, Japan, pp. 268–276 (2020)
5. Sakuraba, A., Shibata, Y., Tamura, T.: Evaluation of performance on LPWA network realizes for multi-wavelength cognitive V2X wireless system. In: The 10th IEEE International Conference on Awareness Science and Technology, (iCAST2019), Morioka, Japan, pp. 434–440 (2019)
6. Sato, G., Sakuraba, A., Uchida, N., Shibata, Y.: A new road state information platform based on crowed sensing on challenged network environments. Internet Things 18, 100214 (2022)
7. Shibata, Y., Sakuraba, A., Sato, G., Uchida, N.: Realtime road state decision system based on multiple sensors and AI technologies. In: The 13th International Conference on Complex, Intelligent, and Software Intensive Systems, (CISIS2019), Sydney, pp. 114–122 (2019)
8. Sakuraba, A., Sato, G., Uchida, N., Shibata, Y.: Performance evaluation of improved V2X wireless communication based on Gigabit WLAN. In: BWCCA-2020, Tottori, Japan, pp. 131–142 (2020)
9. Sakuraba, A., Sato, G., Uchida, N., Shibata, Y.: Field experiment on cognitive wireless V2X communication over high-speed WLAN. In: The 23rd International Conference on Network-Based Information Systems, (NBiS-2020), Victoria, Canada, pp. 450–460 (2020)
10. Shibata, Y., Sakuraba, A., Sato, G., Uchida, N.: A prototype system of social experiment for actual road state information platform based on sensors and V2X communication. In: The 14-th International Conference on Complex, Intelligent, and Software Intensive Systems, (CISIS-2020), Lodz, Poland, pp. 108–119 (2020)

An Architecture Proposal to Support E-Healthcare Notifications

Wagno Leão Sergio, Gabriel di Iorio Silva, Victor Ströele$^{(\boxtimes)}$, and Mario A. R. Dantas

Institute of Exact Sciences, Federal University of Juiz de Fora (UFJF), Rua José Lourenço Kelmer, Juiz de Fora, MG, Brazil
{wagno.leao.sergio,iorio,victor.stroele,mario.dantas}@ice.ufjf.br

Abstract. By constantly focusing on the demands of everyday life, people easily put their health in second place. Such behavior can lead to both physical and psychological risks. This work proposes an architecture capable of sending users notifications regarding their health, thus helping them become aware of their habits. The research was designed to develop a system based on the Fog-Cloud paradigm and machine learning algorithms. The system collects the user's heart rate data and processes it to generate notifications regarding unusual heart rate frequency. A feasibility experiment was conducted for three months, collecting real data from the user. The generated results showed the moments in the day when the user had more instability, thus indicating possible moments for a notification to be sent, demonstrating the proposal's viability.

1 Introduction

As the society we live in grows increasingly larger over time, we can notice that the demand for results over individuals is intensifying. The population's needs are becoming voluminous and complex, making new solutions and products have to reach higher quality, speed, and satisfaction levels. This results in ever-increasing expectations about the degree of productivity of those involved who work in the most varied sectors and institutions. Such pressure caused by these factors can result in long working hours for the individual and the adoption of habits and behaviors that can cause health problems.

Recent research has already warned about the dangers of people's lack of awareness regarding their health [3,23]. It can be seen that the structures of companies and institutions are also being gradually updated to reduce such negative impacts on the people involved. Also, it has been observed that through self-awareness, the individual can improve his condition and adopt healthier daily practices [25]. When analyzing possible efficient approaches to the issues presented, it is possible to observe an increasing adoption of techniques to create tools to support people. In this way, they can identify possible problems, obtain relevant information about their health and behavior, and be assisted in making decisions. Such tools are usually denoted as e-health applications [8], which continuously monitor users and collect data to generate meaningful responses.

L. Barolli (Ed.): AINA 2023, LNNS 661, pp. 157–170, 2023.
https://doi.org/10.1007/978-3-031-29056-5_16

However, it is also important to note that a viable implementation of such applications requires high-quality standards. In our previous researches [13,14], we observed that the application must have high service availability, efficient processing of the data collected to generate the response, and secure access to user information that may be sensitive. Thus, it is necessary to consider several measures when developing applications in this specific area, in addition to modeling an architecture that can support the high volume of data exchange and processing required to make the necessary support resources available to the user.

With the availability of a tool that helps investigate their habits, users have more opportunities to become aware of their well-being and correct their daily routine to a healthier one. There are several different e-health applications, most of them focusing on a specific task related to monitoring the health of individuals [15]. In most cases, sensors are used as sources to collect the necessary data about the users, and then some processing is performed to achieve relevant results for the individuals concerned. These analyses clarify or warn about the health situation of the person in question.

An important factor rarely considered in modeling e-health applications is how to deliver personalized analyses to the monitored user. In most of the applications studied, the information generated was used to support the decision-making of healthcare professionals [15]. However, approaches where the user receives notifications about his activities, may help him become more aware of his health. The lack of attention and care regarding health caused by long working hours or daily physical or mentally unstable activities can cause an individual to acquire behaviors detrimental to their well-being. With this in mind, solutions such as e-health applications are created to address such issues and provide the necessary support in monitoring the user's health. However, solutions that require high scalability and low response time need to adopt new architectures that support the volume of constantly generated data [10].

In addition, we observed that most systems in this category do not present the information acquired during the monitoring in a personalized and relevant way to the user, taking into account only the clinical aspects of the individual. Also analyzed were the various aspects that can bring a negative experience to the user during the use of monitoring systems, as well as data collection methods that are applied and need to offer freedom, comfort, and safety to those being monitored. With each of these points in mind, the research question of this paper is: *How to make use of technologies to help people improve their health?*

This work proposes and develops a computer system to support individuals who need information about their health, focusing on observing and analyzing their daily activities. The specific objectives of this research involve several work steps. The first involves designing a model architecture for data collection, processing, and presentation. The next is to study and select technologies available for collecting individual health data. Then, we aim to structure a model for processing the obtained data and transforming it into relevant information. The following objective is to implement a service that presents the generated analyses to the user. Then we have to perform the implementation of an architecture that is capable of making the operation of the proposed system functional. Finally, a feasibility study is conducted to evaluate the performance and efficiency of the proposed solution.

The article is organized as follows: In Sect. 2, the related works are discussed; in Sect. 3, the functioning and specificities of the proposed architecture are presented; in Sect. 4, the experimental results to validate the proposed architecture are mentioned; and in Sect. 5 we have the conclusions and future work.

2   Background and Related Work

This section presents the main concepts for understanding the development of everyday behavior detection architecture. First, concepts related to our proposed application area are presented: e-health services, ubiquitous computing, and Fog-Cloud architectures. Later, the entire research process is detailed about using different types of sensors to obtain the data. Finally, the techniques used to store and process the collected data are described, including the explanation of Lambda architecture, Data Lake, and Machine Learning algorithms.

Monitoring and support systems for people's health are applications that have been emerging over the years and gaining space both in academic and industry areas [12]. However, some difficulties may occur when incorporating e-health applications into people's daily lives. For example, in [21], it is pointed out that consumers of an e-health application must be familiar with the use of technology to follow the configuration instructions so that they can efficiently use the functionalities made available by the tool.

Despite the statements made, our research focuses on presenting informative analyses that help healthcare professionals in decision-making and a larger set of individuals who lack medical and technical knowledge. Thus, it is essential to develop methods of presenting data that meet both target users' needs.

When thinking about systems that need to process and transfer data between elements of an infrastructure continuously, it is first necessary to consider some important factors. First, most sensors that obtain user signals must have the computing power required to produce the desired information. Furthermore, when dealing with a large number of users using the system, considering that they all have a set of sensors that are taking measurements all the time, it is essential to design a structure capable of storing the volume of data produced [5,24]. Thus, it is necessary to model one or more components responsible for receiving the data collected by the sensors, storing them, and then performing the appropriate computation techniques.

The literature brings several approaches to solving this problem considering Cloud, Fog, and Edge computing [4,20]. By making a comparison between these different approaches, we can reflect on their characteristics [11]. Firstly, we can see that the main difference between them is the network's degree of distribution of computing power. When analyzing Cloud computing, for example, it is noted that there can be problems involving data congestion and an increase in system response time. On the other hand, when focusing on Edge computing, we may encounter difficulties involving connection availability between devices and data synchronization, which can make the system unstable and unreliable. Thus, in this paper, we adopted a hybrid approach to incorporate the best of the paradigms presented.

We considered two points central to our research while identifying works related to this paper: non-invasive and real-time monitoring. The authors in [26] review the different types of ambient sensors used for monitoring the elderly daily activities. It is further pointed out that ambient sensors provide greater independence and comfort for the individuals involved. Nevertheless, it is noted that the target user of the work may differ from what is presented in the article, leading to possible frequent absences from measurements involving individuals who spend a lot of time away from the monitored environment.

Looking from another perspective, the study presented in [22] shows a systematic review of health monitoring systems that use wearable sensors and compares different implementations. The authors analyzed the characteristics of each approach, such as sensor types, processing techniques, and portability. They concluded that wearable sensors might provide better measurement quality and autonomy. However, it is also identified by the authors that there are still some obstacles that must be overcome in the evaluated approaches, mainly interoperability, security, and efficiency. Considering the context in which our research is applied, we decided that, for a better representation of individuals in their daily activities, the use of wearable sensors is a more feasible strategy at the time this research is being developed.

The Data Lake concept, as explained in [19], is a highly scalable data storage repository that holds a vast set of data in its most raw form until processing systems require it. Data Lakes must possess certain properties, such as continuous availability, quality of data indexing, and high traceability of the operations performed. These features proved adequate for registering the measurements made by the sensors of all the users, causing this concept to be used for modeling the proposed architecture.

The lambda architecture generates more availability and reduces the response time of systems that have a time delay between collecting and processing data and its presentation to end users [16]. It is composed of batch and real-time processing. The batch processing stream handles all high-cost computation for the system and stores the results generated for future queries. On the other hand, the real-time processing stream makes already processed analyses available as quickly as possible, ensuring a low response time. Through observing experiments from previous works on implementing this architecture [13,14,18], we concluded that such a framework is viable and fits as a solution for developing the research in progress.

Now that the concepts of the previous topics have been presented, we can focus on one of the most important questions of the problem being worked on: How to use the data collected from sensors to generate knowledge about user behavior? When studying the literature related to this question, we found

proposals involving the use of Machine Learning algorithms for pattern identification and data classification [14]. The idea of a Machine Learning algorithm,
explained in [2], consists of a computational process that uses specified input data
to accomplish a given task that has not been formally programmed to produce a
particular result. In [6], for example, experiments are conducted comparing different Machine Learning algorithms to classify stress signals in individuals. The
research results show that this strategy is feasible and can produce satisfactory
results with the right data selection and algorithms.

3   System Architecture Proposal

In this paper, the implementation of a system to monitor users in real-time and
notify them of possible changes in their health conditions is presented. In order
to develop such a tool capable of monitoring people's activities to improve their
mental and physical health, we decided to collect their data in different settings
and backgrounds. With these aspects in mind, we collected the body data to
create a model that is able to detect the moments in which there is instability
in its behavior in order to warn it of such a situation.

Considering the large volume of data generated continuously by users and
the complexity in the computation flow of the data, we perceived the advantages of using the Fog-Cloud paradigm for data acquisition and processing. This
paradigm reduces the possibility of network bottlenecks, bringing the processing closer to the data extraction nodes and enhancing the overall responsiveness
of the system [7]. The data generated by the sensors are produced by an edge
device and sent to a Fog processing node [11].

It is assumed that both the edge and fog nodes have the required computing
power to handle the data that is constantly being sent and received. Furthermore,
the system design was also based on the concepts of the Lambda architecture,
where the received data is sent simultaneously to two parallel workflows, where
both batch data processing and streaming routine are implemented in a hybrid
way. Finally, the results generated by this process are sent to the cloud to be
stored and used later. Figure 1 provides an overview of the proposed architecture
with its layers and workflow.

The computational flow performed by the presented system starts with the
data extraction from the sensors of one or more users simultaneously. For such
a task, the **extraction layer** has this responsibility, being composed of sensors
from each user that gather the information required for monitoring.

After collection, the data is sent to the Fog node to be pre-processed. When
this data arrives at the Fog node, it is processed in parallel, and the data is
previously stored in a storage service that applies the same principles as a *Data
Lake* [19]. We have adopted a Data Lake to allow storing a large amount of data
at high speed to provide continuous availability of its contents. When performing
research on similar works, we decided to use the Lambda architecture because it
shows satisfactory outcomes in resolving problems concerning both performance
and scalability. As mentioned in [16], the lambda architecture is centered on providing high scalability and having a low delay in both data queries and updates,

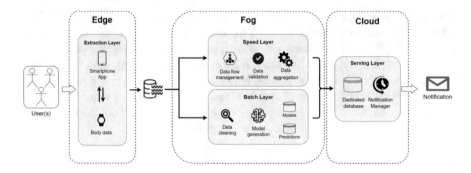

Fig. 1. Proposed architecture with the main components and workflow.

also being able to deal with either human or hardware errors. The presentation of this architecture details a distributed structure of three main layers: the batch layer, the speed layer, and the service layer.

Following the concepts outlined in the Lambda architecture, the **batch layer** performs all the high-cost computing in the system so that the output results produced are stored and are always up to date and available to the other elements. This is done to guarantee that the system will not be slowed down, especially by a delay in response due to possible processing overhead. The batch layer also enables us to have linear scalability of the data capacity that the system can support. After this process, we have a securely stored collection of processed and enriched data that users can access at any time at a low compute cost. For the proposed system, the batch layer takes the body data from a specific user and updates its classification model through machine learning training and evaluation methods to provide increasingly accurate predictions about the individual's state. The generated information is stored in a dedicated database. The model generated using the new data is also stored to ensure the system's tolerance to possible prediction errors or even hardware malfunctions.

When new data is being treated in the batch layer, the same data set is also sent to another stream, a component named **speed layer**. The primary function of the speed layer is to reduce the response time to system requests since it deals exclusively with the most recently sent data from the extraction layer. The speed layer is developed with only low-latency queries and compute functionalities to ensure that monitoring users have real-time information. It can also have the ability to use the enriched data produced by the batch layer. The responsibility of the speed layer in the proposed system is to secure the authentication and validation of the received data while also ensuring the consistency of the data flow management. In addition, this layer performs update procedures at periodic intervals, recovering the data stored in the Data Lake to verify that a user's most recent entries have been changed.

On completion of this processing flow, the produced outputs are sent to the cloud in the **serving layer**. The main task of the service layer is to make the generated information available, present it to end-users, and ensure the data is constantly updated. An automatic notification service has been implemented to perform this task. The service sends messages to users via their email at times when instabilities in their vital signs have been detected, warning them of possible risks to their health. The best times to send messages are estimated from the data collected and are regularly updated as the individual is monitored. At the end of this entire processing flow, the system is intended to provide the appropriate tools to contribute to the decision-making process of monitoring the activities of individuals in real time in a secure and cost-effective approach.

The Speed and Batch components have been configured on the Fog node. We decided on this configuration because the Batch requires more processing time and a higher volume of data access, resulting in latency problems and processing costs if configured in the Cloud. The Speed needs to process the new data quickly, using the model trained and stored by the Batch. So, we chose to configure it on the Fog node too.

As the Serving component is responsible for presenting the data to end-users, we chose to configure it in the Cloud. We believe that in this way, we guarantee greater security of raw data and give access to only permitted information. Also, it facilitates the use of data analysis and query tools.

As can be observed, such architecture requires the modeling and implementation different components that need to communicate with each other. The main idea of this work is, through the proposal presented, to develop a system capable of providing useful information about the individual's behavior throughout the day. Therefore, both the implementation of the system and the carrying out of experiments have as a priority goal to prove the viability of introducing such a project in a real environment. We can then demonstrate, based on data collected from a user throughout his routine, that the system developed according to the proposed architecture is capable of generating appropriate and useful results to identify his behavior and, consequently, possible irregularities in his health.

4   Experimental Results

Given the complexity and scope of data handling, we must consider different scenarios to carry out the proposed research. When dealing with data from corporal sensors, we must collect them to analyze the feasibility of methods to reach a viable solution to the problem at hand. Hence, it is necessary to carefully study the captured signals used as input variables for the machine learning model. As such, we must analyze each of them for possible correlations, missing data, patterns, and intrinsic insights about the data in question [17].

4.1   System Deployment

Figure 2 shows the main technologies used to implement the architecture. Smartwatches designed for physical activities were selected as equipment to collect the vital signs of users. The model used in question has the functionality to connect to a smartphone application, specifically for the android platform. When connected, the device can send data related to several vital signs of the user during the day, such as the number of steps, estimated calories burned, heart rate, blood pressure, and blood oxygenation. It is important to note that despite the device's capabilities to take various measurements, only heart rate was used for this research to maintain complexity criteria. Furthermore, it is necessary to point out that, as an *Edge* device, the smartwatch does not have significant storage capabilities, making the data measured by the sensors have persistence in the device for at most one day. Thus, the role of long-term data logging is transferred to the smartphone application and the rest of the system's processing nodes.

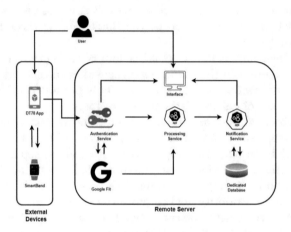

Fig. 2. Infrastructure deployed to implement the architecture proposed.

The smartwatch model used for the research has, by default, an android application for managing data collection. However, the application in question was unable to provide the necessary functionality for the architecture proposed, besides having some instability issues. That said, it was concluded that it was necessary to use a different application to meet the research needs. After researching similar applications, an open-source project was found that implements an alternative application capable of connecting to the smartwatch, named *DT78-App-Android* [9]. Besides offering greater stability of operation, this application gives us the opportunity to develop the updates necessary for the research due to its open-source nature. In addition, the application itself has features for data storage, such as periodic automatic backups. The main modifications made to the application were concerning the time interval between measurements of the device, which were adjusted to occur every 5 min, and in the implementation of sending the collected data via *HTTP* requests to a server that works as a *Fog* node. In this way, we have an infrastructure capable of collecting vital signs in real-time, ensuring control and consistency of the measurements through the modified application, and making a remote connection with the rest of the system.

With the users' vital signs being properly managed by the Edge layer, it is now necessary for this data to be sent to the *Fog* layer so that it can be formatted and processed. The service responsible for this task has been implemented as a micro-service in the form of a *Web API*. The main responsibilities of this service are to receive the data that is sent by the application and store it in the *Data Lake*. The service was implemented using the *Express JS* framework, with the addition of a *SQLite3* database for storing the times when measurements were sent, for data indexing quality issues.

To create an architecture that provides a satisfactory level of security, an authentication flow that makes use of *Google's* authentication system was implemented. First, each new user is inserted into the system by registering through a web application that performs a *Google* authentication login. Upon logging in, *Google* services generate a temporary access code for the user, which is stored by the system for future operations. As the system works, the access codes are updated at periodic intervals so that the system can connect to *Google* services automatically. It is important to note that there is a time limit to how often these access code updates can be made, thus ensuring that the user is required to log into the system on a recurring basis.

Now that valid credentials have been provided by the user, the system can then connect to the *Google* services, specifically to the *API* that is provided to manage health data from *Google Fit*. For quality architecture reasons regarding data availability, protection, and integrity, the *Google Fit* services were used as *Data Lake*, storing the record of the collected measurements in so-called *Data-Sources*. According to the *Google Fit API* documentation [1], a *DataSource* represents a single source of data from one or more sensors and is used to store health data. Every time a new user is registered in the system, a *DataSource* for their vital signs is required to be created. This way, every time the measured data is sent to the authentication service, it is inserted in the user's respective *DataSource* so that it can be queried later.

For the service responsible for processing the collected data, a *Web API* was implemented using the *Flask* framework so that both the received data and the trained prediction model can be controlled remotely. Measurements can be recorded via *HTTP requests*, and then added to the service's database. In addition, a feature to generate predictions about a given input sent via a request has been implemented. This way, information about the user's behavior can be obtained quickly, reducing the system's response time. However, it is necessary to manually train the machine learning model in the Batch layer for the service to work. This is done primarily because we can avoid an increased processing load on the service. The model can then be loaded into the service and used to generate updated predictions in the Speed layer.

Finally, we have performed as a serving layer in the architecture an automated notification service. At the end of processing the user's data, the notification service receives information detailing the moments of the day when there is an instability in the individual's vital signs. The notification service was implemented as a *Web API* using the *FastAPI* framework. The analyses sent to the service are sent via *HTTP* requests so that they can be validated and stored in a dedicated database. In addition, when a new notification request is received, the service uses a scheduling routine to schedule a notification for a specific user. Thus, throughout the system's operation, the service keeps jobs running for each scheduled notification. The message is sent to the user when the scheduled notification occurs during the day. As there is also the persistence of the schedules that must be done, if the system is restarted for some reason, the notification service will still be able to perform the previously scheduled notifications.

4.2   Model Experiments

The experiment conducted to test the feasibility of the proposal was performed using data collected from a volunteer user over a period of about 3 months (06/09/2022–06/11/2022). The user's vital sign selected to demonstrate the collection, processing, and presentation of the analyses was the heart rate. A graph of the monitored individual's heart rate behavior on a specific day is shown in Fig. 3.

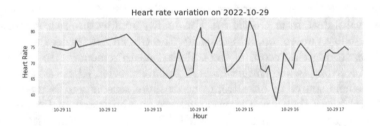

Fig. 3. Heart rate of the volunteer along the day 2022-10-29.

The data collected throughout the monitoring went through a process of validation, cleaning, and formatting to be used in the proposed model. First, all the records in which heart rate was not measured were removed. Then, the data were transformed into a format based on the beginning and end of the measurements that were taken, and within these intervals the maximum and minimum heart rates were calculated. Figure 4 shows the distribution of the heart rate values of the individual that was monitored. We can observe that most of the samples are around 70 beats per minute (bpm). The samples have a high variation within the 60 to 80 bpm range, and there is also a low variability between the values of 100 to 120 bpm.

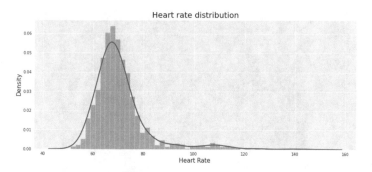

Fig. 4. Distribution of the heart rate processed from volunteer.

By standardizing the data, we have calculated new variables of heart rate increase and the absolute range between maximum and minimum heart rate in the time interval of the recording data. The increase is measured as the difference between two sequential heart rate recordings. With the properly formatted dataset, we can train a model to help identify instability in the individual's heart rate.

The machine learning model built was designed in such a way that the system would produce, over time, a pattern of behavior of the monitored user. In this way, when the system received a heart rate that differed considerably from the prediction made by the trained model, an alert routine would be started, sending a notification to the user, or users, responsible. The data used as input for training the model are the minimum and maximum, the absolute interval between the minimum and maximum, and the increase in heart rate. The value expected to be the model's output is the heart rate that the individual should have at that moment. Part of the data set was separated into a training set, a test set, and a validation set. In all, 1,160 monitoring records were used to construct the sets.

Tests were conducted using different regression models, such as *Decision Tree Regressor*, *K-Nearest Neighboor Regressor*, and *Support Vector Machine Regressor*. The algorithms were evaluated using the R-square score, with their values being 0.97204, 0.98256, and 0.9734, respectively. Based on this, the *K-Nearest Neighboor* regression algorithm was selected to work on the system. Having a properly trained prediction model, we can now use the validation set to analyze the model's behavior. For the validation set, the last 24 h of measurements were selected. The graph in Fig. 5 shows two heart rate curves: one being the actual data collected and the other with the model predictions. The marks on the graph indicate the moments where there was a difference greater than 5% between the two curves.

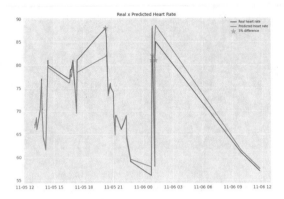

Fig. 5. Comparison between real and predicted heart rate on the last 24 h of monitoring of the user. There are also markers indicating a divergence of 5% or more on the prediction.

As can be seen, there were times when the heart rate received was much more different than that predicted by the model. For example, between the hours of 7 pm and 8 pm, the heart rate received was much higher than that predicted by the model. Such measurements could indicate periods in which the user has heart rate instability, and identify possible beginnings of mental or physical stress. The system then makes a record of the moments in which these instabilities occurred, and schedules notifications to sent to the user at these times of the day.

5   Conclusions and Future Works

This paper proposes an architecture for user monitoring to help them to know themselves better and highlight stressful moments. The architecture was deployed using free technologies, and an experiment was conducted to verify the proposal's feasibility. With the results obtained, it was concluded that the proposed architecture is feasible and capable of identifying moments of instability in people's daily lives through their vital signs, making the presented research a possible solution to the problems.

Some limitations were observed when performing the proposed monitoring experiment. Some instability was identified due to the used device, causing the dataset analyzed to have sparse time intervals of measurements. In addition, in this paper, the only measurement used to recognize the user's behavior was the heart rate for complexity reasons.

In future works, we plan to add the processing of other signals from the individual and data that can contextualize the situation in which he is, such as his location. This way, we could put together a richer and more relevant contextual profile of the user. In addition, there are plans to use more precise and robust equipment in the construction of the system, both in the data collection and processing areas.

There is the additional intention to perform stress experiments on the system to verify how well the proposed architecture handles a very high number of users and sensors. It is also planned to perform experiments to analyze whether the proposed architecture positively impacts the behavior of different individuals, thus improving their health.

References

1. Android api reference — google fit — google developers. https://developers.google.com/fit/android/reference
2. Alpaydin, E.: Introduction to Machine Learning. MIT press, Cambridge (2020)
3. Artazcoz, L., Cortès, I., Escribà-Agüir, V., Cascant, L., Villegas, R.: Understanding the relationship of long working hours with health status and health-related behaviours. J. Epidemiol. Commun. Health **63**(7), 521–527 (2009). https://doi.org/10.1136/jech.2008.082123
4. Cao, K., Liu, Y., Meng, G., Sun, Q.: An overview on edge computing research. IEEE Access **8**, 85714–85728 (2020). https://doi.org/10.1109/ACCESS.2020.2991734
5. Chintapalli, S., et al.: Benchmarking streaming computation engines: Storm, flink and spark streaming. In: 2016 IEEE International Parallel and Distributed Processing Symposium Workshops (IPDPSW), pp. 1789–1792. IEEE (2016)
6. Ciabattoni, L., Ferracuti, F., Longhi, S., Pepa, L., Romeo, L., Verdini, F.: Real-time mental stress detection based on smartwatch. In: 2017 IEEE International Conference on Consumer Electronics (ICCE), pp. 110–111. IEEE (2017). https://doi.org/10.1109/ICCE.2017.7889247
7. Deng, R., Lu, R., Lai, C., Luan, T.H., Liang, H.: Optimal workload allocation in fog-cloud computing toward balanced delay and power consumption. IEEE Internet Things J. **3**(6), 1171–1181 (2016). https://doi.org/10.1109/AINA.2010.187
8. Eysenbach, G., et al.: What is e-health? J. Med. Internet Res. **3**(2), e833 (2001). https://doi.org/10.2196/jmir.3.2.e20
9. Fbiego. Fbiego/dt78-app-android: Alternative app for the dt78 smartwatch. https://github.com/fbiego/DT78-App-Android
10. Firouzi, F., Farahani, B., Ibrahim, M., Chakrabarty, K.: Keynote paper: from eda to iot ehealth: promises, challenges, and solutions. IEEE Trans. Comput.-Aided Des. Integr. Circ. Syst. **37**(12), 2965–2978 (2018). https://doi.org/10.1109/TCAD.2018.2801227
11. Gomes, E., Costa, F., De Rolt, C., Plentz, P., Dantas, M.: A survey from real-time to near real-time applications in fog computing environments. In: Telecom, vol. 2, pp. 489–517. MDPI (2021). https://doi.org/10.3390/telecom2040028
12. Gravina, R., Fortino, G.: Wearable body sensor networks: state-of-the-art and research directions. IEEE Sensors J. **21**(11), 12511–12522 (2021). https://doi.org/10.1109/jsen.2020.3044447
13. Di iorio Silva, G., Sergio, W.L., Ströele, V., Dantas, M.A.R.: ASAP - academic support aid proposal for student recommendations. In: Barolli, L., Woungang, I., Enokido, T. (eds.) AINA 2021. LNNS, vol. 226, pp. 40–53. Springer, Cham (2021). https://doi.org/10.1007/978-3-030-75075-6_4
14. Di iorio Silva, G., Sergio, W.L., Ströele, V., Dantas, M.A.R.: A watchdog proposal to a personal e-health approach. In: Barolli, L., Hussain, F., Enokido, T. (eds.) AINA 2022. LNNS, vol. 450, pp. 81–94. Springer, Cham (2022). https://doi.org/10.1007/978-3-030-99587-4_8

15. Kim, H., Xie, B.: Health literacy in the ehealth era: a systematic review of the literature. Pat. Educ. Counsel. **100**(6), 1073–1082 (2017). https://doi.org/10.1016/j.pec.2017.01.015

16. Kiran, M., Murphy, P., Monga, I., Dugan, J., Baveja, S.S.: Lambda architecture for cost-effective batch and speed big data processing. In 2015 IEEE International Conference on Big Data (Big Data), pp. 2785–2792. IEEE (2015). https://doi.org/10.1109/BigData.2015.7364082

17. Klein, A., Lehner, W.: Representing data quality in sensor data streaming environments. J. Data Inf. Qual. (JDIQ) **1**(2), 1–28 (2009). https://doi.org/10.1145/1577840.1577845

18. Larcher, L., Stroele, V., Dantas, M., Bauer, M.: Event-driven framework for detecting unusual patterns in AAL environments. In: 2020 IEEE 33rd International Symposium on Computer-Based Medical Systems (CBMS). IEEE (2020). https://doi.org/10.1109/cbms49503.2020.00065

19. Miloslavskaya, N., Tolstoy, A.: Big data, fast data and data lake concepts. Procedia Comput. Sci. **88**, 300–305 (2016). https://doi.org/10.1016/j.procs.2016.07.439

20. Munir, A., Kansakar, P., Khan, S.U.: IFCIOT: integrated fog cloud IoT: a novel architectural paradigm for the future internet of things. IEEE Cons. Electron. Maga. **6**(3), 74–82 (2017). https://doi.org/10.1109/MCE.2017.2684981

21. Norman, C.D., Skinner, H.A.: ehealth literacy: essential skills for consumer health in a networked world. J. Med. Internet Res. **8**(2), e506 (2006). https://doi.org/10.2196/jmir.8.2.e9

22. Pantelopoulos, A., Bourbakis, N.G.: A survey on wearable sensor-based systems for health monitoring and prognosis. IEEE Trans. Syst. Man Cybern. Part C (Appl. Rev.) **40**(1), 1–12 (2010). https://doi.org/10.1109/TSMCC.2009.2032660

23. Prasad, B., Thakur, C.: Chronic overworking: cause extremely negative impact on health and quality of life, pp. 11–15 (2019)

24. Shahverdi, E., Awad, A., Sakr, S.: Big stream processing systems: an experimental evaluation. In: 2019 IEEE 35th International Conference on Data Engineering Workshops (ICDEW), pp. 53–60. IEEE (2019)

25. Sutton, A.: Measuring the effects of self-awareness: construction of the self-awareness outcomes questionnaire. Eur. J. Psychol. **12**(4), 645 (2016). https://doi.org/10.5964/ejop.v12i4.1178

26. Uddin, M.Z., Khaksar, W., Torresen, J.: Ambient sensors for elderly care and independent living: a survey. Sensors **18**(7) (2018). https://www.mdpi.com/1424-8220/18/7/2027, https://doi.org/10.3390/s18072027

Sensor Data Integration Using Ontologies for Event Detection

Jefferson Amará[1], Victor Ströele[1(✉)], Regina Braga[1], and Michael Bauer[2]

[1] Graduate Program in Computer Science, Federal University of Juiz de Fora
(UFJF), Juiz de Fora, Brazil
`jnamara@ice.ufjf.br`, {`victor.stroele,regina.braga`}`@ufjf.br`
[2] Department of Computer Science, University of Western Ontario (UWO),
London, Canada
`bauer@uwo.ca`

Abstract. Nowadays, computing applications operate in environments with multiple and heterogeneous data sources, such as data generated from IoT devices. Without contextual information, the information derived from these isolated data sources may cause bias, error, or a lack of correct comprehension. Data integration can help to promote a holistic view of data and support getting the most trustful meaning from the information. This work proposes an architecture in which ontologies help to provide context for data integration. Furthermore, ontologies and complex network concepts enrich context awareness and derive relations among data to identify events of interest. The approach is evaluated in a controlled experiment using real data from hydrological and hydrometric sensors. The results indicate it is possible to detect context and relate events from different data sources to new significant events through ontology and graph network analysis.

1 Introduction

In recent years, data sources, such as sensors and other devices, processed by diverse, ubiquitous, and pervasive applications have increased in number and diversity. The outbreak of IoT is an example of data diversity, heterogeneity, and distribution and where there is growth in day-to-day applications, such as retail, cities, buildings, transportation, agriculture, healthcare, environment, and energy [1].

With so many diverse data sources, an evident need is the capability to integrate these sources to promote better observations, and decision-making [2]. The use of ontologies for data integration is a well-known approach when dealing with sensor data due to its capacity to address syntactic and semantic data heterogeneity, standardization, and sharing [3].

Moreover, integrated data can provide even more meaning when considering their context [4]. According to Abowd et al. [4], context can be defined as any information used to characterize the situation of entities that are considered relevant to the interaction between a user and an application, including the user and the application themselves. Therefore, a system is context-aware if it uses

L. Barolli (Ed.): AINA 2023, LNNS 661, pp. 171–183, 2023.
https://doi.org/10.1007/978-3-031-29056-5_17

context to provide relevant information (including implicit meaning, inference, and things not explicitly said).

Considering the importance of context awareness, this work proposes an architecture to promote data integration based on context and use the integrated data to detect and correlate events before they happen. This work followed five main steps: (I) literature review; (ii) model representation definition; (iii) architecture proposal for data integration and event detection based on context; (iv) feasibility study with real sensor data; and (v) evaluation of results. We seek to answer the following Research Question (RQ): *Can the architecture enable data integration based on context to detect and correlate events?*

The literature review includes works on data integration using ontologies and event detection based on context. The architecture was proposed to enable the data flow from different sensors, event detection, and identification of relationships. Data from hydrometric and hydrological stations were used in the feasibility study to evaluate the proposal. The implementation uses Python as the coding language, NetworkX[1] and OwlReady2[2] libraries to deal with graph and ontology modeling, and Protégé[3] for ontology base construction. The first effort of implementation is available at bitbucket[4].

In addition to this introduction, the article is organized as follows: Sect. 2 presents background and related work; Sect. 3 presents the architecture for data integration and context awareness; Sect. 4 evaluates the architecture through a feasibility study; and in Sect. 5, the final considerations of the work are presented.

2 Background and Related Work

The Internet of Things (IoT) is implemented in many areas and is used by many day-to-day applications. Examples of IoT use can be found in domains such as lifestyle, retail, city, construction, transportation, agriculture, health, environment, and energy; applications exist for smart homes, smart cities, smart energy, and smart industry [1].

With the diversity of domains and applications, data heterogeneity is an intrinsic characteristic of this environment [5]. An IoT environment is characterized by streaming data represented in diverse structures or even unstructured. Sensors are one of the main data sources in this environment, and one of the challenges revolves around integrating data from multiple sources "on the fly" in real-time.

Most of the literature on "data fairness" ignores that data integration is one of the main steps to generating high-quality data [6]. Tan [7] highlights the wealth of "out-of-the-box data" and the ability to consume data from diverse sources as a challenge and an opportunity.

Data integration combines technical and business processes to turn data from different sources into valuable and meaningful information [8]. In integration

---

[1] https://networkx.org/.
[2] https://owlready2.readthedocs.io/en/v0.37/.
[3] https://protege.stanford.edu/.
[4] https://bitbucket.org/JeffersonAmara/integration-system.

approaches, heterogeneity is one of the main challenges considering different data sources. Three main types of heterogeneity must be addressed when integrating data: structural, syntactic, and semantic heterogeneity [9]. Structural heterogeneity refers to the different structures or formats of the data. Syntactic heterogeneity means that table or attribute names may differ in different data sources but refer to the same real-world entities or objects. Semantic heterogeneity is concerned with the occurrence of the same table names or attributes that are in different data sources, but they refer to different entities or objects in the real world.

Ontologies models can play the role of a canonical representation facilitating an integrated view of the data conditions [10]. Moreover, ontology technology allows us to model the data formally and reflect the semantics of that data.

Sensor ontologies and mapping of sensor data to domain ontologies build a solid foundation for sharing data of this nature and reusing and integrating various IoT applications [11]. In addition, using standardization to represent sensor networks also favors data integration. One such standardization is proposed by the World Wide Web Consortium (W3C) and Open Geospatial Consortium (OGC) through the formalization of a semantic sensor network (*Semantic Sensor Network*, SSN) [12].

Once the syntactic and semantic aspects are solved, the data context plays a key role in getting the most trustful meaning of information [4]. Context-aware computing is a core feature of ubiquitous and pervasive computing systems that deals with the real-time environment, dynamic information, need for real use, and making sense to derive future inferences [13]. In this sense, context-aware drives the way to deal with pragmatic data integration [14].

In [4], the authors present some context types, which are: **location, identity, activity,** and **time**, which are capable of answering important questions regarding context-awareness in systems: **who, what, when, and where, and how** related to some event. Although this basis for defining context is not used in its entirety in this work, as it is still an initial effort, its aspects are partially applied and will be evolved in our subsequent work.

Thus, there is an interesting match between data integration and context-aware. So, the use of ontologies can provide knowledge sharing among different data sources (enhancing context-aware needs) and reasoning capabilities due to their well-defined semantics fitting very well to integration needs [15].

Regarding the related work, considering the research areas this paper draws, we selected works addressing data integration, context awareness, and event detection. The purpose was to understand important aspects of the theme, challenges, and limitations of existing solutions.

Researchers are using artificial intelligence for automatic data integration [19,20], using SQL as the standard language to query heterogeneous data sources [22], logic-based techniques [21], and many other approaches using ontology [9, 23,24]. However, these proposals need to use the context in which the data is inserted to promote or use data integration.

In [11], the authors present an interesting mechanism to achieve the association between sensor data and domain ontology, evaluating the similarity of semantic concepts. In that research, there is no intention of integrating data from different sensors nor detecting or relating events from different sources.

The proposal presented in [25] is an Intelligent Context-Awareness Building Energy Management System (ICA-BEMS) that uses ontology to organize smart building knowledge and a new context-awareness mechanism to provide contextual information. However, despite recognizing context and events, there is no effort to relate them or use one to infer another.

Considering the works mentioned above, their contributions and limitations, and the challenges identified by the authors, the approach proposed in this work present a feasible solution to the highlighted points. The main contributions of this work are: (i) allow the integration of data from different sensors, (ii) provide the abstraction of sensor data structure, allowing add new data sources, (iii) detect of events of interest based on ontology definitions, and (iv) relate events from different sensors and relate them based in a context defined.

3 Data Integration and Context Awareness Architecture

This work proposes an architecture to facilitate the integration of sensor data to enable the detection of events of interest. Data sources are sensory devices that we assume provide data from the same knowledge domain. Despite being distributed and independent sources, the data domain is common to all of them. The following subsection outlines the key aspects of our approach.

3.1 Solution Model

We define the elements of our architecture as follows:

$S = \{s_0, s_1, ..., s_n\}$ the set of Sensors in the domain of interest;
$O_{s_i} = \{o_0, o_1, ..., o_p\}$ the set of Observations performed by each sensor s_i;
$A_{o_k} = \{a_0, a_1, ..., a_q\}$ the set of Attributes associated to each observation o_k;
$\epsilon_{s_i} = \{\varepsilon_0, \varepsilon_1, ..., \varepsilon_r\}$ the set of Events associated to a sensor s_i;

As defined with the previous sets, each measure detected by a sensor is an Observation and each sensor has its own set of Events.

An Observation o_k, $\forall o \in O_{s_i}$, is given by:
$o_k = \{(a_0, v_{a_0}), (a_1, v_{a_1}), ..., (a_l, v_{a_q})\}$ such that $v_{a_q} \in \mathbb{R}$
An Event ε_r , $\forall \varepsilon_{s_i} \in \epsilon$, is given by:
$\varepsilon_r = \{(a_0, vi_{a_0}, vf_{a_0}), (a_1, vi_{a_1}, vf_{a_1}), ..., (a_q, vi_{a_q}, vf_{a_q})\}$ such that $vi_{a_q}, vf_{a_q} \in \mathbb{R}$

For example, a given hydrometric sensor can have Observations such as:
$o_0 = \{(\text{date}, 2013\text{-}06\text{-}10), (\text{water\_level}, 2.60), (\text{flow\_rate}, 20.2)\}$
$o_1 = \{(\text{date}, 2013\text{-}06\text{-}10), (\text{water\_level}, 2.85), (\text{flow\_rate}, 21.7)\}$
$o_2 = \{(\text{date}, 2013\text{-}06\text{-}21), (\text{water\_level}, 22.10), (\text{flow\_rate}, 21.0)\}$

And some Event defined as:
$\varepsilon_0 = \{(\text{water\_level}, 0, 3), (\text{flow\_rate}, 0, 15)\}$
$\varepsilon_1 = \{(\text{water\_level}, 3, 4), (\text{flow\_rate}, 15, 30)\}$
$\varepsilon_2 = \{(\text{water\_level}, 4, 100)\}$

Considering vi and vf the initial and final attribute values that limit the classification of an event, and given $o_k = \{(a_0, v_{a_0}), (a_1, v_{a_1}), ..., (a_q, v_{a_q})\}$, and $\varepsilon_r = \{(a_0, vi_{a_0}, vf_{a_0}), (a_1, vi_{a_1}, vf_{a_1}), ..., (a_q, vi_{a_q}, vf_{a_q})\}$, is said to be equivalence between o_k and ε_r $(o_k \equiv \varepsilon_r)$ when:

$\exists\, \mathbf{M} \subset o_k \mid \mathbf{M} = \varepsilon_r$, wich means

$\forall (a_q, vi_{a_q}, vf_{a_q}) \in \varepsilon_r \; \exists (a_q, v_{a_q}) \in o_k \mid vi_{a_q} \leq v_{a_q} \leq vf_{a_q}$

That means that an observation is classified as an event if the values of all attributes in a defined event are detected in an observation. For example, the observation o_2 is classified as an event ε_2 just because its attribute $water\_level$ has a value greater than 4 and less than 100.

In short, a sensor is defined as: $s_i = \{\{o_0, o_1, ..., o_p\}, \{\varepsilon_0, \varepsilon_1, ..., \varepsilon_r\}\}$, such that $p, r \geq 0$.

Be further defined:

$G_{s_i} = \{V, E\}$ the graph associated to the observations of a sensor s_i where:

$V = \{v_0, v_1, ..., v_x\}$ is the set of nodes in G_{s_i}; and

$E = \{e_0, e_1, ..., e_y\}$ the set of edges of the graph G_{s_i};

Each edge $e_y = (v_b, v_c, distance)$, where $distance$ represents the distance between v_b and v_c. Each node $v = \{P, frequency\}$, where $P = \{(p_0, pv_0), (p_1, pv_1), ..., (p_n, pv_n)\}$ is the set of properties of a node and $frequency$ represents the number of ocurrences of a given observation.

$F(o_k, v_t)$ is the functional whose transforms an observation o_k to a node representation v_t. For each observation o_k from a sensor s_i, a new node is inserted in V, as long as the combination of properties and their respective values are not yet represented in V, i.e.:

$$F(o_k, v_t) = \begin{cases} v_t(p_n) = o_k(a_q) \\ v_t(pv_n) = o_k(v_{a_q}), \forall (v(p_n), v(pv_n)) \ni V \end{cases}$$

The distance between two nodes is given by:

$$d\varepsilon = \left(\sum\nolimits_{b,c=0}^{x} (v_{b_{p_b pv_b}} - v_{c_{p_c pv_c}}) \right)^{1/2}$$

Not every observation is represented as a new node in the graph. Only the first occurrence of an observation generates a new node. All repetitions of values for the attributes only result in incrementing the 'frequency' property in the given node. The distance among the nodes is intended to be used for future clustering and network analysis strategies.

3.2 Architecture

The architecture is organized and implemented in layers. Figure 1 presents an overview of the architecture, showing the data flow, its layers, and its components, described below.

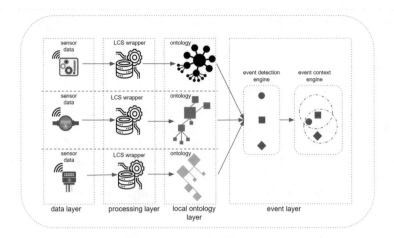

Fig. 1. Proposal Architecture

The **Data Layer** is responsible for data acquisition performed by the sensors. At this point, the data may be heterogeneous in structural, syntactic, and semantic terms and contains only and exactly the values of the attributes read from the entity they monitor.

The **Processing Layer** receives the sensor data from the previous layer and is responsible for creating a canonical representation of this data. *LCS wrapper* components build this representation based on a *Local Concept Schema* (LCS), which defines the rules for transforming the data obtained from the sensor into a canonical model. The processing result is an ontology according to SSN/SOSA model recommended by the W3C and OGC for sensor data.

Based on the LCS definitions, the *LCS wrapper* produces the canonical representation in an ontology, which is stored and has inference and reasoning processes in the **Local Ontology Layer**. Also, in that layer, the rules for event classification are defined.

At this moment, the event classes of each sensor are individually predefined based on historical data from the station. For example, the observations from sensors of hydrometric stations are classified as *Critical events* if the observation result value is greater or equal to the 3° quartile (75%) of the observations noted in the same month of the previous year; and if similar observation values were classified as a *flood* by Global Disaster Alert and Coordination System (GDACS)[5]. All other observations are classified as *Common events*. For the hydrological stations, the classes are the same as the hydrometric stations, and only the quartile condition is observed, based on [18] criteria.

The **Event Layer** uses the result of ontology inference and reasoning for creating a graph representation for each observation already classified. The *Event Detection Engine* adds a new node representation to the graph for each unique event classified in the previous layer. This approach ensures that no duplicated

---

[5] https://www.gdacs.org/flooddetection/ (Accessed at 2022/10/15).

event is represented in the graph, preventing the graph from growing without necessarily adding relevant information.

The *Event Context Engine* selects only the events classified as *Critical event* and, based on the context conditions defined by the ontology, uses one event to trigger an "event alert" related to another sensor station. At this time, the context rules only consider *location* and *temporal* aspects. *Location* context refers to stations located in the same area, and *Temporal* context refers to stations whose influence context can be noted at another station with or without time delay. The Semantic Web Rule Language (SWRL) has been used to define the context rules in the ontology. Table 1 presents some of the context rules definitions. For example, for a station *s2*, the property "is location_context_for" is observed when an event *e* occurred in a station *s1* and *s1* "is_influence_area_for" *s2*.

Table 1. Context rule example

swrl rules
occurred_in_station(?e,?s1) $\wedge$ is_influence_area_for(?s1,?s2) $\Rightarrow$ *is_location_context_for(?e,?s2)*
occurred_at_time(?e,?t) $\wedge$ is_in_range_time_to_influence(?t,?s) $\Rightarrow$ *is_temporal_context_for(?e,?s)*
is_location_context_for(?e,?s) $\wedge$ is_temporal_context_for(?e,?s) $\Rightarrow$ *is_influenced_by(?s,?e)*

4   Feasibility Study

This section describes a feasibility study using real use case scenarios. The architecture was used to detect and relate events from data produced by sensors in hydrological and hydrometric monitoring stations. A feasibility study attempts to characterize a technology to ensure it does what it claims to do and is worth developing [16].

This evaluation follows the Goal Question Metric (GQM), which aims to assess the feasibility of the solution, pointing out evidence that the previously defined objectives were met and that the formulated research question was answered. Thus, the objective is to: **analyze** the architecture **with the purpose of** evaluating the integration proposal **with respect to** the use of ontology and graph network analysis **from the point of view of** decision-makers **in the context of** integrating data for detecting events and correlate them based in context. The research focused on the following research question: **RQ1)** *Can the architecture enable data integration based on context to detect and correlate events?*

4.1   Data Description

We obtained sensor data from hydrological and hydrometric domains. The data from hydrometric stations are available at https://wateroffice.ec.gc.ca/, and data from hydrological stations are available at https://acis.alberta.ca/acis/. Both are public data sources made available by the Government of Canada and Alberta respectively. We also use data to validate event detection. The data is from a public data source available at https://www.gdacs.org/ and is made available by the Global Flood Detection System, which monitors floods worldwide using near-real-time satellite data.

The feasibility study considered data from four hydrological stations (Jumpingpound Ranger, Elbow Ranger, Cop Upper, and Calgary International Airport), which monitor precipitation levels, and five hydrometric stations (05BH004, 05BJ001, 05BJ004, 05BJ008, and 05BJ010), which monitor the water level of the riverbed to which they are associated. The data is from 2012 to 2013. We selected this period given the flooding events that occurred in 2013, popularly called the "flood of floods" [17].

The data provided by the hydrological sensors contains the following features: *"Station Name"*, *"Date (Local Standard Time)"*, *"Precip. (mm)"*, *"Precip. Accumulated (mm)"*, among others. The hydrometric sensors have the features: *ID, PARAM, Date, and Value*, where the *ID* feature represents the station identification and *PARAM* means the parameter monitored by the station (water level or flow rate).

4.2   Experimental Results

We assume the main events of interest are associated with the detection of high-water levels in the hydrometric stations since these kinds of events may indicate flooding and possible danger to people and possible damage to buildings in the affected area. To detect these events, we discuss two approaches. The first one considers only the capacity of event detection based on each sensor individually. The second one uses data from different sensors integrated under the context *'location'* and *'temporal'* aspects of the context when detecting events.

This section presents the results obtained when the sensor data described earlier is submitted to architecture under each scenario. Based on the data and scenarios described above, we want to integrate data from different sensors and detect and relate events. Two secondary research questions were derived from RQ1:

SRQ1) *Can the architecture be used as a tool for integrating data from different sensor data sources?*

SRQ2) *Can the architecture detect and correlate events based on improved context?*

Scenario 1

In the first scenario, sensor data from the five hydrometric stations are submitted to the architecture, which tries to detect events based on pre-defined classifications to the water level defined in the ontology model. So, values above the 3° quartile of water level values observed in the previous years at the current month and classified as flood events by GDACS, are considered *Critical event*. In contrast, the others are classified as *Common event*.

Once the sensors are treated individually, the results are simple and obvious to be expected, as they detect the rise in the station's water level at the moment this rise occurs. Table 2 lists the events where the water level is classified as a *Critical event* from the observations at four hydrometric stations and the events' descriptions.

Table 2. Events descriptions

station	observation	observable property	result time	result value
05BH004	observation21	water_level	2013/06/21	4.093
	observation22	water_level	2013/06/22	3.829
05BJ001	observation21	water_level	2013/06/21	5.213
	observation22	water_level	2013/06/22	3.852
	observation23	water_level	2013/06/23	2.953
	observation24	water_level	2013/06/24	2.805
	observation25	water_level	2013/06/25	2.353
05BJ008	observation21	water_level	2013/06/21	1077.428
	observation22	water_level	2013/06/22	1076.978
05BJ010	observation20	water_level	2013/06/20	28.024
	observation21	water_level	2013/06/21	29.429
	observation22	water_level	2013/06/22	28.886
	observation23	water_level	2013/06/23	27.999

However, this information is not relevant when the intention is to provide quick reaction time for measures to be taken on a certain event since when they are detected, the event has already reached the monitored location. Table 3 shows the days when the events are detected in each station. It is important to note that stations 05BH004, 05BJ001, and 05BJ008 detect the event on day 21; stations 05BJ004 and 05BJ010 on day 20. The number '1' in the table indicates the time of detection in Scenario 1, without data integration and context.

Table 3. Event alert without sensor integration.

station | days	15	16	17	18	19	20	21	22	23	24	25	26	27	28	29	30
05BH004							1									
05BJ001							1									
05BJ004						1										
05BJ008							1									
05BJ010						1										

Scenario 2

In scenario 2, we added rules for integrating data from hydrological sensors, creating *location, and temporal* context. Thus, the hydrological and hydrometric sensors are integrated considering: the location they are installed, if there is influence from one to another, and the time this influence is perceived.

Figure 2 presents the network of integrated sensors in the map. Each one of the dotted lines represent the context relations stablished through context.

For example, when the data from hydrological sensors at Cop Upper and Calgary International stations are integrated into the analysis, the precipitation data can be used to anticipate the water level alert at 05BH004, 05BJ001, 05BJ008, and 05BJ010 considering the amount of precipitation detected. Also, the data from Jumpingpound Ranger and Elbow Ranger can be integrated and considered in the context of these hydrometric stations.

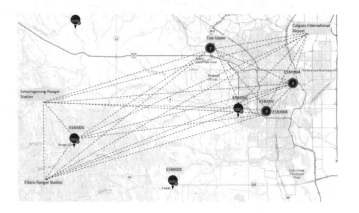

Fig. 2. Context scenario 2

To define the classification of the intensity of rain detected in the hydrological sensors, we used the classification based on the quartiles of the levels of rain observed in the same region in previous periods, as justified in [18], and used as an ontology rule. In this scenario, we consider critical events 'heavy rain' and 'extreme rain', which are equivalent to precipitations above the 0.75 and 1.0 quartiles respectively.

At this point, the architecture is not able to measure how much each hydrological sensor influences the hydrometric sensor. To define the influence context, we arbitrarily defined if there is influence or not based on the previous simultaneity of high precipitation and high-water level. For this, we considered the data collected during June 2012. Then, based on these data, we defined the context rules, as exhibited in Table 1. For example, the station Elbow Ranger is linked to 05BH004, 05BJ004, and 05BJ010 in the context of 'location' and 'temporal' 't < 1'. In other words, the events detected at Elbow Ranger station can influence these other stations on the same day they are detected. The definition of more accurate context rules depends on domain experts or context detection automation, which are points to be improved in future works.

With this integration and context applied, the architecture could anticipate events from the shared context, as we can see schematically in Fig. 3.

In this case, the critical events are detected earlier than in the approach without context. Checking Table 4, for station 05BH004, the detection is anticipated

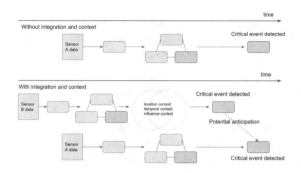

Fig. 3. Schema-integrated event timeline

Table 4. Event alert with sensor integration

station | days	15	16	17	18	19	20	21	22	23	24	25	26	27	28	29	30
05BH004					2		1									
05BJ001						2	1									
05BJ004					2	1										
05BJ008						1										
05BJ010					2	1										

in two days compared to the approach without data integration and context. In stations 05BJ001, 05BJ004, and 05BJ010, the events are detected one day earlier, while station 05BJ008 has no anticipation once no context rule related to Jumpingpound Ranger or Elbow Ranger stations was applied. The number '2' in the table indicates the detection time in Scenario 2.

Considering the results, we can answer the SRQs:

SRQ1) Can the architecture be used as a tool for integrating data from different sensor data sources? *Totally.* We demonstrate that we can integrate data from different sensors using ontologies as a canonical model. Furthermore, the architecture can add new data sources by including a file containing LCS definitions.

SRQ2) Can the architecture detect and correlate events based on improved context? *Partly.* We show that the architecture can detect and relate events from different sensors with expanded context using integrated data and context definitions in the analyzed scenarios. Despite the context attributes are not fully implemented yet, the architecture detected the events and established the relationships. However, some event detection may depend on the context defined, so we intend to improve context attributes in future work. We believe this question can be answered completely in future versions.

Thus, by answering these two secondary research questions, we can answer the main research question once the core of the proposal is *Data Integration* and *Event detection and relation based on context*. **RQ1) Can the architecture enable**

data integration based on context to detect and correlate events? *Totally.*
The architecture promoted data integration and based on context, detected and
correlated events from different data sources.

5   Conclusions and Future Works

This paper proposes using ontologies for data integration based on context and
then the detection and relationship between events. As a first implementation
effort, we used data from monitoring sensors from hydrological and hydrometric
stations. The experimental results indicate the proposal's feasibility once it can
detect and relate events, including anticipating detections based on context.

In future work, we intend to evolve to support syntactic, semantic, and pragmatic integration, automating the context detection attributes and provenance
models for event detection. With these new issues, we will conduct a new cycle
of evaluation with new and different domains and data sources.

References

1. Smys, S.: A survey on internet of things (IoT) based smart systems. J. ISMAC
   2(04), 181–189 (2020)
2. Krishnamurthi, R., et al.: An overview of IoT sensor data processing, fusion, and
   analysis techniques. Sensors **20**(21), 6076 (2020)
3. Sagar, S., et al.: Modeling smart sensors on top of SOSA/SSN and WoT TD with
   the semantic smart sensor network (S3N) modular ontology. In: ISWC 2018: 17th
   Internal Semantic Web Conference (2018)
4. Abowd, G.D., Dey, A.K., Brown, P.J., Davies, N., Smith, M., Steggles, P.: Towards
   a better understanding of context and context-awareness. In: Gellersen, H.-W. (ed.)
   HUC 1999. LNCS, vol. 1707, pp. 304–307. Springer, Heidelberg (1999). https://
   doi.org/10.1007/3-540-48157-5_29
5. Akanbi, A., Masinde, M.: A distributed stream processing middleware framework
   for real-time analysis of heterogeneous data on big data platform: Case of environmental monitoring. Sensors **20**(11), 3166 (2020)
6. Galhotra, S., et al.: Fair data integration. arXiv preprint arXiv:2006.06053 (2020)
7. Tan, W.-C.: Deep data integration. In: Proceedings of the 2021 International Conference on Management of Data, p. 2 (2021)
8. Sreemathy, J., Nisha, S., Rm, G.P., et al.: Data integration in etl using talend. In:
   2020 6th International Conference on Advanced Computing and Communication
   Systems (ICACCS), pp. 1444–1448. IEEE (2020)
9. Asfand-E-Yar, M., Ali, R.: Semantic integration of heterogeneous databases of
   same domain using ontology. IEEE Access **8**, 77903–77919 (2020)
10. Verstichel, S., et al.: Efficient data integration in the railway domain through an
    ontology-based methodology. Transp. Res. Part C: Emerg. Technol. **19**(4), 617–643
    (2011)
11. Liu, J., et al.: Towards semantic sensor data: an ontology approach. Sensors **19**(5),
    1193 (2019)
12. Compton, M., et al.: The SSN ontology of the W3C semantic sensor network incu-
    bator group. J. Web Semant. **17**, 25–32 (2012)

13. Bang, A.O., Rao, U.P.: Context-aware computing for IoT: history, applications and research challenges. In: Goyal, D., Chaturvedi, P., Nagar, A.K., Purohit, S.D. (eds.) Proceedings of Second International Conference on Smart Energy and Communication. AIS, pp. 719–726. Springer, Singapore (2021). https://doi.org/10.1007/978-981-15-6707-0_70

14. Ribeiro, E.L.F., Claro, D.B., Maciel, R.S.P.: Defining and providing pragmatic interoperability: the MIDAS middleware case. Anais Estendidos do XVII Simpósio Brasileiro de Sistemas de Informação. SBC (2021)

15. Malik, S., Jain, S.: Ontology based context aware model. In: 2017 International Conference on Computational Intelligence in Data Science (ICCIDS). IEEE (2017)

16. dos Santos, R.P.: Managing and monitoring software ecosystem to support demand and solution analysis. Ph.D. thesis, Universidade Federal do Rio de Janeiro (2016)

17. Pomeroy, J.W., Stewart, R.E., Whitfield, P.H.: The 2013 flood event in the South Saskatchewan and Elk river basins: causes, assessment and damages. Can. Water Res. J./Revue canadienne des ressources hydriques **41**(1–2), 105–117 (2016)

18. Gouvea, R.L., et al.: Análise de frequência de precipitação e caracterização de anos secos e chuvosos para a bacia do rio Itajaí. Revista Brasileira de Climatologia **22** (2018)

19. Saes, K.R.: Abordagem para integração automática de dados estruturados e não estruturados em um contexto Big Data. Diss. Universidade de São Paulo (2018)

20. Lipkova, J., et al.: Artificial intelligence for multimodal data integration in oncology. Cancer Cell **40**(10), 1095–1110 (2022)

21. Levy, A.Y.: Logic-based techniques in data integration. In: Logic-Based Artificial Intelligence, pp. 575–595. Springer, Boston (2000)

22. Amará, J., et al.: Stream and Historical Data Integration using SQL as Standard Language. Anais do XXXVI Simpósio Brasileiro de Bancos de Dados. SBC (2021)

23. Fathy, N., Gad, W., Badr, N.: A Unified Access to Heterogeneous big data through ontology-based semantic integration. In: 2019 Ninth International Conference on Intelligent Computing and Information Systems (ICICIS). IEEE (2019)

24. Nadal, S., et al.: An integration-oriented ontology to govern evolution in big data ecosystems. Inf. Syst. **79**, 3–19 (2019)

25. Degha, H.E., Laallam, F.Z., Said, B.: Intelligent context-awareness system for energy efficiency in smart building based on ontology. Sustain. Comput. Inf. Syst. **21**, 212–233 (2019)

BANY: An Anycast MAC Protocol Based on B-MAC+ for IoT Systems

Tales Heimfarth[1]([✉]), João Carlos Giacomin[2],
and Gabriel Augusto Lemos Silva[2]

[1] Applied Computing Department, Universidade Federal de Lavras, Lavras, Brazil
`tales@ufla.br`
[2] Computer Science Department, Universidade Federal de Lavras, Lavras, Brazil
`giacomin@ufla.br, gabriel.silva27@estudante.ufla.br`

Abstract. This paper presents BANY, a novel low latency, low power listening MAC protocol for Wireless Sensor Networks. WSNs are employed in data collection side of IoT systems, in order to acquire environmental data and send it to a data-base. BANY is a duty-cycled asynchronous protocol based on preamble sampling with short duration, as B-MAC+. In order to establish communication between two nodes, this protocol uses preamble with the duration of a cycle time to ensure that the forwarder will be awake to receive the data packet. This increases considerably the sleep-delay problem. BANY eases this phenomenon by reducing the preamble duration to a small part of the cycle time. Anycast communication pattern is employed to increase the probability that a possible forwarder is awake after the reduced preamble. Differently from most asynchronous MAC protocols, the channel is probed for activity during a very short interval, reducing the idle network energy consumption. Simulation results shown that BANY achieved smaller end-to-end latency in the majority of the tested scenarios.

1   Introduction

The Internet of Things (IoT) technology has expanded the concept of data communication far beyond the exchange of voice, image and files. Autonomous electronic devices interact with the environment, collecting data and sending them to a processing center. IoT can be described as a multi-domain distributed wireless communication network with many diverse applications and network components [16]. In data collection side is the machine-to-machine (M2M) domain, composed by small autonomous devices which communicate wirelessly and send data by multi hop to a Base Station connected to the Internet. A huge amount of data is stored in the cloud, accessible to consumers, in the application domain.

A Wireless Sensor Network (WSN) takes place in the IoT architecture in M2M domain, collecting data from the environment and sending them to a database. WSNs are composed by a large number of small electronic devices, denominated *sensor nodes*, with low-power consumption components and short-range communication capabilities, as those specified in IEEE 802.15.4 [1] reference model. With limited energy resources, usually small batteries, the main challenge for WSN developers is to maintain network operation for long periods [6].

Medium Access Control (MAC) protocol is a critical element in WSN, since it is responsible for allocating shared communication resources assuring effective connectivity [16]. Another function of WSN MAC is to control the radio operation. The main strategy to save energy is the adoption of duty-cycle [5, 12]. Sensor nodes operate in small periods to perform measurements, processing and communication. Most of the time they are kept inactive, in low power mode (sleep state). As a consequence, the sleep-delay arises [3, 12], enlarging the end-to-end communication.

MAC protocols for WSN can be broadly classified in *contention-based* and *scheduled-based* protocols [12]. Contention-based protocols are preferable for WSNs due to their easy implementation, as well as reduced overhead [16]. Also they do not require centralized control which gives them more scalability. Asynchronous contention-based MACs are simpler and easier to implement. They use preamble sampling to establish peer-to-peer communication [3]. Before data packet the sender node transmits a preamble to advise the intended receiver a data is coming. The preamble transmission must overlaps the sleep state interval of the receiver.

MAC protocols can mitigate the sleep-delay problem exploring opportunistic routing techniques with the use of anycast communication pattern [15]. The sender node selects a group of neighbors with good condition to forward the message towards destination, denominated *Forwarding Candidate Set* (FCS). The routing metrics to select the FCS members are defined previously in the network layer. The sender node starts communication sending a preamble composed by small RTS (Request to Send) packets to advise the FCS members its intention of sending data. The final decision about which FSC member is going to receive data is taken in the MAC layer, observing some conditions at the moment [9]. This procedure reduces end-to-end latency.

In this work, we propose BANY (BMAC+ Anycast Probabilistic), an asynchronous protocol devoted to WSNs, which sends a series of RTS (Request to Send) packets to find a forwarder node. All the nodes employ duty-cycles to save energy and probes the communication channel for activity in the same way of the canonical B-MAC [14] and B-MAC+ [2]. Differently from those works, the preamble length is smaller than the sleep interval. There exists a probability that an FCS member samples the channel during this interval, establishing communication at the end of the preamble. If none of the FCS members probe the channel when RTS are sent, communication will fail. In this case, BANY will try another chance.

2 Related Work

Wireless Sensor Networks plays the role of data collector in the Internet of Things (IoT). They are composed by a large number of electronic devices organized in star or mesh topology. The allocation of transmission resources is normally controlled by the sensor nodes, with a distributed MAC scheme. This give more flexibility and scalability to WSNs [16]. Asynchronous MAC are more easy to implement, more scalable and more power efficient, since no synchronization message exchange is needed [3]. Also, asynchronous Low Power Listening (LPL) protocols are very appealing for multi-hop networks since they consume less energy than scheduled protocols in lightly loaded networks.

B-MAC (Berkeley MAC) [14] was the pioneering LPL protocol developed for WSN. All the nodes follow the same duty cycle but there are not any synchronization in their active periods. During active time, a carrier probe is realized. When no transmission is detected, the node go back to the sleep mode. B-MAC+ [2] is an improvement on B-MAC, aiming at reducing the overhearing of the neighbors of the sender node which are not the intended receiver. Instead of sending a long preamble before data, B-MAC+ sends a long series of short of RTS packets with the same information, the address of the intended receiver and the serial number of the RTS. Each node that receives an RTS can decide by itself to wait for data, if it is the intended receiver, or go back to sleep mode, in the other case.

X-MAC [4] is another asynchronous protocol that uses a preamble composed by a series of RTS packets. Differently from B-MAC+, it introduces listening intervals between RTS, in order to enable the intended receiver to notify its presence. As soon as the sender decode an early Acknowledgment (eACK) packet from the receiver, it stops the transmission of the preamble and starts to transmit the data packet. The authors argue that X-MAC reduces by half the sleep-delay in comparison to B-MAC. In the other hand, the overall energy consumption is largely increased, since the time spent by nodes listening for a possible RTS is greater than that of B-MAC probing channel. That is because the listening time must overlaps the interval between two consecutive RTS.

AGAp-MAC (Adaptive Geographic Anycast Probabilistic MAC) [8] is an asynchronous protocol, which uses strobed RTS packets to achieve rendezvous, but with a small period for channel sampling. It conjugates the small energy cost of Low Power Listening of B-MAC [14] with reduced end-to-end latency of anycast protocols. There is a probability of achieving rendezvous in one cycle time, but there is no warranty. In the case of failure, a mechanism is provided to ensure end-to-end communication.

In this paper we propose BANY-MAC, a cross-layer asynchronous protocol encompassing routing and medium access control functions. This protocol is based on B-MAC+ [2] and benefits from LPL small energy expenditure and reduced end-to-end latency from anycast communication pattern. In a similar way of AGAp-MAC [8], this new protocol achieves rendezvous in one cycle time with a certain probability. The operation process of BANY is explained in details in the next section.

3   The BANY Protocol

BANY is a *cross-layer* asynchronous protocol encompassing functionalities of network (NET) and layer the medium access control (MAC) sub-layer. It is inspired in the B-MAC+ protocol [2], using preamble sampling technique to establish communication. Periodically, each node wakes up, turns its radio on and verify the communication channel activity. This indicates that a neighboring node intends to send a message. In order to avoid costly synchronization among nodes, the schedule of each sensor node is independent of each other. However, all nodes have the same cycle length. When a node is ready to send a packet, it starts to send a series of RTS (Request To Send) packets, acting as preamble. After this series, the node waits for a CTS (Clear To Send) packet from a possible next relay node, in order to achieve rendezvous. This means that the next relay is ready to receive the data packet.

In the original B-MAC+, the series of RTS covers the interval between two consecutive channel probe events. Since it is an unicast protocol, it is necessary to assure that the next relay is awake and waiting for the data packet at the end of the preamble. The RTS has a counter that permits the awaken receiver calculate the time when data will be sent. Then it returns to sleep until this time, saving energy. A drawback of the B-MAC+ protocol is the large latency caused by the sleep-delay problem. At each hop, the sender node transmits the data packet only after the large preamble, even if the receiver is prompt at the beginning.

In order to reduce latency, BANY employs anycast communication combined with a smaller preamble. When a packet should be sent towards a destination, instead of a single possible next hop, any node from a group, called FCS (Forwarding Candidate Set), may be used to relay the data. The rationale here is that, with a large number of possible relays, there is at least one that will receive an RTS earlier, with high probability. Therefore, a shorter preamble may be used. As the one-hop latency is correlated with the extension of the preamble, its reduction has a direct consequence in the message propagation time.

At the beginning of each cycle time, a node probes the channel for activity by a short period, like the Low Power Listening technique. If some activity is detected, the node stays awake longer, hoping to receive an RTS packet from a sending node. The RTS carries the position of the sending node, the position of the destination, a threshold τ and, as in B-MAC+, a counter. When a node receives an RTS, it decides by itself its inclusion in the transmitter's FCS. Threshold $\tau \in (0,1)$ indicates how much advance towards the destination is required ($\tau \cdot R$, where R is the radio range).

When a nodes wakes up, receives an RTS packet and determines that it should participate the communication, it returns to the sleep state until the end of the RTS series. At this moment, the sending node turns the radio to reception mode, in order to receive a CTS from an FCS member, which will become the next forwarder. All members of the FCS, aware of the communication, wake up at the end of the preamble and start a back-off period before sending back a CTS packet. The first node to send a CTS is elected the next forwarder.

Other awake FCS members hear that communication and resign to participate in the transmission, returning to sleep state. After receiving the data packet, the new forwarder restarts the process.

Figure 1 presents an example of communication. The source node S initiates the communication towards node D, the final destination of the message. Node F is a forwarder, which is sending the preamble, called here sender. Nodes a, b and c are members of the FCS formed by τ employed in this example. One node of this group will become the next forwarder, taken the role of receiver in this communication.

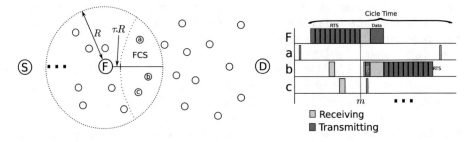

Fig. 1. Example of an ongoing communication using BANY. S is the source node, where the process begins. D is the final destination of the data packet. The forwarder F is a routing node in the path from S to D. Nodes a, b and c are the FCS mebers of the sender F.

At the left part of Fig. 1, the network configuration is presented. Node F has a radio range R, and a minimal advance of $\tau \cdot R$ is required in order to include a neighbor in the FCS. This advance is the shortening in the distance to the final destination D a neighboring node can give when compared to the distance between F and D. In other words, the data packet will be closer to D by an advance larger than $\tau \cdot R$. All the F's neighbors located in the region delimited by the radio range and the arch defined by $\tau \cdot R$ belongs to the FCS. The larger the required advance, the smaller the FCS cardinality. The smaller the required advance, the larger the number of FCS members. When a large number of members is expected, a smaller preamble can be used, speeding up the communication.

At the right side of the figure, a temporal diagram of the underlying communication is exhibited. Node F is willing to forward a message and it starts to send a preamble, comprised as a sequence of RTS packets. The nodes forming the FCS use the same duty cycle but they wake up in different moments. Node a makes a short probe of the channel before the activity, thus it returns to the sleep state. When the probe time of node b arrives, it notices the ongoing preamble from F, and sustains its reception mode until receiving an RTS packet. After this, it returns to the sleep state until the end of the preamble, marked in the figure as moment m. Node c does the same. At m, both b and c wake up in order to answer the RTS with a CTS packet. Nodes b and c contends for the channel making random back-off periods.

At the end of its period, node b makes a carrier sense, identifying a free channel. It starts to send the CTS packet. A small period latter, the back-off time of node c goes off, a carrier sense is made, detecting the communication of b. Node c decides to cancel the communication and returns to the sleep mode, taking its normal duty cycle. After receiving the CTS from b, node F sends the data packet. The new forwarder b restarts the process sending RTSs. The communication ends at node D.

An important remark about the size of the preamble and the value of τ must be made. As presented in results section, different combination of the size of the preamble and threshold values can be used. The density of node in the network combined with τ determines the average cardinality of the FCS. Smaller threshold and dense network result, in average, in larger number of FCS members.

We define the size of the preamble as Pre_{time} parameter. A Pre_{time} of 1.0 means that all the cycle time is covered by the preamble (behavior of B-MAC+), whereas a $Pre_{time} = 0.3$ indicates that just 30% of the cycle is covered by the preamble. Smaller values lead to faster communications, with reduced latency. However, the probability that no node hears the communication is higher in this case. Concluding, for our protocol, a combination of threshold τ and Pre_{time} values that leads to reduced latency and high success delivery rate must be determined.

A final remark is that asynchronous duty-cycled protocols that interleaves RTS packets with listening periods, achieves faster rendezvous, but with an increased network energy cost. This is the case of the canonical X-MAC [4]. In order to guarantee reception of an RTS from a sending node, all nodes in the network must poll the channel for a period long enough to cover the listening period. Differently, B-MAC+ [2] and our protocol perform carrier sense for a very short period of time. This leads to a much smaller energy consumption by nodes in the network that are not participating of a ongoing communication, extending network lifetime.

4   Experimental Results

This section presents the results and discussions related to latency and energy consumption of a WSN with the protocol proposed in this work.

Nodes use the same duty cycle with independent activity time. The new protocol was compared with three asynchronous protocols from literature: B-MAC+ [2], X-MAC [4] and AGAp-MAC [8]. AGAp-MAC and B-MAC+ are important for the comparison since they are also protocols that employ short channel probing. This feature allows to reduce considerably energy consumption for idle nodes, extending the network lifetime. X-MAC is a standard preamble sampling protocol for duty-cycled wireless sensor networks.

The tests were performed with GrubiX, a simulator developed for wireless sensor networks, which is an extension of the Shox simulator [13]. It incorporates the IEEE 802.15.4 protocol stack in a compliant radio.

4.1 Setup

A radio model based on IEEE 802.15.4 with the bit rate of 250 kbps was adopted in our simulation. It was configured for fixed transmission power, bidirectional links and Free Space propagation model for isotropic point source in an ideal medium. For the simulation arena, random node deployment was employed, thus nodes' locations form a Poisson process. X-MAC and B-MAC+ use GPSR [11] as routing protocol. Each node has its own position information. Table 1 presents the set of parameters employed in the simulation.

Each simulation comprises the transmission of a data packet from source S to final destination D, which are 1000 m apart. Each simulation was repeated 50 times in order to obtain statistically relevant results. Three data packet lengths were tested: 6250, 8250 and 12500 bits, whose transmission intervals correspond to approximately 25%, 33% and 50% of cycle time. Other two parameters were important: threshold (τ) and Pre_{time} values. As explained, the FCS members are found in the area defined by a threshold value and the radio range (R) of node F. Pre_{time} parameter configures length (in time) of the preamble, where 1.0 means that it will endure the complete cycle time.

Table 1. Parameters employed in the simulations

Symbol	Parameter	Values
R	Radio range (m)	40
P	Power ON consumption (mW)	60
$\overline{SD}$	Sender to Destination distance (m)	1000
η	Node density ($\frac{\text{nodes}}{m^2}$)	8×10^{-3}
t_{cycle}	Cycle time (s)	0.1
t_{data}	Data transmission interval ($\% t_{cycle}$)	25%, 33%, 50%
t_{RTS}	RTS transmission interval (s)	1.024×10^{-3}
t_{CTS}	CTS transmission interval (s)	0.512×10^{-3}
t_{cw}	Initial Contention Window duration (s) - BANY	2.560×10^{-3}
τ	Threshold values - BANY	$0.2, 0.3, 0.4, 0.5, 0.6$
Pre_{time}	Preamble coverage ($\times t_{cycle}$) - BANY	$0.2, 0.3, 0.4, 0.5, 0.6$
B	Transmission rate ($kbps$)	250

4.2 Results

In order to determine the better combination of preamble length (Pre_{time}) and threshold (τ), we first carry on a series of experiments to explore different combinations of the described parameters. The idea is to find out which combination of the parameters achieve the best performance, in terms of latency.

Table 2 presents the results obtained for a packet size of 6250 bits (25% of cycle time). In order to pick up the best combination of parameters, just experiments whose delivery rate achieved at least 95% were considered. From this appraisal, we could find out that the best combination for this packet size was $Pre_{time} = 0.1\,t_{cycle}$ and $\tau = 0.4$. The same experiments were executed for the other packet sizes. For a packet of 33% of cycle time, the best parameters were $Pre_{time} = 0.1$ and $\tau = 0.4$, which leads to a latency, in average, of 2.34 s. For the large packet (50% of cycle time), we achieved $Pre_{time} = 0.1$ and $\tau = 0.4$ with latency (in average) of 3.04 s.

Table 2. Average latency (s) for a transmission of a packet across 1000 m. Data size: 6250 bits. Delivery rate $\geq$ 95% for white cells.

Pre_{time} $\,^{\tau}$	$\tau = 0.2$	$\tau = 0.3$	$\tau = 0.4$	$\tau = 0.5$	$\tau = 0.6$
0.1	2.25	2.13	**2.02**	2.03	2.12
0.2	2.54	2.36	2.25	2.18	2.22
0.3	3.01	2.72	2.53	2.40	2.39
0.4	3.43	3.11	2.89	2.72	2.65
0.5	3.82	3.50	3.23	3.05	2.91

In order to compare our protocol with the state of the art, an appraisal using the same parameters (described in the Setup section) for B-MAC+, X-MAC, AGAp-MAC and BANY was realized. Figure 2 presents the average latency obtained for a single packet delivery for each protocol.

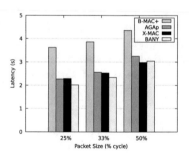
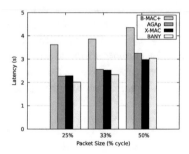

Fig. 2. Average latency (s) of B-MAC+, X-MAC, AGAp-MAC and BANY for a single packet transmission across 1000 m.

Fig. 3. Average energy (J) of B-MAC+, X-MAC, AGAp-MAC and BANY for a single packet transmission across 1000 m. Only the energy spent by the routing nodes was considered.

The results show that our protocol had lower latency, when compared with the state-of-the-art, for almost all packet sizes. The exception is encountered for the protocol X-MAC, with 12500 bit packets. The latency for this configuration is slower than BANY. However, it does not utilizes short channel probing, increasing significantly total network energy consumption, as will be demonstrated latter on. The original B-MAC+, although saving energy due to short channel probing, produce the higher latency of all compared protocols in our experiments. This can be explained by its intrinsic nature of 100% coverage of the cycle time by the preamble.

AGAp-MAC [8] was an improvement in the class of short probing asynchronous MAC protocols. It includes short preamble probing in the AGA-MAC [7] protocol. Due to this feature, there is a chance that the short probing occurs at the moment the sender is waiting for a response (CTS) of the next forwarder, preventing rendezvous. There is a probability of $\frac{t_{wCTS}}{t_{RTS}+t_{wCTS}}$ that an awake receiver does not hear an RTS packet and then goes back to sleep mode. Here, t_{wCTS} is the interval the sender node maintains the radio in reception mode waiting for a CTS packet. To increase the chance of an early encounter of a next forwarder, the anycast communication pattern is employed.

This point is enhanced by our new protocol: as in B-MAC+, BANY preamble do not embody interleaved periods for the CTS answer. The answer is expected only at the end of the complete preamble. Therefore, a short probe performed during the preamble time will always detect an RTS packet. This is an improvement over AGAp-MAC which translates in the lower latency achieved by our protocol for the appraised scenarios. As in AGAp-MAC, there is a probability that an awake receiver hears an RTS packet and sends back a CTS. For our protocol, it is calculated as $\frac{p \cdot t_{RTS}}{t_{cycle}}$. Here, p is the number of RTS packets in the preamble. At the end of a preamble, the sender node (F) waits for a CTS during a t_{wCTS} interval. All FCS members that hear an RTS contend for the channel during t_{cCTS}. If no answer is heard, BANY starts a new preamble. In our simulations, there is a limit of 5 tries and $t_{cCTS} = 5 \cdot t_{CTS}$ and $t_{wCTS} = t_{cCTS} + t_{CTS}$.

Figure 3 presents the average energy spent by the routing nodes for a single packet delivery for each protocol. Only the energy spent by the nodes participating in the communication process was computed. This energy is calculated as the product of the sensor nodes power in active mode and the total active time in transmission and reception modes:

$$E_T = P \sum_{h=1}^{H} \left(t_{cw} + p_h \cdot t_{RTS} + \frac{3}{2} t_{RTS} + 2 \left(\frac{t_{cCTS}}{2} + t_{CTS} + t_{data} \right) \right) \quad (1)$$

where $P = 60\,\text{mW}$ is power of a node in the active state, p_h is the number of sent RTSs in hop h, H is the total hops to send the message from S to D, $\frac{t_{cCTS}}{2}$ models the average time spent by the receiver in contention to send back a CTS to F and t_{data} is the time to transmit a data packet. The total power spent by a node is considered the same no matter if the radio is in idle listening, transmission or reception mode.

The power P is multiplied by the total time of activity, modeled by the rest of the equation. The total time is computed as the sum of the transceivers operating time in both sender and receiver nodes for each hop. For hop h, the sender node spends t_{cw} probing the channel to avoid message collision and then it sends a sequence of p_h RTS packets. The total time of consecutive RTSs is the product of p_h and t_{RTS}. After p_h RTS packets, the next awake relay waits on average $\frac{1}{2}t_{RTS}$ until it starts to receive a complete RTS packet (t_{RTS}). Next it goes into sleep mode until the expected time for the sender to finish the RTS sequence. After the end of the preamble, potential receivers contend for the CTS response, spending up to t_{cCTS}. After that, the receiver sends back a CTS packet and receives the data packet, assuming the role of the next relay, F. The last 3 events involve the sender and receiver nodes.

Figure 3 presents the energy expenditure in a single communication event for the protocols appraised in this article. The energy employed to take a data message from a source node S to a destination D, depends on the total time the transceivers are used for communication. This time depends primarily on the length of preamble in each hop as soon as the number of hops used in the communication. As presented in the figure, B-MAC+ has the greatest energy consumption since its preamble covers a complete cycle time at each hop. Since X-MAC is a unicast protocol, the number of hops is reduced yielding a low energy profile. AGAp-MAC and BANY employ similar routing function. For this reason, they have similar energy consumption profile, with a slight advantage for BANY. The probability of an FCS member to probe the channel during an waiting CTS interval is higher in AGAp-MAC causing a slightly higher energy consumption for small and large packet sizes.

In order to use a WSN in several IoT applications, a very important goal is a long network lifetime. In previous sections, we present the energy expenditure of a single transmission. Since WSN have, for several applications, low traffic profile, the energy spent by all nodes when no transmission is taken place is far more important for the network lifetime. We call this energy expenditure idle consumption (E_{idle}).

In duty-cycled networks, the periodic channel probing is the responsible for idle consumption. All nodes of the network periodically wake up and check the channel for activity. The size of the channel probing (t_{cs}) is very important in this context. We can divide here the protocols in two categories: normal and short channel probing. X-MAC, similarly to most asynchronous MAC protocols, belongs to the first category since the size of the probe must be long enough to enable the reception of an RTS packet. In the short channel probing, the radio is active for a short time that enables the detection of the state of the channel (whether there is some ongoing transmission). In this category we include B-MAC+, AGAp-MAC and BANY.

For the short probing, we considered the following: during sleep time, nodes are kept in power down (PD) mode. Therefore, the crystal oscillator is turned off. Just voltage regulator is working. For the channel probing task, the radio must be put in reception mode. We use the parameters of a CC2420 [10]. The transition time from Power Down (PD) to receiving mode is 1.2 ms. The channel probing size comprises of 5 samples, which are necessary for a trustfully operation [3]. This accounts to another 0.64 ms. Therefore, the whole operation needs a total of $t_{cs} = 1.82$ ms.

The normal probing needs the same 1.2 ms. In addition, there is a necessity to sample entire CTS period plus 9 samples. This can be explained by the following. A node may wake up at the end of an RTS, getting 4 samples and missing one of these RTSs. The total sampling period is $t_{cs} = 2.86$ ms, for a CTS length of 0.512 ms.

With these values, it is possible to calculate the idle energy consumption for our network. It is given by $E_{idle} = N \cdot P \cdot t_{cs} \cdot \frac{t}{t_{cycle}}$ For a network of $N = 2448$ nodes, operational power of 60 mW, $t_{cycle} = 0.1$ s, for a given period of $t = 100$ s, the idle consumption is approximately 420 J for the normal probing and 267 J for short probing. Short probing savings achieves 36% of idle consumption.

For low traffic duty-cycled networks, the idle energy consumption is several orders of magnitude higher than transmission expenditures. For this reason, reducing the node probing period is the key point to enlarge network lifetime. BANY protocol holds down idle energy consumption by employing short channel probing. Moreover, it achieved lower network latency when compared with other protocols that use this method. It employs, for that, anycast communication pattern combined with reduced preamble coverage.

5   Conclusion

A wireless sensor network (WSN) is a important element in an IoT architecture. Environmental data collecting is an example of task performed by such network. Since nodes are deployed in the environment and can't be easily replaced, it is important to support a long network lifetime by reducing energy consumption. WSN MAC protocols are very important in this task, employing in many cases duty-cycling for achieving this objective.

This paper presents BANY, an anycast asynchronous protocol for WSNs. In asynchronous duty-cycled networks, each node have its own independent scheduler. In order to achieve rendezvous, a sender emits a preamble long enough to find an awake next forwarder. B-MAC+ cover the whole cycle with RTS packets to assure next hop reception. This increases the sleep-delay problem. In order to mitigate it, our protocol reduces the preamble coverage. This may lead to communication failure if an unicast communication pattern is employed (as in B-MAC+). To increase success probability, anycast communication method is used.

A very important task is channel probing. Each node wakes up once per cycle to check for incoming communications. Most protocols (e.g. X-MAC) use a large channel probing since RTS packets are interleaved with a reception period for the CTS (clear to send) answer. Our protocol, as B-MAC+, employ short channel probing, which improves considerably the idle energy profile of the system.

Simulations were performed in order to assess the behavior of BANY in comparison to state-of-the-art MAC protocols. For the tested scenarios, our protocol could deliver the data packet with, in average, smaller latency for all packet sizes, with one exception (X-MAC, packet size of 50% of cycle time). The transmission energy consumption was smaller than other MAC protocols with exception of

X-MAC for medium and large packet sizes. However, the idle network energy consumption, which is more important for network lifetime in low traffic networks, was considerably lesser than X-MAC for 100 s of network runtime.

Acknowledgments. The present work was carried out with the support of the Conselho Nacional de Desenvolvimento Científico e Tecnológico (CNPq).

References

1. IEEE standard for low-rate wireless networks. IEEE Std 802.15.4-2020 (Revision of IEEE Std 802.15.4-2015), pp. 1–800 (2020)
2. Avvenuti, M., Corsini, P., Masci, P., Vecchio, A.: Energy-efficient reception of large preambles in MAC protocols for wireless sensor networks. Electron. Lett. **43**(5), 300–301 (2007)
3. Bachir, A., Dohler, M., Watteyne, T., Leung, K.: MAC essentials for wireless sensor networks. IEEE Commun. Surv. Tutor. **12**(2), 222–248 (2010)
4. Buettner, M., Yee, G.V., Anderson, E., Han, R.: X-MAC: a short preamble MAC protocol for duty-cycled wireless sensor networks. In: Proceedings of the 4th International Conference on Embedded Networked Sensor Systems, SenSys 2006, pp. 307–320. ACM, NY (2006)
5. Doudou, M., Djenouri, D., Barcelo-Ordinas, J.M., Badache, N.: Delay-efficient MAC protocol with traffic differentiation and run-time parameter adaptation for energy-constrained wireless sensor networks. Wirel. Netw. **22**(2), 467–490 (2016)
6. Halder, S., Ghosal, A., Conti, M.: Limca: an optimal clustering algorithm for lifetime maximization of internet of things. Wirel. Netw. **25**(8), 4459–4477 (2019)
7. Heimfarth, T., Giacomin, J., De Araujo, J.: AGA-MAC: adaptive geographic anycast MAC protocol for wireless sensor networks. In: 2015 IEEE 29th International Conference on Advanced Information Networking and Applications (AINA), pp. 373–381 (2015)
8. Heimfarth, T., Giacomin, J.C.: A probabilistic preamble sampling anycast protocol for low-power IoT. In: Barolli, L., Woungang, I., Enokido, T. (eds.) AINA 2021. LNNS, vol. 226, pp. 15–27. Springer, Cham (2021). https://doi.org/10.1007/978-3-030-75075-6_2
9. Hong, C., Xiong, Z., Zhang, Y.: A hybrid beaconless geographic routing for different packets in WSN. Wirel. Netw. **22**, 1107–1120 (2016)
10. Inc., T.I.: CC2420 2.4 GHz IEEE 802.15.4 / zigbee-ready RF transceiver (2013). https://www.ti.com/product/CC2420
11. Karp, B., Kung, H.T.: GPSR: greedy perimeter stateless routing for wireless networks. In: Proceedings of the 6th Annual International conference on Mobile Computing and Networking, MobiCom 2000, pp. 243–254. ACM, New York (2000)
12. Kumar, A., Zhao, M., Wong, K., Guan, Y.L., Chong, P.H.J.: A comprehensive study of IoT and WSN MAC protocols: research issues, challenges and opportunities. IEEE Access **6**, 76228–76262 (2018)
13. Lessmann, J., Heimfarth, T., Janacik, P.: Shox: an easy to use simulation platform for wireless networks. In: UKSIM 2008. Tenth International Conference on Computer Modeling and Simulation, 2008, pp. 410–415 (2008)
14. Polastre, J., Hill, J., Culler, D.: Versatile low power media access for wireless sensor networks. In: Proceedings of the 2nd International Conference on Embedded Networked Sensor Systems, SenSys 2004, pp. 95–107. ACM, New York (2004)

15. Qin, X., Huang, G., Zhang, B., Li, C.: Energy efficient data correlation aware opportunistic routing protocol for wireless sensor networks. Peer-to-Peer Netw. Appl. **14**(4), 1963–1975 (2021). https://doi.org/10.1007/s12083-021-01124-3
16. Sotenga, P.Z., Djouani, K., Kurien, A.M.: Media access control in large-scale internet of things: a review. IEEE Access **8**, 55834–55859 (2020)

Technology Factors Influencing Saudi Higher Education Institutions' Adoption of Blockchain Technology: A Qualitative Study

Mohrah S. Alalyan[1,2](✉), Naif A. Jaafari[1], and Farookh Khadeer Hussain[1]

[1] School of Computer Science, University of Technology Sydney, Ultimo, Sydney, NSW 2007, Australia
{mohrah.s.alalyan,naifalajlant.jaafari}@student.uts.edu.au,
malalyan@kku.edu.sa, farookh.hussain@uts.edu.au
[2] College of Art and Science, King Khaled University, Khamis Mushayt, Abha, Saudi Arabia

Abstract. Given the lack of empirical research on blockchain adoption in higher education, this study aims to partially fill this gap in the knowledge by focusing on the technology factors that could play a role in the process. Using Rogers' Diffusion of Innovation Theory, five technology factors were considered: relative advantage, compatibility, observability, complexity, and trialability. The data analysis was based on ten semi-structured interviews with university mid- and senior-level administrators and IT personnel. Relative advantage emerged as the strongest perceived factor. The effect of compatibility and complexity was observed to a lesser extent. The influence of observability and trialability was unclear. The uncovered underlying mechanisms for these effects are discussed and implications for theory, practice, and future research are explored.

1 Introduction

The educational sector plays an important role in any country's development. This is particularly true for countries which are seeking economic diversification and to tap into their enormous human potential. For Saudi Arabia, the development of education, in particular higher education, has become one of the major goals of the Development Plans which commenced in the 1970s [1].

Historically, developments in the educational sector take place alongside technological progress [2, 3]. Given the importance of integrating novel technologies into higher education, administrators and IT specialists in colleges and universities are likely to explore the opportunities and challenges in applying these technologies within their organizations. One of the emerging technologies with a high potential for application in the area of education is blockchain. Introduced primarily as a decentralized cryptocurrency tool [4], blockchain has received widespread attention since Buterin's groundbreaking paper [5], which described its many possible applications in different areas of life.

In the past few years, there has been a dramatic increase in interest regarding blockchain applications in higher education [6, 7]. The range of benefits that blockchain offers is wide, from immutable credentials to academic fraud detection and novel education platforms [8–10]. However, the literature on blockchain applications in education

© The Author(s), under exclusive license to Springer Nature Switzerland AG 2023
L. Barolli (Ed.): AINA 2023, LNNS 661, pp. 197–207, 2023.
https://doi.org/10.1007/978-3-031-29056-5_19

remains predominantly focused on reviews of cases and models for blockchain implementation [6, 9]. Empirical studies are rare, and there is little clarity regarding the underlying mechanisms of blockchain adoption in higher education [11, 12]. To the best of the author's knowledge, studies of blockchain adoption in Saudi higher education institutions (HEIs) are scant.

This study aims to fill this gap in the knowledge by exploring the technology factors relevant to blockchain adoption by Saudi HEIs. Taking the Diffusion of Innovations (DOI) Theory [13] as a basis, the study aims to gain an understanding of the technology attributes relevant to blockchain adoption and to identify the mechanisms behind them.

2 Theory

The DOI Theory [13] is a technology adoption theory that emerged from general observations of the common elements that underlie diffusion research in several disciplines: sociology, psychology, communications, economics, and organizational life. It explains the process of technology adoption based on a number of inherent technology attributes that influence that process. It is one of the few adoption theories that can be applied at the organizational level of research [14, 15].

An innovation in DOI is defined as *"an idea, practice, or object that is perceived as new by an individual or another unit of adoption"* ([13], p. 62). Blockchain is an innovative technology in higher education because of its novelty and its potential to restructure some common operational and service practices. Its advantages include being immutable, it makes it easy to verify education certificates, and it provides approaches for student authentication and the management of education records [16–18].

DOI envisions the process of technology adoption through five stages as shown in Fig. 1. While all stages are important in successful adoption, it is at stage 2 (persuasion) that adopters formulate an opinion about the technology in question, which ultimately determines their decision to adopt or not adopt it. Much of this depends on the perceived attributes of the technology. The first attribute is relative advantage, which is "the degree to which an innovation is perceived as being better than the idea it supersedes" ([13], p. 229). Relative advantage has different expressions, and it could be seen in terms of cost, task completion or even social status. The second attribute is compatibility, which is "the degree to which an innovation is perceived as consistent with the existing values, past experiences, and needs of potential adopters" ([13], p. 15). This innovation aspect is considered important because different groups of adopters are likely to have different values and beliefs. The third attribute is complexity, which refers to "the degree to which an innovation is perceived as relatively difficult to understand and use" ([13], p. 15). This is a negative attribute since technologies which are difficult to learn and apply are likely to have a slower rate of adoption. The fourth attribute is trialability, which is "the degree to which an innovation may be experimented with on a limited basis" ([13], p. 16). Technologies that can be tried have a better chance of being adopted since the potential adopters are likely to see hands-on positive benefits from them. Finally, observability is "the degree to which the results of an innovation are visible to others" ([13], p. 16). This attribute refers to the degree of an innovation's visibility and visibility of the outcomes of its use. Observable innovations that produce visible positive effects are more likely to be adopted.

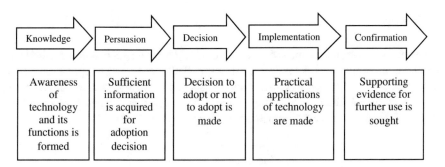

Fig. 1. Technology Adoption Process According to the Diffusion of Innovation Theory [13]

Rogers [13] argued that innovations are adopted faster when there is a stronger presence of all factors except for complexity, which has a negative effect. There is, however, a serious lack of research in higher education institutions investigating such propositions for blockchain adoption [11]. Studies on the effect of the five DOI attributes of blockchain as the drivers for its adoption in other industries also remains inconclusive [19, 20]. In Saudi Arabia, to the best of the author's knowledge, such studies do not exist. To address these gaps, this paper offers an analysis of the influence of blockchain technology attributes on its adoption in Saudi HEIs. The qualitative nature of this research provides a deeper understanding of this phenomenon by uncovering the underlying mechanisms for either the presence or absence of the attributes' effects.

The following research questions were investigated:

1. What blockchain technology factors are relevant for its adoption by Saudi HEIs?
2. What are the mechanisms that support the relevance of these factors in the adoption process?

3 Method

The data collection process involved a series of semi-structured interviews with mid-level and senior-level administrative and technology personnel in Saudi colleges and universities. Semi-structured interviews are commonly used to acquire respondents' insights, views and perspectives on a particular topic [21]. The interview protocol was organized around five topics that engaged the participants to discuss the role of each of the five technology attributes within the DOI theory.

Following the suggestions outlined in the general qualitative research literature [22, 23] and the literature specializing in qualitative interviews [24, 25], a data saturation approach was taken to determine a sample size. Specifically, the process of data collection stopped when no new ideas, themes or insights emerged from the next interview [22]. A directed content analysis technique was used to analyze the collected data since it is considered the best approach to check the assumptions of the existing theories and frameworks [26]. Following Miles and Huberman [27], the analysis involved three steps: data reduction, pattern coding, and data display and interpretation.

4 Results

A convenience sampling approach was used to recruit participants holding middle and senior administrative and technology positions in Saudi HEIs. The sample was balanced in terms of position and type of institution to resemble the distribution of public and private institutions in Saudi Arabia. The saturation data point was reached during interview 11; therefore, the final sample consisted of 10 participants (Table 1).

Table 1. Study Sample

N	Position	Institution	Institution type
P1	IT Support	Department of Education	Ministry
P2	IT Deanship Web Developer	King Khalid University	Public
P3	IT Deanship	Islamic University of Riyadh	Public
P4	Technical Support	King Khalid University	Public
P5	Head of Computer Science Department	Taif University	Public
P6	IT Administrator	KAUST	Private
P7	Service Desk Manager	King Khalid University	Public
P8	Dean of Computer Science and Engineering	Hail University	Public
P9	IT Project Manager	KAUST	Private
P10	e-Learning Specialist	King Khalid University	Public

Relative Advantage. Relative advantage was by far the most consistently and highly rated technology attribute. Five interviewees considered it an "essential technology" for future applications whereas the others called it a "very important" aspect of blockchain. Blockchain's relative advantage was discussed in the context of catering for the growing amount of education which is moving online, the specific needs of HEIs in managing the process, and overall being forward looking, value adding technology with next-level security mechanisms:

> Relative advantage is an essential factor because trust, privacy, and security in blockchain are high. It is essential in terms of the policies that we always work with in relation to cybersecurity and in any conflict. (P2)

> Speaking of the relative advantage of blockchain, first of all, it can cater for specific organizational needs. (P7)

> Blockchain's advantage is in being a purely digital technology. This will be an added value for the University, especially if one of the objectives of the University is to be a digital university. (P9)

Compatibility. With regards to compatibility, the opinions of the study participants were divided. Five interviewees considered it a strong factor in the blockchain adoption process. They spoke about blockchain's compatibility with their institutions' goals and objectives, administrative procedures and educational services:

> I think that in our case, in our university, blockchain is compatible with the University, our vision, message, goals, and so on. (P1)

> Blockchain should be compatible with what we do and, better, with the existing systems we use. (P7)

> Of course, compatibility matters, especially if the technology allows the connection of other tools, I mean, connectibility to other different systems. (P10)

On the other hand, some interviewees expressed the view that compatibility had no meaningful impact or its impact was much weaker in comparison to relative advantage:

> I do not feel that compatibility is necessary for applying blockchain technology because blockchain technology can be applied and adapted to match the environment in which it is applied. (P5)

> With regards to compatibility… Well, in our case, if our school uses a new technology, it must not affect the old systems. It will probably replace them outright. So, it is not an essential factor. (P6)

> I do not think that compatibility or incompatibility is an obstacle or something that may affect the adoption of blockchain technology. (P8)

Accordingly, the following themes emerged: 1) compatibility is sought in all aspects of HEI management: from services to administration thereby making compatibility an important factor in blockchain adoption; 2) from a technical standpoint, compatibility with existing systems may not be that important for the decision makers.

Observability. The opinions regarding this attribute's effect were clearly divided. Some interviewees argued that observability had a strong influence, especially in view of the positive effects in other institutions or in some areas of their own institution:

> If a new technology appears, we must keep pace with it, especially if it is applied in a university. Moreover, suppose I noticed that this university succeeded in using the technology, facilitated many tasks, and reduced costs. In that case, it will encourage the university to apply it. Because I saw, an experiment in other universities in the same region and sector. (P6)

> Hundred percent, observability matters. In our university, we applied many technologies with this. […] Say, this tool is helping you do your job comfortably; you do it efficiently. If there is convincing evidence that it was applied well in some area, then definitely, we will consider its use in other areas. (P7)

Others were somewhat cautious about the effect of the observability factor:

> Ok, I can see, for example, KAUST or the University of Taibah or any university that has adopted the use blockchain. I may start to see what the benefits are, but it

won't be the deciding factor or the only factor influencing our decision to adopt. (P2)

If we see the benefits of blockchain applications, it's not necessarily what we are looking for. We will first consider how blockchain is applied and whether we use the same applications and systems, whether it will benefit us the same. It's a long process, actually, with many variables involved. (P3)

Others expressed the view that observability has a low-to-no impact on blockchain adoption:

I am not sure about the influence of observability. Maybe when the decision-makers see that many universities in Saudi Arabia have successfully applied blockchain, this may help them adopt it in their university because they know its benefits. Again, however, it is not clear right now. (P4)

The adoption of new technologies is based more on politics. [...] Most Saudi universities are public. Much decision making in these institutions depends on bureaucratic procedures and state support. (P8)

In my opinion, observing successful technology adoption may motivate universities to apply it, but it is unnecessary, especially if a university has a strong IT department and implements new technologies regardless. (P5)

Therefore, the following themes emerged during the interviews with regard to observability in relation to blockchain adoption by HEIs: 1) the effect of the observability factor remains unclear; 2) observability will matter if blockchain benefits can be observed both within and outside of a particular HEI; 3) it may take time for a HEI to realise blockchain benefits by observing its effect; and 4) for some HEIs, observability may not matter at all if they have a strong IT development strategy.

Complexity. Opinions regarding the effect of complexity were divided. Those who opined about the influence of the complexity factor agreed that the impact was negative; that is, the more complex the blockchain is considered to be, the worse the administrators' desire to adopt it. However, disagreements arose about the strength of this impact. Those who spoke of a strong impact argued that it would be difficult for users to understand and use:

This factor will negatively impact technology adoption. When the technology is challenging to use and understand and needs a specific infrastructure and competencies, and is financially expensive, it may be a reason for a delay in applying this technology. [...] I see that this is the case with blockchain at the moment. (P4)

I expect complexity is an important factor. Many institutions still do not know exactly what difficulties they may face with blockchain and the weaknesses of this technology. (P5)

Complexity is a significant factor because blockchain is still emerging so it could be hard for users to understand it. (P9)

Several interviewees also acknowledged the impact of complexity, although they thought it would be moderate because of specialist and user training and the learning process:

It is a possible factor, yes. Blockchain is ambiguous. I think the impact of complexity can be negated. (P2)

There is some complexity, perceived complexity barrier in the beginning. But it may not be as strong if users understand it quickly. So, it is a matter of a learning curve. (P10)

Yet others did not believe complexity was a factor in the adoption process. They argued that generally, in the age of technology, complexity is becoming archaic because users are becoming increasingly technology savvy and because it is possible to hire appropriate specialists at the institutional level:

Complexity as a factor? No, no, no. We live in the age of technology. Users, especially young people, they are too well versed with all kinds of emergent technologies. (P2)

No, the complexity of blockchain would not be an obstacle to its adoption because we can use experts who can arrange training for the staff. (P3)

Right now, I think the impact of the complexity factor is not important because each university has an IT department, the staff who work there are rather familiar with new technology, and you can train them easily. Therefore, complexities, even if they exist, will not be an obstacle. (P8)

In the course of the analysis, the following themes emerged: 1) perceived blockchain complexity negatively affects its adoption and it arises from the novelty of the technology and users not being able to understand it outright; 2) the negative impact of complexity can be diminished through training and learning; and 3) the stronger a HEI's IT department is believed to be, the lower the expected impact of complexity.

Trialability. In general, the majority of study participants agreed on the effect of this factor, although to differing degrees. Those who argued in favour of a strong influence connected it to the necessity of trialing any new potentially impactful technology on a limited basis:

Of course, this factor is essential because any new technology must be tested or experimented with to know if this technology is applicable and to what extent and benefit. (P5)

Yes, it is necessary to try out technology. For example, we have three-phase trials in our university tech department for all new technologies. If we see that it has good potential, it has a high chance of being implemented. (P7)

Trialability is very important. Any new technology is always applied on a limited scale before we are sure if it fits our needs. (P8)

Those who believed that trialability only had a moderate degree of influence mostly argued that almost any technology could be trialed, at least from their experience in their HEIs:

> Usually, we need a demo with new technologies. We test and present it to the decision-makers. If they give us their approval, we do not have any problem. (P2)

> We do trial all applications. It is extensive and quick. But, as I said, all applications are triable. (P3)

> Trialability is a factor, to some extent. We, of course, test new technologies to see what they help achieve and how. I do not, however, recollect any difficulties in trying and testing new technologies. (P6)

Finally, one interviewee expressed the view that trialability should not be a concern because all new technologies are tried routinely at her HEI:

> No, trialability is an anachronism. We try all technologies that we consider per-spective, no exception. We test them quickly and then transfer them to the other parties with the most specific features that serve them. (P10)

In the course of the analysis, the following themes emerged: 1) trialability is an influential factor in blockchain adoption, although the strength of the influence is not clear; 2) HEIs that routinely test emerging technologies for school applications are better positioned to ignore the trialability factor.

5   Discussion

The key themes uncovered in relation to the effects of the five technology attributes on blockchain adoption are summarized in Table 2. Three out of five attributes (rel-ative advantage, compatibility, complexity) emerged as influencers in the process of blockchain adoption by Saudi HEIs, while the effect of the remaining two attributes (trialability, observability) is unclear. The effect of relative advantage is considered the strongest which suggests that to blockchain adopters, it is extremely important to under-stand how it is better and more beneficial than the existing systems. This offers some suggestions for blockchain promoters as they should probably focus on this issue first and foremost. It is also important to demonstrate how blockchain will meet the specific needs of each potential organization adopter. However, this may also suggest that there may not be one universal set of benefits that will "sell" blockchain adoption to all HEIs. Overall, blockchain adoption may be a rather holistic process in this regard, given that the effect of the compatibility factor was considered across not only technological but also organizational management, such as its degree of fit with the operational model and strategy.

In relation to the factors that were not considered to have an effect (trialability and observability), an important aspect seems to be organizational innovativeness and resourcefulness. Those respondents who mentioned that their institutions regularly test new technologies, possess resources, and are capable of implementing the required training initiatives quickly seemed to assign lower importance to observability and trialability. According to DOI [13], these organizations could be either innovators or the early majority who are more willing to adopt innovations faster and possess sufficient resources to do this. This could be important given that blockchain is currently in the early stages of adoption in higher education.

Table 2. Summary of Technology Factors

Factors	Themes Related to Blockchain Adoption
Relative advantage	- Strongest perceived influence on adoption; - Catering for the growing amount of education moving online; - Corresponding to specific needs of HEIs is important; - The overall advantage comes from being forward looking and taking advantage of value-adding technology with next-level security mechanisms
Compatibility	- Moderate perceived influence on adoption; - Compatibility is sought in all aspects of HEI management: from services to administration thereby making compatibility an important factor in blockchain adoption; - From a technical standpoint, compatibility with the existing systems may not be that important for the decision makers
Observability	- Unclear effect on adoption; - Observability will matter if blockchain benefits can be observed both within and outside of a particular HEI; - It may take time for a HEI to realise blockchain benefits by observing its effect; - For some HEIs, observability may not matter at all if they have a strong IT development strategy
Complexity	- Moderate perceived effect on adoption; - Complexity arises from the technology novelty and users not being able to understand it outright; - Complexity can be countered with training and learning; - The stronger a HEI's IT department is believed to be, the lower the expected impact of complexity
Trialability	- Unclear strength effect on blockchain adoption; - Trialability is an influential factor in blockchain adoption, although the strength of the influence is not clear; - HEIs which routinely test emerging technologies for school applications are better positioned to ignore the trialability factor

Next, it seems that the technology attributes themselves do not provide a whole picture of the blockchain adoption process. This is because organizational features and factors, as noted above, are present in many mechanisms of each technology attribute's influence on blockchain adoption. At the very least, the results of the study point to the possible effects of factors such as organizational readiness, possession of sufficient resources (financial, technological, human), and sufficient technology knowledge. The possible role of such factors in the blockchain adoption process at the organizational level has been described in other studies [7, 19]. Moreover, hints at the presence of additional contextual variables emerged as well. Examples discussed by the respondents employed in public Saudi HEIs are bureaucracy structures and government support for blockchain adoption. A good approach, therefore, would be to take a more holistic view on the blockchain adoption process by integrating the discussed technology attributes with organizational and, possibly, environmental factors. Technology-Organization-Environment (TOE) framework [28] may serve as a good theoretical foundation for this purpose.

Overall, the practitioners in Saudi HEIs who consider blockchain adoption should 1) understand its functionality and fit with various operational aspects of the organization; 2) clearly understand the benefits over existing systems and whether it is worth implementing; 3) ensure the availability of adequate resources and, possibly, recognize the need to acquire and train staff in the use of new ones.

6 Conclusions and Limitations

This study represents an attempt to uncover the technology factors which play a role in the blockchain adoption process in Saudi HEIs and to understand the mechanisms that could be in play. Due to the qualitative nature of this research, the findings should be considered as exploratory. To confirm the identified relationships, quantitative research is needed. Longitudinal studies could offer an excellent perspective on how these factors influence the blockchain adoption process over time and whether the degree of their influence varies with time. Further, a good approach to create a viable model of adoption would be to test the explored factors within more integrative frameworks. This would offer a bigger picture of adoption and may point to unexpected relationships and combined effects on blockchain adoption with other, non-technological factors. Overall, this study helped to understand certain aspects of blockchain adoption, but it opened many more avenues for continuing the research in the area. Hopefully, the research progress continues with time, and both theoretical and practical knowledge of blockchain adoption in education will expand on par with other industries.

References

1. Allahmorad, S., Zreik, S.: Education in Saudi Arabia. WENR, Riyadh, (2020)
2. Bernacki, M.L., Greene, J.A., Crompton, H.: Mobile technology, learning, and achievement: advances in understanding and measuring the role of mobile technology in education. Contemp. Educ. Psychol. **60**, 101827 (2020)
3. Ratheeswari, K.: Information communication technology in education. J. Appl. Adv. Res. **3**(1), S45–S47 (2018)
4. Nakamoto, S.: Bitcoin (2008). https://bitcoin.org/bitcoin.pdf. Accessed 1 Nov 2020
5. Buterin, V.: A next generation smart contract and decentralized application platform (2014). https://blockchainlab.com/pdf/Ethereum_white_paper-a_next_generation_smart_contract_and_decentralized_application_platform-vitalik-buterin.pdf. Accessed 1 Nov 2022

6. Alammary, A., Alhamzi, S., Almasri, M., Gillani, S.: Blockchain-based applications in education: a systematic review. Appl. Sci. **9**, 2400–2418 (2019)
7. Chen, G., Xu, B., Lu, M.: Exploring blockchain technology and its potential applications in education. Smart Learn. Environ. **5**(1), 1–10 (2018)
8. Bhaskar, P., Tiwari, C.D., Joshi, A.: Blockchain in education management: present and future applications. Interact. Technol. Smart Educ. **18**(1), 1–18 (2020)
9. Awaji, B., Solaiman, E., Albshri, A.: Blockchain-based applications in higher education: a systematic mapping study. In: 5th International Conference on Information and Education Innovations, New York (2020)
10. Lutfiani, N., Apriani, D., Navila, E.A., Juniar, H.L.: Academic certificate fraud detection system framework using blockchain technology. Blockchain Front. Technol. **1**(2), 55–64 (2022)
11. Ullah, N., Al-Rahmi, W.M., Alzahrani, A.I., Alfarraj, O., Alblehai, F.M.: Blockchain technology adoption in smart learning environments. Sustainability **13**, 1801–1818 (2021)
12. El Nokiti, A., Yusof, S.A.M.: Exploring the perceptions of applying blockchain technology in the Higher Education Institutes in the UAE. Proceedings **28**(1), 8–20 (2019)
13. Rogers, E.: Diffusion of Innovations, 2nd edn. Free Press, New York (2003)
14. Lai, P.C.: The literature review of technology adoption models and theories of the novelty technology. J. Inf. Syst. Technol. Manag. **14**(1), 21–38 (2017)
15. Tarhini, A., Arachchilage, N.A.G., Masa'deh, R., Abbasi, M.S.: A critical review of theories and models of technology adoption and acceptance in information systems research. Int. J. Technol. Diffus. **6**(4), 58–77 (2015)
16. Arndt, T., Guercio, A.: Blockchain-based transcripts for mobile higher education. Int. J. Inf. Educ. Technol. **10**(2), 84–89 (2020)
17. Ismail, L., Hameed, H., AlShamsi, M., AlHammadi, M., AlDhanhani, N.: Towards a blockchain deployment at UAE University: performance evaluation and blockchain taxonomy. In: 2019 International Conference on Blockchain Technology, Honolulu (2019)
18. Liu, Y., Li, K., Huang, Z., Li, B., Wang, G., Cai, W.: EduChain: a blockchain-based education data management system. In: CCF China Blockchain Conference, Beijing (2021)
19. Clohessy, T., Acton, T., Rogers, N.: Blockchain adoption: technological, organisational and environmental considerations. In: Business Transformation through Blockchain, vol. 1, pp. 47–76. Palgrave-Macmillan, Cham (2019)
20. Lustenberger, M., Malešević, S., Spychiger, F.: Ecosystem readiness: blockchain adoption is driven externally. Hypothesis Theory **4**, 1–19 (2021)
21. Creswell, J.: Research Design: Qualitative, Quantitative, and Mixed Methods Approaches. Sage, New York (2017)
22. Guest, G., Namey, E., Chen, M.: A simple method to assess and report thematic saturation in qualitative research. PLoS ONE **15**(5), 1–17 (2020)
23. Fusch, P.I., Ness, L.: Are we there yet? Data saturation in qualitative research. Qual. Rep. **20**(9), 1408–1416 (2015)
24. Brinkmann, S., Kvale, S.: Interviews. Sage, London (2015)
25. Saunders, M., Townsend, K.: Reporting and justifying the number of interview participants in organisation and workplace research. Br. J. Manag. **27**(4), 836–852 (2016)
26. Assaroudi, A., Nabavi, F., Armat, M., Ebadi, A., Vismoradi, M.: Directed qualitative content analysis: the description and elaboration of its underpinning methods and data analysis process. J. Res. Nurs. **23**(1), 42–55 (2018)
27. Miles, M.B., Huberman, A.M.: Qualitative Data Analysis: A Methods Sourcebook, 4th edn. Sage, Phoenix (2019)
28. Tornatzky, L.G., Fleischer, M.: The Processes of Technological Innovation. Lexington Books, Lexington (1990)

An Architectural System for Automatic Pedagogical Interventions in Massive Online Learning Environments

Diego Rossi[1], Victor Ströele[1], Fernanda Campos[1], Jairo Francisco de Souza[1], Regina Braga[1], Nicola Capuano[2], Enrique de la Hoz[3], and Santi Caballé[3(✉)]

[1] Department of Computer Science, Federal University of Juiz de Fora, Juiz de Fora, MG, Brazil
[2] School of Engineering, University of Basilicata, Potenza, Italy
[3] Department of Computer Science, Open University of Catalonia, Barcelona, Spain
scaballe@uoc.edu

Abstract. Massive Open Online Courses (MOOCs) are a teaching method that uses Virtual Learning Environments to reach a vast number of students, thus, facilitating access to education by making costs more appealing because of scale economics. Consequently, Tutors' and teachers' interaction is crucial for the successful development of a MOOC. However, due to the size and diversity of the student body in MOOCs, instructors and tutors need help to keep an eye on them carefully and intervene as needed. This work aims to set and validate an architecture for pedagogical interventions in online learning based on how a student feels, using the automatically detected subjective attributes obtained through interactions in the learning management systems. The architecture is based on three layers: (i) the Application layer for managing interaction with the Virtual Learning Environment; (ii) the Knowledge layer for the automatic textual classification, the attributes identification, knowledge representation through ontology and selection of pedagogical intervention actions; and (iii) the Intervention layer carries out pedagogical interventions through an autonomous conversational agent. The proposed architecture can identify the necessary pedagogical intervention, and the conversational agent can make decisions and adopt an approach more suited to the student's needs. The proposed architecture was evaluated using the Stanford MOOC dataset, comprised of 11,042 participants who posted 29,604 messages from eleven courses. The preliminary evaluation results conclude that our approach is able to significantly support the tutor in MOOC environments as 65% of the student posts were automatically managed by the system while only the 35% left needed tutor attention.

1 Introduction

The popularization and advances in Information Technology (IT) have made it increasingly part of our daily lives. Consequently, due to its relevance, it is widely used in the educational environment [20], supporting innovative pedagogical practices, and generating new learning spaces [16]. Furthermore, IT allows

L. Barolli (Ed.): AINA 2023, LNNS 661, pp. 208–221, 2023.
https://doi.org/10.1007/978-3-031-29056-5_20

people to study at distance modalities, which also democratizes access to education since prices are more attractive due to a leaner structure [19]. Another point contributing to increased distance learning is study schedules, which make it more flexible and enable the student to have a full-time job.

Massive Open Online Courses (MOOCs) emerged as an alternative to formal higher education, being made available through technological means, making it possible to serve a wide audience. MOOCs are known as an efficient and important educational tool, but several issues and problems compromise their educational performance. More specifically, the main problems are the number of dropouts during a course, low participation, and lack of student motivation and engagement in general due to several factors, including poor student-student and teacher-student collaboration [3].

According to [10] and [8], one main factor that positively affects learning is motivation. Certainly, tutors play an important role in this process, where they are responsible for interacting with the students to motivate and keep them engaged. To make tutoring more agile and efficient, even automating some tasks, it is necessary to identify which students need specific help. Furthermore, to select the students who will undergo an intervention, it is necessary to identify implicit characteristics in their messages to provide accurate pedagogical intervention.

These challenges demand that tutors accompany students agilely, providing a communication environment capable of answering questions and motivating students [16]. So, we believe that assisted education may contribute to the automatization process of student tutoring. In this context, there are two challenges: the first is sending recommendations and academic topics to students; the second is to predict learning problems and act to minimize their impacts [22].

Discussion forums are among the most popular interaction tools offered by Learning Management Systems (LMS), often used by students to create a sense of belonging and better understand course topics [5]. According to [6] it is possible, through natural language processing (NLP) and predictive models, to detect multiple attributes in post messages, such as Sentiment, Post Type, Urgency, and Confusion. These attributes underlie the choice of pedagogical intervention action. Thus, the semantic detection of forum posts offers implicit information, fundamental for more careful analysis to evaluate the student knowledge, consequently collaborating to moderate and plan interventions. However, students trying to clarify concepts through these forums may not get the attention they need, and the lack of responsiveness often favors dropout [6].

Some computer systems seek to fill this gap through automatic student interactions, such as Conversational Agents (CA), Recommender Systems (RS), Chatbots etc. Usually, these systems use Artificial Intelligence (AI) approaches operating based on a well-defined set of rules that shape their behavior when interacting with humans [9]. Researchers have already applied these computer systems to accomplish various educational goals such as tutoring, question-answering, language learning practice, and the development of metacognitive skills [12].

This works aims to set and validate an architecture for pedagogical interventions in online learning. Knowing students' educational moments is necessary to identify those who need help. We automatically identify subjective attributes in messages from students in virtual learning environments. The attributes identified were Sentiment, Confusion, and Urgency, to determine the most appropriate pedagogical intervention for the student.

The article is structured as follows: Sect. 2 presents concepts that serve as a basis for the development of the proposal of this work. Section 3 presents the proposed architecture and details the main layers that make up the solution workflow. Section 4 brings the evaluation of the proposed architectural system. Finally, Sect. 5 concludes the paper highlighting the gaps and next steps of the research.

2   Background

The term intervention is used in several areas, such as Psychology, Medicine, Administration, and Education. More precisely, in education, it is common to call this intervention a pedagogical intervention. According to [11], these interventions aim to increase motivation, mnemonic skills, self-regulation, study-related skills such as time management, and even general skill itself; create positive attitudes towards content and context; and minimize learning pathologies. Interventions can be broadly classified as cognitive, metacognitive, and affective [1].

- Cognitive interventions are those that focus on developing or enhancing specific task-related skills.
- Metacognitive interventions are those that focus on self-management of learning, that is, planning, implementing, and monitoring a person's learning efforts, and conditional knowledge of when, where, why, and how to use specific tactics and strategies in their activities in appropriate contexts.
- Affective interventions are those that focus on non-cognitive aspects of learning, such as motivation and self-concept.

Students with learning difficulties require pedagogical intervention from teachers, such as offering both oral and written help and allowing extra time to complete the tasks. Learning must be linked to the affective act, it must be pleasant and stimulating [15].

In the work [25], a conceptual framework was proposed offering individualized intervention for blended learning from the perspective of learning analytics, as shown in Fig. 1. The framework describes five steps of implementing the learning intervention: data collecting and screening, data mining and analysis, identifying intervention targets, implementing intervention measures, and evaluating intervention effects.

According to [18], e-Learning platforms store a large amount of data from all student interactions with the platform. These interactions are related to course navigation events, video events (time of the video reproduction, speed, etc.), and exercise logs (attempts, scores, tips used, etc.). In addition, social interactions,

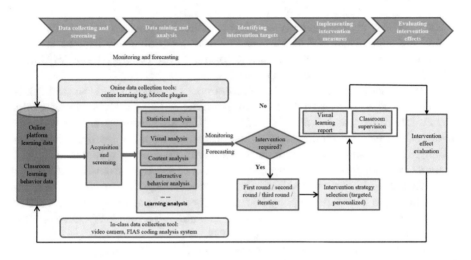

Fig. 1. Conceptual framework for learning intervention

such as forum postings can be used to detect attitudes and sentiments. All of this data can be used to detect problems based on the data collected, but you can also use this information to predict behavior and outcomes. Based on predictions, teachers can anticipate possible problems, providing adaptations to the course or methodology, so instructors or even the institution can rethink the curriculum design or carry out interventions to improve the learning experience.

It is very common for a post to be accompanied by a sequence of events, which makes it even more difficult to get the necessary attention from an instructor in a virtual learning environment (VLE). Trying to solve this problem, students ask colleagues to vote for their posts for the tutor to see and carry out a necessary intervention. Three models were proposed for predicting the instructor's intervention taking into account high-level resources, such as post initiated by an anonymous user, approved, unresolved, or deleted, forum ID, time, last time post, the total number of posts on the topic, word count related to rating, technical, conclusive and request issues, the sum of votes, etc. [7].

According to [13], teacher interventions can be defined as teacher-student assistance that supports individual learning and students' problem-solving process, enabling students to work independently. According to [4] and [23], there are three types of intervention intention. The first is the diagnostic intervention, where an evaluation is carried out throughout the process. The second intention is the tip when something is recommended so that the student can overcome a problem. Finally, the last one is intended to provide feedback, so a comment is provided to the student.

The authors in [14] show an interesting strategy using a predictive system to identify students who are at risk and apply intervention guidelines so that the student can be successful. The authors highlight that existing academic early warning systems have some shortcomings as the fairly common problem is that those systems typically employ a general prediction model that fails to address the complexity of all courses. Another problem is that most early warning systems are designed for online courses or are too dependent on Course Management System (CMS) access data.

Some works use attributes that express how the student feels to avoid data directly linked to the environment. For example, in [24] the detection of confusion and its impact on learning is performed. The educational process frequently involves battles with confusion. An instantaneous response clarifying the confusion can accompany the student to overcome it, causing a beneficial effect. However, the delay in clearing up the confusion can detract from the experience of participating in the course, leading to dropout along the way. Therefore, the confusion experienced during the learning process cannot always be associated with negative results. This confusion does not hinder the student's development and learning and must be resolved promptly.

Sentiment analysis is also important in MOOC environments. According to [17], identifying whether forum messages are positive or negative can give an insight into how students feel about the course, aiming to increase engagement and student satisfaction. Furthermore, message polarity can help identify more complex emotions such as excitement, frustration, or boredom. The analysis carried out in [17] shows a more positive trend towards the messages at the beginning of the course, and the positivity decreases at certain peaks, even days before the deadline for the assessment tasks.

3   The Architectural System

The proposed architecture, PRED-INTER, aims to help tutors and students carry out pedagogical interventions considering the student's educational moment. PRED-INTER is based on how the student feels, using the automatically detected subjective attributes obtained through interactions in the learning management systems. It is important to emphasize that evaluating each student's participation and identifying the necessary pedagogical intervention is essential, as it is impossible to perform this task manually.

The architecture collaborates by monitoring the forums of virtual environments to identify students' needs and automatically help them. So they avoid feelings of abandonment and demotivation. However, some cases are identified as more critical; for example, when the post sentiment is negative, students need more attention. In these cases, our solution alerts the tutors so they can carry out an individualized follow-up. On the other hand, less critical cases, defined according to the attributes, can be monitored automatically through motivational, informative, or thank-you messages or by automatic interaction. These kinds of pedagogical interventions will prevent the student from creating the sentiment of abandonment and help the tutor avoid task overload.

Figure 2 presents the layers and workflow that make up the proposed architecture to carry out the pedagogical intervention. Next, we describe the main three layers:

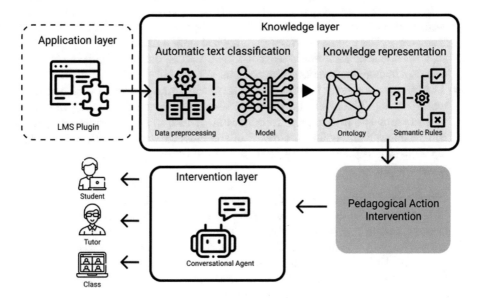

Fig. 2. PRED-INTER Architecture

- The Application layer is responsible for interacting with VLEs and capturing students' messages. These messages can come from chats, forums, or any other form of textual interaction. For the proposed architecture, in addition to the posted message, it is also essential to identify the author.
- The Knowledge layer has a very important part of the work to carry out. It has two main attributions and is divided into two modules. The first module is responsible for automatic textual classification, where the attributes of sentiment, confusion, and urgency are automatically identified in messages from pre-trained models. The second module is responsible for the representation of knowledge. The information is stored so that the domain knowledge is represented, allowing the understanding of the information related to the message. To this end, the Knowledge layer starts with text pre-processing, where unwanted text parts are removed, and then, textual classification is

responsible for identifying the attributes implicit in the messages, where pre-trained models are used to identify these attributes. Then, in the representation module, the post information is stored, along with the identification of the attributes. Storage takes place in an ontology in the second module. Semantic rules are employed to identify the pedagogical intervention necessary to meet the student's post type. These rules guide decision-making, resulting in the choice of which pedagogical intervention is most appropriate to meet the student's needs at that moment. The output of this layer is the selected pedagogical action intervention.

- The Intervention layer is composed of an autonomous conversational pedagogical agent [3]. The agent assists the student and the tutor, sending messages to carry out the pedagogical intervention. It also includes sending messages to the class. The mediation content of the message depends on the identified attributes, which are previously stored in the ontology, as well as the type of intervention.

The type of intervention is also pre-defined and may vary according to the class or course. It may be an automatic message to the student or it may be a message asking the class to help the student. When a post in a discussion forum indicates that the student needs more than a message for pedagogical intervention, the agent may start and conduct an automatic interference (help) or ask the tutor for help. For cognitive intervention, the subject in the message is recognized and pre-selected educational resources are recommended to help the student with the content [21].

This autonomous interference and recommendation of educational resources is based on a recommender component, which defines what will be recommended and the priority that each item will have. The focus of this research is not the integration of a recommendation system with the conversational agent, and in this research, in more complex cases where the agent cannot help the student, the post is forwarded to the tutor.

Before formulating the semantic rules, an interpretable decision model was defined based on subjective attributes, making it possible to identify the appropriate pedagogical intervention. This model is represented through a hierarchical diagram. From the diagram in Fig. 3, it is possible to identify the different types of intervention according to the student's needs at the time of posting the message. Based on the attributes of sentiment, confusion, and urgency, it is possible to assume how the student feels and thus carry out an intervention that is more meaningful to him.

For a pedagogical intervention to take place, it may be necessary to carry out several actions; for example, a Class Help sends a message to ask the classmates to help the student but also asks the tutor to accompany this intervention.

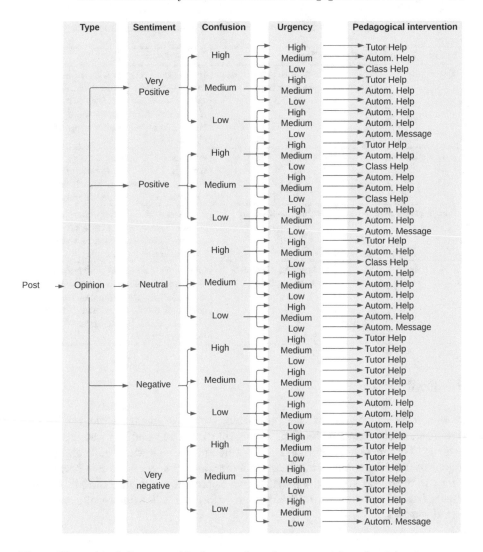

Fig. 3. Hierarchical diagram with the post-formal representation of opinion-type message

Table 1 presents the pedagogical interventions and the actions demanded by them. The Automatic Message intervention can be a thank you, a motivational, or a contact message. The message type is defined according to the student's post-message attributes. For example, a message with a positive sentiment, low confusion, and low urgency receives an automatic message intervention, which can be a thank-you message type.

Table 1. Actions taken for each type of pedagogical intervention

Pedagogical intervention	Actions	Description
Tutor help	Message to student and Message to tutor	In this type of intervention, a message is sent to the students, informing them that they are being monitored. At the same time, the tutor is notified to offer support to the student
Automatic help	Interaction with the student	Automatic help start a dialog and sends a content recommendation to the student, based on the concepts identified in the student's message
Automatic message	Message to student	The automatic message answers the student's post with messages of different purposes, such as thanking, motivating, or contacting
Class help	Message to class and Message to tutor	Class help facilitates dialogue among students in the same class, sending a message encouraging colleagues to comment on the post. The tutor is also alerted to follow up on this intervention

4   Evaluation

In this section, we present the methodology used to evaluate the architecture described in the previous section. A general evaluation method was used in the context of online education research described in [2]. Following this method, first, a goal to describe the specific functional outcome of this research is formulated. Then, information about the participants, apparatus, and procedure to conduct the evaluation are described. Finally, evaluation and validation results are shown and discussed.

4.1   Research Goals

The main research goal is to answer the question: "How should be an architecture of an automatic system to support pedagogical interventions in online learning that maximizes the learning experience?".

4.2   Participant Data and Apparatus

The work was developed using the Stanford MOOC Posts. This dataset has 11,042 participants who posted 29,604 messages, from 11 different courses, in the areas of education, humanities, and medicine. Each message was manually annotated, informing attributes such as post type (opinion, question, and answer), sentiment polarity, level of confusion and urgency.

The data from the students' posts are loaded into the ontological structure, allowing the information to be consistently related and inferences to be made. To accurately determine the pedagogical intervention process that the agent should use, it is necessary to apply semantic rules, which guide us in choosing the pedagogical intervention process, taking into account the attributes related to the post. For constructing the ontology, the Protégé tool was used together with the OWL language, as it is a semantic markup language that allows describing the classes and their relationships. We loaded the data by using the Java language together with the OWL API, which allowed filling the instances after reading the posts and their data in the dataset.

4.3 Procedure

The purpose of this evaluation is to analyze the proposed architecture with respect to its ontological rules from the point of view of professors and tutors in the context of providing useful pedagogical interventions. We thereby derive our general research goal into the following questions: **Q1)** Can the ontology be used to provide semantic rules for the conversational agent? **Q2)** Is the solution able to guide the agent to identify pedagogical interventions to help professors and tutors in MOOCs environments?

For the data to be loaded into the ontology, it was necessary to perform a pre-processing step, where some data needed to be converted to the expected standard. Originally, the dataset has the post type divided into three attributes, opinion (yes/no), question (yes/no), and answer (yes/no), so it was necessary to check which attribute has the answer "yes" and convert it for the post type.

The sentiment attribute also needed a conversion, as the values ranged from 1 to 7, with being 1 the most negative sentiment and 7 being the most positive one. As the framework works with very negative, negative, neutral, positive, and very positive sentiments, the discretization of sentiment is realized into five classes, we have the following distribution: sentiment values less than 2 represent a very negative polarization, values greater than 1.5 and less than 3.5 correspond to negative, greater than 3 and less than 5.5 represent neutral, those greater than 5 and less than 6.5 are positive, and those greater than 6 are very positive. For the attributes confusion and urgency, the low class, which represents little or no presence of this attribute, was assigned to values less than or equal to 3. The class that represents the average presence of the attributes grouped values greater than or equal to 3.5 and less than or equal to 5, while the high was represented by values greater than or equal to 5.5.

4.4 Evaluation Results and Discussion

In the dataset, we found 16.469 (55.6%) posts representing the students' opinions. We created semantic rules to obtain the pedagogical intervention for these types of posts. Some of them were selected as a sample and presented to show the necessary interventions. Table 2 summarizes the results of this step, presenting the percentage of data considering the ontology class "Type".

Table 2. Distribution of interventions generated from the application of semantic rules.

Intervention	Count	Percent
Tutor help	1199	7.23%
Automatic help	12453	75.13%
Class help	949	5.73%
Automatic message	1975	11.91%

In this evaluation step, considering the rules activated by the ontology in relation to the student's post, we analyzed the agent action and the pedagogical intervention. Considering the aggregate number of instances that comprise the dataset, we observed that about 44.7% of the messages are classified as positive or neutral, not confusing, and not urgent. Thus, according to the identified rules, the conversational agent acts autonomously, and automatically answers the post, preventing the tutor from having to waste time answering them.

On the other hand, some messages need immediate intervention, and, in these cases, the tutor or professor must analyze them specifically. Figure 4 shows a sample of students' posts, the conversational agent's action, and the pedagogical intervention. When the student sent the message, it was processed by the ontology, and the agent took action. As it was a negative sentiment message, the conversational agent sent an automatic message in the class forum and asked for tutor help in a private chat. In this way, the student knew that something was done concerning his or her message.

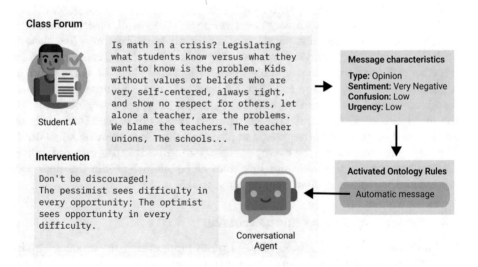

Fig. 4. Sample of one of the negative messages sent by a student in the MOOC and the conversational agent actions.

The messages sent by the conversational agent may vary according to the identified pedagogical intervention process. Therefore, some standard messages are:

1. **Agent help and Tutor help:** Hilary, I noticed that you need help urgently, so I've already notified the pedagogical team. The tutor will contact you shortly. In the meantime, can I offer help to [Concept1]? Thanks for your contact.
2. **Agent thank message:** James, your participation was crucial. Dedication and studies are the secrets to success, so try to contribute as possible. It makes our interaction environment richer. Thank you for your contact.
3. **Agent help:** Oliver, I realized you need help. Do you want to get help for [Concept1]?
4. **Agent class help and Tutor help:** Do you agree or disagree with what was said by Chloe? I would like classmates to help us with this task. The tutor was also invited to this conversation. Thanks for listening.

Considering Table 2, two types of intervention automatically perform the interaction with the student: automatic help and automatic message. Together they represent a considerable number of interventions, approximately 87%. Regarding interventions that depend on human intervention, we have class help, which corresponds to about 5.73%, and tutor help, which corresponds to 7.23%.

5 Conclusions and Future Work

After the ontology was developed, it was possible to load the posts and their data and the links between them. The dataset had 29,604 posts from 11,043 different students, therefore all messages were stored. By loading the post into the ontology, we could make inferences based on the previously defined rules. Thus, the necessary pedagogical intervention was identified, and the conversational agent could make decisions and adopt an approach more suited to the student's needs.

At this point, we were in the position to answer our research questions, as the first question seeks to verify if the ontology can provide semantic rules for the agent. According to the numbers obtained after loading the data, it was possible to conclude that ontology provides the rules. The second question investigated the solution's ability to help the tutor in MOOC environments, as shown above, in addition to identifying the necessary pedagogical intervention, only about 35% of the posts needed the tutor's attention.

These results help us approach a solution to the fundamental problem that motivates our research in the context of massive online learning, where the student/tutor ratio is very high and it is not always possible to respond immediately to all posts in the forums, which can negatively affect the student, making them feel abandoned and often insecure for not being closely monitored by a person able to help them. Therefore, the presented approach of a conversational agent providing automatic pedagogical support can significantly reduce the number of messages that need attention, leaving only the most critical posts for the tutor. It helps the tutor to be more agile and act better in serving the students. In addition, no student remains unanswered, so we greatly reduce the possibility of the students feeling abandoned by the tutor.

To sum up, in this paper we presented an ontology with a set of semantic rules, capable of guiding the conversational agent, indicating the type of pedagogical intervention. This approach can contribute a lot to the quality of teaching, through this help in interactions through student posts, mainly in MOOC environments, where the tutor has a huge difficulty keeping up with all the students, therefore, it also contributes to the tutor as the pedagogical interventions happen automatically.

Some limitations were observed in the development of the work, such as classifying new posts, that is when we do not have the attributes to describe the type, feeling, confusion and urgency. In future work, we intend to resolve these limitations by performing the automatic classification of posts, as also suggested in [6]. This can contribute to expanding posts beyond those present in Stanford MOOC Posts dataset.

References

1. Barbier, K., Struyf, E., Verschueren, K., Donche, V.: Fostering cognitive and affective-motivational learning outcomes for high-ability students in mixed-ability elementary classrooms: a systematic review. Eur. J. Psychol. Educ. **38**, 83–107 (2022)
2. Caballe, S.: A computer science methodology for online education research. Int. J. Eng. Educ. **35**(2), 548–562 (2019)
3. Caballé, S., Conesa, J., Gañán, D.: Evaluation on using conversational pedagogical agents to support collaborative learning in MOOCs. In: Barolli, L., Takizawa, M., Yoshihisa, T., Amato, F., Ikeda, M. (eds.) 3PGCIC 2020. LNNS, vol. 158, pp. 199–210. Springer, Cham (2021). https://doi.org/10.1007/978-3-030-61105-7_20
4. Gurel, Z.C., Bekdemir, M.: The teacher and peer intervention for pre-service mathematics teachers on the validity of mathematical models. Pedagogical Res. **7**(2), em0120 (2022)
5. Capuano, N., Caballé, S.: Towards adaptive peer assessment for MOOCs. In: 2015 10th International Conference on P2P, Parallel, Grid, Cloud and Internet Computing (3PGCIC), pp. 64–69. IEEE (2015)
6. Capuano, N., Caballé, S.: Multi-attribute categorization of MOOC forum posts and applications to conversational agents. In: Barolli, L., Hellinckx, P., Natwichai, J. (eds.) 3PGCIC 2019. LNNS, vol. 96, pp. 505–514. Springer, Cham (2020). https://doi.org/10.1007/978-3-030-33509-0_47
7. Chaturvedi, S., Goldwasser, D., Daumé III, H.: Predicting instructor's intervention in mooc forums. In: Proceedings of the 52nd Annual Meeting of the Association for Computational Linguistics (Volume 1: Long Papers), pp. 1501–1511 (2014)
8. Cuevas, R., Ntoumanis, N., Fernandez-Bustos, J.G., Bartholomew, K.: Does teacher evaluation based on student performance predict motivation, well-being, and ill-being? J. School Psychol. **68**, 154–162 (2018)

9. Demetriadis, S., et al.: Conversational agents in MOOCs: reflections on first outcomes of the ColMOOC project. In: Intelligent Systems and Learning Data Analytics in Online Education, pp. xxxvii–lxxiv (2021)
10. Fandiño, F.G.E., Velandia, A.J.S.: How an online tutor motivates e-learning English. Heliyon **6**(8), e04630 (2020)
11. Hattie, J., Biggs, J., Purdie, N.: Effects of learning skills interventions on student learning: a meta-analysis. Rev. Educ. Res. **66**(2), 99–136 (1996)
12. Khanal, S.S., Prasad, P.W.C., Alsadoon, A., Maag, A.: A systematic review: machine learning based recommendation systems for e-learning. Educ. Inf. Technol. **25**(4), 2635–2664 (2020)
13. Leiss, D.: Adaptive lehrerinterventionen beim mathematischen modellieren-empirische befunde einer vergleichenden labor-und unterrichtsstudie. J. für Mathematik-Didaktik **31**(2), 197–226 (2010)
14. Marbouti, F., Diefes-Dux, H.A., Madhavan, K.: Models for early prediction of at-risk students in a course using standards-based grading. Comput. Educ. **103**, 1–15 (2016)
15. Máximo, V., Marinho, R.A.C.: Intervenção pedagógica no processo de ensino e aprendizagem. Braz. J. Dev. **7**(1), 8208–8218 (2021)
16. Moreno-Guerrero, A.-J., Aznar-Díaz, I., Cáceres-Reche, P., Alonso-García, S.: E-learning in the teaching of mathematics: an educational experience in adult high school. Mathematics **8**(5), 840 (2020)
17. Moreno-Marcos, P.M., Alario-Hoyos, C., Muñoz-Merino, P.J., Estévez-Ayres, I., Kloos, C.D.: Sentiment analysis in MOOCs: a case study. In: 2018 IEEE Global Engineering Education Conference (EDUCON), pp. 1489–1496. IEEE (2018)
18. Moreno-Marcos, P.M., Alario-Hoyos, C., Muñoz-Merino, P.J., Kloos, C.D.: Prediction in MOOCs: a review and future research directions. IEEE Trans. Learn. Technol. **12**(3), 384–401 (2018)
19. Palvia, S.: Online education: worldwide status, challenges, trends, and implications. J. Global Inf. Technol. Manage. **21**(4), 233–241 (2018)
20. Panigrahi, R., Srivastava, P.R., Panigrahi, P.K.: Effectiveness of e-learning: the mediating role of student engagement on perceived learning effectiveness. Inf. Technol. People **34**(7), 1840–1862 (2020)
21. Rossi, D., et al.: CAERS: a conversational agent for intervention in MOOCs' learning processes. In: Guralnick, D., Auer, M.E., Poce, A. (eds.) TLIC 2021. LNNS, vol. 349, pp. 371–382. Springer, Cham (2022). https://doi.org/10.1007/978-3-030-90677-1_36
22. Toti, D., Capuano, N., Campos, F., Dantas, M., Neves, F., Caballé, S.: Detection of student engagement in e-learning systems based on semantic analysis and machine learning. In: Barolli, L., Takizawa, M., Yoshihisa, T., Amato, F., Ikeda, M. (eds.) 3PGCIC 2020. LNNS, vol. 158, pp. 211–223. Springer, Cham (2021). https://doi.org/10.1007/978-3-030-61105-7_21
23. Tropper, N., Leiss, D., Hänze, M.: Teachers' temporary support and worked-out examples as elements of scaffolding in mathematical modeling. ZDM **47**(7), 1225–1240 (2015)
24. Yang, D., Wen, M., Howley, I., Kraut, R., Rose, C.: Exploring the effect of confusion in discussion forums of massive open online courses. In: Proceedings of the Second (2015) ACM Conference on Learning@ Scale, pp. 121–130 (2015)
25. Zhang, J.-H., Zou, L., Miao, J., Zhang, Y.-X., Hwang, G.-J., Zhu, Y.: An individualized intervention approach to improving university students' learning performance and interactive behaviors in a blended learning environment. Interact. Learn. Environ. **28**(2), 231–245 (2020)

Traffic Light Algorithms in Smart Cities: Simulation and Analysis

Artem Yuloskov[1]([✉]), Mohammad Reza Bahrami[1], Manuel Mazzara[1],
Gerald B. Imbugwa[1], Ikechi Ndukwe[1], and Iouri Kotorov[2,3]

[1] Innopolis University, Universitetskaya Str., Innopolis 420500, Russia
`a.yuloskov@innopolis.university`
[2] Karelia University of Applied Sciences, Tikkarinne, 80200 Joensuu, Finland
[3] Institut de Recherche en Informatique de Toulouse (IRIT), Université de Toulouse,
31062 Toulouse, France

Abstract. One of the most important components of a smart city is
smart transport. To design large-scale smart transport systems, simulations are integral to testing the efficacy of various traffic light control
algorithms. The traffic light algorithm designers take advantage of the
simulation software to build reliable and robust algorithms. In this work,
traffic light simulation software was designed, implemented, and tested.
The program runs in a web browser and does not require installation. The
roads and intersections are JSON-configurable, and the algorithms can
be written in JavaScript. The simulation shows real-time statistics of the
algorithms' performance. The result of the work is a working prototype
of traffic light simulation software.

Keywords: Traffic light management · traffic light algorithm ·
simulation · frame work

1 Introduction

A smart city is a city in which many processes are automated and intelligent,
and existence is sustainable [1,2]. Smart cities also have intelligent transportation systems [1,3] that use various sensors to assess traffic conditions. Smart
transportation systems offer numerous benefits, such as lower traffic congestion,
decreased air and noise pollution, savings on fuel consumption and cost, and
less time lost in traffic [4,5]. These benefits lead to improved environmental
conditions and quality of life for road users.

Simulation software can determine the efficacy of different traffic light control
algorithms. Simulation software provides researchers with an interface to control
traffic lights by testing various algorithms in an interactive and intuitive environment. The interactivity is achieved through adaptive transport flow parameters
and the flexible structure of the roads. Some of this software requires installation, while others are web-based. Web-based simulation software needs only a
web browser to work, unlike desktop-based software, which requires installation
on a local machine. Examples of traffic light simulation software are described

© The Author(s), under exclusive license to Springer Nature Switzerland AG 2023
L. Barolli (Ed.): AINA 2023, LNNS 661, pp. 222–235, 2023.
https://doi.org/10.1007/978-3-031-29056-5_21

in Sect. 2. The software implemented in this work runs in a web browser and does not require installation.

Section 3 outlines the main components and the mathematics behind the designed simulation software. In Sect. 4, results obtained from testing the software are reported and discussed. Lastly, in Sect. 5, we conclude stating directions for future work.

2   Related Work

2.1   Existing Traffic Light Simulation Software

SUMO (Simulation of Urban Mobility) is an open source and extremely adaptable traffic light simulation software written in C++ and Python [6]. It can simulate traffic networks, which include automobiles, public transportation, and pedestrians. Many supporting tools come with SUMO, like network import, route estimation, visualization of traffic, and emission calculation. SUMO can be configured to use custom models, and the simulation can be controlled via an Application Programming Interface (API).

SUMO simulator is useful for testing various traffic management algorithms and approaches. For example, Axel Wegener et al. designed a simulation in which vehicles can interact with each other [7]. Such simulations are called vehicular ad-hoc networks (VANETs). The goal of VANET applications is to increase transportation safety and capacity. The authors tested this approach with traffic lights in SUMO. The driver receives the information about traffic lights in advance and can adjust their behavior accordingly. The authors' method is a versatile and efficient way to create a feedback loop that can positively influence road situations and affect the fuel consumption of vehicles.

Multi-model open-source vehicular-traffic Simulator (**MovSim**) [8] is another traffic simulator. MovSim is a web-based traffic simulator. The simulator implements a variety of car-following models from the textbook Traffic Flow Dynamics [9]. The simulator is written in Javascript and has a web interface. It has several preset road configurations like a roundabout, intersection, and ramp. These configurations can be changed dynamically during the simulation.

One drawback of using MovSim is that custom road configurations not included in the preset road configurations cannot be simulated. Also, custom traffic light algorithms cannot be tested in this simulator.

2.2   Traffic Light Simulation for Custom Algorithms

Many works have been published recently to design and study traffic light algorithms. For example, Ach Maulana Habibi and Yusuf Rahadian [10] built a traffic light simulation software with Unity 3D platform using fuzzy logic control. They designed a system with adaptive traffic lights. The results showed that adaptive traffic lights can regulate road intersections more efficiently than non-adaptive traffic lights.

Farooqi et al. [11] also designed and implemented a traffic light simulation software. The authors implemented the program in C# using the Visual Studio .NET 2008. So far, the system can only manage straight roads and requires enhancements. The authors used it because no alternatives were found to test the idea of genetic algorithm application for traffic light management. After simulating for 10 h with 63 chromosomes, the total wait time of the system has decreased by 34 h. This result showed that the use of a genetic algorithm for traffic management can decrease road congestion.

In the work by Marco Wiering et al. [12], the authors suggest a Reinforcement Learning (RL) based adaptive traffic controller and compared it with existing traffic control algorithms. The authors also created the Green Light District (GLD) simulator in Java to simulate an adaptive controller and compared it to other traffic light algorithms on a variety of infrastructures and traffic patterns. In comparison to manually built non-adaptive controllers, simulations revealed that the RL-based algorithm lowered average waiting times by more than 25% in congested traffic.

2.3   Traffic Light Management Algorithms

There are several existing approaches for managing traffic lights like Expert Systems, Fuzzy Logic, Prediction-based optimization, Evolutionary algorithms, and Reinforcement Learning algorithms.

Expert Systems: A program that mimics the decisions of a human expert is called an expert system. Such a system can be used to manage traffic lights. For example, W. Wen proposed a dynamic and automatic traffic light control expert system (DATLCES) [13]. It includes a set of rules to make decisions about traffic lights. For example, some rules depend on inter-arrival time and signal type. Applying these rules results in an efficient algorithm for managing traffic.

Fuzzy Logic: A fuzzy logic control system operates on a continuous set of analog values between 0 and 1. This approach is widely used for traffic light management. For example, Koukol et al. conducted a literature review on the use of fuzzy logic theory to control traffic systems [14]. The study looked at a variety of ways for describing and predicting drivers' behavior and optimizing traffic flow.

Prediction Based Optimization: Prediction based optimization method uses analyzed data to predict the state of the system. This approach can be used to manage traffic flow. Chavhan et al. proposed a technique for traffic management based on predictive information. The inputs for the proposed analysis and prediction model were traffic flow characteristics, historical data, and spatiotemporal correlation data. The proposed algorithm reduced traffic congestion, the growth of traffic density, and travel time.

Evolutionary Algorithms: Evolutionary Algorithms are search methods based on Darwinian evolution that capture global solutions to optimization problems. These algorithms are robust and adaptable. The use of such algorithms is a good fit for traffic light control as it is an optimization problem with constantly changing input. Shaikh et al. gave an overview on employing evolutionary algorithms, swarm intelligence algorithms, and other population-based meta-heuristics to solve the traffic signal control problem [15]. The selected decision variables, the optimized objective, the issue formulation, and the solution encoding were used to evaluate these algorithms. Extracted data demonstrated a growing interest in using evolutionary algorithms and swarm intelligence to control traffic signal problems over the last two decades.

3   Main Components

To implement a physical simulation of vehicle movement we took the formulas from Reynolds's work [16]. The formulas for steering, path following, and arriving are from this work. Now, let's briefly discuss the core classes used in the app including vector, canvas, path, vehicle, path configuration, getting the route from path configuration, traffic light algorithm and main. Firstly, we are going to discuss the utility classes and then the classes that contain the business logic.

Vector: Vectors are useful for building physics simulations. Custom vectors were written which supports addition, subtraction, scaling, normalization, dot product, calculating the angle between two vectors, and distance between two vectors.

Canvas: HTML canvas is a tool for drawing complex Figures on a web page using JavaScript. The component accepts the method "draw" as an argument and calls it at a specified frame rate.

Path: Path class is a core class of the app. It implements the data structure for the path which vehicles will follow. It also implements methods for drawing the path and the traffic lights on the canvas. A Path has *radius* which is a number that describes path width and also it has *points*. Each point is an object, it has properties that describe the coordinates and the point type. A point can be either *road*, *verticalLight* or *horizontalLight*. The separation of the horizontal and vertical lights is needed to properly draw the traffic light on the road. Depending on the type it will be either horizontal - green, yellow, red (from left to right, or vertical - green, yellow, red (from top to bottom).

Besides being a container for storing the points, the *Path* class also has two methods for drawing. The method *draw()* draws the path on the canvas. The method *drawLights()* draws the traffic lights on the path. The methods are separated for the sake of efficiency. The path and the traffic lights are drawn on the different layers because the road should be drawn only once and then it will not change, however traffic lights may change during the execution of a program. Therefore, the program can avoid redrawing the road every time it updates the traffic lights.

Vehicle: The Vehicle is one of the main classes in the app. It is used to determine the autonomous agent's behavior. The logic of the autonomous agents is implemented inside this class. It is also responsible for drawing the vehicle on the screen. The Vehicle has properties, which are used to determine the behavior of the vehicle and it has methods, which are used to change the properties of the vehicle (Fig. 1).

```
constructor(params?: {
  location?: Vector;
  velocity?: Vector;
  acceleration?: Vector;
  color?: string;
}) {
  if (!params) return;

  this.location = params.location || this.location;
  this.velocity = params.velocity || this.velocity;
  this.acceleration = params.acceleration || this.acceleration;
  this.color = params.color || this.color;
}
```

Fig. 1. The vehicle constructor.

The Vehicle class has the implementation of the movement logic. The methods responsible for it are $seek()$, $follow()$, $avoid()$, $arrive()$, $followRoad()$, and $draw()$. First, we will discuss the key method of vehicle movement implementation - $seek(target : Vector)$. This method accepts target point of type Vector and applies the force to the vehicle in the direction of the target point. The formula for steering force according to Reynolds [16] is

$$steeringForce = truncate(steeringDirection, maxSpeed)$$

$Truncate$ here is a function which cuts the force, so that vehicle could not be faster than its maximal speed. The $steeringDirection$ formula is $steeringDirection = target - location$. Then, the flow for this method is as follows: get the steering direction using target point and vehicle location, calculate the steering force, and apply this force to a vehicle. The method implementation is shown on the Fig. 2.

Another method that is needed for vehicle implementation is the method $arrive(target : Vector)$. This method will stop the vehicle if it reached the target point. The method checks if the distance to the target point is less than $nearDistance$ constant and applies the force needed either to stop on this point or to move closer to it.

Now, using the $seek()$ method, we can discuss the $follow(path : Path, vehicles : Vehicle[])$ method. It accepts the $path$ the vehicle should follow as a first argument, and as the second argument, it accepts the other vehicles on this path. The method should steer the vehicle through this $path$ while avoiding other $vehicles$ on it.

```
seek(target: Vector) {
  let desiredVelocity = target.subtract(this.location);
  desiredVelocity = desiredVelocity.normalize();
  desiredVelocity = desiredVelocity.scaleBy(this.maxSpeed);

  // Reynold's formula for steering force
  let steer = desiredVelocity.subtract(this.velocity);
  steer = steer.limit(this.maxForce);

  this.applyForce(steer);
}
```

Fig. 2. The *seek* method.

The implementation of the path following behavior is described in Reynolds's work [16]. The high-level logic is the following. To understand in which direction the vehicle is going and then correct it, the prediction of the future location of the vehicle is needed. The calculation of the prediction vector is straightforward - $prediction = norm(velocity) * lookAheadDistance$. Using the normalized velocity, we can get the *direction* in which the vehicle is moving. Then, multiplying it by *lookAheadDistance*, the location of a vehicle can be predicted. *LookAheadDistance* is just some scalar, which can be set up in the configuration. Now, knowing the prediction vector, we can calculate the normal from the prediction vector to the path segment. This normal point will be the *target* point of the vehicle. Then, the algorithm calls the *seek*() method with the *target* point as an argument. It results in a natural-looking path following behavior.

Another duty of the *follow*() method is to avoid the other vehicles on the same path. It is done through the use of the method *avoid(vehicles : Vehicle[]) : boolean*. This method returns *true* if the vehicle is currently avoiding the other vehicles or *false* if not. The method *avoid*() is called before the method *seek*(). And if the return value is *true*, the method *seek*() is not being called. The method *avoid*() checks the distances to other vehicles on the road in the direction of their movement. If the algorithm finds the vehicle in *vehicles* array the distance to which is less than a constant *interVehicleDistance*, it will call the method *arrive*(). The argument of the method *arrive*() is *vehicleLocation − predictionVector*. The argument is calculated as such because the vehicle must stop before the other vehicle. Therefore, we subtract the *predictionVector* from the *vehicleLocation*.

The method *follow*() can drive the vehicle through the path, which is a collection of points. For the successful implementation, it is needed a method that drives the vehicle through a collection of paths. The method *followRoad(road : Path[], vehicles : Vehicles[])* does this. It accepts an array of paths and an array of vehicles. The array of vehicles passed for the purpose of avoiding them on the road. Each vehicle has a variable *roadSegment* which indicates the road segment the vehicle is currently on. The method *followRoad*() calls the *follow*() method with an argument of $road[roadSegment]$. The $roadSegment$ variable is increased in the $follow()$ method if the vehicle got to the end of the path. Therefore, the vehicle will drive through the collection of paths sequentially.

The method $draw()$ is used to draw the vehicle on the screen. The vehicle is currently being drawn as a square. A function $fillRect()$ - a native function of the canvas context used to draw the square on the canvas.

To conclude, the Vehicle class is used to drive the vehicle through the path or a collection of paths. The class is also responsible for drawing the vehicle on the screen. Another purpose of the class is to contain the parameters of the vehicle. The properties of the vehicle are used to configure the vehicle's physical behavior.

Path Configuration: One of the requirements for the app is to provide a simple road configuration mechanism for the users. The app supports the following format. The user needs to provide the number of roads, the vehicle parameters for each road, the number of vehicles on each road, the spawn points for each road, and the path configuration object. The path configuration object is an array of arrays with the length of $numberOfRoads$ from the configuration. The top-level array should contain the arrays of objects where the keys are probabilities to go to the path and the values are the correspondent segments. The Fig. 3 shows the described structure. In this example road, the vehicle will go the $DHorizontalSeg1$ with the probability of 1. Then it will either go to the $DHorizontalSeg2$ with the probability of 0.6 or to the $LVerticalSeg3$ with the probability of 0.4. The $LVerticalSeg3$ is a final segment of the road. The vehicle will respawn at the spawn point after reaching the end of this segment. However, if the vehicle got to the end of $DHorizontalSeg2$ it will go either to $DHorizontalSeg3$ with the probability of 0.1 or to $RVerticalSeg3$ with the probability of 0.9.

```
[
  {
    1: DHorizontalSeg1,
  },
  {
    0.6: DHorizontalSeg2, 0.4: LVerticalSeg3,
  },
  {
    0.1: DHorizontalSeg3, 0.9: RVerticalSeg3,
  },
],
```

Fig. 3. Example of the road object.

Getting the Route from Path Configuration: The path configuration gives the general routing for a set of vehicles assigned to that routing. However, each of the vehicles has to follow only one path from the configuration. A helper function $getRoute(road : [key : number] : Path[], index : number)$ is used to get the route from the path configuration. It assigns the route for the vehicle based on the vehicle's index and the distribution provided in the configuration. The index of the vehicle is used for the seeded random generation. This method is called on every render, therefore the route assignment for each vehicle has to be deterministic. This means that on every call the vehicle has to get the same route as in the previous one.

Traffic Light Algorithm Configuration. The traffic light algorithm configuration is packed into one method. That method can be called differently for each algorithm, though it has to obey the interface. The method has to accept the *frameCount* : *number* and return the object with the fields *isChanged* : *boolean* and *lights* : *TrafficLight[]*. The *isChanged* variable is responsible for telling the Main component whether to redraw the lights or not. It is *true* when the lights have to be redrawn and *false* otherwise. The *lights* field is an array of type *TrafficLight[]*. Each entry in the array has to be either *"yellow"*, *"green"*, or *"red"* corresponding to the color of the traffic light in the path configuration. The *frameCount* variable is used to get information about the current state of the system. For example, the algorithm can be as such: change the colors sequentially every 200 frames. The only needed piece of information here is the *frameCount* variable. The algorithm may need information about the intensity of the flow on the road. That could be useful for building complex dynamic algorithms. This feature is not implemented yet.

Main: The Main component combines all the logic together in one place. It uses the components and classes described in the previous sections. It has 3 methods for drawing: *drawVehicles*, *drawRoad* and *drawLights*. The method *drawVehicles* is called every frame. This method updates the vehicle's state and draws them on the screen.

```
const drawVehicles = useCallback(
  (ctx: any) => {
    ctx.clearRect(0, 0, ctx.canvas.width, ctx.canvas.height);
    setTotalFrames((prev) => prev + 1);

    for (let i = 0; i < road.numOfRoads; i++) {
      for (let j = 0; j < road.vehiclesNumber[i]; j++) {
        const vehicle = vehicles[i][j];

        if (!vehicle) continue;

        if (isOutOfBounds(vehicle)) {
          if (vehicle.roadSegment > 0) {
            setNumFinished((prev) => prev + 1);
          }
          vehicles[i][j] = new Vehicle(road.vehicleParams[i]);
        }

        vehicle.followRoad(getRoute(road.paths[i], j), vehicles.flat());
        vehicle.update();
        vehicle.draw(ctx, j);
      }
    }
  },

  [road, vehicles, isOutOfBounds, setNumFinished]
);
```

Fig. 4. The *drawVehicles* method.

The implementation of this method can be seen in the Fig. 4. The method has two loops. One for the iteration over the road and another one for the iteration over the vehicles. On each iteration, the vehicle state is being updated using the method *update()* and the vehicle is being drawn using the method *draw*. Also, the method *followRoad* is called.

The method *drawRoad* draws the road on the canvas. It is called only once because the roads are not changed during the execution of a program. The method *drawLight* is executed only when the variable *isChanged* from the traffic light algorithm has a value equal to *true* for optimization purposes.

The Fig. 5 shows the relations between entities.

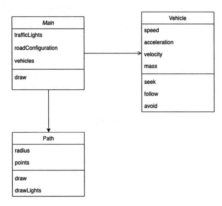

Fig. 5. The class diagram.

4   Results and Discussion

This section describes the test environment and the test results. First, a discussion of the road configuration and traffic light algorithms used in the simulation. Then, go over the comparative Table of test results. Finally, interpret the simulation test results. The three parameters were used to compare the test results. The number of vehicles that passed through the intersection It rises when a vehicle reaches the end of the road. The second metric is vehicles per minute (V/M). The metric is calculated by dividing N by the total time the simulation has been running in minutes. The total number of minutes spent waiting is the third metric. TMW is a cumulative metric that shows how much time all vehicles on the crossroad spend not moving.

4.1   Environment and Algorithms

Road Configurations: The road configuration for these tests is a simple one-lane crossroad as shown on the Fig. 6.

It has one lane in each direction for 4 sides and a traffic light on each side. The vehicles can turn in all directions. This configuration is common in the areas with narrow streets.

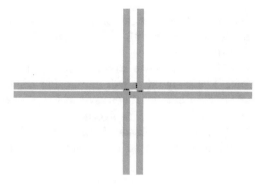

Fig. 6. Simple crossroad configuration.

Traffic Light Algorithms. In this work, we compared four algorithms which work in a similar manner. They all have a fixed cycle and after the cycle finished the signals repeat. The algorithms are: two-way unbalanced (2WU), two-way balanced (2WB), one-way unbalance (1WU), and one-way balanced (1WB). We will describe them in detail in this section. First, we discuss the difference between two-way and one-way algorithms. Then, we will describe the difference between balanced and unbalanced algorithms.

In two-way algorithms, the vehicles are moving towards each other in opposite directions as shown on the Fig. 7. The vehicles on the left and on the right are moving simultaneously. The vehicles on the perpendicular road are waiting for their turn. The vehicles can turn to the perpendicular roads. In case the vehicle needs to make the left turn it has to give way to a vehicle on the opposite direction. The precedence mechanism is not implemented yet, however the vehicles still can avoid each other.

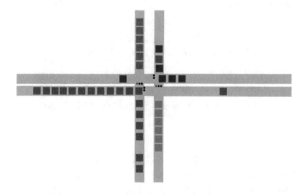

Fig. 7. Two-way traffic light algorithm.

In one-way algorithms, the vehicles can only move in one direction at a time. Only the vehicles on the left are moving. The vehicles on the other directions are waiting for their turn.

Now, we can discuss the difference between the balanced and unbalanced traffic light algorithms. The traffic flow can be unbalanced, meaning there can be more vehicles in one of the directions compared to the other ones. In such cases the traffic light can balance the flow. The algorithm can extend the green signal on the dense directions. Then, the balanced traffic light algorithm is adjusted for the flow density and the unbalanced traffic light algorithm is not adjusted for the flow density.

To sum up, the environment consists of the road configuration and traffic light algorithm. Traffic light algorithm can be balanced or unbalanced and either one-way or two-way.

4.2   Tests

The tests design is the following. We simulated each algorithm for 5 min and calculated the metrics N, V/M, and TMW. To compare the algorithms between each other in different conditions I changed the speed of the traffic flow for different runs.

The results for the speed limit of 100 km/h are in the Table 1. For this road configuration we set the traffic flow on one of the roads to higher density than on the others. Particularly, the leftmost path has the density 3.2 times higher than the others. The balanced algorithms compensate for this density difference by extending the green light signal on the denser road. For these algorithms, the green light signal time on the dense road is twice the green signal time on the less dense road. The speed limit for this test is 100 km/h.

Table 1. 100 km/h speed limit, right balancing test.

Algorithm name	N	V/M	TMW
Two-way balanced	1245	247.56	91.95
Two-way unbalanced	1058	211.59	108.18
One-way balanced	716	143.19	131.35
One-way unbalanced	601	120.19	138.87

Table 2 shows the results of the simulation with the wrong lights balancing. In this example, the denser road does not have an extended green light, however the less dense road has an extended green light signal. For this test, the speed limit of 100 km/h has been set.

Table 2. 100 km/h speed limit, wrong balancing test.

Algorithm name	N	V/M	TMW
Two-way balanced	1010	201.99	92.45
Two-way unbalanced	1076	215.19	91.31
One-way balanced	566	113.19	117.34
One-way unbalanced	596	119.19	115.87

Now, we can test these configurations with lower speed limit. The previous tests were simulating an out of town crossroad, the following ones will be simulating an urban crossroad. Table 3 shows the results for the correctly balanced crossroad. Table 4 shows the results for the incorrectly balanced crossroad.

Table 3. 60 km/h speed limit, right balancing test.

Algorithm name	N	V/M	TMW
Two-way balanced	754	150.96	86.81
Two-way unbalanced	651	128.93	89.94
One-way balanced	414	82.98	101.42
One-way unbalanced	321	63.98	106.22

Table 4. 60 km/h speed limit, wrong balancing test.

Algorithm name	N	V/M	TMW
Two-way balanced	569	113.97	81.76
Two-way unbalanced	634	126.96	80.1
One-way balanced	350	21.39	107.1
One-way unbalanced	381	75.98	105.21

4.3 Results Interpretation

The tests showed that simulation software is able to run the real-world scenario. In total, four configurations were tested and the results were reasonable. For example, the comparison between correct and incorrect balancing showed that the simulation software is able to reflect the correctness of the algorithm. In the Table 1 with the correct balancing the performance is significantly better than in the Table 2 with incorrect balancing. The number of vehicles that passed through the crossroad is higher, and the crossroad has higher throughput with the correctly balanced traffic light algorithm.

5   Conclusions and Future Work

Based on the simulation we can test real-life road configurations with non-uniform traffic flow congestion. There are three categories of features: statistics and parameters, the road configuration mechanism, and traffic light configuration. The ability to vary the vehicle mass and maximal velocity in a particular direction, size and acceleration can improve the environment configuration possibilities. These parameters should be configurable through the interface. Also, various statistical parameters shows users: average velocity in the direction, density, number of people transported, etc. The traffic light algorithms can be programmed with domain-specific language (DSL). DSL usage can smooth out the learning curve of the simulation software and speed up the development of the algorithm. Traffic light-driven route choice, traffic light-dependent congestion, circular dependence between traffic flow, regulation, costs, users' choice, demand flow, and traffic flow will be considered in the future of simulation development. For traffic lights algorithms analysis the different optimum strategies, the choice of the elastic path, and a network of interacting controlled junctions can be considered in the future.

The approximate algorithm design process using the simulation software can be as follows:

– Collect statistics on the particular real-world crossroad. The statistics can include the average speed of vehicles on each direction, density of the flow, number of vehicles per hour, etc.
– Simulate real-world crossroad in the simulation software. The simulation includes configuration of the parameters based on the statistics gathered and road configuration in the software.
– Decide where the traffic light are going to be placed and simulate the decision in the software.
– Design the algorithm using the statistics gathered.
– Run the simulation and tweak the algorithm according to the results of the simulation.

In this work, we designed and implemented the simulation software for traffic lights algorithms. The application can visualize the transport flow in a configurable environment. The software can be useful in traffic algorithm creation and testing. One can configure real-world road environment and test the algorithms designed for the road environment.

References

1. Yuloskov, A., Bahrami, M.R., Mazzara, M., Kotorov, I.: Smart cities in Russia: current situation and insights for future development. Future Internet **13**(10), 252 (2021)
2. Kupriyanovskiy, V.P., Namiot, D.E., Kupriyanovskiy, P.V.: On standardization of smart cities, internet of things and big data. The considerations on the practical use in Russia. Int. J. Open Inf. Technol. **4**(2), 34–40 (2016)

3. Bahrami, M.R.: Modeling of Physical Systems Through Bond Graphs. St. Petersburg Polytechnical University, St. Petersburg (2018)
4. Krzysztof, M.: The importance of automatic traffic lights time algorithms to reduce the negative impact of transport on the urban environment. Transp. Res. Procedia **16**, 329–342 (2016)
5. Bilbao-Ubillos, J.: The costs of urban congestion: estimation of welfare losses arising from congestion on cross-town link roads. Transp. Res. Part A Policy Pract. **42**(8), 1098–1108 (2008)
6. Simulation of urban mobility. https://www.eclipse.org/sumo/. Accessed 20 Oct 2021
7. Wegener, A., Hellbruck, H., Wewetzer, C., Lubke, A.: VANET simulation environment with feedback loop and its application to traffic light assistance. In: 2008 IEEE Globecom Workshops, pp. 1–7. IEEE (2008)
8. MovSim: multi-model open-source vehicular-traffic simulator. www.movsim.org. Accessed 20 Oct 2021
9. Treiber, M., Kesting, A.: Traffic Flow Dynamics. Data, Models and Simulation. Springer, Cham (2013). https://doi.org/10.1007/978-3-642-32460-4
10. Yusuf, A.M.H., Yusuf, R.: Adaptive traffic light controller simulation for traffic management. In: 2020 6th International Conference on Interactive Digital Media (ICIDM), pp. 1–5. IEEE (2020)
11. Farooqi, A.H., Munir, A., Baig, A.R.: THE: traffic light simulator and optimization using genetic algorithm. In: International Conference on Computer Engineering and Applications IPCSIT, vol. 2. Citeseer (2009)
12. Wiering, M., Vreeken, J., Van Veenen, J., Koopman, A.: Simulation and optimization of traffic in a city. In: IEEE Intelligent Vehicles Symposium 2004, pp. 453–458. IEEE (2004)
13. Wen, W.: A dynamic and automatic traffic light control expert system for solving the road congestion problem. Expert Syst. Appl. **34**(4), 2370–2381 (2008)
14. Koukol, M., Zajíčková, L., Marek, L., Tuček, P.: Fuzzy logic in traffic engineering: a review on signal control. Math. Probl. Eng. **2015**, 979160 (2015)
15. Shaikh, P.W., El-Abd, M., Khanafer, M., Gao, K.: A review on swarm intelligence and evolutionary algorithms for solving the traffic signal control problem. IEEE Trans. Intell. Transp. Syst. **23**, 48–63 (2020)
16. Reynolds, C.W., et al.: Steering behaviors for autonomous characters. In: Game Developers Conference, vol. 1999, pp. 763–782. Citeseer (1999)

Scattering with Programmable Matter

Alfredo Navarra[1]([⊠]), Giuseppe Prencipe[2], Samuele Bonini[2], and Mirco Tracolli[1]

[1] Department of Mathematics and Computer Science, University of Perugia, Perugia, Italy
alfredo.navarra@unipg.it
[2] Department of Computer Science, University of Pisa, Pisa, Italy
giuseppe.prencipe@unipi.it, sbonini7@studenti.unipi.it

Abstract. We aim at studying the *Scattering* problem (or *Distancing*) in the context of *Programmable Matter* (PM). This is intended as some kind of matter with the ability to change its physical properties (e.g., shape or color) in a programmable way. PM can be implemented by assembling a system of self-organizing computational entities, called *particles*, that can be programmed via distributed algorithms. A rather weak model proposed in the literature for PM is SILBOT, where particles are all identical, executing the same algorithm based on their local neighborhood. They have no direct means of communication and are disoriented. We aim to achieve Scattering, i.e., all particles are at least two hops far apart from each other. We show that the problem is unsolvable within the pure asynchronous setting whereas we do provide a resolution algorithm for the event-driven case where a particle reacts to the presence of other particles in its neighborhood. Furthermore, we investigate (also by simulations) on configurations where some nodes of the grid can be occupied by obstacles, i.e., immovable but recognizable elements.

1 Introduction

The rise in the last decade of new capabilities to design smart systems, compute and fabricate like never before, has sparked a renewed interest in the performance of materials. In particular, we are now witnessing significant advances in studies dedicated to "active matter", i.e., 3D printing, materials science, synthetic biology, DNA nanotechnology and soft robotics, which have led to the growth of the new field called *Programmable Matter* (PM). This refers to some kind of matter with the ability to change its physical properties (e.g., shape or color) in a programmable way. PM can be implemented by assembling a system of self-organizing computational entities, called *particles* and programmed in a distributed way to collectively achieve global tasks. Several theoretical models for PM have been proposed, ranging from DNA self-assembly systems, (e.g., [17]) to metamorphic robots, (e.g., [15]), to nature-inspired synthetic insects and microorganisms (e.g., [10]), each model assigning special capabilities and constraints to the particles and focusing on specific applications. Among them, the Amoebot model [8,9] is of particular interest from the distributed computing viewpoint. Indeed, in such a model (introduced in [9]), PM is viewed as a swarm of decentralized

Work funded in part by the Italian National Group for Scientific Computation GNCS-INdAM.

L. Barolli (Ed.): AINA 2023, LNNS 661, pp. 236–247, 2023.
https://doi.org/10.1007/978-3-031-29056-5_22

autonomous self-organizing particles, represented by finite automata, moving along a triangular grid.

Recently in [7], the SILBOT model has been introduced where particles' capabilities are very minimal. The movement of a particle in SILBOT is obtained in two steps: first a particle expands toward a neighboring node along a joining edge of the grid, and successively it reaches the desired node, if empty. Basically, each particle may assume two different states: CONTRACTED, i.e., a particle occupies one node; EXPANDED, i.e., a particle occupies one node and an incident edge. We consider a classical problem in distributed computing that is called *Scattering* (see, e.g. [3,14]): starting from any configuration where particles are all CONTRACTED, the aim is to lead the system to a configuration where each particle is still CONTRACTED but admits no neighboring particles.

Outline. The next section provides all the details concerning SILBOT and the SCATTERING problem. In Sect. 3, some impossibility results for SCATTERING under SILBOT are given. Section 4 is devoted to the resolution of SCATTERING when no obstacles are considered. Section 5, instead, refers to the case where configurations may admit obstacles. Finally, Sect. 6 provides conclusive remarks.

2 The Model and the Problem

In this paper, we address the SCATTERING problem within SILBOT, where particles act independently of each other, without explicit communication, in an asynchronous way, based only on local knowledge. SILBOT is a recent variant of the well-established geometric *Amoebot* model (see, e.g., [1,11]).

The Operating Environment. Particles operate on an infinite triangular grid embedded in the plane, where each node has 6 incident edges. Each node can contain at most one particle. There are N particles in the considered system and there might be nodes occupied by *obstacles*, i.e., immovable objects recognizable by the particles.

Particles and Configurations. Each particle is an automaton with two states, CONTRACTED or EXPANDED (it does not have any other form of persistent memory). In the former state, a particle occupies a single node of the grid while in the latter, the particle occupies one single node and one of the adjacent edges. Hence, a particle never occupies two or more nodes at once.

Each particle can sense its surroundings, i.e., if a particle occupies a node v, then it can see the nodes adjacent to v (i.e., nodes at distance one). Specifically, a particle can determine if an adjacent node is occupied by a CONTRACTED or an EXPANDED particle, or occupied by an obstacle, or empty.

Any positioning of CONTRACTED or EXPANDED particles that includes all N particles composing the system plus the obstacles is referred to as a *configuration*.

It is assumed that initially particles are all CONTRACTED.

Movement and States. In order to move to a neighboring node u, a particle in v expands on the edge (v,u). Thus, in the EXPANDED state, the particle occupies one node and one edge. Note that, node u might be occupied by another particle. If the other particle leaves u in the future, the EXPANDED particle will contract into u during its next activation.

A particle commits itself into moving to node u by expanding in that direction, and at the next activation of the same particle, it is constrained to move to node u, if u is empty. A particle cannot revoke its expansion once committed.

Asynchrony and Rounds. SILBOT introduces a fine-grained notion of asynchrony similar to what in the literature is referred to as ASYNC, see, e.g., [4–6, 12, 13]. All operations performed by the particles are non-atomic: that is, there can be delays between the actions of sensing the surrounding, computing the next decision, executing the decision (i.e., change of state, movement, expansion, contraction).

There are no restrictions on the scheduling of these events; thus any possible execution of a physical system can be captured by the model, hence inducing many difficulties for proving the correctness of resolution algorithms (see, e.g. [4,6]).

A *round* is any time window within which all particles have been activated and concluded their activation time at least once. When asynchrony is assumed, it is also required the well-established fairness property by which each particle is activated infinitely often in any execution of the particle system. Hence, the duration of a round is finite but unknown and may vary from time to time.

Orientation and Randomness. Particles are disoriented, we do not make any assumptions about the local coordinate system of a particle. It may even change in each activation of the particle. Moreover, particles only take deterministic decisions. However, we will see in our simulation how randomness may play a central role. It is worth noting that there are cases that cannot be deterministically resolved and are left to the power of the scheduler. For instance, if two CONTRACTED particles decide to expand on the same edge simultaneously, exactly one of them (arbitrarily chosen by the scheduler) succeeds. If two particles are EXPANDED on two distinct edges toward the same node w, exactly one of the particles (chosen arbitrarily by the scheduler) contracts to node w, while the other particle remains EXPANDED.

The Scattering Problem. We can now formally define the SCATTERING (or DISTANCING) problem, introducing two possible variants concerning the behavior of the particles with respect to obstacles.

Definition 1 (SCATTERING). Given an initial (solvable) configuration, possibly with obstacles, an algorithm solves the SIMPLE DISTANCING problem if there exists a time t after which all particles remain CONTRACTED at distance at least two from each other. It solves the STRONG DISTANCING problem if it solves SIMPLE DISTANCING and particles are at distance at least two also from obstacles.

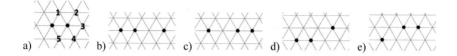

Fig. 1. Basic configurations used in the proof of Theorem 1.

3   Impossibility for SCATTERING **Within** SILBOT

In this section, we show that within SILBOT, the SCATTERING problem cannot be solved. Hence, some more restrictions or particles' capabilities must be added.

Theorem 1. SCATTERING *is unsolvable within* SILBOT.

Proof. To prove the claim, we provide an example where the problem cannot be solved. First of all, it is worth noting that an isolated particle must not move otherwise the time t required in the Definition 1 would be never reached as any CONTRACTED particle can expand as soon as activated. Furthermore, any algorithm that solves the SCATTERING must provide a move for the particles of a (sub-)configuration made by just two neighboring particles like in Fig. 1.a, otherwise the problem would never be solved. Then, for that (sub-)configuration we can basically specify two algorithms in order to guarantee to solve SCATTERING. In the first, say $\mathscr{A}'$, we assume a particle that is neighboring just another one, if activated, moves toward the opposite direction with respect to its neighbor. By referring to Fig. 1.a, the rightmost particle would move toward the position denoted by 3; in the second algorithm, say $\mathscr{A}''$, the same particle, if activated, expands and then moves toward one of the two symmetric positions 2 or 4, the adversary decides. The other possible expansions (toward positions 1 and 5 or even toward the neighboring particle) would always leave the particles neighboring each other, hence we do not consider such cases as they do not resolve the SCATTERING. Let us consider Algorithm $\mathscr{A}'$ and the configuration in Fig. 1.b: the adversary can activate the particles in a (fair) way that brings the system in an infinite loop. In fact, it can activate the particle in the middle, that moves to the right, and the particle on the right, which doesn't move as it is isolated. Once the particle in the middle has moved (Fig. 1.c), the adversary can activate the one on the left, which doesn't move, and the one in the middle that would move back to its original position, hence creating an infinite loop.

When considering Algorithm $\mathscr{A}''$, similar arguments as above can be provided by starting from the configuration shown in Fig. 1.d (or Fig. 1.f).   □

The above theorem confirms that in order to solve SCATTERING within SILBOT, more restrictions to the environment or some more particles' capabilities are required. Enlarging the visibility range as in [7] doesn't seem to be effective.

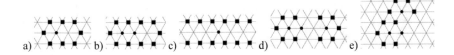

Fig. 2. Basic configurations with obstacles (black squares) used in the proof of Theorem 2.

An interesting case is to consider the asynchronous case with the assumption that a CONTRACTED particle is activated if it is (or becomes) neighboring to at least another particle. In practice, the adversary does not control anymore the activation of CONTRACTED particles which becomes an *event driven* (ED) process. The duration of an activation as well as the time required by a particle to accomplish an expansion or a contraction remains in the decision of the adversary as well as the activation of expanded particles. It turns out that this kind of schedule, that we call ED-ASYNC, is less general than the pure asynchronous one, but it is clearly more general than the synchronous case where all particles are always active. From a practical point of view, it is like the sensing abilities of the particles about their surrounding react to the stimuli given by the presence of neighboring particles.

As we are going to show, this new assumption is enough to allow the resolution of the SCATTERING without obstacles. On the contrary, with obstacles, new difficulties must be faced and the problem turns out to be much harder.

Theorem 2. *Given a configuration with obstacles,* SCATTERING *is unsolvable within* SILBOT, *even when the synchronous schedule is assumed.*

Proof. To prove the claim, we provide an example where the problem cannot be solved for both its variants, even considering the synchronous scheduler which is a special case of both the asynchronous and the ED-ASYNC ones.

We first consider the SIMPLE DISTANCING variant. In that context, it is worth noting that a particle in a situation like the one shown in Fig. 2.a should not move, as otherwise it would move forever back and forth, and the problem cannot be solved. Similarly, two particles "entrapped" in a (sub-)configuration like the one shown in Fig. 2.b should not move concurrently, as otherwise they would move forever back and forth, and again the problem cannot be solved.

Considering instead Fig. 2.d, the particle in the middle should move, whereas its neighboring particle as well as the other one cannot move because of the above arguments. If the allowed move is toward the right (position 3 according to Fig. 1), then a similar configuration would be achieved, from which the same particle would move back to its original position, i.e., an infinite loop occurs. Hence, the move should be toward one of the other two possible directions (positions 2 or 4 according to Fig. 1). However, starting from the configuration shown in Fig. 2.e, similar arguments can be provided, showing that the system can enter in an infinite loop.

Concerning STRONG DISTANCING, it is enough to consider the configuration of Fig. 2.c. In fact, the particle must move in order to exit the tunnel but the adversary can make it move back and forth since the particle is disoriented.   □

As seen above, synchrony doesn't help in the resolution of SCATTERING when obstacles occur. The proof may suggest that a crucial point is about the orientation of the particles. However, even considering particles endowed with chirality, i.e., a common clockwise direction, still there might be unsolvable configurations. It is sufficient to consider the case where a particle needs to traverse a *tunnel*, i.e., as in Fig. 2.c, a corridor of one node width. In fact, in such occurrences, a particle cannot deduce whether it was coming from the left or from the right. This is particularly crucial for the STRONG DISTANCING variant, as even an 'isolated' particle inside a tunnel should move in order to find a position far apart from the obstacles.

In Sect. 5, we conduct some simulations in order to understand how much randomness may affect the solvability of SCATTERING with obstacles.

4   SCATTERING **Without Obstacles**

In this section, we define a very simple algorithm to solve the SCATTERING problem when no obstacle is present. According to Theorem 1, we need to add some restrictions in order to allow the resolution of the problem. We consider the ED-ASYNC schedule where a CONTRACTED particle is activated as soon as it is (or becomes) neighboring to at least another particle. The duration of an activation as well as the time required by a particle to accomplish an expansion or a contraction remains in the scope of the adversary as well as the activation of EXPANDED particles.

- Algorithm $\mathscr{A}$: Given a CONTRACTED particle p, let I be the maximal interval of consecutive adjacent nodes of p where no other particles lie. If $2 < |I| < 6$, then p expands toward the most central node in I – the rightmost one in case of ties.

We remind that particles do not share any common orientation, hence the term "rightmost" used in $\mathscr{A}$ can be interpreted differently by each particle according to its own coordinate system. Moreover, the algorithm does not refer to EXPANDED particles that can only reach their destination as soon as they are activated and the nodes toward which they are expanded get empty, i.e., we have no control on them.

4.1   Aligned Particles

In this section, we consider the case where N particles are all aligned, not necessarily all adjacent within an N-nodes segment. Let S be the smallest segment containing all the particles. It is worth noting how this case is well-related to the Ants on a Stick puzzle according to the specified algorithm, see [18]. In fact, the interaction of the particles reminds the bounce event occurring among ants moving in opposite directions. The main difference is that particles only move if close to each other, i.e., they can reach the end of the stick (segment S) only if stimulated/pushed by other particles. Actually, like for ants, the relative order of the particles never changes as overtaking is not possible.

Lemma 1. *After at most $N-1$ expansions of the leftmost and the rightmost particles of S, such particles never expand again.*

Proof. Let us assume p is the leftmost particle in S. First note that, if p expands, according to our algorithm, it can only expand leftward. Also, according to the assumed scheduler, when p is CONTRACTED, it expands leftward as soon as its (rightward) neighboring node along S is occupied. After p expands and then becomes CONTRACTED, eventually, its neighboring node (i.e., the one previously occupied by p itself) becomes empty, unless the particle p', the one that forced the expansion of p, returns to be neighboring to p because of another expansion forced by another particle, the third one on S, counting from the left bound. Thus, before p can expand again, i.e., enlarging S again, p' must reach the neighboring node of p by means of an expansion generated by the particle to the right of p'.

By iterating the above argument, since each new expansion of p involves one new particle, we can conclude that p can expand at most $N-1$ times. A similar argument can be applied to the rightmost particle. □

Theorem 3. *Given N aligned particles, Algorithm $\mathscr{A}$ solves* SCATTERING.

Proof. We prove the claim by induction on the number N of particles. For $N \leq 2$, the claim trivially holds. Assume the claim for $N-1$ particles.

By Lemma 1, Algorithm $\mathscr{A}$ allows the outermost particles p and p' to expand a finite number of times, upper bounded by $N-1$. Consider the time after which p and p' do not expand anymore. We can restrict our attention to the remaining $N-2$ particles obtained by excluding p and p'. By the inductive hypothesis, Algorithm $\mathscr{A}$, applied on such $N-2$ particles, resolves the SCATTERING. Moreover, by hypothesis, their final positioning cannot induce further expansions for p and p', i.e., p and p' admit no neighboring particles. Hence, the SCATTERING is solved for all the N particles. □

4.2 General Configurations

We consider now the general case where the initial configuration is constituted by any placement of N CONTRACTED particles over any N nodes of the triangular grid. By similar arguments than those applied in the previous section, we prove that Algorithm $\mathscr{A}$ solves the SCATTERING problem also in the general setting.

Instead of the segment S specified in the previous section, given the initial configuration, we consider the smallest regular hexagon H_i centered on a node of the grid and containing all the particles. Let c be the center of H_i; the index i represents the distance in terms of hops between c and the perimeter of H_i (cf. Fig. 3.a). Note that H_i is not necessarily unique; however, we just need to fix one for our analysis purposes. Hence, particles are not required to be aware about H_i nor c.

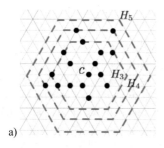
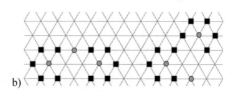

a)   b)

Fig. 3. a) a configuration with H_4 as the smallest enclosing hexagon; b) two possible solutions for SIMPLE DISTANCING obtainable from the configurations of Figs. 2.d and 2.e. Grey circles represent particles that do not move anymore.

As a first observation, a particle on the perimeter of H_i cannot expand toward the inner part of H_i according to Algorithm $\mathscr{A}$. This can be easily verified by means of a case analysis. As for notation, by H_{i+1} we denote the regular hexagon centered in c with one level more with respect to H_i; similarly, by H_{i-1} we denote the regular hexagon centered in c with one level less with respect to H_i.

Lemma 2. *Let H_j be the smallest hexagon centered in c, with $j \geq i$, that contains all particles. Particles on H_j, within finite time, either move to H_{j+1} or they are all isolated (admitting no neighboring particles), eventually.*

Proof. Apart from isolated particles, the only case in which a particle p on H_j cannot move according to Algorithm $\mathscr{A}$ is when it admits two neighboring particles p' and p'' residing on H_j (possibly with other neighboring particles on H_{j-1}) and p does not occupy a corner of H_j. However, such a situation cannot occur for all the non-isolated particles on H_j since p' and p'' must be in the same situation of p, creating a chain of particles on H_j. In order to conclude the proof, it is sufficient to observe that if a corner of H_j is occupied, it admits at least three consecutive neighboring nodes empty, i.e., according to Algorithm $\mathscr{A}$, the particle on the corner either is isolated or it will move, eventually. □

By the above lemma and similarly to Lemma 1, we prove the next result.

Lemma 3. *A particle on H_i can increase its distance from c at most $N-1$ times.*

Proof. A particle p on H_i, in order to move from H_i to H_{i+1}, must be neighboring to at least another particle p' residing either on H_{i-1} or H_i. Successively, once particle p becomes CONTRACTED again, it can move from H_{i+1} to H_{i+2} if:

- it becomes neighboring to a particle on H_i, initially located on a node different from the one previously occupied by p' (hence occupied by a third particle p'');
- or it becomes neighboring to a particle that has also reached H_{i+1}, pushed itself by another particle on H_i (since H_{i+1} has no particles on itself at the beginning).

As the proof of Lemma 1, the claim follows by iterating this argument. □

Theorem 4. *Algorithm $\mathscr{A}$ is correct and terminates.*

Proof. We prove the claim by induction on the number N of particles. For $N \leq 2$ the claim trivially holds. Assume the claim for $N-1$ particles, we prove it for N.

By Lemma 3, Algorithm $\mathscr{A}$ allows the outermost particles to move away from H_i a finite number of times, upper bounded by $N-1$. At that time, all particles on the outermost regular hexagon, say H_j, cannot move anymore; hence, by Lemma 2, all particles on H_j are isolated. Consider the time after which particles on H_j do not expand anymore, nor new particles reach this hexagon. Let us consider now the instance given by all the particles inside H_j and not on it. By the inductive hypothesis, since the considered particles are less than N, Algorithm $\mathscr{A}$ resolves the SCATTERING. Moreover, by hypothesis, the final positioning of the involved particles cannot induce further expansions for the particles on H_j, i.e., particles on H_j are all isolated. Hence, SCATTERING is solved for all the N particles.   □

5   Approaching SCATTERING with Obstacles

In Sect. 3, we have proven that SCATTERING is unsolvable within SILBOT when obstacles are present in the configuration, even in the synchronous setting. Here we aim to approach the problem in the asynchronous setting (not even the ED-ASYNC) but relaxing the power of the adversary. We consider the case where the fairness of the scheduler is randomized, and not addressed to detect the worst case scenario. In practice, although in general the configurations shown in Figs. 2.d and 2.e remain unsolvable, if the scheduler does not play the adversarial role but "fairly" allows the particles to reach symmetric destinations when ties occur, then the situations can be managed differently. Note that, this kind of scheduler is not equivalent to let particles approach random walks. Algorithms are still deterministic but "sometimes" the output, combined with the scheduler, mimics a random choice. To this respect, we have developed a computer simulation framework for approaching both the SIMPLE DISTANCING and the STRONG DISTANCING variants, see [2, 16].

First of all, it is worth noting that when obstacles are considered, an algorithm similar to that provided in Sect. 4 cannot succeed. It is sufficient to consider the configuration of Fig. 2.d where the middle particle may move back and forth forever. Hence we need to define a new algorithm.

5.1   The SIMPLE DISTANCING Problem

Given a particle p, we remind that I is defined as the maximal interval of consecutive neighboring nodes of p where no other particles lie. Note that, in the computation of I, obstacles are treated as empty nodes. Our algorithm works as follows. In general, we fix a direction toward which the particle should expand. Ties are resolved by the local perception of a particle, giving priority to the rightmost direction. We remind that

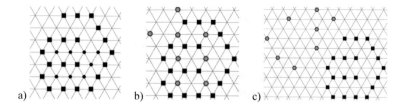

Fig. 4. (a) A constrained configuration and the corresponding solutions obtained when solving (b) SIMPLE DISTANCING or (c) STRONG DISTANCING, respectively.

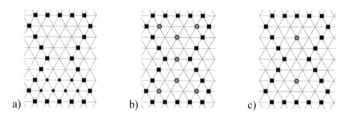

Fig. 5. A constrained configuration (a) and the corresponding solution when solving SIMPLE DISTANCING (b). The solution shown in (c) concerns STRONG DISTANCING with the maximum number of particles allowing to solve the problem within the area confined by the obstacle.

particles do not agree on any direction, and moreover a particle may have a different knowledge/perception with respect to different activations. Once chosen a direction, if an obstacle is present toward that direction, then a neighboring direction is tested until detecting an empty node, if any. Otherwise the particle does not expand. The directions provided by the algorithm are the following:

- If $|I| = 1$, then p tries to expand toward the only node in I;
- If $|I| = 2$, then p tries to expand toward the rightmost node in I;
- If $|I| = 3$ or $|I| = 5$, then p tries to expand toward the rightmost node among the two neighbors of the central node in I;
- If $|I| = 4$, then p tries to expand toward the rightmost node among the two central ones in I.

Figure 3.b shows the obtained solutions when starting from the configurations of Figs. 2.d and 2.e. We have run our simulator on various instances where the resolution of the SIMPLE DISTANCING is rather constrained due to the presence of obstacles. Keeping in mind the commitment of EXPANDED particles, one may expect deadlock situations when the density of the particles is high whereas the moving space is reduced, as in Figs. 4.a and 5.a. Our runs repeated several times always obtained a successful behavior. Figures 4.b and 5.b show two possible solutions obtained from the configurations shown in Figs. 4.a and 5.a, respectively, even though the solution for Fig. 5.a may require a large number of rounds (we experienced about 2200 rounds on average) to converge. In fact, there exists exactly one solution according to the number of particles and the shape of unoccupied nodes.

5.2 The STRONG DISTANCING **Problem**

In this scenario, the goal of the particles is to be distanced also from the obstacles, see Definition 1. The basic algorithm we tested in this case is similar to the one proposed in the previous section, with the additional move required for a particle p when $|I| = 6$ with at least an obstacle neighboring p, i.e., p is not isolated from obstacles:

- if $|I| = 6$ and p is neighboring to at least an obstacle, then p tries to expand toward the rightmost neighboring node.

We remind that if an obstacle is present in the direction chosen by the algorithm, then a neighboring direction is tested until detecting an empty node, if any, otherwise the particle doesn't expand. Clearly, in order to solve STRONG DISTANCING, there cannot exist, in the initial configuration, a particle surrounded by 6 obstacles or a situation like in Fig. 2.a where an area enclosed by obstacles does not allow the distancing of the particles. The proposed algorithm is clearly still able to successfully solve the specific configurations shown in Figs. 2.d and 2.e. Also, it succeeds in configurations similar to that shown in Fig. 5, assuming there is enough space inside the area delimited by the obstacles. In the specific case, only two particles could be placed, see, e.g., the obtained solution in Fig. 5.c. However, the algorithm may find quite some difficulties with the configuration of Fig. 4.a even though we always experienced successful runs requiring about 2500 rounds on average. A corresponding solution is shown in Fig. 4.c.

When considering configurations where the placement of the objects does not constrain much the movement of the particles, the algorithm results to be quite effective. We executed over 1000 runs as follows: at each run, about 60 particles and 40 obstacles were randomly placed in an initial area of diameter of 18 edges. Particles were able to complete the task with an average number of about 240 rounds.

6 Conclusion

We have considered the SCATTERING problem with programmable matter within the very weak SILBOT model. We have shown resolution algorithms and simulations for the different variants of the problem. Several questions are still open, and leave space for further investigations: for instance, to prove whether our algorithms for SCATTERING with obstacles are correct, maybe by introducing some probability arguments. It would be interesting to detect the minimal set of constraints on the obstacle placement under which SCATTERING with obstacles can be solved, in both its variants. Finally, it is worth investigating the case where some particles might be faulty, hence representing non-recognizable obstacles.

References

1. Bazzi, R.A., Briones, J.L.: Brief announcement: deterministic leader election in self-organizing particle systems. In: Izumi, T., Kuznetsov, P. (eds.) SSS 2018. LNCS, vol. 11201, pp. 381–386. Springer, Cham (2018). https://doi.org/10.1007/978-3-030-03232-6_25

2. Bonini, S.: Programmable matter simulator (2022). https://github.com/samul-1/unipi_programmable_matter

3. Bramas, Q., Tixeuil, S.: The random bit complexity of mobile robots scattering. Int. J. Found. Comput. Sci. **28**(2), 111–134 (2017)

4. Cicerone, S., Di Stefano, G., Navarra, A.: Asynchronous arbitrary pattern formation: the effects of a rigorous approach. Distrib. Comput. **32**(2), 91–132 (2019)

5. Cicerone, S., Di Stefano, G., Navarra, A.: Solving the pattern formation by mobile robots with chirality. IEEE Access **9**, 88177–88204 (2021). https://doi.org/10.1109/ACCESS.2021.3089081

6. Cicerone, S., Di Stefano, G., Navarra, A.: A structured methodology for designing distributed algorithms for mobile entities. Inf. Sci. **574**, 111–132 (2021). https://doi.org/10.1016/j.ins.2021.05.043

7. D'Angelo, G., D'Emidio, M., Das, S., Navarra, A., Prencipe, G.: Asynchronous silent programmable matter achieves leader election and compaction. IEEE Access **8**, 207,619–207,634 (2020)

8. Daymude, J.J., Gmyr, R., Hinnenthal, K., Kostitsyna, I., Scheideler, C., Richa, A.W.: Convex hull formation for programmable matter. In: ICDCN 2020. ACM (2020)

9. Derakhshandeh, Z., Gmyr, R., Strothmann, T., Bazzi, R., Richa, A.W., Scheideler, C.: Leader election and shape formation with self-organizing programmable matter. In: Phillips, A., Yin, P. (eds.) DNA 2015. LNCS, vol. 9211, pp. 117–132. Springer, Cham (2015). https://doi.org/10.1007/978-3-319-21999-8_8

10. Dolev, S., Frenkel, S., Rosenblit, M., Narayanan, R., Muni Venkateswarlu, K.: In-vivo energy harvesting nano robots. In: 2016 IEEE International Conference on the Science of Electrical Engineering (ICSEE) (2016). https://doi.org/10.1109/icsee.2016.7806107

11. Dufoulon, F., Kutten, S., Moses Jr., W.K.: Efficient deterministic leader election for programmable matter. In: Miller, A., Censor-Hillel, K., Korhonen, J.H. (eds.) PODC: ACM Symposium on Principles of Distributed Computing, pp. 103–113. ACM (2021)

12. Flocchini, P., Prencipe, G., Santoro, N.: Moving and computing models: robots. In: Flocchini, P., Prencipe, G., Santoro, N. (eds.) Distributed Computing by Mobile Entities. LNCS, vol. 11340, pp. 3–14. Springer, Cham (2019). https://doi.org/10.1007/978-3-030-11072-7_1

13. Flocchini, P., Prencipe, G., Santoro, N., Viglietta, G.: Distributed computing by mobile robots: uniform circle formation. Distrib. Comput. **30**(6), 413–457 (2017)

14. Izumi, T., Kaino, D., Potop-Butucaru, M.G., Tixeuil, S.: On time complexity for connectivity-preserving scattering of mobile robots. Theor. Comput. Sci. **738**, 42–52 (2018)

15. Miyashita, S., Guitron, S., Li, S., Rus, D.: Robotic metamorphosis by origami exoskeletons. Sci. Robot. **2**(10) (2017)

16. Tracolli, M.: Programmable matter simulator (2022). https://github.com/MircoT/programmable-matter-simulator

17. Tucci, T., Piranda, B., Bourgeois, J.: A distributed self-assembly planning algorithm for modular robots. In: Proceedings of the 17th International Conference on Autonomous Agents and MultiAgent Systems, AAMAS 2018, pp. 550–558. International Foundation for Autonomous Agents and Multiagent Systems, Richland, SC (2018)

18. Winkler, P.: The adventures of Ant Alice. In: A Lifetime of Puzzles, pp. 177–185. CRC Press (2008)

The Burden of Time on a Large-Scale Data Management Service

Etienne Mauffret[1,2]($\boxtimes$)(ID), Flavien Vernier[2](ID), and Sébastien Monnet[2](ID)

[1] École Normale Supérieure de Lyon - LIP, 46 Allée d'Italie, 69007 Lyon, France
etienne.mauffret@ens-lyon.fr
[2] Université Savoie Mont Blanc - LISTIC, 5 chemin de bellevue, Annecy-le-vieux, 74940 Annecy, France
{etienne.mauffret,flavien.vernier,sebastien.monnet}@univ-smb.fr

Abstract. Distributed data management services usually run on top of dynamic and heterogeneous systems. This remains true for most distributed services. At large scale, it becomes impossible to get an accurate global view of the system. To provide the best quality of service despite this highly dynamic environment, those services must continuously adapt. To do so, they monitor their environment and store events and states of the system in memory. In this paper, we propose a model to formalize this memory and three strategies to use it. We explain the theoretical difference between those strategies and conduct an experimental evaluation. We show that providing the ability to "forget" old events leads to better performance. However, using fading events (events that progressively disappear) rather than events that suddenly disappear leads to even better performance and is more adequate to detect habits and recurrent behaviors.

1 Introduction

With the advent of large scale services and applications comes the rising of large scale distributed data management systems. Those systems face an ever-changing and heterogeneous environment: new components come and go continuously. Such systems also face dynamic workloads: the number of users may vary, their activity (frequency, type, etc.) their location, etc. Many services are built on top of such volatile conditions and should be able to provide a good quality of service regarding the evolution of the system or the workloads induced by users. There are many works that take past into consideration to adapt to the future environment behavior. Every distributed system that aims to adapt has to grab and store information about its environment. For instance, in [1], the authors present a leader election algorithm in a dynamic network. They combine a wave algorithm with the temporally ordered routing algorithm to face the evolving network. In [2], the authors propose a way to build failures detectors in a Mobile Ad-Hoc Network. In the Linux kernel memory management mechanisms, there is a lot of work concerning this problem (access/dirty bits, double LRU lists, shadow page-cache, etc.). However, even if the weight given to store past events is very important, in the field of distributed systems, to our knowledge, there

L. Barolli (Ed.): AINA 2023, LNNS 661, pp. 248–260, 2023.
https://doi.org/10.1007/978-3-031-29056-5_23

is no work focusing on studying various approaches and their impact on the system performance. For instance, in the context of reputation systems [3], great importance is given to the weight that should be given to the past actions of other nodes according to their freshness, frequency, and regarding the other actions. But this is done in a particular context, where nodes monitor each other and try to attribute a reputation level to others.

In a previous work, we presented CAnDoR [4], a dynamic data placement algorithm taking into account the consistency protocol in use, the workload and an approximate view of system state. In such a service, the local representation of the global environment and the management of past events have a crucial impact on the quality that the service can provide. Indeed, giving the same importance to an "old" event and an event that just occurs may lead to bad adaptation choices and alter the service. This method can also be used to build learning algorithms like proposed by [5] or [6]. For example, let us consider a dynamic data placement service that only consider the workload induced by users. The service will place data such that user interested in those data can access them quickly. To determine which users are interested in a piece of data, the service could compare users accesses. If the service does not differentiate access that was just requested to that access that was requested some time before, it cannot precisely compute which users are interested in the piece of data now. Old requests would still have an important impact on the data placement choice, which will not be as good as it could be. However, only considering the most recent requests could equally lead to bad decisions.

In this paper, we propose a formal model to represent the local memory of a large scale service, and we use this model to compare the impact of three different strategies to use local memory of events to build an adaptable service. In the rest of this paper, we first describe a model to represent a knowledge memory (the set of stored past events and their associated weight) and 3 strategies to use this knowledge (Sect. 2). We then present an experimentation where we implement those strategies in a simulator to observe their impact on a distributed data placement service (Sect. 3). Lastly, we conclude and discuss limitations and future work (Sect. 4).

2   Memory Model

Let S be a dynamic service deployed on a large scale system. This service runs over a period of time $\mathbb{T} = [0; t_\infty]$. Traditionally, a large scale system can be viewed as different nodes that communicate by exchanging messages. Each node in the system participating in the execution has a local instance of this service. From now, only the point of view of a node is considered, thus when we refer to the service S, we refer to the local execution of that service.

During the run, each instance of S builds a local memory that stores the history of events. The memory is composed of all events known by the node and some useful information attached to them. More precisely, an event e is defined by a tuple $e = \langle a_e; s_e; t_e \rangle$ with:

- a_e: the action performed,

– s_e: the source of the event,
– t_e: the time at which the event occurred.

According to the need of services, it may be possible to add components to this tuple, such as a recipient d_e for example. In this paper, we only consider the triplet $\langle a_e; s_e; t_e \rangle$. The set of every event known by a node is denoted $\mathbb{E}$. While building a memory, it is interesting to consider the "importance" of each event. This "importance" can be represented by a weight that depends on the age of the event, its type, its index in the memory both any combination of factors. In order to determine the weight of each event in the memory, we define a weighting function noted w_S, it provides a value between 0 and M that depends on the time t_k, an event e and the elapsed time since e occurred $\Delta_e^k = t_k - t_e$. For example, a service could want to forget or less consider the oldest events, or consider the impact of an event depending on its own history.

In this paper, the focus is on 3 strategies for using the memory. To model these strategies, let us define associated services. To explain the behavior of those services, we use a simple example, illustrated by Fig. 1. In this example, the current time is denoted t_k and past events are denoted from e_0 to e_{i+7}.

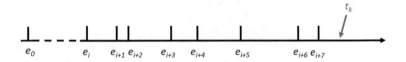

Fig. 1. Events occurrence at time t_k in a service

When an event occurs on a node, the local service adds this event to its memory with the timestamp t_e, the time of occurrence. When the service uses the event e at time step t_k, it computes the weight of this event using its weighting function: $w_S(e, \Delta_e^k)$. This function can define the chosen strategy. We describe here ways of using memory using the weighting function. We consider that the memory always stores every known event with every piece of information. However, the strategy employed dictates how this knowledge can be used. It is important to notice that one service can rely on different strategies for different computations and decision processes.

2.1   Timeless Service

First, consider $\mathcal{T}$, a service that uses all the events the same way without taking their age into account. We call it a *timeless service*. Using the simple example, $\mathcal{T}$ will use every event from e_0 to e_{i+7} with the same weight. This service is illustrated by Fig. 2.

An intuitive way to use the memory is to consider every event with an equivalent weight. For example, if a service needs to use the number of requests, it just needs to count the number of events in the memory. This service does not

Fig. 2. Weight of event at time t_k with timeless service

consider the "age" of events and use a constant function as weighting function. An example of weighting function could be:

$$\forall (e, t_k) \in \mathbb{E} \times \mathbb{T}, w_{\mathcal{T}}(e, \Delta_e^k) = c.$$

One of the main limitation of this strategy is that it could induce inertia and prevent the service from reacting quickly enough. Indeed, by construction, the service does not make any difference between a recent and old event. Therefore, old events will have a non-negligible impact on the decision to make an adaptation or not. For example, let us consider $\mathcal{T}$ to be a data placement service try to place data near users interested in those data. If a user was previously interested in the data but is not active anymore, but a new user is interested in this data, $\mathcal{T}$ will give the same importance to each request. In this situation, illustrated by Fig. 3, $\mathcal{T}$ will keep putting the data closer to the user that was active at the beginning of the execution.

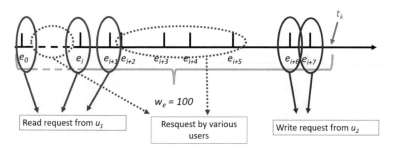

Fig. 3. Weight of event at time t_k with a timeless service

This strategy of memory usage is quite efficient if the sources of events behave roughly in a consistent way during the execution, i.e., if users send the same kind of request periodically during the execution, if the communication between nodes uses the same patterns, etc. If this behavior is expected from the participants, then the service will be able to adapt correctly without being disturbed by "noises" or small irregularity in habits. We expect such behavior in some specific services, such as a personal data service, where a user tends to access the same restricted data over time.

However, this strategy could also lead to an "average solution" for every user. If (a subset of) users have roughly similar habits and activities, the service will tend to consider those users as equivalent and propose a solution equally efficient for all of them. It could be possible to provide a better service for every user by trying to adapt to the users more frequently. Moreover, such a strategy does not take failures into account: every event is weighted equivalently even if one of the sources has failed, crashed or left the system.

2.2 Sliding Window Service

While the timeless strategy seems intuitive and can be efficient in some scenarios, it is usually more efficient to use a finer strategy. Indeed, as we previously saw, a timeless strategy leads to slower adaptations. In many services, it is preferable to provide a service that can adapt itself quickly. For example, a failure detector [7] needs to detect as soon as possible which nodes are still correct, an old event should not be considered. We then consider $\mathcal{W}$, a service that uses the concept of sliding window [8]. A period τ is determined and any event older than τ is ignored.

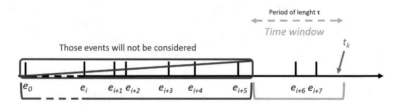

Fig. 4. Weight of event at time t_k with sliding window service

This service is illustrated by Fig. 4. In this example, the events $e_1, \ldots, e_{i+5}$ are older than τ. Thus, those events will not be used in adaptation computations. Events e_{i+6}, e_{i+7} occurred since less than τ time units and are therefore still in the time window: they will be used in adaptation. The service is thus able to adapt quickly, as it only takes into account recent events. Any decision will be made according to recent behavior and any source that became inactive will lost its impact. This can be achieved by considering a step weighting function $w_{\mathcal{W}}$:

$$\forall (e, t_k) \in \mathbb{E} \times \mathbb{T}, w_{\mathcal{W}}(e, \Delta_e^k) = \begin{cases} c, & \text{if } \Delta_e^k \leq \tau, \\ 0, & \text{if } \Delta_e^k > \tau. \end{cases}$$

With such a function, the service will assign a weight of 0 the events older than τ and thus remove any impact they could have on the adaptation. This is illustrated by the example on Fig. 5. In this example, events that occurred before e_{i+6} are given a weight of 0 and thus are ignored by $\mathcal{W}$, while other events are assigned the weight $c = 100$.

When using a sliding window approach, it is important to correctly tune the size of the window, i.e., the period τ. This period can be either static or dynamically evolve through the execution, based on the number and kinds of

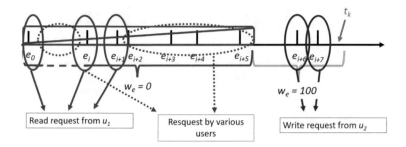

Read request from u_1 Resquest by various users   Write request from u_2

u_2 is considered as the only active user
and must then be favorised!

Fig. 5. Weight of event at time t_k with a timeless service

events, for example. If the period is too short, too few events will be used, important events could be ignored, leading the service to take bad decisions. On the opposite, if the period is too long, too many events will be considered which could also lead to bad decisions, according to the service and its needs.

Some works use a double sliding windows approach. In this case, once an event is out of the first windows, a new weight is assigned to it until it leaves the second windows, with $c_1 > c_2, \tau_1 < \tau_2$ to represent those values:

$$\forall (e, t_k) \in \mathbb{E} \times \mathbb{T}, w_{\mathcal{W}}(e, \Delta_e^k) = \begin{cases} c_1, & \text{if } \Delta_e^k \leq \tau_1, \\ c_2, & \text{if } \tau_1 < \Delta_e^k \leq \tau_2, \\ 0, & \text{if } \Delta_e^k > \tau_2. \end{cases}$$

This approach allows keeping track of events that occurred before the current windows but are not recent enough to be fully considered in computations. This approach can be even more precise and generalized in fading event approach.

2.3 Fading Event Service

Many services aim for a quick adaptation but need to keep information for a longer period than just a period τ, or to give some nuances to events through time. Such service can use a fading event strategy. This strategy consists in making events less important with time. Many weighting functions can be used to implement this behavior, as long as it verifies the following properties:

1. Definition: $w_{\mathcal{F}}$ must be defined for any couple of event and age (e, Δ_e^k) (however, the function does not need to be continuous).
2. Evolution of memory: for a given event e, $w_{\mathcal{F}}$ is not a constant function over time.

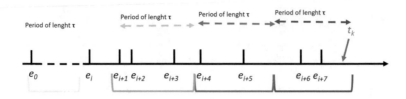

Fig. 6. Weight of event at time t_k with fading event service

3. Fading over time: for any event e, $w_{\mathcal{F}}$ is a pseudo-decreasing function over time, that is that $w_{\mathcal{F}}$ is decreasing over time except for a meager number of outliers.

This service is illustrated by Fig. 6. In this example, t_{i+6} and t_{i+7} will be fully considered, t_{i+4}, t_{i+5} slightly less, $t_{i+1}, t_{i+2}, e_{i+3}$ even less and so on.

There are many relevant functions that can be used to build a memory in a large-scale system. A way to represent such a function is the following:

$$\forall e \in \mathbb{E}, \exists\, t_{k_1}, t_{k_2} \in \mathbb{T}^2 \mid w_{\mathcal{F}}(e, \Delta_e^{k_1}) \neq w_{\mathcal{F}}(e, \Delta_e^{k_2}),$$

$$\forall e \in \mathbb{E}, \forall^*(t_{k_1}, t_{k_2}) \in \mathbb{T}^2 \mid t_{k_1} < t_{k_2} \Rightarrow w_{\mathcal{F}}(e, \Delta_e^{k_1}) < w_{\mathcal{F}}(e, \Delta_e^{k_2}),$$

where $\forall^*$ denotes "for all but a meager of elements". For the clarity of illustrations and explanations, we use a simple function that divides the weight of events by 2 every τ time units:

$$w_{\mathcal{F}}(e, \Delta_e^k) = \frac{100}{2^\delta}, \ \delta = \lfloor \frac{\Delta_e^k}{\tau} \rfloor.$$

This function is a decreasing function over time and thus respects the needed properties while being easily implemented and does not imply a huge computation cost. The philosophy described hereafter still holds with most of the others functions that respects the 3 properties presented.

The impact of this function is illustrated by Fig. 7 where the most recent events have a weight of 100 which is periodically divided by 2. While the service theoretically never truly forget any event, at some point the weight became negligible and the event is almost forgotten. It is possible to provide some variation on purely fading events by allowing the function to occasionally increased the weight of an event, if this event became more relevant again for some reason (e.g., recurring events).

As for a sliding window strategy, it is important to tune correctly the value of τ. Preliminary studies suggest that a dynamic strategy that uses both the frequency and the kind of event allows reaching a well dimensioned period without prior knowledge of the system. This service gives more importance to the recent events, but without *forgetting* them. Instead, the weight is decreased over time.

Most of the time, services only need a decreasing function over time to run smoothly. However, it may be interesting to give more importance to old events if a pattern of events reoccurs. In this case, one may use a pseudo-decreasing

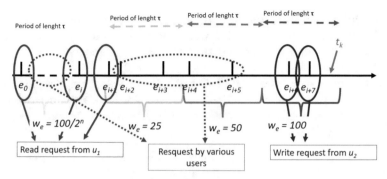

Fig. 7. Weight of event at time t_k with a timeless service

function that will give a bigger weight to old events under some circumstances. Some mechanisms, such as the double LRUs in Linux, are based on this concept.

3   Experimentations

We propose here an evaluation of the three strategies presented in the previous section through simulation of a service that rely on its memory to take some adaptation decisions. CAnDoR is a distributed data placement service that consider both the consistency protocol and the user workload [4]. CAnDoR computes an efficient position for each replica of data requested by users, focusing on requests location of most acceded data in combination of the guaranties associated with the consistency model. In such a service, the computation directly uses the memory and the weighting functions can have a huge impact on the placement solutions found by the service. We consider here 4 different versions of CAnDoR:

- **The static version of CAnDoR** $\mathcal{N}$: which place the replica according to the guaranties but do not try to adapt to users behavior. Therefore, this service does not use any memory model. This service is used as a baseline to compare the efficiency of the memory strategies.
- **The timeless version of CAnDoR** $\mathcal{T}$: this service uses a timeless strategy to weight the requests made on data.

– **The sliding window version of CAnDoR** $\mathcal{S}$: this service uses a time window of length τ to determine which events must be considered for computation. The value of τ is based on the frequency of requests. The weighting function is such that $|t - t_i| < \tau \Rightarrow w(t, t_i) = 100$.
– **The fading event version of CAnDoR** $\mathcal{F}$: this service weights events according to the number of periods τ elapsed since the event occurred. More specifically, the weighting function described in Sect. 2 is used: $w_{\mathcal{F}}(t, t_i) = \frac{100}{2^\delta}$, $\delta = \lfloor \frac{t-t_i}{\tau} \rfloor$.

Experiments presented here are made on CandorSim, a simulator based on peerSim [9]. CandorSim simulates 100 clusters, each of them is considered as a single node due to the fact that the communication inside the cluster is negligible. Users can request access to data to a cluster, while clusters can treat those requests and communicate with each other to periodically compute for better placement. The evaluated metric is the time needed for at least 95% users to get the requested data. The simulator has been calibrated with the result of real-world studies, such as [10,11]: a message sent by a user needs 100 to 300 ms to reach a cluster while a cluster needs 30 to 150 ms to reach another cluster. As the memory strategy used can lead to different results according to the number and the kind of request, experiments have been made using 3 different behaviors for users.

Stable Behavior: In this experiment, some users send many requests during the whole execution and others only send occasional requests. This behavior is to be expected with application using personal data. Each data is acceded by a small group of users (doing regular requests) that can occasionally change locations (and be considered as new users by the service). Some occasional user may also perform some requests as well. We can observe that the strategy in use does not have a huge impact.

For all three services $\mathcal{T}, \mathcal{S}, \mathcal{F}$, the time needed to answer a request drops from 175 to 130 ms. This is due to the lack of adaptation in the system: the set of active users stays the same during the whole execution, and requests are uniformly made during the execution. Thus, considering whole requests or only the recent ones provide the same view of the system. Results of such execution are shown in Fig. 8.

Alternating Behavior: This time, users are divided into 3 groups. The first two groups send many requests but alternatively: while the first group is active, the second is not and reciprocally. The third group always sends occasional requests to add some noise to the computation. These behaviors represent applications where only a defined group of user access the data but regularly change location, or if two groups are sharing the data but at different times.

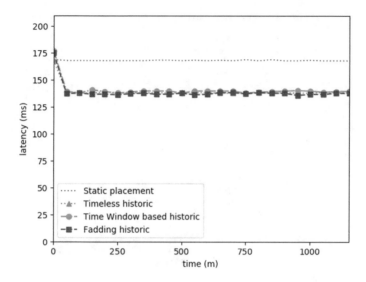

Fig. 8. Time needed to deliver requested data with stable behavior

We can observe that with such a behavior, the timeless strategy provides worst performance. This can be explained by the impact of the inactive group: with a timeless strategy, the inactive group will still be considered with the same importance as the active one. We can thus see a huge latency peak, this peak corresponds to the change of activity. As the two groups alternate the activity, the second will always be less considered than the first one. The use of a sliding window or fading event strategy allows the service to quickly react to this change and while smaller latency peaks can be observed, the service reacts much quicker which provides a better global quality of service. The results of this scenario are shown in Fig. 9.

Unbalanced Alternating Behavior: This scenario is close to the previous one: two groups work consecutively while a third one is working sporadically to add some noise. However, the first two groups do not work during an equivalent period of time in this scenario: the first group is active for a longer period of time and is only temporarily replaced by the second group. The context of such application is similar to the previous one, but with unbalanced loads between the two active groups.

As a consequence, we can observe periodic latency peaks. Those peaks correspond to the period where the second group is active. As this group is less active, it will not be favored a lot by the service. This tendency is especially true with a timeless service, where the second group will never be favored over the first one, hence the periodical peaks. The use of a sliding window or a fading event strategy allows the service to quickly react to this change. Furthermore, the dynamic

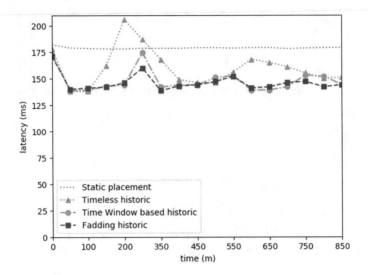

Fig. 9. Time needed to deliver requested data with alternating behavior

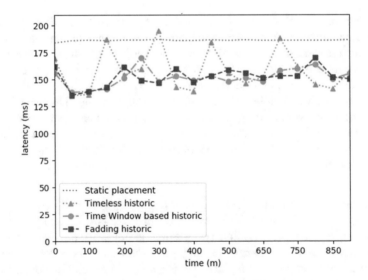

Fig. 10. Time needed to deliver requested data with unbalanced alternating behavior

tuning of τ even allows the service to calibrate it to reduce the impact when the activity changes. The results of this scenario are shown in Fig. 10.

Those simulations show that the strategy of memory usage has a non-negligible impact on the system performance. In a service such as CAnDoR, that relies on its memory for computation, the use of a timeless strategy can lead to bad performance. In our experiments, we observed that a fading event based strategy and a sliding windows one lead to quick adaptation and thus better global quality of service. The simple version of fading events-based strategy seems to provide slightly better performance than sliding windows.

4   Conclusion and Future Works

Large scale services often run on top of dynamic systems and have to face continuously changing workloads. In order to provide a good quality of service and a smooth experience for users, such services must be able to adapt to their dynamic environment. To do so, they usually have to monitor their environment, store past events in their memory, and take decisions to gracefully adapt.

In this paper, we presented a formal representation of the memory of individual nodes of a dynamic system and how they can manage their own memory (set of stored past events). We also propose a theoretical analysis of three strategies to use this memory when a node needs to make an adaptation decision. We used our model and weighting functions to influence the impact of some past events during the computation. More specifically, we proposed a set of restrictions to build *fading events* based memory: a memory that progressively diminish the impact of events with time. We then conduct experiments to evaluate the impact on the quality of service of a distributed data placement service according to the strategy in use. Those evaluations show that it is important to use a strategy that provides the ability to forget, completely to partially, past events. Fading event based strategy seems to provide better results, but not significantly compared to a sliding window strategy.

The strategies presented in this paper mostly rely on the time elapsed since an event occurs. We believe that finer (but more complex) strategies could be used by injecting other considerations, such as taking into account importance or rarity of events. Future works will address such strategies.

References

1. Ingram, R., Shields, P., Walter, J.E., Welch, J.L.: An asynchronous leader election algorithm for dynamic networks. In: 2009 IEEE International Symposium on Parallel & Distributed Processing, pp. 1–12. IEEE (2009)
2. Greve, F., Arantes, L., Sens, P.: What model and what conditions to implement unreliable failure detectors in dynamic networks? In: Proceedings of the 3rd International Workshop on Theoretical Aspects of Dynamic Distributed Systems, pp. 13–17 (2011)
3. Jøsang, A., Ismail, R., Boyd, C.: A survey of trust and reputation systems for online service provision. Decis. Support Syst. **43**(2), 618–644 (2007)
4. Mauffret, E., Vernier, F., Monnet, S.: CAnDoR: consistency aware dynamic data replication. In: 2019 IEEE 18th International Symposium on Network Computing and Applications (NCA), pp. 1–5. IEEE (2019)
5. Iqbal, M., Browne, W.N., Zhang, M.: Reusing building blocks of extracted knowledge to solve complex, large-scale boolean problems. IEEE Trans. Evol. Comput. **18**(4), 465–480 (2013)

6. Li, T., Sahu, A.K., Talwalkar, A., Smith, V.: Federated learning: challenges, methods, and future directions. IEEE Signal Process. Mag. **37**(3), 50–60 (2020)
7. Chandra, T.D., Toueg, S.: Unreliable failure detectors for reliable distributed systems. J. ACM (JACM) **43**(2), 225–267 (1996)
8. Lee, C.-H., Lin, C.-R., Chen, M.-S.: Sliding-window filtering: an efficient algorithm for incremental mining. In: Proceedings of the Tenth International Conference on Information and Knowledge Management, pp. 263–270 (2001)
9. Montresor, A., Jelasity, M.: PeerSim: a scalable P2P simulator. In: 2009 IEEE Ninth International Conference on Peer-to-Peer Computing, pp. 99–100. IEEE (2009)
10. Agarwal, S.: Public cloud inter-region network latency as heat-maps (2018)
11. Popescu, D.A.: Latency-driven performance in data centres. Ph.D. thesis, University of Cambridge (2019)

A Control-Theoretical Approach to Adapt Message Brokers

Nelson S. Rosa[(✉)] and David J. M. Cavalcanti

Universidade Federal de Pernambuco, Recife, Pernambuco, Brazil
{nsr,djmc}@cin.ufpe.br

Abstract. Messaging brokers allow the development of distributed applications whose components are weakly coupled and communicate asynchronously, e.g., publish/subscribe systems. Brokers usually have a set of configuration parameters (e.g., queue size, persistence mode) that can be configured to adjust the broker's behaviour to an expected work-load. However, developers only configure these parameters at deployment time and they are kept unchanged at runtime. This paper presents a step by step to implement a control-theoretical solution that dynamically changes messaging brokers' behaviour. Central to the steps is the definition of controllers, along with their tune methods. Then, different controllers were implemented and tuned using distinct mechanisms following the proposed steps. The controllers have been integrated into a widely adopted open-source messaging service named RabbitMQ. The unique contribution of this paper is to show the steps needed for applying control theory to message brokers, along with a comparative analysis of different controllers and tuning methods. Developers can use similar steps in engineering control solutions for messaging systems.

1 Introduction

Middleware systems have been widely adopted in the development of distributed applications. The middleware drives how the application is built and how its components interact. Meanwhile, different middleware models have become popular such as message-oriented middleware (widely known as messaging brokers) and RPC-based solutions. Whatever the model, they usually hide the complexity of distribution from application developers by implementing distributed transparencies (e.g., access, location) and providing services such as transaction, security and concurrency control.

Adaptive middleware is a particular case of middleware whose behaviour or structure can be adjusted while the middleware executes. For example, a new component can be added, a bug can be fixed, or a runtime parameter can be altered to improve the middleware performance. A great variety of strategies, mechanisms and technologies have been proposed for building this kind of middleware, such as computational reflection [4], process mining [12], and neural networks [5].

© The Author(s), under exclusive license to Springer Nature Switzerland AG 2023
L. Barolli (Ed.): AINA 2023, LNNS 661, pp. 261–273, 2023.
https://doi.org/10.1007/978-3-031-29056-5_24

Even considering the broad number of existing adaptive middleware systems, the adoption of control theory as an adaptation mechanism is minimal [11], or even inexistent in the case of messaging brokers. This scarceness contrasts with the current wave of using control-theoretical elements for adapting software systems [13]. Controllability has been recognized as a recent wave of research in self-adaptive systems [15] and can offer insights into the design of adaptation mechanisms with formal guarantees [7].

In a simplified way, the use of control theory means that the software consists of a *feedback control loop* that includes a plant (e.g., the middleware), plant operation goals (e.g., *throughput > 50 messages/s*) and a controller that sets configurable plant parameters, e.g., size of the message queue used in the middleware. Adjustments (adaptations) are calculated by the controller using monitored data from the plant itself and the defined control targets. In the end, control theory can improve the software execution with a wide variety of adaptation rules having a solid mathematical basis. On the other hand, the adoption of control theory has some associated challenges [13]: how to create a mathematical model of the software and its components, how to choose the controller to use, build software sensors and actuators, and how to ensure the goals of adaptation.

This paper presents a step by step to use control theory elements to adapt an open-source and widely adopted messaging broker named RabbitMQ[1] [6]. Adopting control theory means showing the steps to choosing controllers, selecting tunning methods of these controllers, and implementing and integrating the controllers into the middleware. Ultimately, this paper's unique contribution is to develop a solution that dynamically adjusts the broker behaviour using a controller.

The remaining of this paper is organized as follows: Sect. 2 introduces basic concepts of RabbitMQ and control theory. Next, Sect. 3 presents the steps to control a messaging system. Section 4 presents results on the controllers on action reconfigure the RabbitMQ. Section 5 presents an overview of existing solutions. Finally, Sect. 6 summarises the conclusion and future developments.

2   Background

Before presenting the proposed solution, this section introduces basic concepts related to control theory and messaging systems.

2.1   Control Theory

Control theory is a branch of applied mathematics that uses feedback to impact the behaviour of dynamic systems to achieve an expected goal. The control should also minimise delays, overshoot, and steady-state errors and keep the system as stable as possible.

---

[1] https://rabbitmq.com/.

The central element in control theory is the controller. The controller has a law that defines the action that should be taken to maintain the system (plant) at a given operation goal. Using a system measure output, the controller utilises the error between the goal and the measured output to compute a new value for a given plant configuration parameter, the so-called knob. The change of this parameter acts to bring the system closer to the goal.

The appropriate configuration of the controller can be done through several tuning methods. These methods can adjust gains present in the control laws to adequately capture the plant's dynamic.

The design of controllers has three primary purposes [9]. Firstly, the controller ensures that the system measure output is always as close as possible to a given reference input (regulatory control). Secondly, the controller performs to ensure that disturbances on the system do not significantly impact the system-measured output (disturbance rejection). Finally, the controller acts to optimize the system-measured output (optimization).

2.2   RabbitMQ

RabbitMQ is an open-source message broker widely adopted by enterprises and whose performance has often been evaluated in several ways [6,8]. A distributed application built atop RabbitMQ consists of publishers and consumers whose interactions are intermediated by a messaging broker. The broker stores messages on queues, manages subscribers, and route messages from publishers to consumers.

A vital aspect of these interactions is when the broker can remove messages from the queue, i.e., when the broker can consider the message handled by the consumer. Two approaches are available, and they depend on the acknowledgement mode used by the application. In the first case, a message is removed from a queue after the broker sends it (automatic ack). Another possibility is the message's remotion after the consumer explicitly sends an acknowledgement to the broker (explicit ack).

The broker and consumers can be tunned at runtime in several ways. The broker tunning usually means configuring the Erlang virtual machine, e.g., garbage collection, process scheduler settings, and memory allocation. On the consumer side, it is possible to set a QoS parameter named *channel prefetch count*, which specifies the max number of unacknowledged messages permitted on the channel. In practice, messages are cached in the consumer by the *RabbitMQ Client* until they are processed. It is worth observing that the broker delivers messages to consumers as fast as the network conditions or consumer will allow. Furthermore, by default, the prefetch buffer is unbounded.

The best practice is to keep consumers busy while messages are stored in the broker's queues and available for new consumers. The way to achieve this behaviour is by adequately configuring the prefetch count. This procedure may avoid thousand of messages stored in a given consumer while the queue is empty and new consumers have no access to delivered messages. Meanwhile, the configuration of the prefetch count also may avoid overloaded consumers.

3 Control-Theoretical Steps

As mentioned before, the main contribution of this paper is to show in detail the steps for adopting control theory elements to adapt the RabbitMQ messaging broker. A set of steps was defined to help in this task, as shown in Fig. 1. These steps extend the general ones proposed in [7] and start defining the scope, identifying a quantifiable system's goal to be tracked by the controller, and one or more knobs that change the system's behaviour and impact the goal. After defining the goal and knob(s), the designer must specify the controller(s) that acts on the knobs to close the system to the goal. In the following, the controllers have to be tuned. Depending on the tuned method used, the steps for tuning can variate. Two paths become possible considering the tuned methods adopted in this paper: the long one for Root Locus and the shortest one for the other methods (Ziegler-Nichols, Cohen-Coon and AMIGO). Whatever the method, the last step of the tunning is to compute the controller gains. Finally, the last steps consist of implementing and integrating the controller and testing and validating the system.

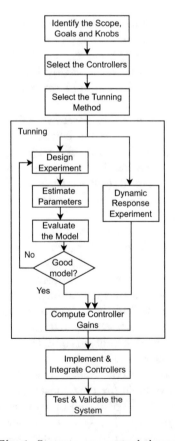

Fig. 1. Steps to use control theory

These steps are detailed in the following sections.

3.1 Identify the Scope, Goal and Knob

Initially, it is needed to define the scope of what is to be modelled clearly. In this work, the *RabbitMQ Client* plays the role of the plant, i.e., the component of the consumer being controlled. As presented in Sect. 2.2, the consumer includes the *RabbitMQ Client* and the *Business Logic*, and the controller acts on the first to regulate the number of messages arriving in the second. A large number of messages can overload the business logic, while few messages lead to poor performance of the business.

Figure 2 is the control theory closed loop associated with the *RabbitMQ Client*. The measurable metric associated with this closed loop is the *arrival rate* that works as the goal. This rate is measured on the consumer and computes the number of messages arriving in the *App* per second after being stored in the *prefetch buffer*. The *arrival rate* should be maintained close to the processing capacity of the business logic to keep the consumer in equilibrium. Finally, the target of the *arrival rate* should be static (defined at deployment time) or dynamic (adjusted at runtime).

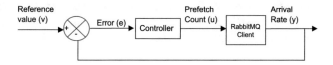

Fig. 2. Closed loop

The most direct way to control the *arrival rate* is to properly define the size of the *prefetch buffer*. This parameter is easily configured in RabbitMQ Clients and substantially impacts consumers' performance. A small value of *prefetch count* usually means fewer messages arriving in the *App* and larger queues in the broker. Meanwhile, a larger *prefetch count* may lead to consumer crashes due to memory run-out and empties queues in the broker.

3.2 Select the Controllers

As the plant has been defined, it is time to select the controller to be used. Due to the absence of previous experiences on using controllers to manage the RabbitMQ, four different controllers are being considered, namely On-Off, P (Proportional), PI (Proportional-Integral), and PID (Proportional-Integral-Derivative). These are very popular controllers whose control laws are defined as follows:

$$u_{on}(t) = \begin{cases} u_{max}, & \text{if } e(t) > 0, \\ u_{min}, & \text{if } e(t) < 0 \end{cases} \tag{1}$$

$$u_p(t) = k_p e(t), \tag{2}$$

$$u_{pi}(t) = k_p e(t) + k_i \int_0^t e(\tau)d(\tau), \tag{3}$$

$$u_{pid}(t) = k_p e(t) + k_i \int_0^t e(\tau)d(\tau) + k_d \frac{de(t)}{dt}, \tag{4}$$

where the control error $e(t) = r - y(t)$ is the difference between the reference signal r and the output of the system y. The integral is merely an abstraction of taking a sum. Then, this integral is a cumulative sum of the error values. Finally, in a discrete-time, the derivative is an approximation of the change of e since the previous time step.

At this step, one or more controllers can be selected to be evaluated. As can be observed in their control laws, they equations involve the definition of the so called controller gains, namely k_p, k_i and k_d.

3.3   Tune the Controllers

As the controllers have been selected, next step consists of tuning them. In practice, the tunning consists of find values to the controller gains (k_p, k_i and k_d). In this case, some well-known and documented tunning methods can be used, such as: Root locus, Ziegler-Nichols, Cohen-Coon, and AMIGO [10].

3.3.1   Root Locus

If the Root Locus is selected, it is necessary to devise a mathematical model able to quantify the effects of given control input (*prefetch count*) on a measured output (*arrival rate*). In practice, the model can be defined through the linear difference equation $y(t + 1) = a.y(t) + b.u(t)$, representing the RabbitMQ Client's dynamics. Then, next prefetch count ($y(t+1)$) depends on its last value ($y(t)$) and the previous arrival rate ($u(t)$).

The construction of an accurate model needs enough data (training data) that allows a reasonable estimation of parameters a and b. The data should be generated through experiments (*Design Experiments*) that produce good quality data. Next, the least squares regression method can be adopted to estimate the parameters [9] (*Estimate Parameters*). Finally, it is necessary to evaluate how good the model is, i.e., how the model explains the data observed. The accuracy of a model can be quantified based on the training data or a separate set of test data. A separate test data and two metrics were used in the evaluation, namely NRMSE (Root Mean Square Error) and R^2 (Coefficient of determination).

3.3.2   Ziegler-Nichols, Cohen-Coon and AMIGO

The Ziegler-Nichols tuning method consists of a set of heuristics devised to perform adequately in various situations. Unlike the previous method, no knowledge of the plant is necessary. This method requires a simple experiment to measure the system's dynamic response to a sudden significant input change. By registering the output generated by the system over time, it becomes possible to extract enough information to calculate the controller gains. It is worth observing that the results produced in this dynamic experiment can also be used to compute the controller gains using the Cohen-Coon and AMIGO tunning formulas [10]. Table 1 shows the controller gains computed to all tunning methods being considered for each adopted controller.

Table 1. Computed controller gains

Controller	Tunning	k_p	k_i	k_d
P	Root Locus	0.02787	–	–
PI	Root Locus	-0.00001	0.00738	–
PID	Root Locus	0,00414	0,00711	0,00225
P	Ziegler-Nichols	0,00077	–	–
PI	Ziegler-Nichols	0,00069	0,00208	–
PID	Ziegler-Nichols	0,00092	0,00463	0,00004
P	Cohen-Coon	0,00104	–	–
PI	Cohen-Coon	0,00133	0,01186	–
PID	Cohen-Coon	0,00123	0,00660	0,00003
P	AMIGO	0,00050	–	–
PI	AMIGO	0,00019	0,00046	–
PID	AMIGO	0,00050	0,00460	0,00001

3.4   Implement and Integrate the Controller

As the controllers were designed, they were implemented in Go language and integrated into a RabbitMQ consumer. Additionally, it was necessary to instrument the consumer to collect the arrival rate metric. This metric is collected from time to time (configurable) and is used to configure the next value of *prefetch count*. The next value of the prefetch count (control variable) is calculated according to controller being used. Next piece of code shows how the arrival rate is computed (Line 1), the value of the prefetch count is calculated (Line 2), and the new prefetch count is configured (Lines 3–6) inside the consumer:

```
1.  s.ArrivalRate = numberOfMessages / monitorInterval
2.  u = controller.Update(s.Ctler, s.ArrivalRate)
3.  err := s.Ch.Qos (
4.          u, // update prefetch count
5.          0, // prefetch size (not used)
6.          false // default
```

3.5   Test and Validate the System

Having implemented the controllers, the consumer's behaviour was observed considering a closed loop in which a given fixed arrival rate (set point) is established. Then each controller tracks it for 10 min.

Publishers and consumers were implemented in Go Language and executed in different hosts, and the RabbiMQ broker was executed in a Docker container. A single consumer tracked by the controller receives messages from 100 publishers individually configured to produce 256-byte messages uniformly distributed with an average of 25 ms and a standard deviation of 2 ms. The set point to be tracked was configured to 400 msg/s.

Figure 3 shows the behaviour of the arrival rate having the goal set to 400 msg/s (red line). These figures present the results[2] of controller OnOff and the

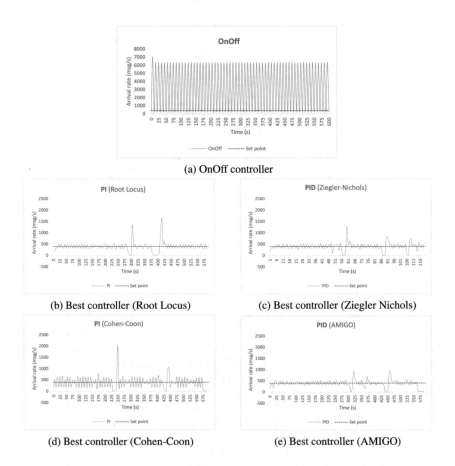

Fig. 3. Controllers tuned with different methods and tracking a *fixed* set point

<hr>

[2] All data of the experiments are avaialable at https://github.com/nsrosa70/control.git.

best controller (P, PI and PID) when different tuning methods are adopted. For example, Fig. 3b shows that the best controller tuned with Root Locus is the PI controller.

The NRMSE (Normalised Root Mean Square Error) was used to define the best tracking. NRMSE reveals how concentrated the data is around the line of best fit, and values close to 0 reveal the best trackings. It is worth noting that the desired fit in all cases is one in which the set point is 400 msg/s. Table 2 presents all values of NRMSE along with the best results for each tuning method (bold). The controllers that produce the best tracking were PI (Root Locus), PID (Zigler-Nichols), PI (COhen Coon) and PID (AMIGO).

Table 2. Controllers in a *fixed* set point scenario

Controller	Tunning	NRMSE
Onoff	–	0,610244118
P	Root Locus	0,318808387
PI	**Root Locus**	**0,120339127**
PID	Root Locus	0,18190188
P	Ziegler Nichols	0,622182286
PI	Ziegler Nichols	0,16327122
PID	**Ziegler Nichols**	**0,129632809**
P	Cohen Coon	0,610349597
PI	**Cohen Coon**	**0,136679983**
PID	Cohen Coon	0,158451381
P	AMIGO	0,622319146
PI	AMIGO	0,291873346
PID	**AMIGO**	**0,157663268**

4   Controllers in Action

Considering the results shown in Table 2, the best controllers were evaluated in a scenario in which the set point changes randomly every 5 s during ten minutes. Figure 4 shows how different controllers track the variable set point. As seen in this figure, the tracking of P, PI and PID controllers is better than the one of OnOff controller.

Table 3 summarises the results of using the best controllers in a variable tracking scenario. As seen in the table, the PI controller tunned with the Cohen Coon method produces the best result, i.e., its NRMSE was the closest to zero. Meanwhile, the controller OnOff was one having the worst tracking. Due to its control law, it presents great instability having spikes and valleys characteristics to this kind of controller.

Table 3. Controllers in a *variable* set point scenario

Controller	Tunning	NRMSE
Onoff	–	0.392572938
PI	Root Locus	0,151959863
PID	Ziegler Nichols	0,216611194
PI	**Cohen Coon**	**0,146376338**
PID	AMIGO	0,187338931

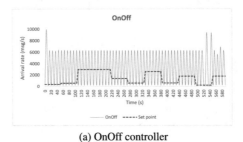

(a) OnOff controller

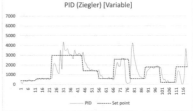

(b) PI controller (Root Locus)   (c) PID controller (Ziegler Nichols)

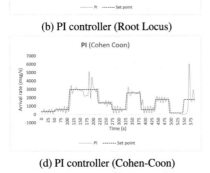

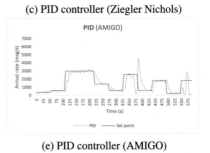

(d) PI controller (Cohen-Coon)   (e) PID controller (AMIGO)

Fig. 4. Controllers tuned with different methods and tracking a *variable* set point

5   Related Works

Li and Nahrstedt [11] proposed a seminal work in this area. In their solution, the authors proposed a middleware capable of reconfiguring parameters and functionalities of a distributed multimedia application taking into account changes in CPU availability and execution environment bandwidth.

The application behaviour is modelled through differential equations, and a PID controller is used. The controller keeps the number of requests arriving at the application close to a defined goal.

Having a focus on QoS (*Quality of Service*), ControlWare [16] is a middleware that uses control theory to provide performance guarantees for services available on the Internet. From the specification of performance goals by the users, the controllers, sensors and actuators necessary for the middleware are automatically generated so that it maintains the desired QoS. A P controller is used and customized according to defined performance goals. In turn, the sensors monitor CPU utilization metrics and queue size of service requests. Adaptations are performed with the controller changing the request queue management policy.

Abdelzaher [1,2] also focuses on server performance control, specifically on Web servers. In this case, controllers are used to avoid overload and guaranteeing server performance when there is an unplanned increase in requests. The solution includes defining the behavioural model of server performance using a PI controller.

Like previous solutions, ACM [14] uses control theory to provide performance guarantees to distributed applications. The solution includes two PID controllers that regulate the *deadlines* for sending responses to client requests. Both controllers are associated with CPU utilization and bandwidth availability.

Unlike previous works, which present middleware systems for service-oriented applications, Banerje et al. [3] propose an adaptive middleware based on control theory for multimedia applications in wireless networks. In this case, the middleware uses an I controller that defines the frequency of sending information (e.g., network traffic and resource availability) from the *streams* server to the application. The adaptations avoid network congestion and guarantee the level of QoS necessary for the application.

An initial difference between the related works and the focus of this paper is that they typically use control theory to reconfigure the resources the application uses. But, they are not intended to reconfigure the behaviour of the middleware itself. Furthermore, they do not focus on messaging systems, which have a very particular dynamic behaviour.

6   Conclusion, Lessons Learned and Future Works

This paper presented the steps to applying control theory to a messaging system. The unique contribution of this work is to show how to incorporate controllers to adapt a messaging consumer's execution. While grounded on mathematical principles, the proposed steps were adopted to implement different controllers for an existing commercial messaging system.

This initial experience have some learning points. Firstly, to properly design the experiments that are essential in the tunning approaches is a great challenge. Messaging systems are complex software systems whose set of configurable parameters is usually very big. Then, to identify the knobs (e.g., prefetch count) that have greater impact on what is being investigated (e.g., arrival rate) needs a good knowledge of the messaging system. Secondly, in relation to the Root Locus tunning method, a first-order linear model is simple but facilitates enourmeously the use of control theory and helps to define the controllers. Thirdly, the integration of control elements into the messaging systems is facilitated by the proximity with the business logic.

Some important points should be taken into account in future works. Firstly, the whole evaluation only involved publishers generating a given workload. Then, it is necessary to evaluate them, considering different workload profiles. Secondly, elements external to the closed loop are usually the source of disturbances that should also be considered, e.g., workload peaks. For example, the controller should also be capable of working on compensating for these disturbances. Thirdly, consumer monitoring may be subject to noise, e.g., delay in collecting the metric. Similarly to disturbances, the controller should also be designed considering measuring noises. Finally, exploring other kinds of controllers and checking their properties should be necessary.

References

1. Abdelzaher, T., Lu, Y., Zhang, R., Henriksson, D.: Practical application of control theory to Web services. In: Proceedings of the 2004 American Control Conference, vol. 3, pp. 1992–1997 (2004)
2. Abdelzaher, T., Shin, K., Bhatti, N.: Performance guarantees for Web server end-systems: a control-theoretical Approach. IEEE Trans. Parall. Distrib. Syst. 13(1), 80–96 (2002)
3. Banerjee, N., Basu, K., Das, S.: Adaptive resource management for multimedia applications in wireless networks. In: Sixth IEEE International Symposium on a World of Wireless Mobile and Multimedia Networks, pp. 250–257 (2005)
4. Beni, E.H., Lagaisse, B., Joosen, W.: Infracomposer: policy-driven adaptive and reflective middleware for the cloudification of simulation & optimization workflows. J. Syst. Arch. 95, 36–46 (2019)
5. Danish, S.M., Zhang, K., Jacobsen, H.A.: Blockaim: a neural network-based intelligent middleware for large-scale IoT data placement decisions. IEEE Trans. Mobile Comput. 22(1), 84–99 (2023)
6. Dobbelaere, P., Esmaili, K.S.: Kafka versus RabbitMQ: a comparative study of two industry reference publish/Subscribe implementations: industry paper. In: Proceedings of the 11th ACM International Conference on Distributed and Event-Based Systems, DEBS 17, pp. 227–238. Association for Computing Machinery, New York, NY, USA (2017)
7. Filieri, A.: Software engineering meets control theory. In: 2015 IEEE/ACM 10th International Symposium on Software Engineering for Adaptive and Self-Managing Systems, pp. 71–82 (2015)
8. Fu, G., Zhang, Y., Yu, G.: A fair comparison of message queuing systems. IEEE Access 9, 421–432 (2021)
9. Hellerstein, J.L., Diao, Y., Parekh, S., Tilbury, D.M.: Feedback Control of Computing Systems. Wiley, Hoboken (2004)
10. Janert, P.K.: Feedback Control for Computer Systems. O'Reilly, Sebastopol (2014)
11. Li, B., Nahrstedt, K.: A control-based middleware framework for quality-of-service adaptations. IEEE J. Sel. Areas Commun. 17(9), 1632–1650 (1999)

12. Rosa, N.S.: Middleware adaptation through process mining. In: 2017 IEEE 31st International Conference on Advanced Information Networking and Applications (AINA), pp. 244–251 (2017)
13. Shevtsov, S., Berekmeri, M., Weyns, D., Maggio, M.: Control-theoretical software adaptation: a systematic literature review. IEEE Trans. Softw. Eng. **44**(8), 784–810 (2018)
14. Shi, X.A., Zhou, X.S., Wu, X.J., Gu, J.H.: Adaptive control based dynamic Real-time resource management. In: Proceedings of the 2003 International Conference on Machine Learning and Cybernetics (IEEE Cat. No.03EX693), vol. 5, pp. 3155–3159 (2003)
15. Weyns, D.: An Introduction to Self-adaptive Systems: A Contemporary Software Engineering Perspective. Wiley, New York (2021)
16. Zhang, R., Lu, C., Abdelzaher, T., Stankovic, J.: ControlWare: a middleware architecture for feedback control of software performance. In: Proceedings 22nd International Conference on Distributed Computing Systems, pp. 301–310 (2002)

A Self-adaptative Architecture to Support Maintenance Decisions in Industry 4.0

Izaque Esteves[(⊠)], Regina Braga, José Maria N. David, and Victor Stroele

Federal University of Juiz de Fora, Juiz de Fora, Brazil
izaque.esteves@estudante.ufjf.br, {regina.braga,
victor.stroele}@ufjf.edu.br, jose.david@ufjf.br

Abstract. The Industry 4.0 era is primarily based on Internet of Things (IoT) devices that generate large amounts of data that can be analyzed to support decisions. However, the system that supports this approach must be highly adaptative and requires knowledge-based and reactive techniques to provide results, characterizing a self-adaptive software solution. In Industry 4.0, preventive maintenance planning is necessary to ensure equipment operation associated with IoT devices. This work presents a self-adaptative architecture to support preventive maintenance. We correlate failure and sensor data and use machine learning and ontologies to analyze these data. We also evaluated the feasibility of the solution in the textile Industry. As a result, we enriched decisionsupport information related to industrial equipment maintenance.

1 Introduction

Software technologies must face several challenges to develop smart services connected to smart products and cyber-physical systems (CPS). In this scenario, the interest in new software development strategies grows, deriving new approaches for complex system development [1]. Self-adaptative software engineering [2], encompassing context awareness and pervasive strategies, is an approach that aims to tackle the complexities from a new perspective. In this scenario, the software development is a knowledge-driven process, encompassing design, integration, and management of complex systems over their life cycles and relies on systems thinking principles to organize the knowledge.

The application domain of Industry 4.0 demands increasingly faster decisions. Therefore, information for decision-making must be built with quality and agility, providing a greater degree of anticipation of operational failures in production processes.

Considering preventive maintenance processes in Industry 4.0, it is possible to improve the decision-making process with self-adaptive systems. Using enabling technologies such as the Internet of Things (IoT) and big data, it is possible to enrich maintenance-related decisions through intelligent technologies, mitigating low criticality failures that usually relate to critical failures.

One of the main players in Industry 4.0 and the maintenance of its status is due to the automation of machines and processes and its continuity is associated with the insertion of sensing for data acquisition and computational intelligence. The strategic application

L. Barolli (Ed.): AINA 2023, LNNS 661, pp. 274–285, 2023.
https://doi.org/10.1007/978-3-031-29056-5_25

of techniques that accelerate decision-making increases productivity and is reflected in maintenance processes.

This paper discusses using sensors and data analysis for predictive machine maintenance through the engineering of a self-adaptative architecture. Predictive software maintenance provides a method to make maintenance processes more efficient since machines equipped with sensors and data processing units can send status signals throughout their use. The architecture can analyze these data to predict failures and improve the equipment's functioning and control system. This way, the planning of maintenance interventions will be more accurate, and we will have less equipment out of use due to failures detected in a predictive way. The strategy is to encourage and leverage the industry productivity based on innovative technologies, i.e., IoT systems, contemporary systems engineering development strategies (self-adaptive SE), and data analysis strategies.

We propose an architecture that collects and analyzes data related to industrial equipment failure events. The architecture must detect the failures through machine learning algorithms (deep learning algorithms), semantic model processing (ontologies), and contextual information. The data analysis results generate alerts and serve as a decision-support approach to improving predictive equipment maintenance.

The prediction of failures in a specific and isolated way is widely discussed in the literature, without analyzing the context in which the failures occur and the possible correlations between them. Therefore, these works do not consider the context as a source for enriching information for decision-making. Moreover, they do not associate the results with a self-adaptative architecture that can automatically alert about possible failures and process changes directly on the devices.

Therefore, we investigate the following research question: How does a selfadaptative architecture using sensor data, contextual information, and data analytics help prevent failures in textile industrial maintenance activities? As a feasibility study, we used the solution in the real-world context of a textile production system. The initial results support the predictive detection of machine failures and the consequent reduction in operating costs. Therefore, from data analysis, the architecture can trigger alerts or process changes directly on the devices, providing an intelligent and context ware solution for Industry 4.0.

The paper is organized into the following sections, besides the Introduction. Section 2 presents the theoretical background, and Sect. 3 discuss related work. In Sect. 4 the solution is detailed. Section 5 presents a feasibility study. Section 6 presents the conclusions and future work.

2 Background

Complex software systems are expected to operate under uncertain conditions without interruption. Possible causes of uncertainties include changes in the operational environment, availability of resources, and variations in user goals. Self-adaptation aims to let the system collect additional data about the uncertainties during operation. The system uses this data to resolve uncertainties, to reason about itself and, based on its goals, reconfigure or adjust itself to satisfy the changing conditions or, if necessary, to degrade gracefully [2].

All these changes led to a set of critical challenges to software development. The key idea is to let the system gather new knowledge at runtime to resolve uncertainties, reason about itself, its context, and goals, and adapt to realize goals. A self-adaptive system is a system that can handle uncertainty in its environment and its goals autonomously (or with minimal human interference). It consists of two distinct parts, i.e., a component that interacts with the environment and the domain, i.e., Actual System (AS), and a component that interacts with the first component and has adaptation concerns, i.e., Managing System (MS). The AS exists within its environment and produces data related to different aspects of the system. The MS captures these data and uses models to conduct different operations/actions on the AS.

Considering the challenges of digitizing industrial processes, their efficiency can be increased through predictive maintenance, based on sensing and processing operational status signals. The collected signals can present the behavior of a machine and allow a causal link between failures, producing context information. This way, it is possible to plan maintenance interventions with greater precision, reducing machine downtime for corrective maintenance.

Machine Learning (ML) techniques can contribute to finding important relationships but not naturally detected. In this vein, Artificial Neural Networks (ANNs) [4] favor learning from experience. Learning takes place by analyzing pre-established patterns or the analysis of results. In the textile industry scenario, the predictive model based on deep neural networks can use sensor data, which, combined with the dynamic adjustment capacity of the model, favors decision-making in maintenance processes in a more anticipated and assertive way.

Understanding the scenario where failures occur involves assimilating information that shapes the context of the decision. According to [5] "Context" is any piece of information that may be used to characterize the situation of an entity, and "contextawareness" is a dynamic property of the system that can affect the overall software system behavior when realizing interaction between an actor and the system. It is possible to build an information model about failure events from the data collected from machines connected through sensors and IoT devices, including information about the context (environment, agents, and other related information). For example, Table 1 shows sensor data and context information related to a textile machine failure.

Table 1. Textile operational failure data.

What?	Who?	When?	How?	Where?	What temperature?	What humidity?
Wire Breakage	Machine #54	2021/11/21 16:45:22	Production order #4521	Rewire	38 °C	47%

In Table 1, (i) "what" represents the failure type, (ii) "who", is the machine where the failure occurs, and (iii) "when" means the hour and data of the failure. The "how" represents the production process executed when the failure occurs. The "where" represents the place where the machine are installed. The "what temperature" represents the environment temperature from the machine's location, and the "what humidity" means the humidity in this same environment.

One of the components of a self-adaptive system is the adaptation goals. Different models can represent the adaptation goals, including ontological models [6], that can infer specific situations by analyzing data through processing semantic rules. Then, alerts can be generated automatically, and the MS can process changes directly on the devices.

3 Related Work

Some works consider preventive maintenance as a key issue in Industry 4.0 systems. However, they need to consider the complexity of software systems that Industry 4.0 demands. Grabot [7] proposes a new approach related to association rule mining in industrial data. The work proposes a semantic (based on a priori knowledge) and objective data analysis (considering the numerical characteristics of the rules).

Syafrudin et al. [8] proposed a solution for real-time decision-making to reduce losses due to unforeseen failures. Min, Qingfei, et al. [3] discuss a framework for building an IoT-based DT for the petrochemical industry. The authors stand out for integrating machine learning and industrial big data in real-time to train and optimize digital twin models.

Carbery et al. [9] present a framework that combines data acquisition and training of learning models to detect events that influence manufacturing processes. The authors used AI to support process decisions and increase equipment performance. Costa et al. [10] present a framework for transforming machine documentation data or log data into highly structured data. Despite using data and techniques potentially contributing to predictive maintenance, the work is limited to structuring machine data and processing natural language. Finally, Ruschel et al. [11] seek to estimate process cycle time variations by applying process mining techniques from event log information.

Table 2. Comparison between related works.

Related work	Processing data about process/equipment	Application of machine learning techniques for data classification	Ontology for information enrichment	Use digital twin in process monitoring and control
Syafrudin et al.	Yes	Yes	No	No
Carbery et al.	Yes	Yes	No	No
Costa et al.	Yes	No	Yes	No
Grabot et al.	No	Yes	No	No
Qingfei et al.	Yes	Yes	No	Yes
Out work	Yes	Yes	Yes	Yes

Table 2 compares the works considering the techniques used to process industrial data to support maintenance processes. Comparatively, our proposal differs in i) using intelligent techniques and semantic models to support preventive maintenance in Industry 4.0; ii) using self-adaptative techniques; and iii) considering the context a source of information capable of supporting decision-making in preventive maintenance through a self-adaptive system. As a result, we provide information that can trigger preventive actions.

4   Self-adaptative Architecture

Contemporary industrial production systems support high connectivity through related industrial sensors or IoT devices. In this type of system, we have a scenario composed of machines that share information about events, sensors capable of detecting context information, and computer systems, as shown in Fig. 1.

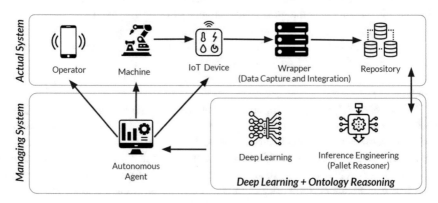

Fig. 1. Failure events communication infrastructure.

It is possible to connect machines through IoT devices, connect sensors, wired or not, and transmit status signals or data related to the machine's productivity and production process. It is also possible to receive command signals via the communication network, such as the MQTT[1] protocol. Our solution uses these available resources to collect, store and process data.

Figure 2 presents the main components of the self-adaptative architecture, showing the scenario where the devices communicate and the infrastructure provided for collecting, storing, and processing information. The Actual System (AS) component is responsible for the system's day-by-day operation and data capture. It has two main modules, i.e., environment and knowledge acquisition. The Environment connects the devices to the data network and broadcasts telemetry data. Older machines typically lack connectivity to industrial networks, not communicating sensor data. IoT devices connected to machines allow telemetry data, such as failures, to be computed and transmitted through existing communication sockets. This module is also responsible for

---

[1] http://mqtt.org/.

real-time capturing the context data from the environment and the closest operators and maintenance managers so that Managing System (MS) can create specific alerts for quick decision-making by the operators. The data communication protocol is MQTT. In this model, we have the devices connected to a broker that acts as a dispatcher for IoT devices and connected equipment and then the subsequent data storage on a server. The Knowledge Acquisition has the Pre-processing component responsible for data cleaning and formatting. After pre-processing, the data is processed by the Analysis module.

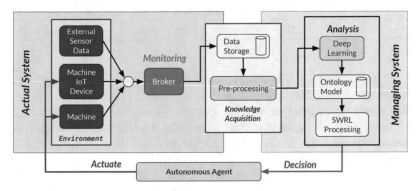

Fig. 2. Self-adaptative solution main components.

The MS component performs intelligent processing using deep learning techniques and ontological inference rules to provide strategic information to support the decision process. The Deep Learning module is responsible for running machine learning algorithms and knowledge discovery. This module uses Neural Networks to analyze machine failure data, environment data and historical maintenance data to estimate the probability of a data set of machine failure configuring a critical failure. The data relating to possible critical failure identified in the Deep Learning module is sent to the Ontology Model module. This module processes the data related to critical failure, together with contextual information acquired by the environmental sensors. Based on logical rules, the SWRL Processing module provides new information and relations that can be used to activate the Autonomous Agent module. The Autonomous Agent can trigger alerts or modify the AS, based on information provided by the previous two MS modules. Therefore, the Autonomous Agent can search for the available devices that must receive the alerts and/or process changes directly on the devices or in the AS functioning, such as turning off a specific machine.

4.1 Deep Learning Component

Generally, in an industrial production environment, the criticality of a failure is not exclusively determined by the type of failure. Other information, such as the repair time, the cost of the solution, or the failure history, must also be tractable. The combination of these components defines the criticality of a failure.

The goal of Deep Learning technique is to analyze the components that characterize the occurrences of a failure and how the interaction between such components defines

Table 3. Example of failure data from textile industry.

Machine ID	Type of failure	Timestamp	Time repair	Cost	Criticality	Hum	Temp	Label
81	7	3637419	1583193600	522	499	61	97	0
66	5	297	1593388800	485	307	63	112	0
32	9	3831058	1600128000	75	798	6	40	1
43	4	300732	1597708800	553	531	48	141	0
62	2	127	1585699200	335	367	74	148	0

a failure as critical or not. The data, acquired from the sensors and pre-processed by the Knowledge Acquisition module, contains information about the type of failure, the machine where the failure occurred, the moment (timestamp) of the failure, the time taken to resolve the failure, and the maintenance cost. A positive class mark also signals that that dataset represents a critical failure in the training data set. Table 3 presents this data considering the textile industry.

From this neural network, we used deep learning algorithms to build a model to predict the criticality of a failure, based on information from maintenance records with positive class indication for criticality. The gain in the use of neural networks is associated with learning from synapse weight adjustments. The model adjustment is made during its training. The adjustment includes the distance between the predicted value by the model and the supervision value. Thus, synapse weights are readjusted at each iteration until the model can be used without supervisory data and can predict the output class, that is, classify a failure as critical or not.

4.2   Ontology Model

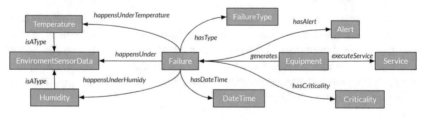

Fig. 3. SmartMaintenanceOntology main classes and associations.

From the data classified by the Deep Learning module, the ontology[2] is instantiated to process information related to failure events and context information. From SWRL.

(Semantic Web Rule Language) rules and inference algorithm processing, it is possible to derive new relationships between data and discover new information, which

---

[2] https://github.com/izaqueesteves/ontology_self_adapt_systems.

may require MS automatic actions related to more critical failure prevention. Figure 3 presents an example of an ontology used in this domain.

5 Feasibility Study

The objective of this feasibility study was to "analyze the self-adaptative solution to assist in decision making from the perspective of managers and operators in the context of textile industry production system." The following research question was derived: How does a self-adaptative architecture using sensor data, contextual information, and data analytics help prevent failures in textile industrial maintenance activities?

Once defined the goal and its subsequent research question, the evaluation methodology was carried out. The solution was in a real-world textile industry context to answer the research question. The activities carried out were the collection of data, the analysis, and the validation of the actions taken by the self-adaptative architecture.

Machines from the textile industry, more specifically from the "meshing" sector, produce failures such as needle breakage that are resolved quickly, without operational loss. The occurrence of several small failures can result in a machine breakdown or even a qualitative loss of a batch of products. The analysis of this information together with environmental data can help in predictive maintenance of equipment.

Temperature and humidity, for example, interfere with the properties of the wires and the operation of the machines, contributing to the occurrence of critical failures and, therefore, are important contextual information in this scenario.

Generally, the environment's temperature and humidity substantially influence the equipment's productivity. In addition, the proximity and expertise of a particular employee can be crucial when detecting the probability of a failure or analyzing the environment for a failure to occur.

The self-adaptative architecture favors this possibility through data availability from deep learning, combined with context data and processed through ontologies and autonomous software agents. To support the architecture, a specific ontology was constructed.

The main classes and associations of SmartMaintenanceOntology are presented in Fig. 3. The Failure class represents the failure event which has information such as the equipment failure code, the sector to which the equipment belongs, and the code that identifies the failure event. FailureType represents the type of failure.

Associated with the Failure class we have the DateTime class that provides the timestamp of the failure event. The registered failures occur in a production environment where the humidity and temperature variables are controlled, which are properties of the Failure class. The Criticality class represents the criticality degree of a failure estimated by the "Deep Learning module".

Finally, the Alert class is associated with the Failure class through the gas alert property, which is dynamically defined from SWRL rules that consider specific characteristics of failure occurrences, such as the amount of a certain event classified as critical, and environmental data such as temperature and humidity, which are types of environment data captured by sensors. Figure 4 shows the main classes and their properties in the

Protegé[3] tool, and Fig. 6 details one of the SWRL rules, which evaluates the humidity and temperature indices in the environment.

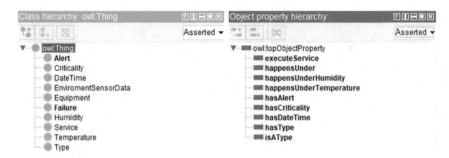

Fig. 4. Classes and properties.

To evaluate the viability of the solution in analyzing data from failure events to trigger autonomous agents from ontological processing, a dataset with 10,000 failure records was used. The set has machine and production order data, environmental data and a criticality label as shown in Fig. 6. Early warning of criticality, according to industry knowledge, facilitates supervised learning of the classification model.

We implemented a failure classification model in Python, and it was expected that failures would be automatically classified as critical or not, according to supervised learning. Using the Scikit-Learn library, the dataset with 10,000 fault records was organized into two parts, one with 75% of the data for training the model and the remaining 25% to evaluate the model's accuracy in class predictions. Machine learning was implemented through a neural network based on the Percetron[4] multilayer classifier with one or several hidden layers with a certain number of neurons. We trained it through a backpropagation algorithm. In this way, the new failure records could be classified automatically by the neural network and serve as input to the ontological model. The result of processing via the neural network is a new dataset with defined classes. After using the neural network to process the test dataset, the resulting dataset with the assigned criticality values was imported by the ontology (Table 1).

Fig. 5. Failure instance and its attributes.

---

[3] https://protege.stanford.edu/.

[4] https://www.sciencedirect.com/topics/computer-science/perceptron-algorithm.

The ontology processed the combination of failure data already processed by the neural network with the environmental data and qualitative and quantitative failure metrics that must be observed. Figure 6 presents an example of SWRL rules that helped to infer alerts from failure and environment data. We had an instance defined as "Failure1" with some failures and failure type attributes respectively set to "4" and "5" (Fig. 5).

Repetitions of events and environmental conditions can worsen critical failures. In the textile industry, the temperature and humidity, for example, can assert machine failure or even make the break of one of its components imminent, resulting in the stoppage of the equipment.

6   Results

Thus, from the environmental information (temperature and humidity), the SWRL rules evaluated the scenarios where a technical intervention was urgent to mitigate the risk. Figure 6 shows one of the rules, which generates an alert code with a value of 100 if the ambient temperature is greater than 35 °C and the ambient humidity is less than 25%.

Similarly, another rule was created to trigger the autonomous agent if the number of "4" type faults exceeds 4 and the ambient temperature exceeds 30 °C. In this case, the alert code generated has a value of "200", as shown in Fig. 7.

Moreover, to illustrate the validity of the SWRL rules, the instance "Failure1" is taken as an example, defined with the following attributes and values: temperature: 45 °C, humidity: 20%, fault type:4, and the number of fault occurrences: 5. After the execution of the inference engine, the ontology was able to process two alerts with codes "100" and "200", as shown in Fig. 7.

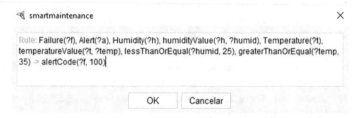

Fig. 6. SWRL Rule – Assessing Temperature and Humidity.

Therefore, the ontology processed the inferences from these data (machine and environmental data) inserted in the ontology and geolocation data captured from the mobile device of a nearby operator. In the latter case, alert could be sent directly to machines or IoT devices responsible for machines, according to the generated alert, to preserve the equipment or guarantee safety conditions for operation.

In this way, based on machine failure data, neural network processing, and semantic models, the solution was able to assess the failure criticality associated with the environmental data jointly and the qualitative and quantitative characteristics of the failures enriching the information. The action of the autonomous agent was processed by sending alert messages via MQTT protocol with directives to the machines. Alert messages

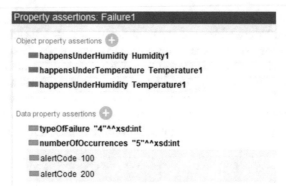

Fig. 7. Inferences produced by SWRL rules.

for operators were also sent to mobile devices closest to the equipment. Thus, it was possible to indicate that the combined use of intelligent techniques and data mining, and autonomous software agents can detect maintenance failures, answering the proposed research question. However, more detailed experimental studies must be conducted to confirm this perspective in other real-world contexts and with a broader data volume.

7 Conclusions

One of the challenges of Industry 4.0 is the agility in decision-making related to preventive maintenance. The ability to acquire data on failures, mine knowledge, and enable decision-making are challenges we can overcome through information technology. This work presented a self-adaptative architecture that supports decisionmaking related to predictive maintenance in Industry 4.0. From the analysis of machine failure data and contextual information processed from intelligent techniques, it was possible to anticipate failures, making maintenance planning decisions more assertive. A feasibility study was described, illustrating how the solution supported decisionmaking activities. The study considered collecting machine failure data through sensors and IoT devices, storing data in the cloud, and processing contextual information, supported by data mining and ontological analysis techniques.

As a result, targeted alerts were generated, considering contextual information, and geographically distributed operators made decisions. The possibility of accessing fault data and association rules mining results favors teams with multiple competencies, and without geolocation restrictions for problem analysis, since both the storage and processing of fault data can be performed remotely. As a result, specialists can interact to tackle complex problems in Industry 4.0. As future work, we are discussing how to extend the solution to provide an integrative platform to support all production phases. Studies to develop the improvements are being conducted.

Acknowledgements. This work was partially funded by UFJF/Brazil, CAPES/Brazil, CNPq/Brazil (grant: 311595/2019-7), FAPEMIG/Brazil (grant: APQ-02685-17) and (grant: APQ-02194-18).

References

1. Weyns, D., Iftikhar, M.U., De La Iglesia, D.G., Ahmad, T.: A survey of formal methods in self-adaptive systems. In: Proceedings of the Fifth International C* Conference on Computer Science and Software Engineering, pp. 67–79 (2012)
2. Eramo, R., Bordeleau, F., Combemale, B., van Den Brand, M., Wimmer, M., Wortmann, A.: Conceptualizing digital twins. IEEE Softw. **39**(2), 39–46 (2021)
3. Min, Q., Lu, Y., Liu, Z., Su, C., Wang, B.: Machine learning based digital twin framework for production optimization in petrochemical industry. Int. J. Inf. Manage. **49**, 502–519 (2019)
4. Basheer, I.A., Hajmeer, M.: Artificial neural networks: fundamentals, computing, design, and application. J. Microbiol. Methods **43**(1), 3–31 (2000)
5. Abowd, G.D., Dey, A.K., Brown, P.J., Davies, N., Smith, M., Steggles, P.: Towards a better understanding of context and context-awareness. In: International Symposium on Handheld and Ubiquitous Computing, pp. 304–307 (1999). Springer
6. Buneman, P., Khanna, S., Wang-Chiew, T.: Why and where: A characterization of data provenance. In: International Conference on Database Theory, pp. 316–330 (2001). Springer
7. Grabot, B.: Rule mining in maintenance: Analysing large knowledge bases. Comput. Ind. Eng. **139**, 105501 (2020)
8. Syafrudin, M., Alfian, G., Fitriyani, N.L., Rhee, J.: Performance analysis of IoT-based sensor, big data processing, and machine learning model for real-time monitoring system in automotive manufacturing. Sensors **18**(9), 2946 (2018)
9. Carbery, C.M., Woods, R., Marshall, A.H.: A new data analytics framework emphasising preprocessing in learning ai models for complex manufacturing systems. In: Li, K., Fei, M., Du, D., Yang, Z., Yang, D. (eds.) ICSEE/IMIOT -2018. CCIS, vol. 924, pp. 169–179. Springer, Singapore (2018). https://doi.org/10.1007/978-981-13-2384-3_16
10. Costa, R., Figueiras, P., Jardim-Gonçalves, R., Ramos-Filho, J., Lima, C.: Semantic enrichment of product data supported by machine learning techniques. In: 2017 International Conference on Engineering, Technology and Innovation (ice/itmc), pp. 1472–1479 (2017). IEEE
11. Ruschel, E., Santos, E.A.P., Loures, E.D.F.R.: Mining shop-floor data for preventive maintenance management: integrating probabilistic and predictive models. Procedia Manuf. **11**, 1127–1134 (2017)

DCANon: Towards Distributed Certification Authority (CA) with Non-fungible Token (NFT)

Rafael Descio-Trineto[✉], Maurício Pillon, Guilherme Koslovski,
and Charles Miers

Graduate Program in Applied Computing, Santa Catarina State University, Joinville,
Brazil
rafael@descio.com.br,
{mauricio.pillon,guilherme.koslovski,charles.miers}@udesc.br

Abstract. Distributed applications rely on Certification Authorities (CAs) entities associated with a few institutions around the world. There are countless trusted institutions on the networks, *e.g.*, universities, governments, banks, making effective use of CAs. Although largely used, the decentralized management of CAs is an administrative challenge. In this context, the joint emergence of digital ledgers and Non-Fungible Tokens (NFTs) compose a new opportunity to effectively decentralize single assets safely. Towards Distributed Certification Authority with Non-Fungible Token (DCANon) is a distributed Certification Authority (CA) built to trusted institutions and supported by Non-Fungible Token (NFT) concepts. DCANon changes the way revocation may be find, permitting that all consortium member crawler the blockchain for revoked certificates, independently from which CA this revocation occurs. A proof-of-concept prototype was implemented with Hyperledger Fabric demonstrating the benefits of DCANon.

1 Introduction

The use of Non-Fungible Tokens (NFTs) has become increasingly common, with large investments in research and development of this concept. With the announcement of the use of NFT generating sales of digital assets for millions of dollars, the benefits of this technology quickly arousing the interest of several researchers and industry to explore which areas this concept can be applied. Currently, a large part of the use of NFT has been associated with digital images, certifying their uniqueness and, consequently, adding financial value to them. However, their designs and applications cover several domains, including, among others, the type of asset used to generate the NFT, how this it will be generated, what information will be included, the platform on which it will be inserted, whoever may have access and manipulate this information, etc. The particularity of this concept lies in its non-fungibility, that is, an NFT must represent a single, irreplaceable and indivisible asset.

L. Barolli (Ed.): AINA 2023, LNNS 661, pp. 286–298, 2023.
https://doi.org/10.1007/978-3-031-29056-5_26

Fungible tokens represent assets that can be replicated, exchanged and shared. An NFT must be immutable from the moment of its conception, and there cannot be more than one NFT with the same value. In this case, asset is understood as the input object that is used to generate the token, which can be a binary string, a file, a document, or any other set of information that can be digitized to generate a unique and indivisible token. To guarantee immutability, the concept of NFT extends the use of the blockchain, since after its creation, the token is stored as a transaction within a blockchain's data block and distributed among the entire blockchain network. The use of NFT still needs to be explored in other contexts. New platforms are emerging in different fields of human activity, and many types and formats of assets can take advantage of this arising, taking in count immutability, security, guarantee of ownership, traceability and other the benefits that the use of NFT combined with blockchain can bring. Motivated by this, new applications for the use of NFT are coming up in different sectors of industry, services and entertainment exploring this concept deeply. Given the facts, this work proposed the use of NFT to implement a decentralized CA. Specifically, this work presents a proof-of-concept through an application using Hyperledger blockchain to issue trusted Secure Sockets Layer (SSL) certificates in a Public Key Infrastructure (PKI), providing some unexplored solutions that can bring NFT benefits to identify business entities to a blockchain-based Certification Authority (CA). In summary, the contributions of this work are observed in different parts of a CA infrastructure: (i) The transport part of the PKI infrastructure is performed through a blockchain consortium; (ii) The use of the NFT to keep non-fungibility of a SSL certificate; (iii) The use of Chaincode to query and update SSL certificates; (iv) The use of Interplanetary File System (IPFS) to solve blockchain's block size limitation; (v) Improves the overall communication performance between Online Certificate Status Protocol (OCSP) and Certificate Revocation List (CRL) relying on blockchain speed and security processing; and (vi) Proposes a trustworthy and accessible CA structure.

It is important to mention that many sources emphasize that a NFT cannot have multiple values, and that is true, however a NFT can represent a state of other existing NFT, *i.e.*, an artist can create a limited number of copies of the same *art-work* NFT and sell them individually. This concept can be applied to our need, using the same certificate to validate different resources, *i.e.*, subdomains, but using the same Chaincode. This paper is organized as follows. Section 2 addresses background, prior concepts of NFT and a set of related works found in literature. Section 3 presents DCANon, while the experimental analysis is presented in Sect. 4. Section 5 concludes and proposes future work.

2   Background and Motivation

2.1   NFT Platforms and Implementations

A Non-Fungible Token (NFT) is a single encrypted record that represents unique, non-replicable information stored in a digital ledger (blockchain) [13]. In theory, any information can be stored in an NFT. They can only have one official owner;

however, ownership is transferable and are protected by the blockchain. With this protection, modifying the ownership record or altering an existing NFT is severely hampered, or even impossible[1]. On the other hand, fungibles tokens store interchangeable, divisible, and non-unique values, that is, two fungible tokens of the same type have the same value. A NFT creation is called Tokenization or Mint/Minting, the term refers to the creation of the non-fungible *token* and insertion into the blockchain, *i.e.*, the transformation of the asset into a *token* with information that can be inserted into the blockchain so that it is unique. Once a NFT is created there are two ways to control it in a blockchain, with and without a smart contract. The concept of contract involves an agreement of obligation between one or more parties, in which all the participants predispose to fulfill the terms of the agreement. In Smart Contracts, this concept is transported to the digital medium, in which this agreement is carried out from machine to machine, using a program to ensure that all contract participants comply with the terms of the contract. Finally, there are two ways to store the NFT asset in the blockchain [7]: (i) In On-Chain approach, the data is stored in the chain itself, however when used with limited number of blocks or with large amount of information, this way becomes impractical. (ii) On the contrary, Off-Chain stores the information outside the block that will be inserted into the blockchain and creates a reference to the data in the block that is stored.

Theoretically, NFTs can be used with any blockchain, however not all blockchain platforms support the unique features of NFT such as uniqueness and indivisibility. There are two main blockchain platforms widely used in academia that support NFTs: Hyperlerger Fabric[2] and Ethereum[1]. But other blockchain platforms such as, Cardano[3], Tezos[4], IBM Blockchain[5], and Cosmos NFT[6] are also supporting and expanding the use of NFTs. More recent developments use the Off-Chain concept to create NFT in a blockchain *i.e.*, Hyperledger Fabric. Developed by the Hyperledger Foundation, Hyperledger Fabric is an enterprise-grade distributed accounting platform that offers modularity and versatility for a broad set of industry use cases. The Hyperledger Foundation is a non-profit organization that brings together all the resources and infrastructure necessary to ensure thriving and stable ecosystems around open-source software blockchain projects[2]. The modular architecture for Hyperledger Fabric accommodates a variety of business use cases by means of plug and play components such as consensus, privacy, and adherence services[2].

Supporting the use of NFT with Hyperledger Fabric, the IBM Research has recently designed and implemented two blockchain components that facilitate the exchange of tokens in the enterprise context, the *Fabric Smart Client* and the *Fabric Token SDK*. The Fabric Smart Client allows off-chain exchanges between

---

[1] Available: https://ethereum.org/en/nft/.
[2] Available: https://hyperledger-fabric.readthedocs.io/.
[3] Available: https://docs.cardano.org/.
[4] Available: https://tezos.com/.
[5] Available: https://www.ibm.com/blockchain.
[6] Available: https://hackmd.io/@okwme/cosmos-nft.

clients, application/client-side state management, and more flexible transaction creation. The Fabric Smart Client also allows Fabric applications to integrate a wider range of technologies that enhance privacy and/or reflect a variety of business processes as needed for the use case at hand[5]. The Fabric Token SDK is a library - leveraging the Fabric Smart Client - that allows the exchange of tokens in Fabric with configurable corporate privacy properties. More specifically, the Fabric Token SDK can be configured to provide mechanisms where token exchange participants' privacy and value are preserved and coexist with auditability. At the same time, the Fabric Token SDK can offer clear asset exchange, for example, without privacy considerations[5].

In turn, the Ethereum platform became quite popular for NFTs, as it was the first decentralized computing platform based on blockchain that standardized NFTs in a well-defined interface, called ERC-721 [6]. Ethereum Request for Commentss (ERCs) are application-level standards and conventions, including contract standards such as token standards, name records, URI schemes, library/package formats, and formats portfolio [19]. Ethereum is a global open-source platform for Decentralized Applications (DApps), being the leading blockchain based platform for smart contracts [16]. In addition, Ethereum has its own Off-Chain propose called Plasma [12]. As NFT is commonly used with large files, *e.g.*, images, and storing these files on the blockchain is inefficient, Off-Chain approach are the most usually found, mainly because block size limitations require files to be split and reassembled outside the blockchain [15]. A common Off-Chain solution has been to use the protocol IPFS in conjunction with NFT. IPFS is a decentralized protocol and peer-to-peer network for storing and sharing data in a distributed file system[7]. Once the local file is transmitted to the IPFS system, it is available to the entire network and can be tracked and identified by its hash content [18]. This hash can be referenced in the metadata at NFT creation time.

2.2 Public Key Infrastructure

PKI provides the core framework for a wide variety of components, applications, policies and practices to combine and achieve the three principal security functions (integrity, authentication and non-repudiation). A PKI is a combination of hardware and software products, policies and procedures. It provides the basic security required for secure communications so that users who do not know each other or are widely distributed, can communicate securely through a chain of trust. Digital certificates are a vital component in the PKI infrastructure as they act as *digital passports* by binding the user's digital signature to their public key [9].

Essentially, CA is a trusted authority in a network that issues and manages security credentials and public keys for message encryption [1]. A CA provides the trust basis for a PKI as it manages public key certificates for their whole life cycle. The CA will: (1) Issue certificates by binding the identity of a user

---

[7] Available: https://ethereum.org/en/developers/docs/.

or system to a public key with a digital signature; (2) Schedule expiry dates for Certificate Revocation List (CRL); and (3) Ensure certificates are revoked by publishing Reputation [9]. In turn, Public Key Cryptography Standards (PKCS) standards are a set of standards, called PKCS #1 through #15. These standards cover RSA encryption, RSA signature, password-based encryption, cryptographic message syntax, private-key information syntax, selected object classes and attribute types, certification request syntax, cryptographic token interface, personal information exchange syntax, and cryptographic token information syntax [17].

Registration Authority (RA) is an optional but common component of a PKI. An RA is used to perform some of the administrative tasks that a CA would normally undertake. The main purpose of an RA is to verify an end user's identity and determine if an end entity(*Relying Party*) is entitled to have a public key certificate issued [1]. *Relying Party* is an entity that relies on the certificate and the CA that issued the certificate to verify the identity of the certificate owner and the validity of the public key, associated algorithms, and any relevant parameters in the certificate, as well as the owner's possession of the corresponding private key [3]. Another important component, Certificate Revocation List (CRL) is a list of revoked public key certificates created and digitally signed by a certification authority [4].

The Online Certificate Status Protocol (OCSP) was created as an alternative to the CRL protocol. Both protocols are used to check whether an SSL certificate has been revoked. The CRL protocol requires the browser to download potentially large amounts of SSL certificate revocation information: certificate serial numbers and status of each certificate's last publication date. The problem with the CRL protocol is that it can increase the time spent completing the SSL negotiation. The OCSP protocol does not require the browser to spend time downloading and then searching a list for certificate information. With OCSP, the browser simply posts a query and receives a response from an OCSP responder (a CA's server that specifically listens for and responds to OCSP requests) about the revocation status of a certificate[8]. *OCSP Stapling* can be used to enhance the OCSP protocol by letting the webhosting site be more proactive in improving the client (browsing) experience. OCSP stapling allows the certificate presenter (*i.e.*, web server) to query the OCSP responder directly and then cache the response. This securely cached response is then delivered with the Transport Layer Security (TLS)/SSL handshake via the *Certificate Status Request* extension response, ensuring that the browser gets the same response performance for the certificate status as it does for the website content. OCSP stapling addresses a privacy concern with OCSP because the CA no longer receives the revocation requests directly from the client (browser). OCSP stapling also addresses concerns about OCSP SSL negotiation delays by removing the need for a separate network connection to a CA's responders[8].

Finally, Certificate Transparency (CT) Logs is a system for logging and monitoring the issuance of TLS certificates. CT greatly enhances everyone's ability to

---

[8] Available: https://digicert.com/kb/enabling-ocsp-stapling.htm.

monitor and study certificate issuance, and these capabilities have led to numerous improvements to the CA ecosystem and Web security. As a result, CT is rapidly becoming critical infrastructure[9].

2.3 Related Work

Multiples researchers explore the use of blockchain to implement PKI applications, not only for public applications, but also for private CAs ecosystems, like Internet of Things (IoT) and Vehicular Ad hoc NETworks (VANETs). In IoT context, the authors [5] propose a certificate audit scheme based on blockchain, however, the possibility of using Smart Contracts or Chaincodes in the certificate creation and verification process was not explored. As regards to VANETs, other research group proposed a lightweight threshold CA for consortium blockchain along with a privacy-preserving location-based service protocol in blockchain enforced Vehicular Social Networks [14], but no revocation systems are explored. Finally, a decentralized CA was implemented using a generic two-party secure computation protocol with an evaluation of a prototype implementation that uses signed certificates. They suggest a new models for certificate generation, where multiple CAs would need to agree and cooperate before a new certificate can be generated, or even where certificate generation would require cooperation between a CA and the certificate recipient (subject) [11], despite the proposed idea being aligned with the blockchain concept, the proposed solution used other formats of distributed systems.

In the scope of CRL, different solutions are used to handle it, *e.g.*, Google Chromium Browser uses the CRLSet for revocation checking. A CRLSet is a client saved list of revoked certificates which is updated by crawling CRLs published by CAs[10], important note that by default no OCSP check is done. Mozilla Firefox uses a solution called OneCRL with is a list of intermediate certificates that have been revoked by CAs in Mozilla's root program[11]. Apple Safari Browser uses a similar approach based on the solution showed in the Worldwide Developers Conference 2017 (WWDC2017), but in addition to the Chrome solution, Apple makes a OCSP checks in certificates that Apple already thinks are revoked.

3 DCANon Architecture

All CAs ecosystems from Subsect. 2.3 used different sources of trust, which means that the same certificate may be valid in one application and invalid in another one at the same time. Furthermore, these solutions are dependent on a centralized CA. Our proposal, DCANon, change the way revocation may be find, permitting that all consortium member crawler the blockchain for revoked certificates, independently from which CA this revocation occurs. Although the

---

[9] Available: https://letsencrypt.org/docs/ct-logs/.

[10] Available: https://www.chromium.org/Home/chromium-security/crlsets/.

[11] Available: https://blog.mozilla.org/security/.

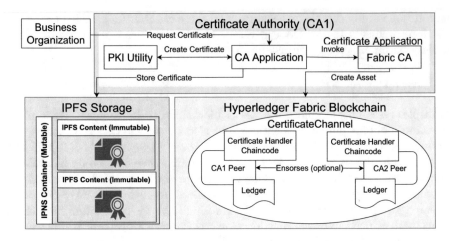

Fig. 1. Distributed CA Architecture supporting by NFT concepts.

proposed structure can be used in any environment, it has great potential to be explored in small consortium, such as, communication between a large industrial manufacturer and its supply chain.

Centralized CAs are controlled by a few business groups like happened with all coins until creation of bitcoin (national governments). A distributed ledger allows decentralize the control guaranteeing the security. This work proposes a Distributed CA supporting by distributed ledger and NFT concepts (summarized in Fig. 1). Considering that we are dealing with a consortium blockchain, each member must be properly authenticated, and for that we rely on Fabric CA module[2]. The CA is the trusted entity member of the consortium that issues the certificates, and holds a Fabric CA membership that is used to secure access to the ledger. The Business Organization is the entity requesting the certificate, the relying party, and within the proposed architecture is considered to be known and trusted by CA. Therefore, to start the ledger, in an initial flow, some administrative calls are necessary. An initial peer is created, called Peer0, which belongs to our first root CA (CA1) and is owned by administrator. The administrator is the user responsible for creating credentials for the other members, all credentials are stored and used in an Membership Service Providers (MSP). In each new CA organization is installed a client application that connects to the Hyperledger Fabric service and transact with the blockchain network. They have a single wallet[2] that held the credentials supplied by the administrator, these credentials are used as organization as MSP identifier and included in the asset.

Once we have all the members properly identified, we can start the NFT creation cycle. The project was performed considering a common public key certificate standard, X.509 standard[12]. For testing purpose, we used the project

---

[12] Available: https://www.itu.int/rec/T-REC-X.509/.

EasyRSA[13], which is a utility for managing X.509 PKI and it is utilized as CA for treat incoming Certificate Signing Requestss (CSRs). The Business Organization asset metadata structure information is shown in Table 1, inside each asset are stored all the certificates issued by the organization, theses certificates (Certificate Asset Metadata Structure (CAMS)) are formatted following the structure present in the Table 2.

Table 1. Business organization asset metadata structure

Field	Type	Description
ID	String	Unique identifier (D-U-N-S)
owerOrg	String	Organization identifier (MSP)
ipnsCID	String	Mutable CID to the IPFS directory
commonName	String	Business Organization Common Name
countryName	String	Business Organization Country Name
stateName	String	Business Organization State Name
localityName	String	Business Organization Locality Name
organizationName	String	Business Organization Name
Certs	Array of object	A list of certificate issued, in format CAMS
Revoked	Boolean	A flag indicating the revocation status

Table 2. Certificate Asset Metadata Structure (CAMS)

Field	Type	Description
ID	String	Serial number
ipnsCID	String	Immutable CID to the certificate stored in the IPFS
notBefore	Date	Validation start date
notAfter	Date	Validation end date
Revoked	Boolean	Flag indicating certificate revocation status
ipnsCRL	String	IPNS CID pointer to the CRL file

To grant the uniqueness we proposed the use of a D-U-N-S number as asset identification in the NFT metadata. The D-U-N-S number is a unique nine-digit number that identifies business entities on a location-specific basis, widely used as a standard business identifier[14]. Because of its universal recognition and unique assignment, the D-U-N-S number can also serve as a primary data key

---

[13] Available: https://easy-rsa.readthedocs.io.
[14] Available: https://developer.apple.com/support/D-U-N-S/.

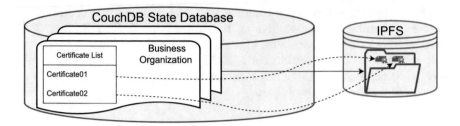

Fig. 2. Database and IPFS relation

within an organization's Master Data Management architecture. In this manner, the D-U-N-S number enables a single source of truth in the capture and storage of information related to a company's customers, partners, and suppliers[15].

Our proposal considers a mid-to-end service, whereupon the Business Organization provides the subject to generate a Certificate Signing Requests (CSR) in PKCS#10 format, the CA Application is responsible for handle this subject, process, and return Privacy Enhanced Mail (PEM) format file. In sequence, this file is imported as request into EasyRSA CA and signed resulting in a Security Certificate File (CRT) format file that is stored on the blockchain as part of the NFT. It's commonly known that sometimes may exists multiples certificates from same issuer and domain, so to handle this issue the certificate is stored in a Interplanetary File System (IPFS) directory, this approach let it store multiple certificates and cover situations like when a certificate it is about to expired and the provider wants point user to two different certificates, to avoid downtime.

Based on this, each individual business organization will have a IPFS directory (Showed in Fig. 2). This solution has a bias, because the IPFS immutability every time the directory content change a new Content Identifier (CID) will be provide, to solve this the asset stores each certificate file IPFS CID inside a vector structure and points the directory to a Interplanetary Name System (IPNS) CID, which is a mutable way to point to a IPFS addressed content. The asset creation smart contract gets the MSP identifier (MSPID) of the organization that submitted the request as the CA owner identifier, this data is stored in the asset metadata and used to control asset updates and transfers. Also in creation, the Smart Contract uses a Chaincode level endorsement policy that requires an endorsement from any channel member. This allows any root CA to create a certificate that they own without requiring an endorsement from other channel members. Transactions that update or transfer existing certificate assets will be governed by state-based endorsement policies. In scenarios with intermediate CA, the root CAs will act as issuing authority to also endorse create transactions. In case when an certificate must be revoked, the revocation information are store in the Certificate Asset as describe in Table 2, and for traceability the CRL file handle by the CA it attached as a IPNS CID. This allow have a unique CRL directory in the IPFS structure, that all CAs may share. Currently,

---

[15] Available: https://www.dnb.com/duns-number.html.

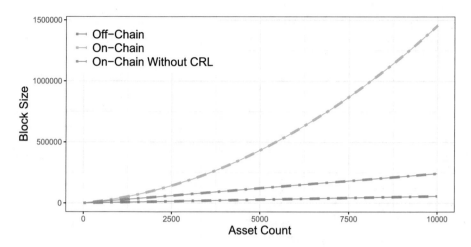

Fig. 3. Block Size Results

Hyperledger Fabric provides two ways to store peer state data, LevelDB and Apache CouchDB. LevelDB is the default key-value state database embedded in the peer process. CouchDB allows store data in JSON format, issue JSON queries against data, and use indexes to support your queries [2]. Our proposal to database relation with IPFS storage (Fig. 2) uses a CouchDB instance in each peer, this allow indexing issued certificate and speed up the search/retrieve process at the Chaincode level.

4   Experimental Analysis

Two metrics were defined to evaluate the results, the block size growth, and the response time to search for an asset. The block size is an important metric in blockchain-based systems, as this value will define the blockchain transactions throughput. And as a PKI-oriented solution, it should be considered search response time, owing to the overhead of adding a new component must be as small as possible. The systems were built in 10 docker containers running on Oracle VM Virtualbox 6.1 hosted in a desktop with processor i9-9900KF, 32G RAM and Windows 10 OS. The guest virtual machine (VM) runs Debian 11 64bits with 4 cores processor and 4G RAM. As a proof-of-concept, DCANon was implemented and tested based on Hyperlegder Fabric version 2.4.6, IPFS 0.16.0, RAFT Consensus of etcd 3.3.10[16], CouchDB database 3.1.1, EasyRSA 3.0.8 and NodeJS 16.17.1 with Typescript 3.9.10. Results were acquired in two different scenarios with data stored On-Chain and Off-Chain, and an extra scenario variation On-Chain without CRL. To ensure the reliability of the data, before each round the whole environment was reset to the initial state, erasing IPFS, CouchDB database and the blockchain ledger. To compare results,

---

[16] https://github.com/etcd-io/etcd.

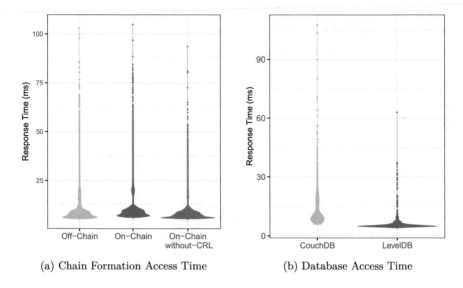

(a) Chain Formation Access Time   (b) Database Access Time

Fig. 4. Response time of chain formation and database

an extra application was developed to capture the on-chain results, the main difference is that the certificates were store in plain text and the CRL file was moved to an external NFT asset to store in the blockchain.

In our test plan, 10000 business organization were created, with 2 certificates per organization, and one in four certificates were revoked and included in the CRL file. Figure 3 depicts the results based on the block size growth. To 10000 certificates, the block size chain with Off-Chain has gone up from 56 MB to 1.4 GB, if On-Chain is choice. Nonetheless, whether On-Chain without CRL is used, the block size Chain reaches 238 MB. In addition to block size, response time is also observed in Fig. 4. A violin plot of Fig. 4a shows response time of three type of Chain formation. It possible to conclude that On-Chain without CRL has the shortest response time and the median difference for greatest (On-Chain) was 22%. Finally, database access time between CouchDB and LevelDB is compared by violin plot (Fig. 4b). The time spent to search assets by ID is tested and, as seen the use of CouchDB has a penalty, the response time are better when we have a small number of assets. But these times differences are small compared to the benefits that CouchDB provides. LevelDB is designed for a key-value architecture and has no indexing system, which limits complex searches that the application and auditing system may require, a more in-depth study can be found at [10].

5   Considerations and Future Work

The main difference of the proposed architecture is that multiple CAs can participate in the consortium and independently from which CA a certificate is issued, it can be retrieved by any consortium member. In addition, the use of CouchDB in conjunction with IPFS to store the certificate and the CRL permits find, inspect, and audit the certificate from any *permissioned* member. Furthermore, by storing the certificates grouped by Business Entity we guarantee a better management of integrity, security, and space for multiple domains in the blockchain environment.

Future security upgrades may be explored. Share secure files between CA, *e.g.*, using a secure IPFS network as suggested by [8]. Also, for the current scenario may be an interesting approach alter the Chaincode to include solution as OCSP stapling or Certificate Transparency (CT) Logs in the certificate validation method as alternative for security reinforcement.

Acknowledgements. This work was funding by the National Council for Scientific and Technological Development (CNPq), the Santa Catarina State Research and Innovation Support Foundation (FAPESC), UDESC, and developed at LabP2D. This work received financial support from the Coordination for the Improvement of Higher Education Personnel - CAPES - Brazil (PROAP/AUXPE) 0093/2021.

References

1. Al-Janabi, S.F., Obaid, A.K.: Development of certificate authority services for web applications. In: 2012 International Conference on Future Communication Networks, pp. 135–140. IEEE (2012). https://doi.org/10.1109/ICFCN.2012.6206857
2. Anderson, J.C., Lehnardt, J., Slater, N.: CouchDB: the Definitive Guide: Time to Relax. O'Reilly Media, Inc., Sebastopol (2010)
3. Barker, E.: Guideline for using cryptographic standards in the federal government:. Technical Report, National Institute of Standards and Technology, Gaithersburg, MD (2020). https://doi.org/10.6028/NIST.SP.800-175Br1
4. Bidgoli, H.: Handbook of Information Security. Wiley, USA (2005)
5. Chen, J., et. al.: CertChain: public and efficient certificate audit based on blockchain for tls connections. In: IEEE INFOCOM - IEEE Conference on Computer Communications, pp. 2060–2068 (2018). https://doi.org/10.1109/INFOCOM.2018.8486344
6. Chirtoaca, D., et al: A framework for creating deployable smart contracts for non-fungible tokens on the ethereum blockchain. In: Proceedings of the IEEE International Conference DAPP, pp. 100–105 (2020). https://doi.org/10.1109/DAPPS49028.2020.00012
7. Hepp, T., Sharinghousen, M., Ehret, P., et al.: On-chain vs. off-chain storage for supply- and blockchain integration. It Inf. Technol. **60**(5-6), 283–291 (2018). https://doi.org/10.1515/itit-2018-0019
8. Huang, H.S., et. al.: A secure file sharing system based on IPFS and blockchain. In: Proceedings of the 2020 2nd International Electronics Communication Conference, pp. 96–100. ACM, New York, NY, USA (2020). https://doi.org/10.1145/3409934.3409948
9. Hunt, R.: Technological infrastructure for PKI and digital certification. Comput. Commun. **24**(14), 1460–1471 (2001). https://doi.org/10.1016/S0140-3664(01)00293-6

10. Javaid, H., et. al.: Optimizing validation phase of hyperledger fabric. In: IEEE 27th International Symposium on Modeling, Analysis, and Simulation of Computer and Telecom. Systems (MASCOTS), pp. 269–275 (2019). https://doi.org/10.1109/MASCOTS.2019.00038

11. Jayaraman, B., Li, H., Evans, D.: Decentralized certificate authorities. CoRR (2017)

12. Poon, J., Buterin, V.: Plasma: scalable autonomous smart contracts (2017)

13. Sakız, B., Gencer, A.H.: Blockchain beyond cryptocurrency: non-fungible tokens. In: International Conference on Eurasian Economies, pp. 154–161 (2021). https://doi.org/10.36880/C13.02527

14. Shen, H., et. al.: Blockchain-based lightweight certificate authority for efficient privacy-preserving location-based service in vehicular social networks. IEEE Internet Things J. 7(7), 6610–6622 (2020). https://doi.org/10.1109/JIOT.2020.2974874

15. Steichen, M., et al.: Blockchain-based, decentralized access control for IPFS. In: IEEE International Conference on Internet of Things (iThings) and IEEE Green Comp. and Comm. (GreenCom) and IEEE Cyber, Physical and Social Comp. (CPSCom) and IEEE Smart Data (SmartData), pp. 1499–1506 (2018). https://doi.org/10.1109/Cybermatics_2018.2018.00253

16. Tikhomirov, S., et. al.: SmartCheck. In: Proceedings of the 1st International Workshop on Emerging Trends in Software Engineering for Blockchain, pp. 9–16. ACM, New York, NY, USA (2018). https://doi.org/10.1145/3194113.3194115

17. Wang, Y.: Public key cryptography standards: PKCS. CoRR abs/1207.5 (2012)

18. Xu, Q., et. al.: Building an ethereum and IPFS-based decentralized social network system. In: 24th International Conference on Parallel and Distributed Systems (ICPADS), pp. 1–6 (2018). https://doi.org/10.1109/PADSW.2018.8645058

19. Zhang, P., Xiao, F., Luo, X.: A framework and dataset for bugs in ethereum smart contracts. In: 2020 IEEE International Conference on Software Maintenance and Evolution (ICSME), pp. 139–150 (2020). https://doi.org/10.1109/ICSME46990.2020.00023

Development of a Building Tool Combining Building Information Modeling and Digital Twin

Markus Aleksy[✉] and Philipp Bauer

ABB Corporate Research Center, Ladenburg, Germany
{markus.aleksy,philipp.bauer}@de.abb.com

Abstract. Modern building automation systems are subject to major challenges. Their architecture becomes increasingly diversified. Recent developments, such as increased digitization and Internet of Things (IoT) enable new implementation and application opportunities. The provided capabilities increase at every system level starting from more powerful embedded devices over development of more advanced integration concepts and networking solutions to modularization at application level and distribution at edge/cloud level. These changes are tightened on the information modeling level. The main modeling method regarding the generation and management of digital representations of physical and functional characteristics of buildings is referred to Building Information Modeling (BIM). However, new developments in context of Industrie 4.0, such as the Digital Twin (DT) require considerations how both approaches can be combined.

In this paper, we present various use cases in building automation systems that benefit from DT-approach, discuss different architectural approaches that were proposed to implement DT-based systems, and finally describe the tool that implements our approach combining BIM and DT to provide real time information about installed assets in a building in an overview as well as on detailed-level.

1 Introduction

Digital twin is a central concept of Industry 4.0. use cases. Since the first definition of the term digital twin published in 2010 by NASA [18], a myriad of definitions has been proposed. Boyes et al. [4] state that the number of digital twin definitions has more than doubled between 2017 and 2019. Moreover, the authors provide a comprehensive literature review regarding digital twin definitions. According to Grieves [10], it contains three main parts: physical products in real space, virtual products in virtual space, and the connections of data and information that ties the virtual and real products together.

On the other hand, Building Information Modeling (BIM) is the most common approach for the design, construction and maintenance of buildings. It improves information flow between different stakeholders involved at various stages of building creation and in some countries, the introduction of the BIM and BIM methods is already quite advanced and supported by various software tools.

Both concepts provide digital representations of physical entities that can be used to address various use cases. Moreover, the included models can be used for different types of simulations. E.g., BIM models can be used as input for various simulation and analysis tools, including structural analysis, energy performance simulation, daylight analysis, computational fluid dynamics, etc. [5] This is also true for digital twins (DT). A digital twin may contain a variety of computational and presentational models regarding its real-world entity including natural laws, statistical data, machine learning/artificial intelligence, geometrical and material models and visualization-oriented approaches [6].

Coupry et al. [8] observe that the concept of using a BIM to develop a DT has been widely explored in the literature. Using BIM as a basis for creating a DT for the construction industry addresses topics such as usage of a BIM-based DT, creation of a DT using a BIM, benefits of a BIM-based DT for lifecycle management, and improvements for data management.

The paper is structured as follows: Sect. 2 describes digital twin use cases in building management. Section 3 discusses software architectures proposed for the implementation of cyber-physical systems (CPS) and digital twins. Section 4 presents design considerations that guided our development of the proposed tool as well as some selected implementation aspects. Finally, Sect. 5 concludes the paper.

2   Digital Twin Use Cases in Building Management

Various use cases related to the utilization of digital twins in building management have been identified in the literature:

- Smart asset management
  Lu et al. [14] present several limitations and gaps of current research and standards for developing smart asset management in operations and maintenance phases related to various categories, such as technology related issues, information related issues, organization related issues, and standard related issues. Therefore, the authors propose a digital twin enabled asset management framework to overcome these gaps.
- Predictive maintenance
  Hosamo et al. [13] propose a digital twin predictive maintenance framework of air handling units to overcome the limitations of facility maintenance management systems that are used in buildings nowadays.
- Monitoring
  Coupry et al. [8] identify global monitoring of the building and its equipment as the first improvements highlighted by researchers. According to the

authors, the centralized database provided by the DT enables different stake-holders obtaining easier access to data related to the equipment they want to monitor.

- Asset anomaly detection
  Xie et al. [21] propose a building asset anomaly detection framework based on a digital twin platform that integrates multiple fragmented data sources.

Additionally, Sharma et al. [19] describe general scenarios and cases where digital twin can be majorly beneficial.

The goal of our development was to create a tool combining BIM and DT to address asset monitoring by providing an overview of available assets conditions as well as detailed presentation of each asset depending on users preferences. Moreover, sustainability and energy efficiency topics should be supported by the tool, e.g. by providing details about energy efficiency of ABB room automation system using heatmap visualization.

3 Architectural Considerations

A digital twin system can be implemented based on various software architectures. A software architecture refers to the basic organization of a system represented by its components, their relationships to each other and to the environment, and the principles that determine the design and evolution of the system.

Since there is no silver bullet regarding the architecture of digital twin systems, different approaches were proposed to describe architectures of cyber-physical systems (CPS) [12] and digital twins. They usually consist of a particular number of layers. Haße et al. [11] propose a three layer architecture composed of a data visualization layer, data processing layer, and a semantic layer with an optional data acquisition layer. Aheleroff et al. [2] describe a three layer model consisting of four parts: physical layer, digital layer, and cyber layer supplemented by communication for data exchange among the three layers. Zheng and Sivabalan [22] propose a generic architecture consisting of four different layers (physical layer, data extraction and consolidation layer, cyberspace layer, and the interaction layer). Lu et al. [15] present an architecture comprised of five layers: data acquisition layer, transmission layer, digital modeling layer, data/model integration layer, and service layer. Xie et al. [21] discuss system architecture for building digital twin falling back on the layers proposed by Lu et al. but utilize simplified structure with less details in their approach. In conclusion, there is a big variety and little consistency between the proposed approaches.

3.1   Patterns

A design sample describes a comprehensive and good solution for a frequently recurring task in program design, as it was gained from experience. The engineering approach to designing object-oriented software is to identify one or more design patterns [17] for a specific programming task, which solve this task. The primary use of a design pattern is in describing a solution for a particular class of problems. Further benefits result from the fact that each pattern has a name, which simplifies the discussion among software architects and developers since one can talk abstractly about a software structure. Thus, design patterns are basically programming language-independent. Unlike design patterns, architectural patterns document the fundamental organization and interaction between the components of an application system. The known architecture patterns include, for example, the client/server architecture pattern or the layer architecture [7].

Tekinerdogan, and Verdouw [20] propose a design pattern catalog for developing digital twins. It is based on a conceptual model of control systems and includes a total of nine different design patterns that address different problems and that can be applied to different systems engineering life cycle stages to leverage the development of high quality digital twin-based systems.

4   Design Considerations

The architecture of our building tool corresponds to a three tier architecture [16]. We used different development platforms and libraries to facilitate the development of our application:

- Forge
  Forge [3] is a cloud-based developer platform from Autodesk that enables access of design and engineering data in the cloud. Utilizing the Forge application programming interface (API) one can automate processes, connect teams and workflows, or visualize data.
- React
  React (a.k.a. React.js or ReactJS) is an open source JavaScript library for building user interfaces [9]. React can be used as a base in the development of single-page, mobile, or server-rendered applications with frameworks like Next.js. However, React is only concerned with state management and rendering that state to the Document Object Model (DOM), so creating React applications usually requires the use of additional libraries for routing, as well as certain client-side functionality.

Figure 1 shows a screenshot of the proposed building tool combining building information modeling and digital twin. It presents the BIM of the ABB Campus in Mannheim that has been created by our partner ATP Architekten Ingenieure. Additionally, two sprites are representing two assets located inside the building. The color of the sprites depends on the temperature of the asset (green color

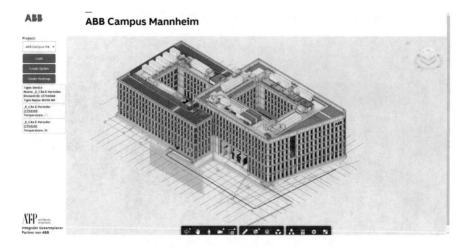

Fig. 1. Proof of concept of the building tool combining building information modeling and digital twin

means temperature is ok while red color corresponds to too high temperature). It is possible to connect to the digital representation of each asset by clicking on the corresponding link to obtain detailed information about it. The BIM model is hosted in the Autodesk Construction Cloud[TM] while the digital twin can be hosted elsewhere, e.g. using ABB Ability[TM] [1] solutions, such as ABB Ability[TM] condition monitoring for electrical systems (cf. Fig. 2).

By clicking on the "Create Heatmap" button one can see details about energy efficiency utilizing heatmap visualization (cf. Fig. 3).

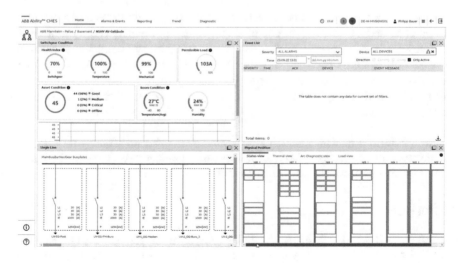

Fig. 2. ABB Ability[TM] condition monitoring for electrical systems (CMES)

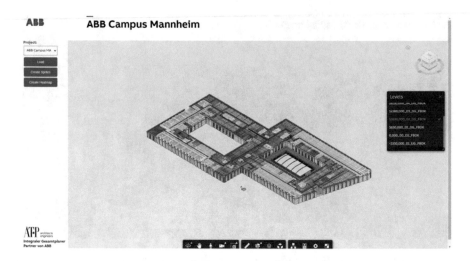

Fig. 3. Heatmap visualization regarding energy efficiency of a building floor

5   Conclusions

In this paper we have elaborated various use cases in building automation systems that can benefit from utilizing digital twins. Afterwards, we have discussed different architectural approaches that were proposed to implement such systems. Finally, we have described our tool that combines BIM and DT to enrich the digital representation of physical and functional characteristics of a building with real time information provided by installed assets in it. This information is shown in an overview manner or can provide detailed insight into the asset itself depending on users' needs. Colored sprites are used to show an overview over existing assets in the building and their conditions while detailed information can be obtained by clicking on one of them. This can be achieved by linking the BIM objects to a connected system, such as ABB Ability Condition Monitoring for electrical systems (CMES) or other industrial internet of things (IIoT) system. Additionally, details about energy efficiency of room automation systems can be obtained using heatmap visualization that is included in the proposed tool.

Acknowledgments. This research is supported by the Federal Ministry for Economic Affairs and Climate Action (BMWK). under grant number 03EN1002D. The responsibility for this publication lies with the authors.

References

1. ABB: ABB Ability^{TM}. https://global.abb/topic/ability/en Accessed 03 Nov 2022
2. Aheleroff, S., Xu, X., Zhong, R.Y., Lu, Y.: Digital twin as a service (DTaaS) in Industry 4.0: an architecture reference model. Adv. Eng. Inf. **47**, 101225 (2021). https://doi.org/10.1016/j.aei.2020.101225

3. Autodesk: a cloud-based developer platform from Autodesk https://forge.autodesk.com/ Accessed 02 Nov 2022
4. Boyes, H., Watson, T.: Digital twins: an analysis framework and open issues. Comput. Ind. 143 (2022). https://doi.org/10.1016/j.compind.2022.103763
5. Borrmann, A., König, M., Koch, C., Beetz, J.: Building information modeling - why? What? How?: Technology foundations and industry practice. Book Chapter (2018). https://doi.org/10.1007/978-3-319-92862-3_1
6. Boss, B., et al.: Digital twin and asset administration shell concepts and application in the industrial internet and industrie 4.0. An industrial internet consortium and plattform industrie 4.0 joint whitepaper (2020). https://www.plattform-i40.de/PI40/Redaktion/EN/Downloads/Publikation/Digital-Twin-and-Asset-Administration-Shell-Concepts.html
7. Buschmann, F., Meunier, R., Rohnert, H., Sommerlad, P., Stal, M.: Pattern-Oriented Software Architecture: A System of Patterns, Wiley, New York (1996)
8. Coupry, C., Noblecourt, S., Richard, P., Baudry, D., Bigaud, D.: BIM-based digital twin and XR devices to improve maintenance procedures in smart buildings: a literature review. Appl. Sci. (2021). https://doi.org/10.3390/app11156810
9. Fedosejev, A.: React.js Essentials: A Fast-Paced Guide to Designing and Building Scalable and Maintainable Web Apps With React.js. Packt Publishing Ltd, Birmingham (2015)
10. Grieves, M.: Digital twin: Manufacturing Excellence Through Virtual Factory Replication. White paper (2014)
11. Haße, H., Li, B., Weißenberg, N., Cirullies, J., Otto, B.: Digital twin for real-time data processing in logistics. In: Proceedings of the Hamburg International Conference of Logistics (HICL) (2019). https://doi.org/10.15480/882.2462
12. Hoffmann, M.W., et al.: Developing industrial CPS: a multi-disciplinary challenge. Sensors 2021(21), 2021 (1991). https://doi.org/10.3390/s21061991
13. Hosamo, H.H., Svennevig, P.R., Svidt, K., Han, D., Nielsen, H.K.: A digital twin predictive maintenance framework of air handling units based on automatic fault detection and diagnostics. Energy Buildings 261 (2022). https://doi.org/10.1016/j.enbuild.2022.111988
14. Lu, Q., Xie, X., Heaton, J., Parlikad, A.K., Schooling, J.: From BIM towards digital twin: strategy and future development for smart asset management. In: Borangiu, T., Trentesaux, D., Leitão, P., Giret Boggino, A., Botti, V. (eds.) SOHOMA 2019. SCI, vol. 853, pp. 392–404. Springer, Cham (2020). https://doi.org/10.1007/978-3-030-27477-1_30
15. Lu, Q., et al.: Developing a digital twin at building and city levels: case study of west Cambridge campus. J. Manage. Eng. 36(3) (2020). https://doi.org/10.1061/(ASCE)ME.1943-5479.0000763
16. Orfali, R., Harkey, D., Edwards, J.: Client/Server Survival Guide, 3 Edn, Wiley, New York (1999)
17. Reussner, R., Hasselbring, W. (Hrsg.): Handbuch Software-Architektur. Dpunkt Verlag (2006)
18. Shafto, M., et al.: DRAFT modeling, simulation, information technology & processing roadmap technology area 11. National aeronautics and space administration. https://www.nasa.gov/pdf/501321main_TA11-MSITP-DRAFT-Nov2010-A1.pdf Accessed 18 Oct 2022
19. Sharma A., Kosasih. E., Zhang, J., Brintrup, A., Calinescu, A.: digital twins: state of the art theory and practice, challenges, and open research questions. J. Ind. Inf. Integr. (2022). https://doi.org/10.1016/j.jii.2022.100383

20. Tekinerdogan, B., Verdouw, C.: Systems architecture design pattern catalog for developing digital twins. Sensors **20**(18), 5103 (2020). https://doi.org/10.3390/s20185103

21. Xie, X., Lu, Q., Parlikad, A.K., Schooling, J.M.: Digital twin enabled asset anomaly detection for building facility management. In: 4th IFAC Workshop on Advanced Maintenance Engineering, Services and Technologies - AMEST, vol. 53, no. 3 (2020). https://doi.org/10.1016/j.ifacol.2020.11.061

22. Zheng, P., Sivabalan, A.S.: A generic tri-model-based approach for product-level digital twin development in a smart manufacturing environment. Robot. Comput. Integr. Manuf. **64**, 101958 (2020). https://doi.org/10.1016/j.rcim.2020.101958

An Efficient Approach to Resolve Social Dilemma in P2P Networks

Avadh Kishor[1(✉)] and Rajdeep Niyogi[2]

[1] Department of Computer Science and Engineering, ABV-Indian Institute
of Information Technology and Management Gwalior, Gwalior, India
akishor@iiitm.ac.in
[2] Department of Computer Science and Engineering,
Indian Institute of Technology Roorkee, Roorkee, India
rajdeep.niyogi@cs.iitr.ac.in

Abstract. The premise of a peer-to-peer (P2P) system is based on the voluntary contribution of peers. However, an inherent conflict between individual rationality and social welfare engenders a new situation called the free-rider problem, in which some peers consume the resource without contributing to return. The tendency of free-riding in a P2P system is a critical issue, as it threatens the viability of the whole system. This can be reduced by employing an effective mechanism that deters the free-riding by incentivizing peers to contribute to the system. In this paper, we consider a social planner that acts as a decision-maker among individual agents. We cast the problem as a constrained optimization problem. A metaheuristic algorithm called the artificial bee colony algorithm is used to solve the problem. The empirical results show that each peer contributes to its full capacity and receives a fair (equal) profit share.

Keywords: Social dilemma · Public goods game · P2P network · Free riding

1 Introduction

In our daily lives, we often encounter situations where our interests are at odds with the welfare of society. Such conflicting situations are called social dilemmas. Social dilemmas describe situations in which every individual in a group faces a conflict between personal gain and collective welfare [1,2]. In a social dilemma, each individual behaves selfishly in the pursuit of their short-term profit, even though all the individuals would be better off acting in the group interest. The underlying fact of the dilemma is that being socially aware (considering the benefit of others) may lead to a better outcome than being individually rational (self-interested). In [3], it is suggested that with regard to the focus of the dilemmas, social dilemmas can be classified in two ways, i.e., (i) resource dilemma and (ii) public goods dilemma. Resource dilemma refers to a situation where each individual has to decide how much to harvest from a shared resource that is freely accessible to all members. Public goods dilemma refers to a situation in which each individual must decide whether and what to contribute to the common pool

L. Barolli (Ed.): AINA 2023, LNNS 661, pp. 307–318, 2023.
https://doi.org/10.1007/978-3-031-29056-5_28

of resources since these resources are equally shared among all group members, irrespective of their contribution. In this situation, those who do not contribute have a more advantageous position than those who contribute. As a result, there is the temptation to enjoy the goods without contributing. Those who get tempted and do not contribute are referred to as free-riders, and this problem is referred to as the public goods Provision problem (PGPP) ([4]).

The PGPP is a well-known problem that has been extensively studied in anthropology, biology, economics, sociology, psychology, political science, mathematics, and computer science cf, e.g.,([1,5–7]) and references cited therein. To illustrate the PGPP, we consider a distributed file-sharing system where all the files are freely available to all the users. Each user has two choices: (i) to access the information and contribute(share the files) in return, and (ii) to shirk and free ride on the effort of others. A user who contributes increases the amount of information a bit, but that increased amount is shared equally between all others, regardless of their contribution. Since the information in the system is assumed to be available to all users without any cost, each user has a clear incentive to minimize her contribution. If each user reasons in the same way, however, there will be severe underprovisioning of the resource, and the whole system will collapse. This is a classic example of PGPP.

In this paper, in light of the observations made above, we propose an optimization model to solve the PG dilemma. This model focused on balancing fairness and efficiency in the PG dilemma. For efficiency, the utilitarian welfare function is used, and an inequality index called the Gini index is used to maintain fairness. We consider that agents operate in a continuous strategy space. To validate the effectiveness and performance of the proposed approach, we apply a metaheuristic algorithm – artificial bee colony (ABC) algorithm.

The rest of this paper is organized as follows. In Sect. 2, related work is discussed. Problem modeling and formulation is discussed in Sect. 3. Proposed approach is discussed in Sect. 4 followed by simulation results in Sect. 5. Finally, in Sect. 6, conclusions are drawn.

2   Related Work

The work presented in this paper aims at providing a generalized solution to the public goods provision problem. Hence, it necessitates discussing the relevant works that have addressed this problem. This problem is ubiquitous all around the real world, ranging from economic systems to communication and computation systems. However, different attempts have already been made to resolve this problem from different perspectives. The following paragraphs describe them in detail.

An inequity aversion model, which is based on Homo Egualis society [8], is propsed in [9]. In that, agents dislike inequitable outcomes. More precisely, an agent enjoying the highest payoff may reduce her payoff to promote equality in the group. On the other hand, if she has the lowest gain, she strongly urges others to reduce their payoffs and maintains a degree of equality. Despite its wide acceptability by game theorists and computer researchers, this model is not able to predict the implication of group size on PGG. In [10], the resource allocation problem with severe communication constraints is considered a generic model to capture the problem of under-provisioning in public goods. To find an optimal contribution to the public good, they [10] proposed an ex-post approach. Halpern and Pass [11], inspired by the regret minimization concept of decision theory, introduced a new solution concept – iterated regret minimization – to predict the NE of strategic form games. In that, regret for an action a is the difference between the best possible utility value and the utility of acting a. In this approach, as the name suggests, all the strategies that do not minimize regret are deleted iteratively. However, this concept does not make it suitable for PGG because it predicts that if there is a dominant strategy, then play it. Such a prediction in PGG leads to a unique NE where everyone will contribute nothing, which is undesirable.

In [12], a computational model of achieving fairness is suggested. The authors [12] incorporated the element of human-decision making for learning in a multi-agent system. Their computational model combines the Homo Egualis function and continuous action learning automata. To analyze the impact of inequality in the provisioning of public goods, a voting-based scheme is suggested in [13].

Further, several works have considered the public goods provision problem as a cooperation problem between agents. In this view, researchers [14–16] proposed different types of peer punishment schemes to solve the problem. In these schemes, free-riders are punished so they can be discouraged from free-riding, and cooperation can be established. Interestingly, these schemes give rise to another severe problem called second-order free-rider problem [17].

3   Problem Modeling and Game Formulation

In this section, we provide some formative concepts and definitions that we will use all along this paper.

3.1   Public Goods Provision Problem

We consider a multiagent society with a finite set of rational agents, $N = \{1, 2, \ldots, n\}$, and a common pool (i.e., public good). Each player i ($i \in [n]$) has an initial endowment ϑ_i, and each agent is independent to decide her level of contribution, x_i ($0 \leq x_i \leq \vartheta_i$), to the common pool. Total contribution to the common pool ($\sum_{i=1}^{n} x_i$) is multiplied by a synergy factor r ($r > 1$) and then equally distributed among all the agents. We use $\rho = r/n$ to denote the marginal per capita return (MPCR) to the common pool. The profit accrued by agent i is given by

$$u_i(x_1, x_2, \cdots, x_n) = \vartheta_i - x_i + \rho \sum_{i=1}^{n} x_i \tag{1}$$

where $1/n < \rho < 1$.

3.2   Game Formulation

From Eq. (1), it can be seen that the decisions of other agents influence the resource provisioning decision of an agent. Each agent is rational in the sense that she is only interested in maximizing her level of satisfaction. In fact, in this resource provisioning scenario, agents are competing with each other to maximize their level of satisfaction. From a game-theoretic point of view, such a scenario is considered an instance of a non-cooperative game [18]. Thus, we formulate the public goods provision problem as a non-cooperative game called public goods game (PGG) as follows.

Definition 1 (Public goods game). Public goods game (PGG) with n players is defined as the tuple $\mathscr{G}_P = \langle N, (x_i), (u_i) \rangle$, where $N = \{1, 2, \ldots, n\}$ is the set of players, x_i is is the set of strategies of player i ($i \in [n]$), and u_i denotes the utility function of player i.

We use the notation $\mathscr{X} = (x_1, x_2, \cdots, x_n)$ to denote the players' joint strategy profile. Also, let $x_{-i} = (x_j)_{j \in [n] \setminus i}$ denotes this same strategy profile except the strategy of player i, so that (x_i, x_{-i}) forms a joint strategy profile.

Definition 2 (Nash equilibrium). A strategy profile $\mathbf{x}^* = (x_1^*, x_2^*, \cdots, x_n^*)$ is said to be Nash equilibrium (NE) of game $\mathscr{G}_P$, if $\forall i \in [n]$:

$$x_i^* \in \underset{x_i}{\operatorname{argmin}}\ u_i(x_i, x_{-i})$$

At the NE, it is assured that no player has the incentive to change her own strategy unilaterally.

We next define social welfare in the public goods game.

Definition 3 (Social welfare). The social welfare of a strategy profile $\mathbf{x} = (x_1, x_2, \cdots, x_n)$ is defined as the sum of utilities derived by each player:

$$W(x_i, x_{-i}) = \sum_{i=1}^{n} u_i(x_i, x_{-i}) \tag{2}$$

3.3   Analysis of PGG ($\mathscr{G}_P$)

In this section, we establish two propositions to show that there is a conflict between individual and social interest of players in $\mathscr{G}_P$. Next, we prove that if the agents are self-interested, then in equilibrium each agent makes zero contribution and this equilibrium is unique.

Proposition 1. *In PGG ($\mathscr{G}_P$) with n players, the optimal strategy (best response) for player i is to choose $x_i = 0$.*

Proof. Consider an arbitrary contribution vector $\mathbf{x} = (x_i)_{i=1}^{n}$ of the players, where $0 \le x_i \le \vartheta_i$. The payoff of the player i is given by $u_i(x_i, x_{-i}) = \vartheta_i - x_i + \rho \sum_{k=1}^{n} x_k$. The first derivative of the function is $\frac{\partial u_i}{\partial x_i} = (\rho - 1) < 0$. The value of the first derivative of the $u_i(.)$ is negative, implying that function $u_i(.)$ is strictly decreasing. Thus, the maximum value of the function will be obtained only at minimum contribution level, i.e., $x_i = 0$. □

Proposition 2. *In PGG ($\mathscr{G}_P$) with n players, the maximal social welfare can be achieved if each player chooses to contribute its full endowment.*

Proof. Given a contribution vector $\mathbf{x} = (x_i)_{i=1}^n$ of the players, the social welfare function is

$$W(x_i, x_{-i}) = \sum_{i=1}^n u_i(x_i, x_{-i}) = \sum_{i=1}^n \vartheta_i + (n\rho - 1)\sum_{i=1}^n x_i$$

For the optimality, the partial derivatives are $\frac{\partial W}{\partial x_1} = \frac{\partial W}{\partial x_2} = \cdots = \frac{\partial W}{\partial x_n} = (n\rho - 1) > 0$. The values of derivatives are positive, implying that the maximum value of the function will be obtained only at maximum contribution level, i.e., $x_i = \vartheta_i$. □

Lemma 1. *The PGG ($\mathscr{G}_p$) with linear utility function $u_i(x_i, x_{-i})$ has unique NE; and in the NE the strategy profile is $\mathbf{x}^* = (x_1^*, x_2^*, \cdots, x_n^*) = (0, 0, \cdots, 0)$.*

From Proposition 2, it is evident that the overall welfare of the players of $\mathscr{G}_P$ is derived from their collective utilities. Thus, to maintain social welfare, all the players must contribute their full endowment. On the other hand, Lemma 1 shows that no player will contribute to the shared pool at equilibrium. In summary, in $\mathscr{G}_P$, there exists the free-rider problem.

4 Proposed Approach: Solution to PGG

In this section, we discuss the proposed solution to the PGG. Our goal is to maximize the utilities of each agent. Moreover, according to [19], it is evident that fair provisioning of resources and inflicting punishment on the defector is not the credible remedy to eradicate the free-rider problem. Thus, in this paper, we assume that a benevolent social planner is responsible for making all the decisions in the common pool provisioning and distributing the profit to each agent. The planner is benevolent because she cares about the agents' well-being. The graphical representation of the planner and agents is shown in Fig. 1.

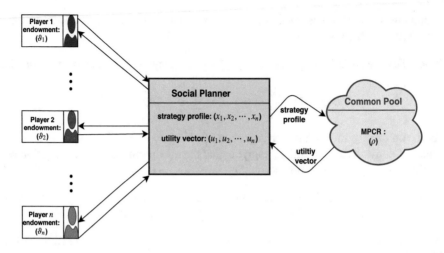

Fig. 1. Working model

4.1 The Objective Function

In order to maximize the social welfare (efficiency), we adopt utilitarian model, i.e., the sum of utilities as our objective function. The mathematical model can be formulated as

$$\text{maximize } f(\mathscr{X}) = \sum_{i=1}^{n} u_i(x_i, \mathbf{x}_{-i}), \ \forall i \in [n] \tag{3}$$

4.2 The Constraints

Constraints define the limits of decisions. Here, our rationale behind defining constraints is to maintain the fairness criterion. Due to the subjective nature of fairness and various interpretations of equitability, there are, however, no universally accepted criteria for evaluating the fairness. Obviously, adopting an inequity index is a subjective matter. However, the well known and perhaps a more commonly used measure of relative inequity is *Gini index* ([20]). Keeping our problem in mind, we use Gini index to measure the relative inequity in the contribution between the players, whose mathematical formula is given below.

$$\xi(\mathscr{X}) = \frac{\sum_{i=1}^{n} \sum_{j=1}^{n} |x_i - x_j|}{2n \sum_{i=1}^{n} x_i} = 0 \tag{4}$$

where $\xi \in [0,1]$; $\xi = 0$ indicates that all players contribute perfectly equal.

The constrained objective function, that social planner will solve is as follows.

$$\text{maximize } f(\mathcal{X}) = \sum_{i=1}^{n} u_i(x_i, \mathbf{x}_{-i})$$

$$\text{s.t.} \quad \frac{\sum\limits_{i=1}^{n} \sum\limits_{j=1}^{n} |x_i - x_j|}{2n \sum\limits_{i=1}^{n} x_i} = 0 \tag{5}$$

$$\sum_{i=1}^{n} x_i \leq n\vartheta \ \& \ x_i \geq 0$$

To optimize the (5), a meta-heuristic optimization algorithm called Artificial Bee Colony (ABC) algorithm is used. The ABC algorithm, proposed by Karaboga [21], is a swarm based meta-heuristic optimization algorithm, which imitates the foraging behavior of honey bees. ABC algorithm has been successfully applied to many other constrained optimization problem [22–24] as well. Please note that algorithmic details of the ABC algorithm are omitted to conserve the space. However, we refer the interested reader to the original study of ABC algorithm [21] for getting the algorithmic concept as well as for a detailed exposition. A source code (in C, Java, and MATLAB) of ABC algorithm is also available from http://mf.erciyes.edu.tr/abc/.

5   Simulation Results

There are two essential requirements that must be met before a meta-heuristic algorithm can be used for a particular problem. First, we need the representation of candidate solution to the problem, and second, we need the way of evaluating the solution – fitness evaluation method.

In fact, the solution of the problem (5) is a contribution vector (i.e., a joint strategy of n players). Though, when applying ABC algorithm to the problem, the position of a food source is considered analogous to a contribution vector, where a parameter of the vector represents the contribution of an agent. Therefore, the dimensionality of solution vector is equal to the number of agents.

Fitness Evaluation
To deal the constraints and determine the feasible solutions, we adopt dynamic penalty method suggested by [25]. In this method, as the generation count increases, the value of the panelity factor also increases. The fitness of a solution i is evaluated as following:

$$\mathcal{F}(\mathcal{X}_i) = \begin{cases} f(\mathcal{X}_i) & \text{if } \xi(\mathcal{X}) = 0 \\ f(\mathcal{X}_i) - [(c*t)^\alpha * (\xi(\mathcal{X}))^\beta] & \text{else} \end{cases} \tag{6}$$

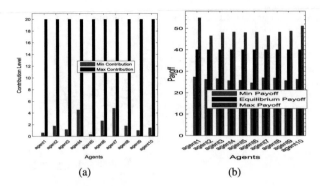

Fig. 2. (a) Bar graph representing the maximum and minimum contribution of each agents and (b) Bar graph representing the minimum, maximum, and equilibrium payoff agents

where t denotes the current generation count; α and β are user defined constants. The following parameter settings are adopted for all the experiments. The number of food sources, NP (one half of the colony size) is equal to the number of agents, n. *limit* is set to $(NP * n/2)$ as in [21]. The values of the both α and β are set to 1. We fix the value of the maximum number of generations. For robustness, the ABC algorithm is simulated 20 times independently. The stopping criterion is that the value of constraint $\xi(\mathscr{X}_i^*) = 0$, or the maximum number of generations is reached.

Further, to make a qualitative assessment of obtained solutions, we use two following performance metrics.

- Price of Fairness (PoF): In order to quantify the loss of social welfare of a fair solution compare to the system optimum, we adopt the metric called *Price of Fairness* (PoF) ([26]). It is a measure of the relative social welfare loss under a fair contribution scheme. we use the PoF in a similar way to [27].

$$PoF = \frac{f^* - f_t(\mathscr{X}_i)}{f^*} \qquad (7)$$

Obviously, PoF $\in [0,1]$, where f^* is the optimal solution of problem (5) and $f_t(\mathscr{X}_i)$ is the value of social welfare for solution i at generation t. Here, $\mathscr{X}_i$ is the final solution obtained by the ABC algorithm in generation t.

- Relative Percentage Deviation (RPD): This metric is used to measure the deviation of an achieved solution form best solution. In this paper, we adopt this metric to measure the variation of total contribution. It is measured as

$$RPD_{TC} = \frac{\mathscr{X}_{Total}^* - \mathscr{X}_{Total}(t)}{\mathscr{X}_{Total}^*} \qquad (8)$$

where $\mathscr{X}_{Total}^* = n * \vartheta$ and $\mathscr{X}_{Total}(t)$ is the total contribution achieved at the end of iteration t.

5.1 Results Discussion

In order to determine whether our proposed model possesses the traits that are of our interest, we carry out the simulations using the ABC algorithm. The required parameters for this simulation is as follows. The number of agents is 10; initial endowment for each agent $\vartheta = 20$; and value of $\rho = 0.2$. We fix the value of maximum number of generations to 200. We perform 20 simulations using different initial populations and obtained similar result. Here, we only show the result of a typical simulation run which takes the highest number of generations, i.e., 190.

In Fig. 2a the minimum and maximum contribution level of each agent in a run is shown. Figure 2b reflects minimum, maximum, and equilibrium payoff of each agents. Figure 2 reflects that in our proposed model, all the agents contributes equally and enjoys the equal profit. In Fig. 3a, we show the progress of agents towards the optimality. For clarity, the progress of only two agents, 1 and 10, is demonstrated. From Fig. 3a, it can be seen that initial agents contributes randomly, as time increases they learn impact of defection and finally converge towards full cooperation. Figure 3b show the how the values of PoF, RPD_{TC}, and Gini index vary over the generation. Figure 3b depicts that, at the equilibrium, agents contribute equally. Finally, PoF becomes zero, which confirms the optimal outcome.

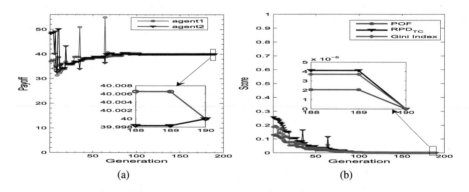

Fig. 3. (a) Demonstrating convegence towards full cooperation and (b) Deviation of PoF, RPD_{TC}, and Gini index over the generation

5.1.1 Effect of MPCR

A key parameter in the PGG, marginal per capita return (MPCR), is the profit that each participant receives from each unit of contribution to the common pool by any group member. A higher MPCR decreases the cost of contribution to the common pool and may encourage the agents for higher contribution.

To investigate the effect of MPCR in our proposed model, we carry out the experiments over varying MPCR, i.e., $\rho = 0.4, 0.6$, and 0.8. The number of agents and the maximum number of generations for each value of MPCR are kept same, i.e., 10 and

200, respectively. We determine (a) the point of convergence, i.e., value of total contribution and social welfare and (b) the final performance of the system, i.e., how system respond to the change in MPCR.

From Table 1 and Fig. 4, it is easy to observe that variation in the contribution level as well as social welfare decreases as the value of MPCR increases. The reason for this low variation is as follows. It is clear from (1) that for the higher value of MPCR, the payoff for an agent always will be higher. Therefore, as the simulation starts, the agents come to know that if the value of MPCR is high, then more contribution to the common pool will be more beneficial than less contribution.

Table 1. Effect of the MPCR (ρ): (a) minimum, maximum, mean (average), and std (standard deviation) of total contribution in a typical run and (b) minimum, maximum, mean (average), and std (standard deviation) of social welfare in a typical run; here CV: coefficient of variance = std/mean

(a)

ρ	Min	Max	Mean	Std	CV
0.4	122.75	200	190.38	17.93	0.0942
0.6	134.57	200	194.47	12.26	0.0631
0.8	159.41	200	193.19	10.37	0.0537

(b)

ρ	Min	Max	Mean	Std	CV
0.4	568.26	800	771.14	53.80	0.0698
0.6	872.85	1200	1172.30	61.31	0.0523
0.8	1315.90	1600	1552.30	72.61	0.0468

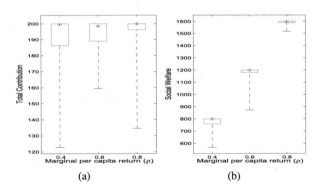

(a)   (b)

Fig. 4. Effect of varying value of MPCR: (a) Total contribution and (b) Social welfare

6   Conclusion

This paper discussed the public goods provision problem, one of the fundamental topics in p2p networks. In this problem, there is a clash between the social and individual objectives of agents. The socially optimal strategy is to contribute, while the individual optimal strategy is to free-ride. We formalized this problem as a constrained optimization problem. We use an artificial bee colony algorithm to solve the optimization problem. The experimental results revealed that the proposed model can effectively solve the problem. Our future plan is to study the public goods provision problem in heterogeneous settings and use this model to solve real world problems, such as resource allocation, health care system and networks.

Acknowledgements. The second author was in part supported by a research grant from Google.

References

1. Paul, A.M., Lange, V., Joireman, J., Parks, C.D., Van Dijk, E.: The psychology of social dilemmas: a review. Organizational Behav. Human Decision Processes **120**(2), 125–141 (2013)
2. Messick, D.M., Brewer, M.B.: Solving social dilemmas: a review. Rev. Personal. Social Psychol. **4**(1), 11–44 (1983)
3. Komorita, S.S., Parks, C.D.: Social dilemmas. Brown & Benchmark (1994)
4. Friedman, D.: Problems in the provision of public goods. Harv. JL & Pub. Pol'y **10**, 505 (1987)
5. Hua, J.S., Huang, S.M., Yen, D.C., Chena, C.W.: A dynamic game theory approach to solve the free riding problem in the peer-to-peer networks. J. Simul. **6**(1), 43–55 (2012)
6. Kishor, A., Niyogi, R., Chronopoulos, A., Zomaya, A.: Latency and energy-aware load balancing in cloud data centers: a bargaining game based approach. IEEE Trans. Cloud Comput. 2168–7161 (2021)
7. Kishor, A., Niyogi, R., Veeravalli, B.: Fairness-aware mechanism for load balancing in distributed systems. IEEE Trans. Serv. Comput. **15**(4), 2275–2288 (2022)
8. Gintis, H.: Game theory evolving: A problem-centered introduction to modeling strategic behavior. Princeton university press (2000)
9. Fehr, E., Schmidt, K.M.: A theory of fairness, competition, and cooperation. The Quart. J. Econ. **114**(3), 817–868 (1999)
10. Sanghavi, S., Hajek, B.: A new mechanism for the free-rider problem. IEEE Trans. Autom. Control **53**(5), 1176–1183 (2008)
11. Halpern, J.Y., Pass, R.: Iterated regret minimization: a new solution concept. Games and Economic Behavior **74**(1), 184–207 (2012)
12. De Jong, S., Tuyls, K.: Human-inspired computational fairness. Auton. Agent. Multi-Agent Syst. **22**(1), 103–126 (2011)
13. Colasante, A., Russo, A.: Voting for the distribution rule in a public good game with heterogeneous endowments. J. Econ. Interact. Coord. 1–25 (2016)
14. Masclet, D., Noussair, C.N., Villeval, M.C.: Threat and punishment in public good experiments. Econ. Inq. **51**(2), 1421–1441 (2013)
15. Kishor, A., Niyogi, R.: A game-theoretic approach to solve the free-rider problem. In: 2017 Tenth International Conference on Contemporary Computing (IC3), pp. 1–6 (2022)
16. Kishor, A., Gargt, T., Niyogi, R.: Altruistic decision making approach to resolve the tragedy of the commons. In: 2016 International Conference on Advances in Computing, Communications and Informatics (ICACCI), pp. 1858–1863 (2016)
17. Okada, A.: The second-order dilemma of public goods and capital accumulation. Public Choice **135**(3), 165–182 (2008)
18. Killingback, T., Bieri, J., Flatt, T.: Evolution in group-structured populations can resolve the tragedy of the commons. Proc. Royal Society London B: Biol. Sci. **273**(1593), 1477–1481 (2006)
19. Fehr, E., Gächter, S.: Cooperation and punishment in public goods experiments (1999)

20. Ceriani, L., Verme, P.: The origins of the gini index: extracts from variabilità e mutabilità (1912) by corrado gini. J. Econ. Inequality **10**(3), 421–443 (2012)
21. Karaboga, D., Basturk, B.: A powerful and efficient algorithm for numerical function optimization: artificial bee colony (abc) algorithm. J. Global Optim. **39**(3), 459–471 (2007)
22. Karaboga, D., Akay, B.: A modified artificial bee colony (abc) algorithm for constrained optimization problems. Appl. Soft Comput. **11**(3), 3021–3031 (2011)
23. Akay, B.B., Karaboga, D.: Artificial bee colony algorithm variants on constrained optimization. Int. J. Optim. Contr. Theor. Appl. (IJOCTA) **7**(1), 98–111 (2017)
24. Kalayci, C.B., Ertenlice, O., Akyer, H., Aygoren, H.: An artificial bee colony algorithm with feasibility enforcement and infeasibility toleration procedures for cardinality constrained portfolio optimization. Expert Syst. Appl. **85**, 61–75 (2017)
25. Joines, J.A., Houck, C.R.: On the use of non-stationary penalty functions to solve nonlinear constrained optimization problems with ga's. In: IEEE World Congress on Computational Intelligence, pp. 579–584 (1994)
26. Bertsimas, D.: Vivek F Farias, and Nikolaos Trichakis. The price of fairness. Operations Res. **59**(1), 17–31 (2011)
27. Nicosia, G., Pacifici, A., Pferschy, U.: Price of fairness for allocating a bounded resource. Eur. J. Oper. Res. **257**(3), 933–943 (2017)

A Reduced Distributed Sate Space for Modular Petri Nets

Sawsen Khlifa[1]([✉]), Chiheb Ameur Abid[2], and Belhassen Zouari[1]

[1] Mediatron Lab, Sup'Com, University of Carthage, Tunis, Tunisia
{sawsen.khlifa,Belhassen.Zouari}@supcom.tn
[2] Mediatron Lab, Faculty of Sciences of Tunis, University of Tunis El Manar,
Tunis, Tunisia
chiheb.abid@fst.utm.tn

Abstract. This paper presents a modular verification structure of modular Petri nets. We go a step forward in improving the modular state space of a given modular Petri net. The proposed structure allows the creation of smaller modular graphs by reducing on the fly the number of its meta-states. Each modular graph draws the behavior of the corresponding module and outlines some global information. Experiments shows that this version helps to overcome the explosion problem by using less memory space. In this condensed structure, the verification of some generic properties concerning one module is limited to the exploration of its associated graph.

1 Introduction

Currently, computer systems are more and more distributed, that is to say that they are composed of several sequential systems which perform parallel calculations and which communicate with each other. These systems are involved in all critical fields such as health, transport, aeronautics, nuclear center and others. Given their importance and their sensitive role, it is crucial to verify their correctness and to maintain a high level of their performance. It is therefore essential to precisely define the behavior of such systems in a formal way and to distinguish as well between the expected behaviors and the forbidden ones. In this context, we can take advantage of Petri nets (RdPs) as a powerful tool to perform system modeling and verification. One of the common approaches for verifying a Petri net is the construction and the exploration of its state space. However, this method habitually suffered from the state explosion problem, especially, in systems which are composed of many parallel processes: the size of the state space grows exponentially with the number of processes that it often becomes impossible to continue the analysis task. From an other side, a concurrent system is usually, seen as the collection of several components or modules, semi-autonomous, which interact with each other. Considering this aspect, modular verification approaches are developed to overcome the problem of the state explosion: instead of building the state space of the whole system,

L. Barolli (Ed.): AINA 2023, LNNS 661, pp. 319–331, 2023.
https://doi.org/10.1007/978-3-031-29056-5_29

these approaches consist of building the state spaces of each module, and then applying verification techniques. These methods allow not only to parallelize the construction but also to use less memory space. Moreover, the verification of certain properties is enabled by taking advantage of their location within the modules. In this work, we propose first, to reduce the distributed state space [1] presented as a new modular verification approach by proposing a condensed version which hides for each module the synchronized actions of other modules of which it is not concerned and eliminating on the fly the redundant meta-states . This version allows to overcome the explosion problem and to use less memory space. Then, we evoke algorithms allowing local generic properties verification. After that we present some preliminary results.

The paper is scheduled as follows: in Sect. 2, we recall the definitions of modular Petri nets; in Sect. 3, we list several works exploiting the modular representation of the state space; in Sect. 4 we introduce the reduced distributed state space structure and we recall some generic properties; Sect. 5 illustrates the results of our preliminary experiments and the final section outlines the conclusions and the perspectives of this work.

2   Preliminaries

To deal with modular verification using Petri nets, the definitions of modular Petri net need to be reminded.

Modular Petri nets is a structure which allows modules to be described independently. In this paper, only modules that communicate through shared transitions are considered. The definition of the modular Petri net is taken from [2,3]

Definition 1 A **modular Petri net** is a pair $MN = (S, TF)$, satisfying:

1. S is a finite set of **modules** such that:
   a. Each module, $s \in S$, is a Petri net $s = (P_s, T_s = T_{sync,s} \cup T_{l,s}, W_s, M_{0_s})$, $T_{sync,s}$ denotes the **synchronised** transitions set of module s. $T_{l,s}$ denotes the **local (internal)** transitions set of module s.
   b. The sets of nodes corresponding to different modules are pair-wise disjoint: $\forall s_1, s_2 \in S, s_1 \neq s_2 \Rightarrow (P_{s_1} \cup T_{s_1}) \cap (P_{s_2} \cup T_{s_2}) = \varnothing$,
   c. $P = \bigcup_{s \in S} P_s$ and $T = \bigcup_{s \in S} T_s$ the set of all places and all transitions of all modules.
2. $TF \subseteq 2^T \setminus \{\varnothing\}$ is a finite set of non-empty **transition fusion** sets.

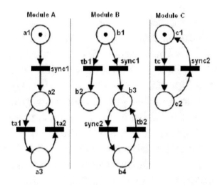

Fig. 1. Example of a modular Petri net

In the following, T_{sync} denotes the set of all synchronised transitions, i.e. $T_{sync} = \bigcup_{tf \in TF} tf$. T_l denotes the set of all local transitions (non synchronised transitions), i.e. $T_l = T \setminus T_{sync}$.

A transition fusion set is a set of synchronized transitions belonging to different modules that its crossing represents the execution of a synchronized action. The notion *transition groups* allows to consider the enabling and the occurrence rules for both local and synchronized transitions.

Lets consider as an example the modular Petri net of Fig. 1. This net is composed of three modules A, B and C. Module A is synchronised on transition $Sync1$, module B is synchronised on transitions $Sync1$ and $Sync2$ while modules C is synchronised on transition $Sync2$. These transitions are assumed to constitute three transition fusion sets.

3   Modular and Distributed Verification Methods: Related Work

Several works exploiting the modular representation of the state space have been proposed for the verification of concurrent systems. Among them we cite Algebra process, Incremental technique by modular decomposition and modular analysis of Petri nets. The algebra process methods [4,5] consists on the generation of a condensed version of the state space of a system in a compositional manner: Respecting an equivalence relationship, the local state spaces are progressively reduced and composed until the obtaining of only one reduced graph. The obtained graph is then used to check preserved properties by the execution of classic algorithms based on its exploration.

In order to verify an $LTL - X$ property, the incremental and modular techniques [11] begin by choosing a subnet according to structural information taking into account the property in question. The verification is thus carried out on this subnet. If this fails, another subnet more extended is chosen.

The modular analysis approaches proposed in [2,6] generate first, for each module of the system, a local graph composed of its reachable local states. Then, a synchronization graph comprising the information on the nodes reachable during the crossing of the fusion transitions is produced. The synchronization graph and the local state spaces define a structure called modular state space allowing all the properties of the system to be preserved. The main contribution of the modular state space development is the ability to directly check certain properties without having to compute the global markings graph. Nonetheless, using modular verification is still hard because verifying a property of module in isolation often requires to introduce its environment, so that the module is not completely free in its interaction with environment. That is why, the modularity of the verification process using modular state spaces has been improved in [16] by limiting it to the exploration of local and some global information. Moreover, the modular state space in this work, is distributed on every machine associated with one module. A new modular representation of the state space of a modular Petri net where only one kind of graphs are associated with modules is proposed in [1]. This modular representation avoids switching the exploration of two types of graphs during the verification of a property related to one module. The Symbolic Observation Graph (SOG) is defined in [7] as a condensed representation of the state space based on a symbolic encoding of the nodes.It has the advantage to be much reduced comparing to the original state space graph while being equivalent with respect to linear time properties.The performances of the SOG construction is improved by using parallelization techniques. By the way, there has been increased interest in parallel and distributed verification approaches in the last decade [10]. These distributed approaches offer more available memory for storage of the state space and has the potential to speed-up the process of verification in time. Generally, parallel approaches can be distinguished into two main categories: *multi-core approaches* [12–14] where parallelization is performed on several cores on the same machine with a shared memory platform and *cloud approaches* where model checking is distributed on multiple machines in the cloud [15].

A state-space analysis technique was adopted in [8]. A case study of an elevator model was evoked to explain the proposed process of developing and composing Colored Petri net-based model components and verifying the composed model.

To guarantee the safety and the effectiveness for insulin infusion pump systems, a formal Model-Based Approach (MBA) with a Coloured Petri nets (CPN) named MBA/CPN was proposed in [9]. A case study on a commercial insulin infusion pump system is exposed to verify and validate the reference model. The proposed method evaluates the quality of such systems in a cost-effective and time-efficient manner and provides re-usable project artifacts.

4   Reduced Distributed State Space

Before introducing the reduced version of the distributed state space, we call back the definitions related to the distributed sate space mentioned in [1].

4.1   Distributed State Space

Given a system composed of a set S of modules modelled by a modular Petri net, the distributed state space (DSS) as it was proposed in [16] consists of a set S of graphs, where each graph is associated with a module s which is called a *meta-graph* represents the local behaviours of s and the firings(crossing) of synchronisation actions. Every node of a meta-graph, called a *meta-state*, is a set of markings linked by local transitions. The arcs linking meta-states correspond only to the firing of synchronised transitions and are called *synchronized arcs*.

The definition of a meta-state is presented as follows

Definition 2 Let M be a reachable marking from M_0. A meta-state associated with marking M in module s, denoted by $\widehat{M_s}$, is a graph representing local markings of module s locally reachable from M. It is a graph defined as follows:

1. $\widehat{M_s}.id$ denotes the label of the meta-state s.t.
   $\widehat{M_s}.id = M^{\phi}$,
2. $\widehat{M_s}.Q$ is the set of nodes s.t.
   $\widehat{M_s}.Q = \{M'_s | M'_s \in [[M_s\rangle_s\}$,
3. $\widehat{M_s}.A$ is the set of arcs s.t.
   $\widehat{M_s}.A = \{(M1_s, t, M2_s) | M1_s, M2_s \in [[M_s\rangle_s \wedge t \in T_{l,s} \wedge M1_s[t\rangle M2_s\}$.

A synchronised arc in a module s corresponding to the firing of a transition fusion set $tsync$ from a marking M is labelled with M_s^c and the transition fusion set $tsync$.

Definition 3 Let MPN be a modular Petri net. The Distributed State Space of MPN is a set of meta-graphs $DSS = \{G_s = (\widehat{N_s}, \widehat{A_s}) | s \in S\}$ s.t. $\forall s \in S$, $G_s = (\widehat{N_s}, \widehat{A_s})$ is the meta-graph associated with module s, where:

1. $\widehat{N_s}$ denotes the set of meta-states of the meta-graph associated with module s.
$$\widehat{N_s} = \{\widehat{M_{0s}}\} \cup \{\widehat{M_s} | \exists M' \in [M_0\rangle, \exists tf \in TF : M'[tf\rangle M\}$$

2. $\widehat{A_s}$ denotes the set of synchronisation arcs of the meta-graph associated with module s.
$$\widehat{A_s} = \{(\widehat{M'}_s(M^{\phi}, tf), \widehat{M"}_s) | M \in [[M'\rangle \wedge tf \in TF \wedge$$
$$M[tf\rangle M"\}$$

Figure 2 illustrates the DSS of the MPnet of Fig. 1. Initial meta-states of meta-graphs are built from the initial marking a1b1c1.

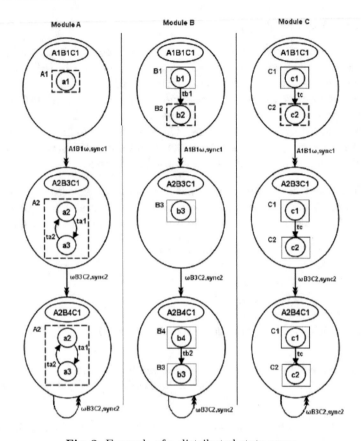

Fig. 2. Example of a distributed state space

4.2   Reduced Version of the Distributed State Space

4.2.1   Definitions

The reduced distributed state space is presented as a condensed version of the
distributed state space. In fact, instead of representing for each module all the
synchronized transitions, only the crossing sequences which belong to the module
are explicitly represented. The others ones are masked under the label τ without
taking into account the marking that activated them and their nomenclature.
The presence of τ is intended to indicate that the other modules have crossed
a synchronized transition by which the module in question is not concerned.
Thereafter, for a given global marking, all the synchronized transitions that the
module is not concerned with, and which can be crossed by the other modules
are represented by the same arc labeled by τ.

Let $T_{sync,s}$ be the set of synchronised transitions of the module s. The definition of the meta-state in the reduced distributed state space is kept as it is
mentioned above (Definition 9). Changes take place at the meta-graph level.
The set of markings reachable from M by firing a sequence of local transitions

and/or synchronised transitions of the other modules which are not shared by the module s is denoted by $[M\rangle\rangle_{\bar{s}}$, i.e. $[M\rangle\rangle_{\bar{s}} = \{M'/\exists\sigma = ((\sigma1)^*(\sigma2)^*)^*, \sigma1 \in (T_{sync} \setminus T_{sync,s})^* \wedge \sigma2 \in (T_{l,s})^* : M[\sigma\rangle M'\}$

The definition of the reduced distributed state space is given as follows:

Definition 4 Let MPN be a modular Petri net.
The reduced distributed State Space of MPN is a set of meta-graphs $RDSS = \{RG_s = (\widehat{N}_s, \widehat{A}_s)|s \in S\}$ s.t. $\forall s \in S$, $RG_s = (\widehat{N}_s, \widehat{A}_s)$ is the reduced meta-graph associated with module s, where:

1. $\widehat{N}_s$ denotes the set of meta-states of the reduced meta-graph associated with module s.

$$\widehat{N}_s = \{\widehat{M_{0s}}\} \cup \{\widehat{M_s}|\exists M' \in [M_0\rangle, \exists t_{sync} \in T_{sync,s} : M'[t_{sync}\rangle M\}$$

2. $\widehat{A}_s$ denotes the set of synchronisation arcs of the reduced meta-graph associated with module s.

$$\widehat{A}_s = \{(\widehat{M'}_s(M^{\phi}, t_{sync}), \widehat{M''}_s)|M \in [[M'\rangle \vee M \in [M'\rangle\rangle_{\bar{s}}$$
$$\wedge (t_{sync} \in T_{sync,s}) \wedge M[t_{sync}\rangle M''\} \cup \{(\widehat{M'}_{s\prime}, \tau), \widehat{M''}_s)|$$
$$M \in [[M'\rangle \vee M \in [M'\rangle\rangle_{\bar{s}} \wedge (t_{sync} \notin T_{sync,s})\}$$

It remains to prove that the projection of the language generated by the graph of a module on its own transitions is the same as the language generated by the entire system projected on the transitions of this module.

Proofs.

Let Σ be the language generated by the Modular Peti net MPN over the set of its transitions, $\Sigma_{/s}$ the projected language of Σ on the transitions proper to the module s, Σ_s the language generated by the meta-graph of the module s and $\Sigma_{s/s}$ the projection of Σ_s on the transitions that concerns the module s.
Let σ be an infinite sequence of Σ formed on the alphabet of transitions. It describes a series of transitions to be activated. In our case, the sequence of transitions can be made up of local transitions and / or fusion transitions for the different modules of the system. σ can be written in the following form $\sigma = ((t_l)^*(t_{sync})^*\tau^*)^*$ where $t_l \in T_l$ and $t_{sync} \in T_{sync}$. As T_l is the set of the local transitions its composed of all the local transitions of different modules: $T_l = \bigcup_{s \in S} T_{l,s}$. The same for T_{sync} which is the set of all synchronised transitions $T_{sync} = \bigcup_{s \in S} T_{sync,s}$.
So,

$$\sigma = ((t_{l,s1})^*(t_{l,s2})^*...(t_{l,sn})^*(t_{sync,s1})^*(t_{sync,s2})^*...(t_{sync,sn})^*\tau^*)^*.$$

The projection of σ on the transitions of the module sk denoted by $\sigma^{\Sigma/sk}$ consists in removing all the synchronized transitions not shared by sk and keeping only the transitions relative to sk whether local or synchronized $\sigma^{\Sigma/sk} = ((t_{l,sk})^*(t_{sync,sk})^*)^*$. We conclude that $\sigma^{\Sigma/sk} \in \Sigma_{sk}$ and subsequently $\Sigma_{/sk} \subset \Sigma_{sk/sk}(1)$.

Similarly and according to the definition of the meta-graph in the reduced distributed state space $\forall \sigma \in \Sigma_{sk}$ σ is built on the allowed transitions in the sk module which are none other than its local transitions, its synchronized transitions and τ. Therefore σ can be written as $\sigma = ((t_{l,sk})^*(t_{sync,sk})^*\tau^*)^*$. the projection of σ on the transitions related to module sk denoted by $\sigma^{\Sigma_{sk}/sk}$ is obtained by eliminating the synchronized transitions not shared by the module sk which were represented by τ. Consequently, by removing the transitions labled by τ, $\sigma^{\Sigma_{sk}/sk}$ is expressed as follows $\sigma^{\Sigma_{sk}/sk} = ((t_{l,sk})^*(t_{sync,sk})^*)^*$ which is an element of $\Sigma_{/sk}$. So $\Sigma_{sk} \subset \Sigma_{/sk}(2)$.

(1) and (2) give $\Sigma_{/sk} = \Sigma_{sk/sk}$.

4.2.2   Construction Algorithm of the Reduced Distributed State Space

The proposed formal definition of RDSS makes it possible to deduce the algorithm for the construction of the meta-graph of each module. The initial meta-state of a meta-graph is built from the initial marking. Other meta-states are determined iteratively from existent meta-states by firing the fusion transitions of the considered module. Algorithm 1 traces the construction of the reduced distributed state space of a given Modular Petri net.

$Waiting = \varnothing$
$S' = \varnothing$
$\widehat{\mathcal{N}}_s = \varnothing$
$\widehat{\mathcal{A}}_s = \varnothing$
forall $s \in S$ **do**
    Build the meta-state $\widehat{M_0}_s$
    $\widehat{\mathcal{N}}_s \longleftarrow \widehat{\mathcal{N}}_s \cup \{\widehat{M_0}_s\}$
    $element[s] \longleftarrow M_0^q$
$Waiting \longleftarrow \{element\}$
repeat
    Select $element = \prod_{s \in S} q(s)$ not marked in $waiting$
    Mark $element$ in $waiting$
    Let M be the marking s.t. $M_s = q(s)_s$
    forall $tf \in TF$ **do**
        forall M' such that $\exists M' \in [[M\rangle : M'[tf\rangle M''$ **do**
            forall $s \in S$ such that $tf \cap T_{sync,s} \neq \varnothing$ **do**
                $element[s-1] \longleftarrow M'^q$
                $S' \longleftarrow S' \cup \{s\}$
            forall $s \in S'$ **do**
                $\widehat{\mathcal{N}}_s \longleftarrow \widehat{\mathcal{N}}_s \cup \{\widehat{M'_s}\}$
                $\widehat{\mathcal{A}}_s \longleftarrow \widehat{\mathcal{A}}_s \cup \{(\widehat{M}_s, (M''^q, tf), \widehat{M'_s})\}$
            forall $s \in S \setminus S'$ **do**
                $\widehat{\mathcal{N}}_s \longleftarrow \widehat{\mathcal{N}}_s \cup \{\widehat{M}_s\}$
                $\widehat{\mathcal{A}}_s \longleftarrow \widehat{\mathcal{A}}_s \cup \{(\widehat{M}_s, (M^q, \tau), \widehat{M}_s\}$
            $S' = \varnothing$
    forall $s \in S$ **do**
        forall $\widehat{N}_s \in \widehat{\mathcal{N}}_s$ **do**
            if $output(\widehat{M}_s) = output(\widehat{N}_s)$ and $M_s^c = N_s^c$ **then**
                $fusion(\widehat{M}_s, \widehat{N}_s)$
                $element[s] = N_s^q$
        if $element \notin Waiting$ **then**
            $Waiting \longleftarrow Waiting \cup \{element\}$
until all elements of $Waiting$ are marked;

Algorithm 1: Construction algorithm of a reduced distributed state space

Figure 3 shows the reduced distributed state space of the Modular Petri net of Fig. 1.

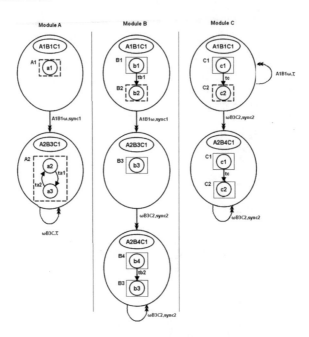

Fig. 3. Example of a reduced distributed state space

4.3   Sync-closure Property and Local Properties Verification

In order to check the local properties of a module without the need to explore the local graphs of other modules, the properties of *Sync-closure* introduced in [16] is used in conjunction with the structure of the reduced distributed state space. The Sync-closure property indicates whether a module of the system reaches a state from which it evolves only by executing local transitions. This property is defined for terminal SCC of local markings in meta-states.

The RDSS structure presented in this work helps to verify the properties related to one module in a modular way. This section evokes some local generic properties. We verify that the algorithms cited in [1,16] remains valid for the proposed structure.

For the following local properties, we checked that the algorithms depicting the verification of some local generic properties remain applicable for the the reduced structure proposed. In fact, hiding synchronized transitions that a particular module is not involved in has no effect on the sync-closure properties. Subsequently, it does not change the way to check the general properties discussed. All the propositions proofs and examples cited in the previous works [1,16] are still valid.

- *Local deadlock markings*: Deadlock is an important property of Petri net. In fact, checking the presence of the deadlock allows knowing if a module can still be active or not. That is to say if it is always possible to find a path allowing local or synchronized transitions to be fired in the concerned module.
- *Local liveness*: A transition in a module is live if it could be fired after each reachable marking of this module.
- *Locally home space*: A set of local markings X is said locally home space if it is always possible from any reachable local marking of the module to reach a marking of X

5   Experiments and Preliminary Results

In the current section, we present the preliminary results of our experiments using **the Distributed State Space Builder**[1] which is a distributed verification implementation based on modular Petri nets. It allows to build the Reduced distributed state space presented in   4. The Philosophers, RobotManipulation and ERK which are famous academic examples are used in our experimentation. They are taken from the Model Checking Contest benchmark[2].

Table 1. Experimental results for the construction of a modular Petri net(RDSS) in comparison of the flat one

Model	flat petri net		Modular petri net (RDSS)			
Net	Transitions	markings	local arcs	syn arcs	total arcs	markings
ERK1	30	13	10	13	23	16
Philo3	768	325	141	540	681	273
Robot1	274	110	25	18	43	23

Th results presented in Table 1 show that the new structure helps to lessen the size of the state space as it could reduce the total arc number and the reachable marking number especially for systems whose modules are not all synchronized on the same fusion transitions (case of the example ERK1). For the two other examples, there is a considerable decrease in the number of possible markings (Fig. 3).

---

[1] see https://github.com/chihebabid/DSS-Checker.
[2] https://mcc.lip6.fr/2023/models.php.

6   Conclusion

In this paper, we have proposed a reduced modular structure of the state space for modular Petri nets. This structure builds a set of reduced graphs representing the conduct of each module. The verification of some properties concerning could be done just by exploring to the reduced graph of the said module.

Comparing to the modular state space proposed in ([1]), we hide for each module the synchronized actions of other modules of which it is not concerned. This allows to reduce the size of the graph representing each module. For that reason, our proposed structure could show good results in coping with the explosion problem especially with systems whose modules are not all synchronized on the same fusion transitions.

Our work opens up several perspectives: as our structure reinforces the modularity of the verification process, we intend to proceed to the verification of an LTL/X propriety related to a module which is formed by its local and synchronized transitions. The verification will be done by exploring only the graph of the concerned module which avoids switching between the graphs of the different modules.

References

1. Ouni, H., Abid, C.A., Zouari, B.: A distributed state space for modular Petri nets. In: 7th International Conference on Modelling, Identification and Control (2015)
2. Christensen, S., Petrucci, L.: Modular analysis of Petri nets. Br Comput. Soc: Comput. J. **43**, 224–242 (2000)
3. Lakos, C., Petrucci, L.: Modular analysis of systems composed of semiautonomous. In: International Conference on Application of Concurrency to System Design, pp.185-196 (2004)
4. Valmari, A.: Compositional state space generation, Lecture Notes in Computer Science, pp. 427-457 (1993)
5. Valmari, A., Tienari, M.: An improved failures equivalence for finite state systems with a reduction algorithm, North-Holland Publishing Co.: Proceedings of the IFIP WG6. In: International Symposium on Protocol Specification, Testing and Verification XI, pp. 318 (1991)
6. Ouzara, F., Boukaka, M., Petrucci, L.: Verification of modular systems. In: Proceedings of the International Conference on Information Technology for Organization Development, pp. 240-245 (2014)
7. Ouni, H., klai, K., Abid, C.A., Zouari, B.: Parallel Symbolic Observation Graph. In: IEEE International Symposium on Parallel and Distributed Processing with Applications and IEEE International Conference on Ubiquitous Computing and Communications (ISPA/IUCC), pp. 770-777 (2017)
8. Mahmood, I., Askari, S.H., Sarjoughian, H.S.: Composability Verification of Complex Systems Using Colored Petri Nets, IEEE Publisher: Winter Simulation Conference (WSC) (2021)
9. Costa, T.F., et al.: Coloured petri nets-based modeling and validation of insulin infusion pump systems. Appl. Sci. **12**(3), 208–223 (2022)

10. Barnat, J., et al.: Parallel model checking algorithms for linear-time temporal logic. In: Handbook of Parallel Constraint Reasoning, pp. 457–507. Springer, Cham (2018). https://doi.org/10.1007/978-3-319-63516-3_12
11. klai, K., PEtrucci, L., Reniers, M.: An incremental and modular technique for checking ltl-x properties of petri nets. In: FORTE '07 : Proceedings of the 27th IFIP WG 6.1 International Conference on Formal Techniques for Networked and Distributed Systems, (Berlin, Heidelberg), pp. 280-295 (2007)
12. Holzmann, G.J.: Parallelizing the spin model checker. In: International SPIN Workshop on Model Checking of Software, pp. 155-171 (2012)
13. Filippidis, I., Holzmann, G.J.: An improvement of the piggyback algorithm for parallel model checking. In: Proceedings of the 2014 International SPIN Symposium on Model Checking of Software, pp. 48-57 (2014)
14. Ouni, H., klai, K., Abid, C.A., Zouari, B.: A parallel construction of the symbolic observation graph: the basis for efficient model checking of concurrent systems. In: The 8th International Symposium on Symbolic Computation in Software Science, pp. 107–119 (2017)
15. Holzmann, G.J.: Proving properties of concurrent programs, International SPIN Workshop on Model Checking of Software, pp. 18–23 (2013)
16. Abid, C.A., Zouari, B.: Local verification using a distributed state space. IOS Press: Fundamenta Informaticae **125**, 1–20 (2013)

Efficient Heuristic for Broadcasting in Chordal Networks

Hovhannes A. Harutyunyan and Narek Hovhannisyan[✉]

Department of Computer Science and Software Engineering, Concordia University,
Montreal, QC H3G 1M8, Canada
haruty@cs.concordia.ca, narek.hovhannisyan@concordia.ca

Abstract. Broadcasting is one of the information dissemination primitives where a message is passed from one node (called originator) to all other nodes in the network. With the increasing interest in interconnection networks, an extensive amount of research was dedicated to broadcasting. Two main research goals of this area are finding inexpensive network structures that maintain efficient broadcasting and finding the broadcast time for well-known and widely used network topologies.

In the scope of this paper, we will mainly focus on determining the broadcast time and near-optimal broadcasting schemes in networks. Determination of the broadcast time of a node x in arbitrary network G is known to be NP-hard. Polynomial-time solutions are known only for a few network topologies. There also exist various heuristic and approximation algorithms for different network topologies. In this paper, we will consider a large subset of chordal networks. We will present an efficient heuristic for the broadcast time problem in these networks and will present some empirical results.

Keywords: Interconnection networks · Broadcasting · Heuristic

1 Introduction

Broadcasting is one of the most important information dissemination processes in an interconnected network, that has been studied over the past decades. The broadcasting process starts from a single node of the network and ends when all nodes have the information. Over the last four decades, a large amount of research work has been published concerning broadcasting in networks under different models. These models differ in the number of originators, the number of receivers at each time unit, the distance of each call, the number of destinations, and other characteristics of the network. In the context of this paper, we are going to focus on the classical model of broadcasting. In the context of broadcasting, the network is modeled as an undirected graph $G = (V, E)$, where $V(G)$ and $E(G)$ denote the vertex set and the edge set of a given graph G correspondingly. The classical model follows the below-mentioned basic assumptions.

1. The broadcasting process is split into discrete time units.
2. The only vertex that has the message at the first time unit is called *originator*.

© The Author(s), under exclusive license to Springer Nature Switzerland AG 2023
L. Barolli (Ed.): AINA 2023, LNNS 661, pp. 332–345, 2023.
https://doi.org/10.1007/978-3-031-29056-5_30

3. In each time unit, an informed vertex (*sender*, a vertex that has the message) can *call* at most one uninformed neighbor (*receiver*).
4. During one time unit, all calls are performed in parallel.
5. The process halts as soon as all the vertices in the graph are informed.

We can represent each call in this process as an ordered pair of two vertices (u, v), where u is the sender and v is the receiver. The set of calls that are placed in time unit i is denoted by C_i.

Definition 1. ***Broadcast scheme*** *is the order of calls made by each vertex during a broadcasting process and can be represented as a sequence* $(C_1, C_2, ..., C_t)$, *where* C_i *is the set of calls performed in time unit* i.

An informed vertex v is *idle* in time unit t if v does not make any call in time t. Given that every vertex, other than the originator, can be informed by exactly one vertex, the broadcast scheme forms a directed spanning tree (*broadcast tree*) rooted at the originator. We are also free to omit the direction of each call in the broadcast tree. Note that if multiple vertices independently inform the same vertex in the same time unit, one of them can be selected as a sender to construct a broadcast tree.

Definition 2. *The **broadcast time** for a vertex* v *in a given graph* G *is the minimum number of time units required for broadcasting in* G *if* v *is the originator and is denoted by* $b(v, G)$.
The broadcast time for a given graph G, *is the maximum broadcast time from any originator in* G, *formally* $b(G) = max_{v \in V(G)}\{b(v, G)\}$.

A broadcast scheme for an originator v that uses $b(v, G)$ time units is called optimal broadcast scheme and is denoted by $\mathcal{S}(v, G)$. Obviously, by the Assumption 3, the number of informed vertices after each time unit can be at most doubled. Meaning, in general, the number of informed vertices after time unit i is upper bounded by 2^i. Therefore, it is easy to see that $b(v, G) \geq \lceil \log n \rceil$, where n is the number of vertices in G, which implies that $b(G) \geq \lceil \log n \rceil$.

The general *broadcast time* decision problem is formally defined as follows. Given a graph $G = (V, E)$ with a specified set of vertices $V_0 \subseteq V$ and a positive integer k, the goal is to determine if there is a sequence $V_0, E_1, V_1, E_2, V_2, \ldots, E_k, V_k$ where $V_i \subseteq V$, $E_i \subseteq E(1 \leq i \leq k)$, $E_i = \{u, v\}, u \in V_{i-1}, v \notin V_{i-1}$, $V_i = V_{i-1} \cup \{v\}$ and $V_k = V$. Here k is the total broadcast time, V_i is the set of informed vertices at round i, and E_i is the set of edges used at round i. It is obvious that when $\mid V_0 \mid = 1$ then this problem becomes our broadcast problem of determining $b(v, G)$ for an arbitrary vertex v in an arbitrary graph G.

Generally, the broadcast time decision problem in arbitrary graphs is NP-complete ([14,34]). However, the broadcast time problem was proved to be NP-Complete even for some specific topologies, such as 3-regular planar graphs [28]. There is a very limited number of graph families, for which an exact algorithm with polynomial-time complexity is known for the broadcast time problem. Exact linear time algorithms are available for the broadcast time problem in trees ([31,34]), in connected graphs with only one cycle (unicyclic graphs [18,19]), and in necklace graphs ([26]). For a more detailed introduction to broadcasting, we refer the reader to [12,17,23,24].

A network is chordal [8] if its underlying structure can be represented as a chordal graph. Recall that an edge is a chord of a cycle if it joins two nodes of the cycle but is not itself in the cycle. A graph is chordal if and only if every cycle of length greater than three has a chord. Chordal graphs play a central role in techniques for exploiting sparsity in large semidefinite optimization problems [37] and in related convex optimization problems involving sparse positive semidefinite matrices. Chordal graph properties are also fundamental to several classical results in combinatorial optimization, linear algebra, statistics, signal processing, machine learning, and nonlinear optimization ([36,38]). Split graphs are a subclass of chordal graphs and are exactly those chordal graphs whose complement, i.e., the same graph in which edges and nonedges are swapped, is also chordal [15]. Therefore, all problems which are polynomial-time solvable for chordal graphs are also solvable for split graphs ([14,15]). Moreover, in [4], the authors showed that in the limit as n goes to infinity, the fraction of n-vertex chordal graphs that are split approaches one. Less formally, they showed that almost all chordal graphs are split graphs, thus making split graphs an important area of research.

A split graph is a graph in which the vertices can be partitioned into a clique and an independent set [27]. Split graphs were first studied in [10], and independently introduced in [35]. Besides being a large subfamily of chordal graphs, split graphs are widely used as an interconnection network topology. Split graphs can be used as the topology of networks that have an important group of tightly coupled nodes (or in other words the *core* of the network), and a number of independent nodes that are only connected to the network core. The same structure is possible in dynamic networks, where nodes can join and leave the network by connecting to the physically closest nodes in the core. In terms of social networks, split graphs correspond to the variety of interpersonal and intergroup relations [2]. The interaction between the cliques (socially strong and trusty groups) and the independent sets (fragmented and non-connected groups of people) is naturally represented as a split graph. Different optimization problems were studied in split graphs due to their many important characteristics ([3,9]). Some of the problems that are NP-complete in the general case are fairly trivial in split graphs. However, finding a Hamiltonian cycle, a Minimum Dominating Set, or a λ-coloring remain NP-complete for split graphs ([5,6,29]). Given all of the above, it is interesting to study the problem of broadcasting on split graphs.

Currently, there exist a number of heuristics for the broadcast time problem in arbitrary graphs that achieve good results and performance ([1,13,16,20–22, 32]). However, for the specific case of split graphs, none of the above algorithms takes advantage of the characteristics of the input graph. In this paper, we devise an algorithm that is designed considering the characteristics of split graphs and achieves near-optimal results.

The rest of this paper is organized as follows. In Sect. 2 we introduce an efficient heuristic for the broadcast time problem in split graphs for an arbitrary originator. Further, in Sect. 3, we will discuss the experimental results of the proposed heuristic, and finally, we will conclude the paper in Sect. 4.

2   Broadcasting Heuristic

In this section, we will introduce a heuristic for the broadcast time problem in split graphs. Consider a split graph $G = (V, E)$ such that the vertex set can be partitioned into a clique K on n vertices and an independent set I on m vertices. Note that a given split graph G may have more than one partitioning into a clique and an independent set. Clearly, we can assume that the clique K is a maximal (and also maximum) clique, otherwise, we could change the partitioning by adding some vertices from I to K to make it maximal.

Let t be a positive integer. A *t-star-matching* of a graph G is collection of mutually vertex disjoint subgraphs $K_{1,i}$ of G with $1 \leq i \leq t$ [25]. A *perfect t-star-matching* of a graph G is a t-star-matching that covers every vertex of the graph G. A *proper t-star-matching* of a split graph G is a perfect t-star-matching such that any vertex $v \in K$ belongs to the smaller partition of a bipartite subgraph (is a center of a star).

Given a connected split graph G, a clique K, an independent set I, and an arbitrary originator $v \in V(G)$ we can consider the following two cases.

1. The originator is from the clique: $v \in K$.
2. The originator is from the independent set: $v \in I$.

However, for both of these cases, the algorithm strategy is the same. The only difference is that for an originator in the independent set, the originator will not be considered in the step of finding a star-matching. Before proceeding to the actual algorithm, we will describe a procedure for finding a proper star-matching, such that the maximum degree (*maxdegree*) of the subgraph induced by that matching is the minimum out of all proper star-matching.

2.1   Finding a Proper Star-Matching with Minimum Maxdegree

In this section, we will introduce an algorithm for finding a proper t-star-matching for a given split graph G such that t is the minimum out of all possible proper star-matchings.

In order to find the minimum positive integer t, for which there exists a proper t-star-matching, we will reduce the problem to an instance of a maximum flow problem ([11,33]). An instance (G, s, t, m) of the maximum flow problem consists of a flow graph $G = (V, E)$, the source vertex s, the sink vertex t, and a natural number m. The goal of the problem is to decide if the maximum flow for graph

---

Algorithm 1. Reduction Algorithm

    Input A split graph G, a clique K, an independent set I, and an integer t
    Output An instance of the maximum flow problem
1: **procedure** MAXFLOWREDUCTION
2:      Create a copy G' of the graph G
3:      Remove all edges between a pair of vertices u and v, where $u, v \in K$
4:      Assign capacity 1 to each edge (u, v), where $u \in K$, $v \in I$
5:      Add a source vertex s and connect it by an edge to every vertex in K. Assign a capacity t to those edges
6:      Add a sink vertex t and connect it by an edge to every vertex in I. Assign a capacity 1 to those edges
7:      **return** $(G', s, t, |I|)$
8: **end procedure**

G from s to t is greater than or equal to m. The reduction algorithm is described in Algorithm 1.

It is easy to see that there exists a proper t-star-matching for the graph G if and only if the maximum flow in the flow graph G' (Algorithm 1) is equal to m, where m is the cardinality of the independent set I. When there exists a proper t-star-matching, the same edges can be used in graph G' to achieve flow with value m. And similarly, if the maximum flow is equal to m, the paths used to construct the flow can be used to find a proper star-matching for the graph G. Since each vertex in K has an incoming edge capacity of t, it will result in a proper t-star-matching for the graph G.

Thus, we can find the minimum integer t for which there exists a proper t-star-matching. Further, the maximum flow algorithm will be used in combination with a binary search to find the minimum degree for which there exists a proper star-matching, whereas, the edges used in the max flow will be used to induce the actual matching. Algorithm 2 presents a procedure to verify the availability of a proper t-star-matching. It can be used in a binary search on t, which can be in the range from 0 to $|I|$.

The procedure *MaxFlowReduction* can be implemented with complexity $\mathcal{O}(n) + \mathcal{O}(m) = \mathcal{O}(|V|)$. The time to find the maximum flow in a graph highly depends on the selected algorithm. One of the best algorithms for maximum flow has a complexity of $\mathcal{O}(|E||V|)$ [30]. Thus, considering the binary search time, the overall time complexity of this algorithm would be $\mathcal{O}(|E||V|\log|I|)$.

2.2   Algorithm Description

In this section, we will devise a heuristic for the broadcast time problem in split graphs from an arbitrary originator. The main goal of the algorithm is to avoid vertex idling. Moreover, the broadcasting scheme also prioritizes informing vertices in the clique, and will only start broadcasting to the independent set as late as possible. The vertices in the clique will be given some priority based on their degree in the proper star-matching. Vertices that have more neighbors

---

Algorithm 2. Algorithm for finding a proper star-matching

    Input A split graph G, a clique K, an independent set I and an integer t
    Output A proper t-star-matching M for G with minimum t

1: **procedure** FINDMATCHING(t, M)
2:    Use $MaxFlowReduction$ to create a flow graph G'
3:    Run a maximum flow algorithm on G'
4:    $M \leftarrow$ the edges used in the maximum flow
5:    $V(M) \leftarrow$ the maximum flow number
6:    **if** $V(M) = |I|$ **then**
7:        **return** M
8:    **else**
9:        **return** $NULL$
10:   **end if**
11: **end procedure**

---

to inform in the independent set will be informed earlier. However, instead of directly finding a broadcast time for the given split graph we will try to find an answer to the broadcast time decision problem. Given a graph $G = (V, E)$, the originator $v \in V$, and a natural number b, the goal of the broadcast time decision problem returns "Yes" if $b(v, G) \leq b$ and "No" otherwise.

Let M be a proper t-star-matching, and let $d_i, 1 \leq i \leq n$ be the degree of vertex $v_i \in K$ in M. Without loss of generality assume that $d_1 \geq d_2 \geq ... \geq d_n \geq 0$. Every vertex $v_i \in K$ will start broadcasting to his neighbors in the independent set at time unit $b - d_i + 1$, as that is the latest point of time when it will be able to inform all of the vertices in the independent set. Thus, we can guarantee that the independent set will be fully informed by the time unit b, and we will only need to make sure that the vertices in the clique are also informed. The pseudocode for this algorithm is presented in Algorithm 3.

Additionally, note that the algorithm can be easily adapted for dynamic networks. The only modification required is to update the list of uninformed vertices d and their corresponding deadlines whenever a node joins or leaves the network.

The time complexity of the $FindBroadcastTime$ procedure in Algorithm 3 consists of several components. First, as we already discussed, finding the star-matching will have a complexity of $\mathcal{O}(|E||V|\log|I|)$. Next, sorting of the degree array will have a complexity of $\mathcal{O}(|K|\log|K|)$. Any operation in the sub-procedure $Broadcast$ can be implemented with constant time complexity. Thus, overall, the procedure $Broadcast$ will have complexity $\mathcal{O}(b)$ caused by the loop on line 16. Hence, overall complexity of the heuristic will be $\mathcal{O}(|E||V|\log|I|)+ \mathcal{O}(|K|\log|K|)+ \mathcal{O}(b\log|V|) = \mathcal{O}(|E||V|\log|I|)$. Lastly, the Algorithm 3 will be invoked during a binary search on range $[\log|V|, |V| - 1]$.

Algorithm 3. Algorithm for Broadcasting in Split Graphs

Input A split graph G, a clique K, an independent set I, an originator $v \in V(G)$, and a natural number b

A broadcast time b.

1: **procedure** FINDBROADCASTTIME
2:     $G' \leftarrow G$
3:     **if** $v \in I$ **then**
4:         Remove v from G'
5:     **end if**
6:     Find a star-matching of G' (Algorithm 2)
7:     $d \leftarrow []$
8:     $d[i] \leftarrow$ degree of $v_i \in K$ in M
9:     Sort the degree array d in a decreasing order
10:    Do a binary search on range $[\log |V|, |V| - 1]$ by invoking *Broadcast* procedure
11: **end procedure**
12: **procedure** BROADCAST($d[]$)   ▷ **Input**: sorted degree array d
13:                                    ▷ **Output** "Yes" if $b(v, G) \le b$; "No" otherwise
14:    $q \leftarrow v$ ▷ Set of informed vertices
15:    $i \leftarrow 1$
16:    **while** $i \le b$ **do**   ▷ Iterate through rounds
17:        Remove $d[0, |q| - 1]$ from d ▷ Inform first $|q|$ vertices in the list d
18:        Add $d[0, |q| - 1]$ to q
19:        **if** a vertex v_i with $d_i = t - i + 1$ is not informed **then**
20:            **return** "No"   ▷ Vertex v_i fails to start broadcasting to I
21:        **else**
22:            Remove v_i from q
23:        **end if**
24:    **end while**
25:
26:    **if** d is not empty **then**
27:        **return** "No"   ▷ Not all vertices of K are informed
28:    **end if**
29:    **return** "Yes"
30: **end procedure**

3   Experimental Results

In this section, we will discuss the results achieved after different experiments with the proposed heuristic. Generally, the experiments that we conduct can be divided into two groups: structured and unstructured. In the first group of experiments, we test the results of the heuristic on random graphs with no predefined structure. The main purpose of this group of experiments is to validate the motivation of the heuristic by comparing our results with one of the best heuristics for the general case of the problem. In the second stage, we study the behavior of the proposed heuristic on graph instances that were generated following some predefined structure.

3.1   Unstructured Graphs

For the first round of experiments, we create random split graphs with different densities. Firstly, the vertex set of the graph is randomly divided into two partitions: the clique and the independent set. After that, edges are added between every pair of clique vertices. Lastly, with given probability p (density parameter), an edge is added between every pair of vertices, where one vertex is from the independent set and the other vertex is from the clique.

We run our algorithm up to 50 times for different graph sizes and for three values of the density parameter: 0.2 (sparse graphs), 0.5 (dense graphs), and 0.9 (near-complete graphs). The broadcasting is initiated from a randomly selected originator. The results for different values of the density parameter p and different number of vertices N are presented in Table 1. The algorithm is executed on up to 50 generated graph instances for each pair of input parameters (N, p), and for each execution the resulting broadcast time $b(X)$ is retrieved. In Table 1, we present the minimum (Min $b(X)$), the maximum (Max $b(X)$), and the average (Avg $b(X)$) of all resulting broadcast times. The above-mentioned values are compared with the broadcast time lower bound of $\log N$. Moreover, we executed one of the best heuristic algorithms for broadcasting in arbitrary graphs (*DH* [16]) and presented the results in the table.

Table 1. Experimental results for random split graphs of different number of vertices (N) and density of edges (p).

N	p	$\lceil \log N \rceil$	DH	Min $b(X)$	Max $b(X)$	Avg $b(X)$
1000	0.2	10	12	10	11	10.5
500	0.2	9	10	9	10	9.4
100	0.2	7	9	7	8	7.46
80	0.2	7	9	7	8	7.24
50	0.2	6	7	6	8	6.52
32	0.2	5	7	5	7	5.8
20	0.2	5	6	5	8	5.58
1000	0.5	10	12	10	11	10.6
500	0.5	9	11	9	10	9.6
100	0.5	7	8	7	8	7.3
80	0.5	7	9	7	7	7
50	0.5	6	8	6	8	6.46
32	0.5	5	8	5	7	5.56
20	0.5	5	6	5	8	5.2
1000	0.9	10	13	10	11	10.5
500	0.9	9	10	9	10	9.3
100	0.9	7	9	7	8	7.46
80	0.9	7	9	7	7	7
50	0.9	6	9	6	8	6.3
32	0.9	5	8	5	7	5.46
20	0.9	5	7	5	8	5.08

We can see that the proposed heuristic returns broadcast times that are rather close to the lower bound of $\lceil \log N \rceil$. In some cases, we can see that the maximum broadcast time of the scheme X returned by the algorithm is equal to the lower bound. Additionally, we can see that the proposed heuristic outperforms the heuristic for the general case of the problem in the vast majority of cases. Given that we do not know the exact value of the minimum broadcast time, the results achieved by the algorithm are considered successful. Moreover, one can notice that the average broadcast times for the same graph size are slightly different based on the density of the graph. In order to understand the effect of density on broadcast time achieved by our algorithm we designed a chart to study the behavior of the algorithm. Figure 1 provides the comparison between average broadcast times for different density parameters. Based on the experimental data that we pose, there is no visible trend change between different density graphs. Hence, we conclude that our algorithm achieves equally good results for randomly generated split graphs regardless of the density of the graph instance.

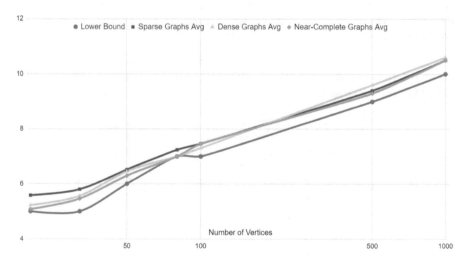

Fig. 1. Comparison between lower bound of the optimal time and average broadcast times of sparse, dense, and near-complete graphs on different number of vertices.

3.2   Bounded Degree Split Graphs

In this section, we will present experimental results on split graphs where vertices in the independent set have bounded degrees. We call such graphs *k-restricted split graphs*, where k is the maximum degree out of all vertices in the independent set. This subfamily of graphs represents networks where nodes have a limit on the number of connections to the network core. In order to simulate such networks, we create split graph instances with a given number of vertices (N), and a predefined number of vertices in the clique (n) and independent set (m)

partitions. After the partition is defined, vertices in the independent set are connected by edges to a randomly selected set of vertices in the clique, such that the size of the set is limited by a given input value d (degree bound). Later, we run our heuristic on this graph instances with random originators and report the results in a way similar to the previous section: the minimum (Min $b(X)$), the maximum (Max $b(X)$), and the average (Avg $b(X)$) of all resulting broadcast times for each input parameter set.

Tables 2 and 3 present experimental results on 5-restricted and 15-restricted split graphs respectively. We can see that in most cases the average broadcast time is equal to the lower bound of the broadcasting, which means that the algorithm found the minimum broadcast time in all executions. Moreover, from the tables, it is clear that the average broadcast time grows as the size of the clique decreases while keeping the number of vertices constant. Moreover, in some cases, we can see that the minimum reported broadcast time increases as well. As in the majority of the cases, the algorithm correctly found a broadcast scheme with the optimal broadcast time, it could mean that the optimal broadcast time of the graph is not equal to the lower bound anymore. Overall, we can conclude that the algorithm performed very well for k-restricted split graphs.

Table 2. Experimental results for 5-restricted split graphs of different number of vertices (N), and different cardinalities of the clique (n) and the independent set (m).

N	n	m	$\lceil \log N \rceil$	Min $b(X)$	Max $B(X)$	Avg $b(X)$
130	85	45	8	8	8	8
130	75	55	8	8	8	8
130	65	65	8	8	9	8.04
130	55	75	8	8	9	8.02
130	45	85	8	8	10	8.3
70	45	25	7	7	7	7
70	40	30	7	7	7	7
70	35	35	7	7	8	7.06
70	30	40	7	7	7	7
70	25	45	7	7	8	7.1
40	30	10	6	6	6	6
40	25	15	6	6	6	6
40	20	20	6	6	7	6.16
40	15	25	6	6	8	6.1
40	10	30	6	7	8	7.04

3.3   Threshold Graphs

In the case of k-restricted split graphs, vertices in the independent set are connected to their neighbors with uniformly random probability and share the same

Table 3. Experimental results for 15-restricted split graphs of different number of vertices (N), and different cardinalities of the clique (n) and the independent set (m).

N	n	m	$\lceil \log N \rceil$	Min $b(X)$	Max $B(X)$	Avg $b(X)$
550	300	250	10	10	10	10
550	250	300	10	10	10	10
550	220	330	10	10	10	10
550	190	360	10	10	11	10
550	180	370	10	10	11	10.45
550	150	400	10	11	11	11
550	100	450	10	12	12	12
260	170	90	9	9	9	9
260	150	110	9	9	9	9
260	130	130	9	9	9	9
260	110	150	9	9	9	9
260	90	170	9	9	10	9.04
260	70	190	9	9	10	9.74
130	85	45	8	8	8	8
130	75	55	8	8	8	8
130	65	65	8	8	8	8
130	55	75	8	8	8	8
130	45	85	8	8	8	8
130	30	100	8	9	9	9

set of limitations. Since these graphs represent fairly balanced networks, it is still interesting to study graphs with imbalanced adjacency between vertices.

Our next group of experiments was performed on threshold graphs. Threshold graphs are constructed from a single vertex graph by addition of vertices. Each added vertex is either not connected to any existing vertex (*isolated*) or is connected to the rest of existing vertices (*dominating*). Threshold graphs are introduced in [7] and are a subfamily of split graphs. Let every vertex that was selected as isolated during the construction process be in the set I and every vertex that was selected as dominating be in the set K. From the construction of the graph G we know that no pair of vertices in I is connected by an edge, hence, making it an independent set. Now, let $u, v \in K$ be an arbitrary pair of vertices in K, such that, without loss of generality, v was added later than u. Since v was selected to be dominating then it would be connected to every vertex that was already in the graph, including u. Hence, K will be a clique in G. Thus, we showed that for any threshold graph G, there exists a decomposition into a clique and an independent set, making it a split graph.

For the sake of our experiments, we construct threshold graphs, where each added vertex is randomly selected to be isolated or dominating with a uniform probability. However, the last added vertex is always a dominating vertex, because otherwise the graph would not be connected. Notice that this process will generate sparse graphs with an approximately equal number of vertices in two partitions. Moreover, the process generates graphs with a wide distribution of vertex degrees, where vertices in the independent set have more neighbors if added earlier, and vice versa for the vertices in the clique.

Table 4. Experimental results for randomly generated threshold graphs on different number of vertices N.

N	Lower bound		$b(X)$	N	Lower bound		$b(X)$
	$\lceil \log N \rceil$	Min maxdegree			$\lceil \log N \rceil$	Min maxdegree	
20	5	2	5	320	9	1	9
36	6	3	7	516	10	3	12
48	6	1	6	528	10	1	10
68	7	1	7	576	10	1	10
80	7	1	7	768	10	3	12
132	8	5	10	1028	11	2	12
144	8	1	8	1040	11	1	11
192	8	2	9	1088	11	2	12
260	9	1	9	1280	11	2	12

The results of the heuristic execution from randomly selected originators are provided in Table 4. For a given graph with N vertices, we analyze the broadcast time achieved by the heuristic ($b(X)$) by comparing it with two lower bounds of the minimum broadcast time: trivial lower bound of $\lceil \log N \rceil$ and the minimum maxdegree of any proper star-matching. Note that the minimum broadcast time from a single originator cannot be less than the broadcast time when all vertices in the clique are informed, hence the maximum degree of the proper star-matching is a lower bound for the broadcast time.

We can conclude that the heuristic returns results that, in the majority of cases, are equal to the lower bound. Moreover, the broadcast time $b(X)$ differs from the $\lceil \log N \rceil$ when the minimum maxdegree of all proper star-matchings is bigger. This behavior is explained by the fact that vertices in the clique will need to spend more time placing calls toward the independent set. Overall, experiments show that the broadcast times returned by the proposed algorithm are rather close to the lower bound.

4   Conclusion and Future Work

We designed an efficient heuristic, which, unlike existing general heuristics, takes advantage of the characteristics of split graphs. After numerous experiments, we conclude that the heuristic has acceptable performance and returns broadcast times with a distribution close to the lower bound. Unfortunately, as of right now, there do not exist exact values or better lower bounds for the broadcast time in split graphs. Hence, we are limited to the comparisons with the lower bound. Although for some cases the resulting broadcast time was greater than the lower bound, it still could be equal (or close) to the minimum broadcast time.

In the future, we plan on improving the procedure for finding a proper star-matching. We believe that the algorithm performance is greatly affected by this step. Moreover, there is an open problem of finding better bounds for the broadcast time in split graphs. Additionally, the technique used in our proposed heuristic can, in theory, be adapted for other graph families, such as bipartite graphs and other graphs with known decomposition into two subgraphs.

References

1. Beier, R., Sibeyn, J.F.: A powerful heuristic for telephone gossiping. In: The Seventh International Colloquium on Structural Information and Communication Complexity (SIROCCO '00), pp. 17–36. Carleton Scientific (2000)
2. Belik, I.: The analysis of split graphs in social networks based on the k-cardinality assignment problem. Int. J. Netw. Sci. **1**(1), 53–62 (2016)
3. Belmonte, R., Kim, E.J., Lampis, M., Mitsou, V., Otachi, Y., Sikora, F.: Token sliding on split graphs. Theor. Comput. Syst. **65**(4), 662–686 (2021)
4. Bender, E.A., Richmond, L.B., Wormald, N.C.: Almost all chordal graphs split. J. Aust. Math. Soc. **38**(2), 214–221 (1985)
5. Bertossi, A.A.: Dominating sets for split and bipartite graphs. Inf. Process. Lett. **19**(1), 37–40 (1984)
6. Bodlaender, H.L., Kloks, T., Tan, R.B., van Leeuwen, J.: Coloring of Graphs. In: Reichel, H., Tison, S. (eds.) STACS 2000. LNCS, vol. 1770, pp. 395–406. Springer, Heidelberg (2000). https://doi.org/10.1007/3-540-46541-3_33
7. Chvtal, V., Hammer, P.: Aggregation of inequalities in integer programming. Ann. Discrete Math. **1**, 145–162 (1977)
8. Cifuentes, D., Parrilo, P.A.: Chordal networks of polynomial ideals. SIAM J. Appl. Algebra Geometry **1**(1), 73–110 (2017)
9. Collins, K.L., Trenk, A.N.: Finding balance: split graphs and related classes. arXiv preprint arXiv:1706.03092 (2017)
10. Földes, S., Hammer, P.L.: Split graphs. In: Proceedings of the Eighth Southeastern Conference on Combinatorics, Graph Theory and Computing. vol. XIX, pp. 311—315. Winnipeg: Utilitas Math (1977)
11. Ford, L.R., Fulkerson, D.R.: Flows in networks. In: Flows in Networks. Princeton university press (2015)
12. Fraigniaud, P., Lazard, E.: Methods and problems of communication in usual networks. Discret. Appl. Math. **53**(1–3), 79–133 (1994)
13. Fraigniaud, P., Vial, S.: Comparison of heuristics for one-to-all and all-to-all communications in partial meshes. Parallel Process. Lett. **9**(01), 9–20 (1999)
14. Garey, M.R., Johnson, D.S.: Computers and intractability, vol. 174. freeman San Francisco (1979)
15. Golumbic, M.C.: Algorithmic graph theory and perfect graphs. Elsevier (2004)
16. Harutyunyan, H.A., Hovhannisyan, N., Magithiya, R.: [full] Deep heuristic for broadcasting in arbitrary networks. In: 2022 21st International Symposium on Parallel and Distributed Computing (ISPDC), pp. 1–8. IEEE (2022), Best Paper Award

17. Harutyunyan, H.A., Liestman, A.L., Peters, J.G., D., R.: Broadcasting and gossiping. In: Handbook of Graph Theory, pp. 1477–1494. Chapman and Hall (2013)
18. Harutyunyan, H., Maraachlian, E.: Linear algorithm for broadcasting in unicyclic graphs. In: Lin, G. (ed.) COCOON 2007. LNCS, vol. 4598, pp. 372–382. Springer, Heidelberg (2007). https://doi.org/10.1007/978-3-540-73545-8_37
19. Harutyunyan, H.A., Maraachlian, E.: On broadcasting in unicyclic graphs. J. Comb. Optim. **16**(3), 307–322 (2008)
20. Harutyunyan, H.A., Shao, B.: A heuristic for k-broadcasting in arbitrary networks. In: Seventh International Conference on Information Visualization (IV '03), pp. 287–292. IEEE (2003)
21. Harutyunyan, H.A., Shao, B.: An efficient heuristic for broadcasting in networks. J. Parallel Distributed Comput. **66**(1), 68–76 (2006)
22. Harutyunyan, H.A., Wang, S.: Efficient multicast algorithms for mesh-connected multicomputers. In: Tenth International Conference on Information Visualisation (IV '06), pp. 504–510. IEEE (2006)
23. Hedetniemi, S.M., Hedetniemi, S.T., Liestman, A.L.: A survey of gossiping and broadcasting in communication networks. Networks **18**(4), 319–349 (1988)
24. Hromkovič, J., Klasing, R., Monien, B., Peine, R.: Dissemination of information in interconnection networks (broadcasting & gossiping). In: Combinatorial network theory, pp. 125–212. Springer (1996). https://doi.org/10.1007/978-1-4757-2491-2_5
25. Lin, W., Lam, P.C.B.: Star matching and distance two labelling. Taiwan. J. Math. **13**(1), 211–224 (2009)
26. Maraachlian, E.: Optimal broadcasting in treelike graphs. Ph.D. thesis, Concordia University (2010)
27. Merris, R.: Split graphs. Eur. J. Comb. **24**(4), 413–430 (2003)
28. Middendorf, M.: Minimum broadcast time is np-complete for 3-regular planar graphs and deadline 2. Inf. Process. Lett. **46**(6), 281–287 (1993)
29. Müller, H.: Hamiltonian circuits in chordal bipartite graphs. Discret. Math. **156**(1–3), 291–298 (1996)
30. Orlin, J.B.: Max flows in o(nm) time, or better. In: Proceedings of the Forty-Fifth Annual ACM Symposium on Theory of Computing, pp. 765–774 (2013)
31. Proskurowski, A.: Minimum broadcast trees. IEEE Trans. Comput. **30**(05), 363–366 (1981)
32. Scheuermann, P., Wu, G.: Heuristic algorithms for broadcasting in point-to-point computer networks. IEEE Trans. Comput. **33**(09), 804–811 (1984)
33. Schrijver, A.: On the history of the transportation and maximum flow problems. Math. Program. **91**(3), 437–445 (2002)
34. Slater, P.J., Cockayne, E.J., Hedetniemi, S.T.: Information dissemination in trees. SIAM J. Comput. **10**(4), 692–701 (1981)
35. Tyshkevich, R.I., Chernyak, A.A.: Canonical partition of a graph defined by the degrees of its vertices. Isv. Akad. Nauk BSSR, Ser. Fiz.-Mat. Nauk (in Russian) **5**, 14–26 (1979)
36. Vandenberghe, L., Andersen, M.S.: Chordal graphs and semidefinite optimization. Foundations and Trends® in Optimization **1**(4), 241–433 (2015)
37. Zhang, R.Y., Lavaei, J.: Sparse semidefinite programs with near-linear time complexity. In: 2018 IEEE Conference on Decision and Control (CDC), pp. 1624–1631. IEEE (2018)
38. Zheng, Y.: Chordal sparsity in control and optimization of large-scale systems. Ph.D. thesis, University of Oxford (2019)

FLOWPRI-SDN: A Framework for Bandwidth Management for Prioritary Data Flows Applied to a Smart City Scenario

Nilton José Mocelin Júnior and Adriano Fiorese[✉]

Department of Computer Science (DCC), Santa Catarina State University (UDESC),
Santa Catarina, Brazil
nilton.junior@edu.udesc.br, adriano.fiorese@udesc.br

Abstract. Applications have different requirements in terms of bandwidth and delay to deliver the expected quality of service. Furthermore, these requirements can change dynamically, as well as the importance of data traffic flows in the service context. Addressing this dynamism is one of the greatest challenges in the management of conventional networks. For this and other reasons, new paradigms such as software-defined networking have emerged. SDN decouples network control from forwarding, allowing development of new applications for dynamic programming of the forwarding devices in a centralized way. The FLOWPRI-SDN Framework is proposed in this work as an application built on top of the Ryu controller for priority resource reservation. Finally, the Framework is applied to a Smart City scenario, where it is shown that there are benefits in its use.

1 Introduction

One of the biggest challenges of communication networks in providing Quality of Service (QoS) is to deal with the dynamic nature of application data streams [1]. The QoS requirements of the application's data streams can vary over time, due to the environment and customer they serve. In this sense, smart cities are examples of scenarios where QoS and flow priority have a lot of impact.

Smart city applications are usually developed for monitoring and controlling the quality of life of citizens, by means of solutions for public safety, energy efficiency, emergency response, among others [2]. In this kind of environment, there are traffic flows that may need guaranteed resources due to the importance of their use. For example, applications whose data flows monitor places in an emergency situation. It means that, without proper control, smart city application flows compete for network resources with each other and with different background data flows, risking not having their QoS and prioritization requirements met [3]. In this sense, guaranteeing QoS is a complex task in environments based on conventional networks. However, interest has been increasing in more modern applications, which motivate the emergence of new network paradigms, such as Software-Defined Networking (SDN).

SDN is a network architecture that seeks to control networks as logical entities, separating the data plane from the control plane and centralizing network management using a component called controller [4]. This device makes it possible to have a global view of the managed network, so that it is possible to dynamically orchestrate data elements by means of controller built-in applications or that communicate with it.

When analyzing the related works presented, it is verified that there are many works that propose solutions to guarantee QoS. However, some of them are not oriented towards multi-controller scenarios and mainly implement flow shaping by creating HTB queues that individually limit throughput. This way, to deal with the dynamic resource provisioning and the priority delivery needs of general flows, this work proposes the FLOWPRI-SDN Framework. Flowing with Priority in SDN (FLOWPRI-SDN) is a Framework consisting of an implementation of the SDN Ryu controller with the objective of providing hierarchical resource reservation for flows that declare their requirements in the form of a QoS contract, through the programming of managed network switches. In addition, the Framework is able to find other Frameworks and distribute contracts, being compatible with networks that do not support the Framework.

The remainder of the paper is organized as follows. In Sect. 2, a bibliographic review is presented. The architecture and implementation of FLOWPRI-SDN are presented in Sect. 3. Section 4 presents the experimental results and analyses. Finally, the Sect. 5 provides the conclusion and finalizes the work.

2   Related Work

A difficulty of QoS provisioning applications is to discover the necessary requirements of each data traffic flow and the related works present their solutions to the problem. The work presented in [8], develops a resource management system for multi-controller environments. In that work, the controllers query the same database to find out the flows' resource requirements, which need to be configured beforehand. The mechanism to control the bandwidth usage of each link is based on querying the states of the switches involved in the data delivery path, instead of saving the states in the Framework. However, it is not explained how the bandwidth is actually reserved and managed.

HiQoS is a Framework proposed in [5], which tries to provide QoS for flows classified as video streams, interactive video/audio and best-effort. In this approach, when a flow is accepted, a queue is created that limits the bandwidth by its configuration. Furthermore, it uses a best-path-choice mechanism to avoid congestion and deliver packets of interactive flows with less delay. In [6], a Framework for bandwidth reservation is also presented, where QoS requirements are obtained from a file, and where bandwidth consumption is limited by individual queues. This application also monitors bandwidth usage and calculates routes based on bandwidth usage.

FlowQoS is a system proposed in [9] that runs over a gateway in a home network to communicate with a controller and to provide QoS. In this system, customers define application requirements using an WEB page. From the data on this page, the network controller identifies the type of application and its QoS requirements to create forwarding rules on network devices.

In this way, it is clear that the works [8,9] are the ones that provide multi-domain solutions, with bandwidth reservation according to the requirements of the flows by creating specific priority queues. However, synchronization between controllers involves third-party tools and the bandwidth managing is performed by different traffic flow queues.

3   FLOWPRI-SDN Architecture

The FLOWPRI-SDN is a Framework for allocation and management of bandwidth in a prioritized and distributed way, that runs as an SDN Ryu built in controller application for SDN OpenFlow networks. Figure 1 presents the architecture of the FLOWPRI-SDN Framework. This architecture shows the most important actions applied when OpenFlow events happen and each of them is described in the Sect. 3.1.

Using OpenFlow, FLOWPRI-SDN can program OpenFlow switches on your domain. In this work, a domain is assumed as a group of switches with one-hope connection to the FLOWPRI-SDN, which represent the network transmission devices. In addition, hosts can communicate with the FLOWPRI-SDN, through a TCP port in the Framework, which should be used for requesting flow requirements.

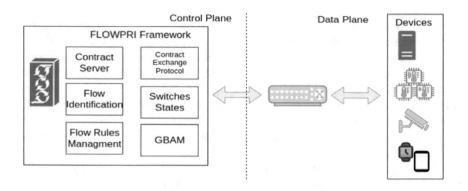

Fig. 1. FLOWPRI-SDN architecture.

At first, FLOWPRI-SDN requires the bandwidth division at the ports of the switches of the domain to be pre-configured using Hierarchical Token Bucket (HTB) queues, as defined in [10]. The recommendation is that, no more than 33% of bandwidth be used for real-time applications, 25% for applications that do not have specified requirements (best-effort), 7% for control traffic (like ICMPs) and the rest 35% for non-real-time data applications. This is because, if real-time applications are the only ones that consume link bandwidth, the performance of non-real-time applications can be significantly degraded.

This work combines the described resource reservation strategy with the concept of prioritizing flows and bandwidth borrowing mechanisms, as shown in Fig. 2. The Generic Bandwidth Allocation Method (G-BAM) model controls the bandwidth borrowing between the real-time and non-real-time classes with guaranteed resource reservation, if the flow is accepted [11]. Whereas, the HTB queuing model lends bandwidth unused by all classes, without resource reservation, to the queue of the best-effort class [12].

In this Framework, is proposed to have three possible priorities for flows that need resource reserving: low priority (1), medium priority (2) and high priority (3). On the other hand, best-effort flows are low priority and control flows are high priority. Therefore, each class priority is implemented as a HTB queue, with their priority associated. For the HTB queues, the priority instructs the kernel to try sending packages first for the highest priority queues, as well as, it allocates not consumed bandwidth from other classes to this queue. Whereas, G-BAM uses priority to determine which flow has the right to be allocated.

The minimum bandwidth for real-time and non-real-time queues is the sum of both reserved bandwidth for those classes. This way, the G-BAM method can manage the bandwidth usage for these classes, sharing bandwidth between them and reserving bandwidth for flows following their priority. All these bandwidth configurations are needed to be done using the OpenFlow tool "ovs-vsctl". Ryu does not implement messages for queue configuration, only implements queue stats query.

In addition to class and priority, flows can have a throughput requirement. Thus, to represent all these flow requirements, it is proposed to use specific Differentiated Services Code Points (DSCP) for flows, based on traffic studies from [13]. Therefore, DSCP represent priority, class, and bandwidth of a flow according to Table 1 and Table 2, including DSCP 60 for best-effort flows and DSCP 61 for control flows. From these values, meter band rules are created in the switches, which can be associated with forwarding rules to limit the bandwidth usage per flow. FLOWPRI-SDN takes advantage of these rules to control bandwidth usage by classes of services and then, by matching DSCP code, switches can decide to which queue the flow is supposed to be forwarded.

Thus, devices that have QoS requirements can tell them to FLOWPRI-SDN via a QoS contract over a connection with 4444 TCP port. The QoS contract informs the source and destination IP addresses, the bandwidth, priority and class of the flow that is intended to be sent by the device, displayed in Fig. 3. This way, the Framework can create rules in switches to identify the flow and provide their resources, using the right DSCP to sending it through the corresponding queue.

Table 1. DSCP traffic class codes for real-time flows.

Bandwidth	Priority 1 DSCP: Queue	Priority 2 DSCP: Queue	Priority 3 DSCP: Queue
32 kbps	1:0	11:1	21:2
64 kbps	2:0	12:1	22:2
128 kbps	3:0	13:1	23:2
500 kbps	4:0	14:1	24:2
1 Mbps	5:0	15:1	25:2
2 Mbps	6:0	16:1	26:2
5 Mbps	7:0	17:1	27:2
10 Mbps	8:0	18:1	28:2
25 Mbps	9:0	19:1	29:2

Table 2. DSCP traffic class codes for non-real-time flows.

Bandwidth	Priority 1 DSCP: Queue	Priority 2 DSCP: Queue	Priority 3 DSCP: Queue
32 kbps	31:3	41:4	51:5
64 kbps	32:3	42:4	52:5
128 kbps	33:3	43:4	53:5
500 kbps	34:3	44:4	54:5
1 Mbps	35:3	45:4	55:5
2 Mbps	36:3	46:4	56:5
5 Mbps	37:3	47:4	57:5
10 Mbps	38:3	48:4	58:5
25 Mbps	39:3	49:4	59:5

Fig. 2. Switch port configuration

Fig. 3. Contract model

3.1 FLOWPRI-SDN Implementation

This way, FLOWPRI-SDN has functionalities that take advantage of communication with switches, hosts and Frameworks. The communication with switches are triggered by OpenFlow events, more specifically "switch_features", "packet_in" and "flow_removed", that allow switches to call the FLOWPRI-SDN for some action. Whereas, Host functionality is enabled by establishing connections as a server over a TCP port, in order to deal with the application's sent prioritizing flows contracts.

3.1.1 Switches States

FLOWPRI-SDN stores the configurations of OpenFlow switches in its domain in the form of object-oriented classes. There is an object-oriented class to represent switches, one to represent switch ports, a class to represent active switch rules, and an object-oriented class to represent OpenFlow commands to be executed on a switch. This way, every switch port class has one list for each priority for the classes real-time and non-real-time, that stores copies of active flow rules that are consuming bandwidth. Each active flow rule is stored as a class, which has information about its DSCP code and whether it is borrowing bandwidth from another class. With this, FLOWPRI-SDN can manage bandwidth usage and availability, avoiding allocating invalid amounts of bandwidth and reducing communication between switches and Framework.

Knowing that, the first event handled by FLOWPRI-SDN is "switch_ features". It activates when an OpenFlow switch starts connecting to the controller to establish a connection. Then, it allows the switches to be proactively configured. Therefore, the Framework is able to apply pre-established definitions according to the identification number of each switch, identified from the event packet. This way, the Framework create the instances of the switches, also, each switch port is stored as a class with the information about the amount of total bandwidth and the bandwidth consumed by these classes, along with the lists of active flow rules. For each port, routes with destination prefixes are configured (stored in the class), according to the topology of the previously known network, and stores the switch identifier for next hop in the domain. This enables the Framework to keep the state of the switches without having to query them directly and get the list of switches of a route at any time.

In order for hosts to reach the Framework, it configures routes for itself on each switch so that they forward using the control queue. Furthermore, FLOWPRI-SDN configures its OpenFlow switches with two flow tables: table 0 for classification rules and Table 1 for forwarding rules. The classification rules are created only on the closest switches to the traffic issuing edge to remark flow packages with their respective DSCP and send them to the Table 1. The default rule for Table 1 is to send the package to FLOWPRI-SDN as a "packet_in" event, in case of there is no other matching rule in these tables.

3.1.2   G-BAM

The G-BAM bandwidth allocation model is implemented to find space in one of the classes (real-time or non-real-time) to reserve bandwidth for a flow. It aims to allocate the traffic flow in the priority queue of the class defined in the QoS contract for that flow, if there is enough bandwidth available in the link of the output port for that class (checked in the port class) and then, the actions are created to install the new flow. If there is not enough bandwidth, it checks to see if there are enough flows borrowing bandwidth in the class by checking all priority lists. If they represent enough bandwidth, actions are created to remove these rules and install the new flow.

If there are no borrowing flows in the original class of the QoS contract, it checks whether there is available bandwidth in the other class with resource reservation (real-time or non-real-time). If so, the action to install the new flow is created, tagging that is a flow that borrows bandwidth. However, if there is not enough bandwidth in that class, it goes back to the original class and tries to find flows that have a lower priority than the flow trying the G-BAM routine. Then, if there are lower priority flows that are using enough bandwidth for the new flow, create the actions to remove the lower priority flows to free up enough space and create the action to install the new flow. If there is not enough bandwidth to allocate to the new flow, it rejects the new flow.

After all switches in the route accept the new flow and get their actions from the G-BAM routine, the actions are executed. If any switch rejects the flow, all actions are also rejected. This way, when the actions are executed, first the meter rule is created in each switch, with a unique identification and the forwarding rate determined by the QoS contract. Then, the classification rule is created in the OpenFlow switch closest to the issuing edge, to mark the DSCP field of the flow packets with the corresponding code and send the packets to the forwarding table. In the forwarding table of each switch, the flow rule matching the IP addresses and the DSCP code of the packets is created to apply the previously created meter rule and forward to the corresponding queue through the correct output port.

Furthermore, a copy of the forwarding rule is stored in the list of priority rules corresponding to the class whose bandwidth is being used. In this way, the amount of bandwidth available to the class is deducted from the bandwidth used for the new flow. If the flow is borrowing bandwidth, it will be stored in the queue of the other class (real-time or non-real-time) with the same priority. In case of any delete rule action, the forwarding rule is removed in the OpenFlow switch and in the list of rules of the class, increasing the available bandwidth for the class of service.

3.1.3   Client/Server for QoS Contracts

Data senders must state their flow QoS requirements if they intend to receive them. The Framework has two TCP sockets for receiving QoS contracts from data senders, one listening for connections on port 4444 for domain hosts and another on port 8888 for controllers in other domains. When a connection is established, a contract is received. With a contract, it checks if there is already a corresponding contract for the same IP addresses, if so, it obtains the route switches between these IP addresses and removes the rules referring to that contract in the stored rules lists and in the switches' OpenFlow tables. If there is no corresponding contract already stored, it goes straight to G-BAM to allocate and create the rules. This closes the routine of the controller's contract server, the difference is that in the host version, after creating the rules, this contract is announced to other domains along the route.

3.1.4   Contract Exchange Protocol

The contract exchange protocol allows FLOWPRI-SDN to extend the QoS support beyond its switch domain, by disseminating QoS contracts along the route the flow will take. This protocol involves three stages: the announcement of QoS contracts via ICMP 15, the request of a contract via ICMP 16 and the server receiving contracts in another FLOWPRI-SDN. The first step occurs when a new QoS contract is established or a "packet_in" event identifies a flow with a corresponding contract in some Framework. Then, the Framework announces the contract by generating an ICMP code 15 package, called Information Request

[14]. The packet carries the advertised contract source IP address in its data field, while the destination IP is settled up in the ICMP destination field. The ICMP source field is set to the IP address of the FLOWPRI-SDN that announced the contract.

The next step is to receive the ICMP 15 on another domain. The first switch in a domain that receives this packet, generates the packet_in event to handle it in its FLOWPRI-SDN. The advertised contract source IP address is obtained from the data field of the packet, while the destination IP address is obtained from the destination field of the packet. Next, the FLOWPRI-SDN checks if it already has a contract with the advertised addresses already stored and gets its DSCP code. After that, an ICMP 16 packet, Information Reply, is created [14]. In its source field, the IP address of the controller (which received the ICMP 15) is defined, while in the destination field, the source IP of the ICMP 15 packet is defined. In the data field, the DSCP found is settled (or nothing if not found), the source IP address and destination of the requested QoS contract (taken from ICMP 15). Then, the created packet is injected into the switch closest to the sender ICMP 15, while the received ICMP 15 is re-injected into the switch closest to the destination ICMP 15.

On the other hand, when some FLOWPRI-SDN receives an ICMP 16 packet from a packet_in event, it has two processing behaviors: 1) the ICMP 16 arrives at the destination controller or 2) it arrives at a controller that is not the final destination. In the first case, FLOWPRI-SDN must send the requested QoS contract. Thus, the IP addresses of the requested QoS contract along with the DSCP code (when defined) are obtained from the ICMP 16 data field. Then, the corresponding QoS contract is fetched and its DSCP code is compared to that of ICMP 16, if they are the same nothing is sent. Otherwise, the route switches between the domain Framework and the ICMP 16 source IP are obtained to create the classification and forwarding rules, which mark the packets with control DSCP to send through the control queue of the correct output port of the switch, the same to the return traffic. After all, the contract is sent to the source IP of the ICMP 16, to the TCP port 8888 of that FLOWPRI-SDN.

If the controller is not the destination of ICMP 16, then it is in a domain that is part of the route and therefore, the contracts must pass through the switches of that domain through the control queue. Thus, the switches of the route between the destination and the source of the ICMP 16 package are obtained for the creation of classification and forwarding rules in their OpenFlow tables, with DSCP control, so that the contracts can travel through the domain. After that, the ICMP 16 packet is injected into the switch closest to the destination edge of its destination.

3.1.5 "Packet_in" Handler

The "packet_in" event corresponds to the main function of FLOWPRI-SDN. Packets that do not match with the switch table rules are sent to the Framework, that handles them according to their type.

For this packets, a match is sought in the stored QoS contracts. If so, the switches that are part of the route between the contract IP addresses are obtained and each one is tested in the G-BAM routine. If all switches accept the flow, the actions generated in the process are executed and the flow rules are created.

However, if the flow has no correspondence with the QoS contracts, then it is a flow of the best-effort service class. In this case, the route switches between the source and destination addresses of the packet are obtained, to create the classification rule in the switch closest to the issuing edge, to mark them with DSCP for best-effort, and the forwarding rule in each switch, to send the packets through the best-effort queue through the corresponding output port.

3.1.6   Flow Rules Management

Every time a forwarding rule is created, it sets an idle_timeout value, in the case of real-time and non-real-time flows, or hard_timeout in other cases. The first determines the time a rule can remain inactive without being removed from the OpenFlow table; hard_timeout, on the other hand, determines the absolute time a rule is active and then removes it. This allows flows that depended on rules to need to go through a new packet_in event. Furthermore, the forwarding rules of the real-time and non-real-time classes have an OpenFlow flag turned on called "send_flow_removed'. This instructs the switch to send a copy of the expiring rule to the FLOWPRI-SDN so that it can synchronize the stored active rules. In this way, FLOWPRI-SDN obtains the DSCP value of the rule and removes the copy stored in the instance of each switch of the route defined in the contract, as well as the associated meter rule, freeing up space for the class.

4   Results and Experiment Analysis

This section presents the experimental tests and results obtained regarding FLOWPRI-SDN evaluation. The experiments were performed in a virtual machine with 4 CPUs and 2 GB of RAM. The operating system used was Ubuntu 20.04.1 LTS, Kernel 5.4.0-42-generic. The experimental scenarios were set up using OpenFlow network simulator Mininet [15] version 2.3.0b2. The OpenFlow controller Ryu version 4.34 was used as well as Open Virtual Switch (OVS) 2.13.1.

In the case of the Smart Cities scenario, there are several devices that send and receive data that are part of the Smart City's application, but there are also devices that do not transmit data from this type of applications. The Fig. 4 presents the Smart City network topology scenario, where a police officer equipped with a video camera is working on an important operation that needs to be monitored and commanded based on the received video by a central command office. The officer is located in the domain of controller 5 and he is able to access the network through the switch of this domain. In this way, the camera sends real-time video to the cloud server in domain 1, where the command center can access and monitor the operation. The camera sends a UDP stream of Full HD video to the cloud. As the network is not used exclusively for the camera, another generic device consumes the bandwidth for best-effort traffic that is not part of the city's services, towards the same cloud server.

In this scenario, the environment is compared being managed with FLOWPRI-SDN vs without it. Switch links have 15 Mb of available bandwidth, split between the service classes: 4.95 Mb reserved for real-time flows, 5.25 Mb reserved for non-real-time flows, 1.05 Mb for control flows and 3.75 Mb for best-effort flows.

In the no-Framework scenario, forwarding rules are pre-configured on the switches so that hosts are able to communicate, without QoS. Whereas, in the FLOWPRI-SDN scenario, the rules are created within the QoS contract establishment. According to the studies of QoS requirements of [13], the camera flows need about to 5 Mbps, and by the context the priority is low (1) for real-time class, that computes the DSCP code 7. In this case, for a scenario that implements FLOWPRI-SDN, the camera creates a QoS contract with the nearest controller ($C5$) that distributes it towards other FLOWPRIs. In this case, G-BAM has to borrow bandwidth from the non-real-time class because the real-time class does not have 5 Mbps of bandwidth, creating the rules in the OVS and storing the rules with the borrowing flag in the borrowing queue list of the borrowing class.

The generic device that simulates the bandwidth consumption for the best-effort generates an UDP flow (DSCP 60). This flow is represented by the communication between an Iperf server on the host that represents the cloud and an Iperf UDP client on the generic device, which tries to use all the link bandwidth.

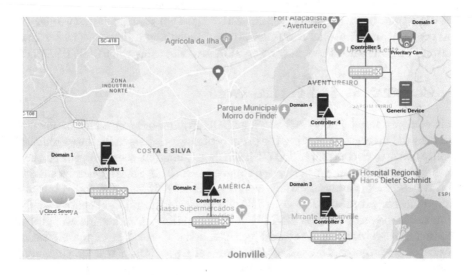

Fig. 4. Smart city scenario.

Figure 5 shows the behavior of priority video streaming traffic in a scenario with intense background traffic (best-effort). In the results, it can be observed that with FLOWPRI-SDN, the priority flow has reserved bandwidth and is not influenced by the best effort flow, remaining constant. However, in the no-Framework case, the priority application is very compromised when the Iperf traffic becomes active. In this scenario, UDP flows start to compete for resources, alternating who can get the expected bandwidth.

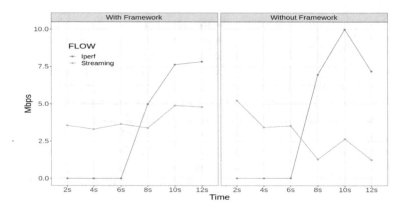

Fig. 5. Priority flow over time

It is important to point out that there is a setup time for the domain controllers to receive the QoS contracts and create the corresponding rules. In this case, the time between the establishment of the QoS contract in controller $C5$ and the creation of rules in controller $C1$ was on average 19ms, that is, for flows shorter than this time, there may be no advantages in using the FLOWPRI-SDN.

5   Conclusions and Future Work

This work proposed FLOWPRI-SDN, an SDN Framework to provide bandwidth in a reserved manner, respecting the priority of flows as defined by the established application's QoS contract. FLOWPRI-SDN is aimed at scenarios where there are flows with QoS and priority requirements, such as Smart Cities scenarios. FLOWPRI-SDN is an application that runs on the SDN Ryu controller, but also communicates with hosts to receive the application's (hosts) QoS contracts. The main mechanisms of FLOWPRI-SDN are the handling of packet_ins events with creation of remarking and forwarding rules using G-BAM resource allocation modeling, and the contract exchange protocol to approach the QoS handling of an end-to-end approach in compatible and non-congested scenarios. Thus, as the Framework's functionality experiment demonstrates, it efficiently manages network resources for Smart City applications. As future work, FLOWPRI-SDN can be extended to use more DSCP codes and support IPv6 flows. In addition, developing application instrumentation methods to automate the establishment of contracts or perform resource discovery efficiently are also future goals.

Acknowledgments. This work received financial support from the Coordination for the Improvement of Higher Education Personnel - CAPES - Brazil (PROAP/AUXPE) 0093/2021.

References

1. Guleria, A.: Traffic engineering in software defined networks: a survey. J. Telecommun. Inform. Technol. **4**, 3–14 (2016)
2. Costa, D.G., de Oliveira, F.P.: A prioritization approach for optimization of multiple concurrent sensing applications in smart cities. Future Gen. Comput. Syst. **108**, 228–243 (2020)
3. Ndiaye, M., Hancke, G.P., Abu-Mahfouz, A.M.: Software defined networking for improved wireless sensor network management: a survey. Sensors **17**(5), 1031 (2017)
4. Tank, G.P., Dixit, A., Vellanki, A., Annapurna, D. Software-Defined Networking-The New Norm for Networks (2012)
5. Yan, J., Zhang, H., Shuai, Q., Liu, B., Guo, X.: Hiqos: An sdn-based multipath qos solution. China Commun. **12**(5) 123–133 (2015)
6. Tomovic, S., Prasad, N., Radusinovic, I.: SDN control framework for QoS provisioning. In: 2014 22nd Telecommunications Forum Telfor (TELFOR). (2014)
7. Leonardi, L., Lo Bello, L., Aglianó. S.: Priority-based bandwidth management in virtualized software-defined networks. Electronics **9**(6), 1009 (2020)
8. Aglianó. S., Ashjaei, M., Behnam, M., Lo Bello, L.: Resource management and control in virtualized SDN networks. In: 2018 Real-Time and Embedded Systems and Technologies (RTEST). IEEE, 2018. pp. 47–53 (2018)
9. Seddiki, M. S., Shahbaz, M., Donovan, S., Grover, S., Park, M., Feamster, N., Song, Y.: Flowqos: Per-flow quality of service for broadband access networks. Georgia Institute of Technology (2015)
10. QoS Design Principles and Best Practices. Cisco Press (2018). https://www.ciscopress.com/articles/article.asp?p=2756478
11. Reale, R.F., Bezerra, R.M.D.S., Martins, J.S.: G-BAM: A Generalized Bandwidth Allocation Model for IP/MPLS/DS-TE Networks. arXiv preprint arXiv:1806.07292 (2018)
12. Ren, S.: A service curve of hierarchical token bucket queue discipline on software defined networks based on deterministic network calculus: An analysis and simulation. J. Adv. Comput. Netw. **5**(1), 8–12 (2017)
13. UNION, I.T. Itu-t g. 1010: End-user multimedia qos categories. G SERIES: Transmission Systems and Media, Digital System and Networks-Multimedia Quality of Service and Performance Generic and User-Related Aspects (2001)
14. POSTEL, Jon. Internet control message protocol (1981)
15. Bholebawa, I.Z., Dalal, U.D.: Design and performance analysis of openflow-enabled network topologies using mininet. Int. J. Comput. Commun. Eng. **5**, 419–429 (2016)

Unsupervised Learning via Graph Convolutional Network for Stock Trend Prediction

Mingxuan Sun, Rongbo Chen, Jianfei Zhang, and Shengrui Wang[✉]

University of Sherbrooke, Sherbrooke, Canada
{Mingxuan.Sun,Rongbo.Chen,Jianfei.Zhang,Shengrui.Wang}@USherbrooke.ca

Abstract. Stock trend prediction has received a significant amount of attention in recent years. Existing methods could not exploit the peculiar trends for prediction, which are valuable in rising-falling trend analysis for short-term or long-term investments. In this paper, we propose an integrated model that can discover peculiar trend patterns for stock trend prediction. Our proposed model is mainly divided into two parts: the clustering and prediction processes. In the clustering process, we use a Graph Convolutional Network (GCN) model to explore the trend patterns groups from a set of subsequences of time series. In the prediction process, an Long Short-term Memory (LSTM) model will be trained based on the discovered patterns for predicting future stock trends. Experimental results on real-world financial datasets demonstrate that our model yields better performance in terms of stock trend prediction, and outperforms state-of-the-art forecasting models for long-term investment.

1 Introduction

The stock market has always attracted the attention of many investors and institutions due to its high returns and high risk. However, the stock price fluctuates greatly over time, and the trend of price change truly and instantly reflect the stock value. The significant price fluctuation indicates the reaction to emerging important stock-related information, usually in the form of events. Investors also aspire to predict the price movement accurately and get the desired benefits. In order to increase the profit and reduce the risk in investment, stock trend prediction has been investigated to forecast stock price tendency in the long run [12]. On the other hand, the stock market is inherently complex, dynamic and chaotic due to the joint actions of many factors such as the political and economic environment, company conditions, supply and demand, investor psychology, etc., which makes the prediction of stock market price movement challenging.

There have been a considerable number of methods reported for stock forecast, e.g., autoregressive model, Support Vector Machine (SVM) [1,16], LSTM [26], and Convolutional Neural Network (CNN) [23]. Although these models may yield accurate forecasts, they usually oversimplify the complexity of the financial market in one or more assumptions:

© The Author(s), under exclusive license to Springer Nature Switzerland AG 2023
L. Barolli (Ed.): AINA 2023, LNNS 661, pp. 358–369, 2023.
https://doi.org/10.1007/978-3-031-29056-5_32

- dichotomous stock trend: existing methods have focused exclusively on either rising or falling stock trends in an individual stock;
- uniform distribution of trends: existing methods are not effective in dealing with dynamic stock trends, e.g., a stock price may cease rising and start dropping at some turning points; this is important for investors;
- independence between various stocks: existing methods consider each stock as an isolated entity and ignore the relationship/interactions between various companies' stocks during different periods

To address the above issues, in this paper, we propose a novel GCN-LSTM method that integrates GCN [30] and LSTM for stock trend prediction. GCN-LSTM leverages a GCN to learn trend patterns in an unsupervised way and an LSTM model to forecast the stock trends based on the GCN clustering results. We transform time series into a set of segments first and then cluster these segments according to the DTW [9] distance between them. By such clustering process, GCN-LSTM can discover various underlying trend patterns at each time in time series. Then, we label and serialize the corresponding patterns of the original sequences according to the time series partition. The experiments on real-world stock data demonstrate that our method can yield accurate forecasts. The main contributions of this paper are as follows:

- We develop a GCN clustering method to render it suitable to discover trend patterns from the time series segments;
- Our method is capable of dealing with cross-company, cross-period, and unbalanced data;
- Experimental results shows that our method can earn great financial benefit on real-world data by our trend forecasts.

2   Related Work

Stock Trend Prediction. Traditional stock prediction models are generally based on sequence modeling with sequence data input. The authors of [14,28] leveraged Markov chain Monte Carlo to carry out simulation smoothing and Bayesian posterior analysis of parameters, and on importance sampling to estimate the likelihood function for classical inference and provide a likelihood analysis of an extension of the usual Gaussian state space form. [1,16] developed a hybrid feature selection to predict the trend of stock markets, and [11] proposed an end-to-end hybrid neural network for forecast. Although some traditional methods can achieve good results in time series prediction, the principal structure of these methods leads to a bottleneck. The potential information of time series data can not be learned in completed by these previous methods.

In recent research, researchers like to adopt machine learning methods for their development model. For example, Recurrent Neural Networks (RNNs) [20] are powerful in discovering the dependency of sequence data, and LSTM has been used in [5] to predict the price movement for 300 CSI constituent stocks. In [19], the restricted Boltzmann machine (RBM) was utilized for short-term stock market trend prediction. The [27]'s researchers proposed a novel approach that converts 1-D financial time series into a 2-D image-like data representation in order to be able to utilize the power of the deep convolutional neural network for an algorithmic trading system. These machine learning methods with time-series values input are adapted to the relatively stable stock market, but they rarely can react to the various changes of the stock market.

Graph Convolutional Neural Network (GCN). A family of linkage-based clustering methods makes no assumption on data distribution and achieves higher accuracy. In previous research, GCN is used in many fields for prediction. GCN, which generalises well-established CNN to arbitrarily structured graphs, has widely been applied in lots of applications, such as image classification [8], document classification [21], and semi-supervised learning [18]. In many machine learning problems, the input can be organized as graphs. We could organize the series set into a graph. The data in each series will be regarded as the nodes, and the link between the nodes will be regarded as the relationship between the series.

According to the definition of convolution on graph data, GCNs could be categorized into spectral methods and spatial methods. Spectral-based GCNs [4, 7,15] generalize convolution based on Graph Fourier Transform, while in spatial-based GCNs [10,29], graph nodes and their neighbors play an important role on the process of convolution. In spatial-based methods [2,17], the authors have used the Gaussian distance adjacency matrix, multi-graph representation, and the adaptive adjacency matrix to construct the adjacency matrix required by the graph convolution. These methods generally make all the data in a single cluster, because they converge the data feature of graph notes into the same value. In this work, we propose a spatial-based GCN method to deal with the series clustering issues. The designed GCN performs graph node clustering in the inductive setting.

3   Proposed Method

Problem Statement. Assume that we have the stock sequence segments of a collection of stock time series $X = [x_1, ...x_N]^T \in R_{N*D}$ where N is the number of sequence segments and D the length a segment. The goal of sequence segments clustering is to assign a pseudo label y_i to each $i \in [1, 2, ..N]$ so that instances with the same pseudo label form a cluster. To this end, we adopt the link-based clustering method, which aims at predicting the likelihood of linkage between pairs of objects. The objects linked to each other form clusters.

Motivation. This work focuses on the efficiency and accuracy of stock sequence segments clustering for mining their trends. We will create the linkage likelihood between two objects if they satisfy two conditions at the same time: (a) one belongs to the k nearest neighbors of the anther; (b) the similarity of them is beyond a cap α. With different k and α, we directly connect each pair of objects when they satisfy conditions (a) and (b). The clustering performance is quite dependent on various values of k and α. This indicates the potential effectiveness of predicting linkages between objects, which satisfy conditions (a) and (b), rather than among all potential pairs. The advantage of adopting such a strategy is that we could obtain reasonably high clustering accuracy while the system has the benefit of being efficient.

Framework. Figure 1 presents the framework of our data pre-processing, GCN learning (i.e., adjacency matrix building), GCN clustering, and LSTM forecast. This work focuses on the efficiency and accuracy of stock sequence segments clustering for mining their trends. Our aim is to mine the natural patterns in these serial data to provide the basic information for predicting future stock movements.

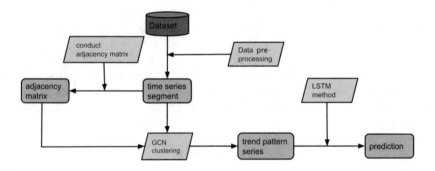

Fig. 1. Overview of the proposed method

3.1   Building GCN Adjacent Matrix

Linkage-based clustering methods (i.e., GCN) can achieve higher accuracy without making assumptions about data distribution. Due to the complexity of the stock market, it is more favorable to use the GCN method since it can capture features of the stock market by studying data that has a topological structure.

Traditional Gaussian kernel function graphs could result in the problem of unbalanced clusters. If the weighted graph formed by the primitive Gaussian kernel function graph construction algorithm is directly used to establish the connection matrix, it will lead to too many links between sequence segments, so that the final clustering results tend to form large clusters, and erode the smaller clusters.

We improve the traditional gaussian kernel composition method by pruning the distance matrix. The similarity matrix calculated by pairwise distance between time series segments is set as D, where $D_{i,j}$ is the DTW distance between the current segment of time series i and j. Our method employs two threshold parameters whose values are set via an empirical model selection. The threshold α is used as a cap on $D_{i,j}$ beyond which no link is allowed between the segment of time series i and j. We also employ an upper limit γ for the maximal number of links to a vertex. Let $N_i(k)$ represent the k nearest vertices (neighbors) of vertex i. The adjacency matrix $A = (A_{i,j})$ of the graph is defined as follows :

$$A_{ij} = \{ \begin{matrix} 1, & D_{ij} \leq \alpha, j \in \mathcal{N}_i(\gamma) \\ 0, & otherwise \end{matrix} \tag{1}$$

To determine the value α and $\mathcal{N}_i(\gamma)$, we use the S-Dbw index [25] to measure the validity of choices of values for α and γ on the clustering results. S-Dbw is an index that can measure the clustering validity of a clustering algorithm. S-Dbw can fully reflect the degree of separation and the compactness of the clusters. The smaller the S-Dbw index value, the better the clustering results. We should therefore choose the combination of parameter values with the smallest S-Dbw.

3.2   GCN Clustering

The context introduced the Adjacent Matrix for GCN will play a important rule in the process of clustering. In this section, we are going to introduce how to use the Adjacent Matrix on GCN.

The essence of GCN is multiple layers of neural networks. By iterating GCN, the adaptive system can be trained. A graph convolution layer takes as input a node feature matrix X together with an adjacency matrix A and outputs a transformed node feature matrix Y. In this paper, GCN obeys the following formula :

$$Y = \sigma(GXW^l) \tag{2}$$

where $X \in R^{N*d_{in}}$, $Y \in R^{N*d_{out}}$ N is the number of nodes, and d_{in}, d_{out} are the dimension of input / output node features. G is an aggregation matrix of size $N * N$ and each row is summed up to 1. W^l is the weight matrix in the lth layer in the graph convolution layer. $\sigma(\cdot)$ is the non-linear activation function.

The graph convolution operation can be separated into two steps. In the first step, by left multiplying X by G, the underlying information of nodes' neighbors is aggregated. Then, the input node features X are concatenated with the aggregated information GX along the feature dimension. In the second step, the concatenated features are transformed by a set of linear filters, whose parameter W is to be learned. In our method, we used the Mean Aggregation which is shown as follows.

$$G = \widetilde{D}^{-1/2} \widetilde{A} \widetilde{D}^{-1/2} \tag{3}$$

where $\widetilde{A} = A + I_N$, A is the adjacency matrix of the graph, I_N is identity matrix of graph; $\widetilde{D}_{i,i} = \sum_j \widetilde{A}_{i,j}$, $\widetilde{D}$ is the degree matrix of the adjacency matrix; The mean aggregation could performs average pooling among neighbors.

The GCN used in this paper is the stack of multi-graph convolution layers activated by the ReLU function. Then we use the information entropy as the objective function for optimization. In addition, ADAM [22] method will be used to optimize the GCN algorithm. The final output Y denotes the label prediction for all data in which each row X_i denotes the label prediction for the ith node. The optimal weight matrices trained by minimizing the information entropy object function as:

$$L = -\sum\sum Y_{ij}logY_{ij} \qquad (4)$$

where Y_{ij} represents the label information of the data. Here we use the information entropy as object function, it's because in information entropy doesn't need the label data to supervise the clustering process.

The process of the GCN clustering is shown as algorithm 1:

Algorithm 1: GCN clustering

Input: Time-series segment set X, Adjacent matrix A, degree matrix D
Output: The probability clusters of each segment X_i belonging to Y
begin

    Initialize GCN weight matrix:W;
    $\widetilde{D} = D + I$;
    $\widetilde{A} = A + I$;
    /* Matrix I is identity matrix   */
    while $n < epoch$ **do**
        $n = n + 1$;
        $Y = \sigma(\widetilde{D}^{-1/2}\widetilde{A}\widetilde{D}^{-1/2}XW^{(l)})$;
        /* Y saves all of probability of clusters that each segment
           X_i belonging to   */
        W is updated by ADAM;

3.3   LSTM for Prediction

By the GCN clustering on the time series segments, we obtain multiple sequences of stock patterns. Next, we will use these patterns to train an LSTM model for predicting the patterns of stock trends in the future.

Assuming that the time series segment result of GCN clustering is X=[x_1, x_2, x_3, ..., x_N], we use the method mentioned in [6] to train an LSTM model. Here is an example:

$$(x_{n-k}..., x_{n-2}, x_{n-1}) \rightarrow x_n \qquad (5)$$

where n represents the ordinal of the training examples, and k represents the length of the input training samples. In the formula, $(x_{n-k}, ..., x_{n-2}, x_{n-1})$ is a training sample, while x_n is the label of that sample. The state data will be transformed into a one-hot vector. This can resolve the problem of the classifier being unable to deal with attribute data, and expand features to a certain extent.

The classification prediction results indicate how the state of a time series segment changes in the short-term future: specifically, the probability that the next sequence will be assigned to each state. In LSTM, we use the softmax activation function [13] and the cross entropy loss function. We use Adam to update the gradient in the prediction part.

4   Experiments

4.1   Data

In order to verify the validity of our proposed stock trend pattern clustering method, we selected 150 stocks from the Yahoo Finance website which fall into five categories (time period 2017–2020 for all stocks): bank stocks, biotech stocks, medical stocks, oil and gas stocks, and semiconductor stocks. In the financial research domain, it is customary to process stock data using the closing price. The factors that can affect stock trends are very complex, including unexpected factors such as the global financial crisis, hence, we only considered stock trends under a relatively stable external environment.

4.2   Stock Segments Clustering Analysis and Evaluation

To search for the optimal number of clusters and evaluate the clustering performance of GCN, we measure the effect of clustering via the S-Dbw index. We did experiments with various numbers of clusters, from 2 to 6. We also use k-means, DBSCAN, and k-Shape algorithms to make forecasts and compare these forecasts to the GCN method. These algorithms have been widely used in time series segment research [3,24]. The k-means algorithm relies on the data distance, while DBSCAN relies on the data density and distribution and k-Shape on the shape of time series segments.

Table 1. The S-Dbw index value in different methods and number of clustering

Number of clusters	GCN	K-means	DBSCAN	k-shape
2	2.7636	24.2409	22.3032	2.3381
3	1.9523	16.5827	8.6872	2.0443
4	**1.3017**	11.4227	7.3704	1.9971
5	3.6308	24.0388	9.1742	2.3458
6	4.1158	38.3207	10.2397	3.5239

The result is preformed in the Table 1. According the change in S-Dbw index, we finally determined to have 4 clusters which achieves the lowest S-Dbw. As shown in Table 1, the GCN algorithm is obviously superior to k-means and DBSCAN, because in this project, with the help of an appropriate adjacency matrix, the GCN algorithm takes into account the particularities of time series segment data and makes appropriate adjustments in the definition of similarity between time series segments. Rather than considering only whether they are close in terms of physical distance, more consideration is given to whether the shape is similar. k-Shape also considers whether the shapes of time series segments are similar in the clustering process. However, the problem of data size balance is not taken into account, so the performance of GCN in the clustering process is better than that of k-Shape.

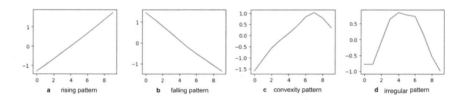

a rising pattern   b falling pattern   c convexity pattern   d irregular pattern

Fig. 2. Average value of four types of stock trend patterns

Each subfigure in Fig. 2 shows a learned trend pattern from a cluster – i.e., how the daily average normalized closing price (y-axis) changes in 10 d (x-axis). The four patterns here are: (a) the rising trend pattern, (b) the falling pattern, (c) the rising-falling convexity pattern, and (d) mixture patter that the trend rises first and then falls. Note that, for the mixture pattern, the curve hardly changes, because compared with the degree of change for the other three clusters, the ordinate range of the fourth is obviously smaller. Therefore, this curve in fact represents an irregular pattern.

4.3   Analysis of the Prediction Results

To observe the performance of our proposed method, we compared it with classic methods and more recent ones, including LSTM, CNN-TA, SVM, and MLP. It is not clear for us whether there is a consensus on the existence of a state-of-the-art in the area. The LSTM method in our comparative experiment is that reported in [31], and the CNN-TA method is proposed by [27]. With the exception of CNN-TA, all of the comparison methods share the same strategy for stock series labeling. In these methods, series data that has a higher price at the end of the series than at the beginning is labeled as a rising pattern and the other series as a falling pattern. The labeling strategy in CNN-TA is referred to [27].

We used Bank, Biotech, Medical, Oil and Gas, and Semiconductor stocks from Yahoo Finance as data for training models. We trained the models and observed the accuracy of the rising and falling patterns. We attempt to use the historical data of the stock to predict its future target.

Table 2. Accuracy of the stock pattern prediction in single stock training

Stock categories	GCN-LSTM	LSTM	CNN-TA	SVM	MLP
banks	**0.7483**	0.7012	0.5469	0.6766	0.6127
biotech	**0.8048**	0.6693	0.5592	0.6153	0.5765
medical	**0.7310**	0.7133	0.5631	0.6767	0.6171
Oil and Gas	**0.6809**	0.6601	0.5448	0.5795	0.5695
semiconductor	0.6819	**0.7523**	0.5549	0.6844	0.6356

Table 2 shows the single-stock forecast accuracy. It can be seen that the proposed GCN-LSTM method outperforms on the training sets. It is worth noting that GCN-LSTM did better despite the fact that it is harder for a multi-trend detection method to increase its accuracy than for a binary trend detection method. The other methods only need to detect two types of trends, whereas GCN-LSTM needs to detect four.

4.4   Financial Evaluation

To further demonstrate the usefulness of the proposed method, we perform the financial evaluation. In this evaluation, we simulated stock trading based on some common stock trading strategies while taking advantage of trend predictions by different methods. The initial capital for financial evaluation is 50,000 dollars.

We adopt a method presented in the work [6] to simulate the process of stock trading. Noticing that we predict the four types of pre-generated trend patterns, i.e. the rising pattern, the falling pattern, the convexity pattern, and the irregular pattern. If the stock trend is predicted as a rising pattern, the stock will be bought at the closing price of that day with all of the currently available capital. If the stock trend is predicted as a falling pattern, the stock will be sold out at the closing price of that day. If the stock trend is predicted as a convexity pattern, half of the stock will be sold at the closing price of that day. If the stock trend is predicted as an irregular pattern, no action will be taken on that day.

Moreover, we make use of two commonly used stock trading strategies as a benchmark to compare with the proposed method. Similar to the paper [6], we chose the Simple Moving Average signal (SMA) strategy as one of the comparison methods. In this momentum strategy, it is hypothesized that the use of the immediate past period predicts the future. To verify the effectiveness of our method, we also added a no action (NA) strategy. Just buy and hold the stocks until the experiment end. This passive strategy provides an idea of overall changes in the profit rate over the testing period. We calculate the profit margin of all of the stocks in 52 weeks under each strategy. For illustration purposes, we calculate the average profit margin over each category of stocks, and the results are shown in Fig. 3.

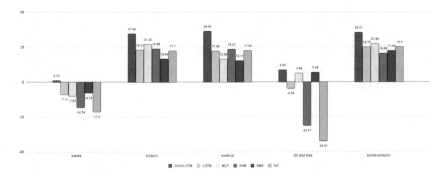

Fig. 3. The profit rate (%) of stocks in each method

From Fig. 3, we can find out that the GCN-LSTM method obviously outperforms the other methods in terms of profit rate. Due to the page limit, we show
here only one profit situation of stock in Figure 4. We choose the stock which
is coded "NG", and observe the profit situation from Nov 2019 to Nov 2020
by each strategy. Further investigation into the investment actions reveals that
the GCN-LSTM method allows adopting a more conservative investment strategy compared to other methods due to our accurate stock trend prediction. This
helps the investors avoid investment risks and explains why the proposed method
is more effective than the others in achieving stable and consistent returns.

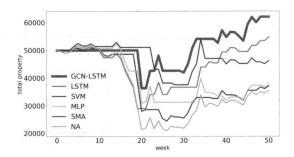

Fig. 4. Profits of stock code "ING" from Nov 2019 to Nov 2020

5   Conclusions

This paper proposes a stock trend prediction method based on the combination
of GCN and LSTM to detect some peculiar trends and predict stock trends in the
short-term. By making use of advanced techniques for pre-processing and comparing time series, our method can discover the trend types of stock sequence
segments through a clustering approach. The novelty of the proposed method

resides in the combination of these techniques to improve the performance of GCN for trend discovery and performance of LSTM for trend prediction. Experimental results show that our method can effectively discover and predict trends while dealing with the unbalanced data in clustering, and can help investment decision-making while avoiding risks.

References

1. Abonyi, J., Feil, B., Nemeth, S., Arva, P.: Modified gath-geva clustering for fuzzy segmentation of multivariate time-series. Fuzzy Sets Syst. **149**, 39–56 (2005)
2. Alfke, D., Stoll, M.: Semi-supervised classification on non-sparse graphs using low-rank graph convolutional networks. arXiv preprint **1905**, 10224 (2019)
3. Birant, D., Kut, A.: ST-DBSCAN: an algorithm for clustering spatial-temporal data. Data Knowl. Eng. **60**(1), 208–221 (2007)
4. Bruna, J., Zaremba, W., Szlam, A., LeCun, Y.: Spectral networks and locally connected networks on graphs. arXiv preprint arXiv:1312.6203 (2013)
5. Chen, M.Y., Liao, C.H., Hsieh, R.P.: Modeling public mood and emotion: stock market trend prediction with anticipatory computing approach. computers in human behavior. Comput. Human Behav. **101**, 402–408 (2019)
6. Chen, W., Jiang, M., Zhang, W.G., Chen, Z.: A novel graph convolutional feature based convolutional neural network for stock trend prediction. Inform. Sci. **556**, 67–94 (2021)
7. Defferrard, M., Bresson, X., Vandergheynst, P.: Convolutional neural networks on graphs with fast localized spectral filtering. NIPS **29** (2016)
8. Fu, S., Liu, W., Tao, D., Zhou, Y., Nie, L.: Hesgcn: Hessian graph convolutional networks for semi-supervised classification. Inform. Sci. **514**, 484–498 (2020)
9. Guan, X., Huang, C., Liu, G., Meng, X., Liu, Q.: Mapping rice cropping systems in vietnam using an NDVI-based time-series similarity measurement based on DTW distance. Remote Sensing **8**, 19 (2016)
10. Hamilton, W., Ying, Z., Leskovec, J.: Inductive representation learning on large graphs. In: Advances in Neural Information Processing Systems **30** (2017)
11. Hao, Y., Gao, Q.: Predicting the trend of stock market index using the hybrid neural network based on multiple time scale feature learning. Appl. Sci. **10**, 3961 (2020)
12. Idrees, S.M., Alam, M.A., Agarwal, P.: A prediction approach for stock market volatility based on time series data. IEEE Access **7**, 17287–17298 (2019)
13. Jang, E., Gu, S., Poole, B.: Categorical reparameterization with gumbel-softmax. arXiv preprint arXiv:1611.01144 (2016)
14. Johansen, A., Sornette, D.: Evaluation of the quantitative prediction of a trend reversal on the japanese stock market in 1999. Int. J. Modern Phys. (C) **11**, 359–364 (2000)
15. Kipf, T.N., Welling, M.: Semi-supervised classification with graph convolutional networks. arXiv preprint arXiv:1609.02907 (2016)
16. Lee, M.C.: Using support vector machine with a hybrid feature selection method to the stock trend prediction. Expert Syst. Appl. **36**, 10896–10904 (2009)
17. Li, R., Wang, S., Zhu, F., Huang, J.: Adaptive graph convolutional neural networks. In: AAAI. vol. 32 (2018)
18. Li, S., Li, W.T., Wang, W.: Co-gcn for multi-view semi-supervised learning. In: Proceedings of the AAAI Conference on Artificial Intelligence. vol. 34, pp. 4691–4698 (2020)

19. Liang, Q., Rong, W., Zhang, J., Liu, J., Xiong, Z.: Restricted boltzmann machine based stock market trend prediction. In: IJCNN, pp. 1380–1387 (2017)
20. Lin, T., Guo, T., Aberer, K.: Hybrid neural networks for learning the trend in time series. In: Proceedings of the Twenty-Sixth International Joint Conference on Artificial Intelligence, pp. 2273–2279. No. CONF (2017)
21. Lin, Y., Meng, Y., Sun, X., Han, Q., Kuang, K., Li, J., Wu, F.: Bertgcn: Transductive text classification by combining gcn and bert. arXiv preprint arXiv:2105.05727 (2021)
22. Liu, Z., Shen, Z., Li, S., Helwegen, K., Huang, D., Cheng, K.T.: How do adam and training strategies help bnns optimization. In: International Conference on Machine Learning, pp. 6936–6946. PMLR (2021)
23. Livieris, I.E., Pintelas, E., Pintelas, P.: A CNN-LSTM model for gold price time-series forecasting. Neural Comput. Appl. **32**, 17351–17360 (2020)
24. Paparrizos, J., Gravano, L.: k-shape: Efficient and accurate clustering of time series. In: SIGMOD, pp. 1855–1870 (2015)
25. Rini, D., Novianti, P., Fransiska, H.: Internal cluster validation on earthquake data in the province of bengkulu. In: IOP Conference Series: Materials Science and Engineering, vol. 335, p. 012048. IOP Publishing (2018)
26. Roondiwala, M., Patel, H., Varma, S.: Predicting stock prices using LSTM. IJSR **6**, 1754–1756 (2017)
27. Sezer, O.B., Ozbayoglu, A.M.: Algorithmic financial trading with deep convolutional neural networks: Time series to image conversion approach. Appl. Soft Comput. **70**, 525–538 (2018)
28. Shephard, N., Pitt, M.K.: Likelihood analysis of non-gaussian measurement time series. Biometrika **84**, 653–667 (1997)
29. Veličković, P., Cucurull, G., Casanova, A., Romero, A., Lio, P., Bengio, Y.: Graph attention networks. arXiv preprint arXiv:1710.10903 (2017)
30. Wang, Z., Zheng, L., Li, Y., Wang, S.: Linkage based face clustering via graph convolution network. In: CVPR, pp. 1117–1125 (2019)
31. Yao, S., Luo, L., Peng, H.: High-frequency stock trend forecast using LSTM model. In: ICCSE, pp. 1–4 (2018)

A Novel Clustering Model TEC for Station Classification

Lei Yuan[1], Lijuan Liu[1,2(✉)], Shunzhi Zhu[1,2], and Long Chen[1]

[1] College of Computer and Information Engineering, Xiamen University of Technology,
Xiamen 361024, China
`2022031502@s.xmut.edu.cn, {ljliu,szzhu}@xmut.edu.cn`
[2] Fujian Key Laboratory of Pattern Recognition and Image Understanding, Xiamen 361024,
China

Abstract. Station classification plays an important role in urban intelligent transportation systems. Recently, most of studies usually choose different features to analyze the function of stations. Moreover, they usually choose K-means as the clustering method, which may lead to unstable clustering results. Therefore, a novel clustering model (Transformer Encoder Clustering, TEC) has been proposed for station classification by only using passenger flow data. Firstly, transformer encoder has been applied to deeply extract the features from original passenger flow. Then, an improved Canopy + has been proposed to determine the initial centers for K-means. Finally, a novel evaluation metric (NR) has been introduced to verify the stability and volatility of station classification results. Experiments are conducted on Xiamen bus rapid transit ridership dataset. The results show that the 44 stations have been classified into three categories, and the NRs in all stations are 0.75, which verifies the stability of results for station classification.

Keywords: Deep feature extraction · Transformer encoder · Improved Canopy + · K-means · Station classification

1 Introduction

Station classification in urban metro is one of the most fundamental and crucial research, which is very helpful for many downstream tasks in enhancing urban transportation systems. Tang et al. [1] proposed a new train operation scheme in suburban urban rail transport lines, which based on station classification. Li et al. [2] proposed a flex-route transit design based on station classification in a bus network.

Furthermore, it is well known that the station classification can be affected by many different features. And these features can be used not only for station classification, but also for other applications in intelligent transportation. Wu et al. [3] proposed a recommendation method of limited-stop bus stops based on multicriteria collaboration which comprehensively considers the passenger volume at bus stops, potential passenger flow and the pivotal role of bus stops based on passenger flow data. Hu et al. [4] employed a hybrid combination of the multilayer dilated convolution, long short-term memory

L. Barolli (Ed.): AINA 2023, LNNS 661, pp. 370–379, 2023.
https://doi.org/10.1007/978-3-031-29056-5_33

network (LSTM) and attention mechanism, to extract the spatio-temporal features of the traffic data for traffic monitoring stations classification, which provided an important reference for the planning and construction of urban transportation infrastructure. Du et al. [5] proposed a SOM neural network and fuzzy classification algorithm for urban rail station classification by selecting the features of passenger flow, transportation capacity and topological location of stations, which was significant for the congestion propagation under larger passenger flow.

With the aim to different applications, the commonly used features for station classification can be divided into four categories: (a) passenger flow [6]; (b) Point of Interest (POI) [7]; (c) land use, location and surrounding environment [8]; (d) mixed features [9]. Most of the previous studies usually chose different features, which more preferred to analyze the function of stations, instead of passenger flow prediction. Moreover, they usually choose K-means as the clustering method, which might lead to unstable clustering results due to its inherent deficiency, such as setting initial center randomly, and assigning the value K artificially. In addition, visualizing the passenger flow for the same type of stations is the commonly used way to evaluate the classification results, which is not objective enough. Therefore, a novel clustering model (Transformer Encoder Clustering, TEC) has been proposed to make a more stable clustering result for station classification by only using passenger flow data, which may be more significant for multiple station passenger flow prediction with the similar flow in the same type of stations. In summary, the main contributions are summarized as following:

(1) Transformer encoder has been applied to effectively extract the deeper features from pure passenger flow data to obtain more important information.
(2) An improved Canopy + has been proposed to determine the value of K and the initial centers, which can effectively solve the limitations of uncertain value of K and unstable clustering results in K-means.
(3) A novel evaluation metric has been introduced to quantitatively verify the stability and volatility of station classification results.

2 Methodology

2.1 Input Feature Description

$P_{FeatureData}$ Represents the original input feature used in this paper, P_{Day} represents the daily passenger flow, and P_{Hour} indicates the hourly passenger flow. All of them considered the inbound flow and outbound flow, respectively, which are shown in Eqs. (1)–(3). P_{Day} and P_{Hour} are extracted as the daily and hourly passenger flow from $P_{featureData}$ according to the column index, respectively.

$$
P_{FeatureData}=
\begin{bmatrix}
p_{1\_1}^{in,day_1\_hour_i} & \cdots & p_{1\_1}^{in,day_m\_hour_n} & p_{1\_1}^{out,day_1\_hour_i} & \cdots & p_{1\_1}^{out,day_m\_hour_n} \\
p_{1\_2}^{in,day_1\_hour_i} & \cdots & p_{1\_2}^{in,day_m\_hour_n} & p_{1\_2}^{out,day_1\_hour_i} & \cdots & p_{1\_2}^{out,day_m\_hour_n} \\
\vdots & \ddots & \vdots & \vdots & \ddots & \vdots \\
p_{1\_w}^{in,day_1\_hour_i} & \cdots & p_{1\_w}^{in,day_m\_hour_n} & p_{1\_w}^{out,day_1\_hour_i} & \cdots & p_{1\_w}^{out,day_m\_hour_n} \\
\vdots & \ddots & \vdots & \vdots & \ddots & \vdots \\
p_{s\ w}^{in,day_1\_hour_i} & \cdots & p_{s\ w}^{in,day_m\_hour_n} & p_{s\ w}^{out,day_1\_hour_i} & \cdots & p_{s\ w}^{out,day_m\_hour_n}
\end{bmatrix}
\tag{1}
$$

where s represents the number of stations, w represents the number of weeks, in represents the inbound flow, out represents the outbound flow, day_1 represents the first day per week, and day_m represents the last day per week. $hour_i$ Represents the started service hour per day, and $hour_n$ represents the ended service hour per day.

$$P_{Day} = \left[p_{day_1} \; p_{day_2} \cdots p_{day_m} \right]^T \tag{2}$$

$$P_{Hour} = \left[p_{hour_i} \; p_{hour_{i+1}} \cdots p_{hour_n} \right]^T \tag{3}$$

where $p_{day_m} \in \mathbb{R}^{(s \times w) \times [2 \times (hour_n - hour_i + 1)]}$, $p_{hour_n} \in \mathbb{R}^{(s \times w) \times [2 \times (day_m - day_1 + 1)]}$.

2.2 Overall Framework

The framework of proposed TEC is shown in Fig. 1. It is divided into two modules: extraction module based on Transformer encoder, and clustering module based on improved Canopy + and K-means.

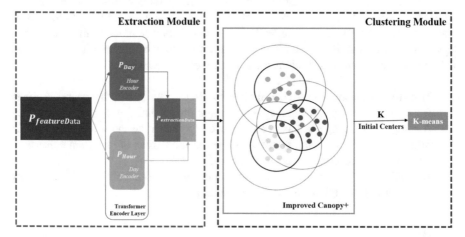

Fig. 1. Framework of proposed TEC.

Firstly, z-score is used to standardize the input features ($P_{FeatureData}$) shown in Eq. (4).

$$Z = \frac{(x_{ij} - \mu_j)}{\sigma_j} \tag{4}$$

where Z is the standardized result; x_{ij} is the actual value of the input feature in low i and column j; μ_j is the mean value of all input feature in column j; σ_j is the standard deviation of all input feature in column j.

Transformer encoder has been used to extract the deep features ($P_{extractionData}$) from P_{Day} and P_{Hour}. Then, $P_{extractionData}$ is used as the inputs in improved Canopy +, which is the pre-clustering module to obtain the two important parameters of K and initial centers. Finally, these two parameters will be applied to K-means to obtain the final clustering results of station classification.

2.3 Extraction Module

This module is mainly based on the encoder in Transformer [10]. There are two different encoders to extract the deep features from P_{Hour} and P_{Day}. First of all, we follow the positional encoding technology in the original Transformer model [10] to realize position encoding of P_{Hour} and P_{Day} in Eqs. (5)–(6).

$$PE_{(pos,2i)} = sin\left(pos/10000^{2i/d_{model}}\right) \tag{5}$$

$$PE_{(pos,2i+1)} = cos\left(pos/10000^{2i/d_{model}}\right) \tag{6}$$

where d_{model} in Eqs. (5)–(6) are set as $2 \times (day_m - day_1 + 1)$ in P_{Hour}, which 2 represents the passenger flow in both directions, and $(day_m - day_1 + 1)$ represents the number of days during a week. d_{model} in Eqs. (5)–(6) are set as $2 \times (hour_n - hour_i + 1)$ in P_{Day}, which 2 represents the passenger flow in both directions, and $(hour_n - hour_i + 1)$ represents the total service hours per day.

Then, we use multi-heads attention and the corresponding encoder to achieve the feature extracting task.

2.4 Clustering Module

2.4.1 Pre-clustering

The detailed steps of improved Canopy + algorithm has been shown in Table 1. As shown in Table 1, Canopy_clusters is the result of pre-clustering, which contains the elements and its relevant center for each cluster. How to determine the values of T1 and T2 is the key problem in the improved Canopy + algorithm. T1 is calculated in line 4 in Table 1, and T2 is set in the following Sect. 3.2.

2.4.2 K-means Clustering Results and Evaluation Metric

The center of each cluster in canopy_clusters shown in Table 1 will be used as the initial centers in K-means.

As shown in Eq. (1), the inbound and outbound passenger flow during more than one week for the same station will be used as the inputs, and different weekly passenger flow in the same station may be quite similar. It means that if the same station with different weekly passenger flow has been clustered as the same cluster, it will be a desirable result. Inspired by this hypothesis, a novel evaluation metric named non-repetition rate (NR) has been proposed in Eq. (7).

$$NR = 1 - \frac{N_i}{w} \tag{7}$$

where N_i represents the number of the clusters for the i th station with different weekly passenger flow, and w has been defined in Eq. (1). The larger the value is, the better the result is.

Table 1. Improved Canopy + algorithm.

```
Algorithm: Improved Canopy+ algorithm.
Input: canopy_clusters=[],
       removal_elements=[],
       canopy_elements=[],
       D_currentDataset=P_extractionData
1.  average_point=Mean(P_extractionData)
       #Calculate the mean value of all samples in P_extractionData
2.  L1=Max(average_point,D_currentDataset)
       #Obtain the maximum distance between average_point and every sample in D_currentDataset
3.  L2=Min(average_point,D_currentDataset)
4.  T1=L1-L2, initial value of T2 is described in Section 3.2
5.  while (D_currentDataset != ∅):
6.      average_point=Mean(D_currentDataset)
           #Calculate the mean value of current dataset
7.      center_index=Get_index(Min(average_point,D_currentDataset)
           #Obtain the index of sample data which is nearest to average_point
8.      removal_elements.append(D_currentDataset[center_index])
9.      delete D_currentDataset[center_index]
           #Delete such center point data
10.     while (D_currentDataset != ∅):
11.         for i=1…len(D_currentDataset) do:
12.             if Dist(D_currentDataset[center_index], D_currentDataset)<T1:
13.                 canopy_elements.append(D_currentDataset[i])
14.             if Dist(D_currentDataset[center_index], D_currentDataset)<T2:
15.                 removal_elements.append(D_currentDataset[i])
16.                 delete D_currentDataset[i]
                       #Delete the sample data from the current dataset
17.             if len(removal_elements)>2:
18.                 canopy_clusters.append(D_currentDataset[center_index],
                    canopy_elements)
19. Return canopy_clusters
Output: canopy_clusters
```

3 Experiment

3.1 Data Description

We use the passenger flow dataset of Xiamen bus rapid transit (BRT) from March 4th to March 31st, 2019, which has totally four weeks. The total number of stations in Xiamen BRT is 44. The data are aggregated into 1-h interval. The daily service hours is from 5:00 to 23:00 for each station, and the total daily service hours are 18.

The parameters of $P_{featureData}$ in Eq. (1) are set as follows: $s \in [1, 44], w \in [1, 4], m \in [1, 7], i = 5, n \in [5, 22]$. For example, $p_{44\_3}^{out,7\_22}$ represents the outbound hourly passenger flow of at the 44th station from 22:00 to 23:00 on Sunday in the third week. It is note that the passenger flow on both workdays and weekends is all considered in the experiment.

The parameters of P_{Day} and P_{Hour} in Eqs. (2)–(3) are set as follows: $m \in [1, 7], i = 5, n \in [5, 22]$. For example, p_{day_2} represents the 18 hourly inbound and outbound passenger flow on every Tuesday during the four weeks for the 44 stations, respectively. p_{hour_8} Represents the 7 hourly inbound and outbound passenger flow in the period of 8:00–9:00 during the four weeks for the 44 stations, respectively.

The number of multi-heads attention is set as 2.

3.2 Pre-clustering

Firstly, we follow the threshold T1 set in [11]. Then, we found that when the threshold T2 set too small, it will lead to too many clustering results. So, we determine the threshold T2 according to the experimental results step by step shown in Table 2. The detailed experimental procedure has been introduced in Table 1, and the experimental results are shown in Table 2.

Table 2. Determination of pre-clustering parameters and K-values.

Items	T1	T2	Removal_rate	Removal_amount	Canopy_amount	Canopies	K
Ideal initial value	11.5	9.8	0.88, 0.79, 0.77, 0.67, 1,1	129, 33, 10, 2, 1, 1	146, 42, 13, 3, 1, 1	6	4
Result 1	11.5	9.9	0.9, 0.82, 0.91, 1,1	132, 32, 10, 1, 1	146, 39, 11, 1, 1	5	3
Result 2	11.5	10	0.91, 0.87, 1, 1	133, 33, 9, 1	146, 38, 9, 1	4	3
Result 3	11.5	10.2	0.93 0.91, 1	136, 32, 8	146, 35, 8	3	3
Result 4	11.5	10.4	0.94, 0.91, 1	137, 31, 8	146, 34, 8	3	3
Result 5	11.5	10.6	0.95, 0.94, 1	138, 31, 7	146, 33, 7	3	3
Final Value	*11.5*	*10.2*					

As shown in Table 2, canopy_elements refers to all the overlapping sample elements in the current cluster, and removal_elements refers to the sample elements that only belong to the current cluster. The total amounts of these two types of elements are called removal_amount and canopy_amount, respectively. Removal_rate is calculated by dividing removal_amount by canopy_amount. Canopies refers to the number of clusters, and the value of K is the number of clusters we have obtained.

As shown in Table 2, the number of pre-clustering clusters is determined as 3, which is consistent to the value of K-values. It is noted that a small increase in T2 had no effect on the final experimental results. Then, the parameters of the value of K and the initial clustering centers have been sent to K-means to obtain the final clustering results.

3.3 Final Clustering Results

Based on the clustering results of K-means, we obtain the evaluation value of each station, and the results show that the values of NR of all stations are 0.75, which all has the largest value. This indicates that all the 44 stations during the different four weeks have been clustered in the same categories. The clustering results are quite stable.

Moreover, Table 3 shows the detailed information of the final station clustering results. There are total 28 stations in the first cluster, whose volume of passenger flow is small. 9 stations have been clustered in the second cluster with the medium volume of passenger flow, and 7 stations have been clustered in the third cluster with the large volume of passenger flow.

Table 3. Station clustering results.

Cluster	Stations	Number	Passenger flow
1	1st Wharf, Douxi Rd, Longshanqiao, Dongfang Shangzhuang, *Airport Terminal 4*, Chengyi University College, Huaqiao University, University Town, Chinese Academy of Sciences, Dongzhai, Tiancuo, Institute of Technovation, Gaoqi Airport, Fenglin, Dong'an, Houtian, Dongting, Meifeng, Caidian, Pantu, Binhai Xincheng (Xike) Junction, Guanxun, Light Industry and Food Park, Sikouzhen, Industrial Zone, 3rd Hospital, Chengnan, Tong'an Junction	28	Small
2	Kaihe Intersection, Sibei, Ershi, *Jinbang Park*, Railway Station, Lianban, Municipal Administrative Service Center, Xianmen North Railway Station, Hongwen	9	Medium
3	Wolong Xiaocheng, Caitang, *Jinshan*, Shuangshi Middle School, Xianhou, Tan Kah Kee Stadium, Qianpu Junction	7	Large

3.4 Visualization

The four weekly passenger flow of the typical stations in the three different clusters have been visualized in Fig. 2. The stations set in italics in Table 3 are the selected typical stations, which are the centers of each cluster. There are seven days used in every week, and the daily service hours are 18 in Xiamen BRT.

In Fig. 2, time span from 1 to 126 represents the inbound flow, and time span from 127 to 252 represents the outbound flow, respectively. Take the inbound passenger flow as an example, the time span from 1 to 18 indicates the 18 hourly passenger flow on Monday, and the time span from 91 to 108 represents the hourly passenger flow on Saturday.

In Fig. 2(a), we can see that the overall passenger flow at Airport Terminal 4 station is relatively small, whose highest passenger flow is above 800. The trends of daily inbound and outbound flow are opposite, especially on weekdays. The passenger flow on weekdays is generally higher than that on weekends. Therefore, we define this type of station as the low-ridership station.

In Fig. 2(b), compared with Airport Terminal 4 station, the overall passenger flow at Jinbang Park station is moderate, whose highest passenger flow is above 2000. The trends of daily inbound and outbound flow are similar, and the outbound flow is slightly higher than the inbound flow. The passenger flow on weekdays is generally higher than that on weekends. Thus, we define this type of station as the station with medium ridership.

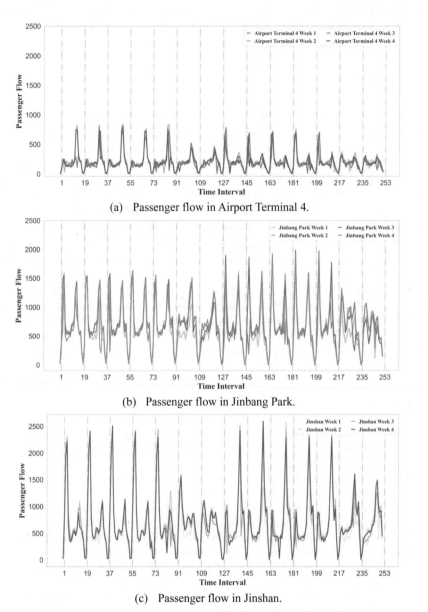

(a)   Passenger flow in Airport Terminal 4.

(b)   Passenger flow in Jinbang Park.

(c)   Passenger flow in Jinshan.

Fig. 2.   Passenger flow of typical stations in the three clusters during 4 weeks.

In Fig. 2(c), compared with Jinbang Park station, the overall passenger flow at Jinshan station is larger, whose highest passenger flow is above 2500. The trends of daily inbound and outbound flow are opposite, especially on weekdays. It's different from that the inbound flow during the morning peak hours is larger than the inbound flow during the evening peak hours compared to the inbound flow at Airport Terminal 4 station. The outbound flow has the opposite characteristics with the inbound flow. Moreover, passenger flow on weekdays is also generally higher than that on weekends. Hence, we define this type of station as the large-ridership station.

4   Conclusions

A novel station clustering model TEC has been proposed in this paper. The proposed TEC is consisted of the extraction module based on Transformer encoder, and clustering module based on improved Canopy + and K-means. Moreover, a new evaluation metric NR for such unsupervised clustering task has been introduced. The evaluation results show that the clustering results in all stations are stable at 1-h time granularity, and all the 44 stations in Xiamen BRT are divided into three categories according to the volume of passenger flow: small-ridership station, medium-ridership station, and large-ridership station. In future, we will apply the station clustering results in passenger flow prediction task for multiple stations. In addition, the proposed TEC can be further applied to the station classification in other public transport modes, such as metro, and normal bus.

Acknowledgement. This work was partly supported in part by the National Natural Science Foundation of China (No. 62103345), Fujian Provincial Natural Science Foundation of China (No. 2022H0023, No. 2020J02160), Xiamen Youth Innovation Fund Project (No. 3502Z20206067), High-Level Talents Research Launched Project of Xiamen University of Technology (Grant No. YKJ190112R), and Key Laboratory of Fujian Universities for Virtual Reality and 3D Visualization.

References

1. Tang, L.H., Xu, X.F.: Optimization for operation scheme of express and local trains in suburban rail transit lines based on station classification and bi-level programming. J. Rail Transp. Plann. Manage. **21**(100283), 1–12 (2022)
2. Li, J.Y., He, Z.C., Zhong, J.M.: The multi-type demands oriented framework for flex-route transit design. Sustainability **14**(9727), 1–23 (2022)
3. Wu, Q.Y., Wan, Y.P.: Stop selection of limited-stop bus services based on multi-criteria collaboration. J. Transp. Syst. Eng. Inf. Technol. **21**(1), 162–168 (2021)
4. Hu, Z.Q., Sun, R.C., Shao, F.J., Sui, Y., Lv, Z.H.: Traffic station classification based on deep spatio-temporal network. Comput. Electr. Eng. **97**(107558), 1–15 (2022)
5. Du, C.L., Li, X.L., Sun, R.R., Zhang, P., Zhu, G.Y.: Classification of urban rail stations based on passenger flow congestion propagation. J. Beijing Jiaotong Univ. **45**(1), 39–46 (2021)
6. Yuan, F.T., Chen, T.J., Wei, J.B.: Research on classification of rail stations based on AFC data. Transp. Eng. **21**(1), 48–52, 57 (2021)
7. Wang, H.D., Ma, H.W.: Classification method of urban rail transit stations based on POI. Traffic Transp. **36**(4), 33–37 (2020)
8. Jiang, Y.S., Yu, G.S., Hu, L., Li, Y.: Refined classification of urban rail transit stations based on clustered station's passenger traffic flow features. Transp. Syst. Eng. Inform. Technol. **22**(4), 106–112 (2022)

9. Xia, X., Gai, J.Y.: Classification of urban rail transit stations and points and analysis of passenger flow characteristics based on k-means clustering algorithm. Modern Urban Transit **4**, 112–118 (2021)
10. Vaswani, A., et al.: Attention is all you need. In: 31st Conference on Neural Information Processing Systems (NIPS 2017), CA, USA, pp. 1–11 (2017)
11. Wang, H.Y., Cui, W.C., Xu, P.D., Li, C.: Optimization of Canopy on K selection in partition clustering algorithm. J. Jilin Univ. (Science Edition) **58**(3), 634–638 (2020)

CVEjoin: An Information Security Vulnerability and Threat Intelligence Dataset

Francisco R. P. da Ponte$^{(\boxtimes)}$, Emanuel B. Rodrigues, and César L. C. Mattos

Federal University of Ceará (UFC), Fortaleza, Brazil
fco.rparente@gmail.com, {emanuel,cesarlincoln}@dc.ufc.br

Abstract. The risk of exploiting information security vulnerabilities should not be determined solely by a single metric, such as the Common Vulnerability Scoring System (CVSS). This approach disregards the global threat landscape and the vulnerable asset. Therefore, in addition to using traditional Vulnerability Management (VM) tools, analysts and researchers must manually curate datasets containing threat intelligence and context-specific information about security flaws. However, this activity is non-trivial and error-prone. To aid this endeavor, we developed a fully automated tool capable of gathering data about the intrinsic characteristics of vulnerabilities available in the National Vulnerability Database (NVD) and augmented it with information collected from multiple security feeds and social networks. Altogether, we collected data on more than 200,000 vulnerabilities that can be used for various research topics, e.g., analyzing the risk of exploiting security flaws, vulnerability severity prediction, etc. In this paper, we present a detailed description of the methodology used to create our dataset with its attributes. Additionally, we perform an exploratory analysis of the data gathered, and finally, we present an illustrative example of how analysts could use the data collected. The CVEjoin dataset and the scripts used for its construction are publicly available on GitHub.

1 Introduction

Given the large volume of information security vulnerabilities published in the last years [1] and the shortage of skilled professionals [2], it is clear the need to identify those critical flaws that must be fixed promptly. That way, analysts can prevent cybercriminals from exploiting the security vulnerabilities that can cause greater damage to the organization's infrastructure.

Traditionally, analysts employ Vulnerability Management (VM) tools that analyze and prioritize vulnerabilities based on the Common Vulnerability Scoring System (CVSS), which is an open-source framework that defines the severity of the vulnerabilities based on their characteristics [3]. The CVSS is a score from zero to ten that divides the vulnerabilities into four classes: LOW, MEDIUM, HIGH, and CRITICAL.

© The Author(s), under exclusive license to Springer Nature Switzerland AG 2023
L. Barolli (Ed.): AINA 2023, LNNS 661, pp. 380–392, 2023.
https://doi.org/10.1007/978-3-031-29056-5_34

Nonetheless, researchers claim that vulnerability remediation policies based on a single metric, such as those that use only the CVSS, are not ideal [4]. Thus, analysts should assess the risk of exploiting vulnerabilities using threat intelligence, which analyzes and processes real-time information to identify threats and malicious users that can harm computers, applications, or networks [6]. Additionally, analysts must consider the environment in which the vulnerabilities appear, i.e., their context, as this information can help guide their prioritization and remediation process [5].

Therefore, besides using VM tools, analysts need to collect and process several additional sources of information to successfully estimate the risk of exploiting security flaws. This process is time-consuming and is prone to errors caused by fatigue, since it is a nontrivial task often done manually [7].

To the best of our knowledge, none of the works in the literature offers a single comprehensive database containing information about vulnerabilities and threat intelligence. Thus, we propose to overcome this limitation by defining a methodology for the automatic collection and curation of a dataset containing public data that can help analysts assess vulnerability risks.

In summary, this paper makes the following contributions:

(a) We survey existing security vulnerability datasets and discuss their strengths and weaknesses;
(b) We propose a method that automatically collects and process information from different security feeds;
(c) We implement the proposed approach in a tool that is made publicly available on GitHub;
(d) We share a version of the dataset that contains information about the characteristics and threat intelligence of the more than 200,000 vulnerabilities published between 2002 and 2022;
(e) We present a script that can help analysts to easily incorporate context information from their network into the dataset.

The rest of this paper is organized as follows: Sect. 2 describes the related datasets and how our work differs from them; Sect. 3 describes the process of creating the vulnerability dataset; Sect. 4 discusses the results of the exploratory analysis of the data; Sect. 5 discusses the application of the dataset; and finally, Sect. 6 concludes the paper and presents ideas for future work.

2   Related Work

This section describes the related works that developed vulnerability-related datasets over the last decade. We discuss their strengths and weakness, and how they differ from our work. Finally, Table 1 summarizes our findings.

Bhandari *et al.* [8] created a tool called CVEfixes that automatically fetches all available Common Vulnerabilities and Exposures (CVE) records from National Vulnerability Database (NVD) with an associated open-source repository. The CVE is a unique identifier for publicly known vulnerabilities. The authors analyze the commits in those repositories and enrich the dataset with metadata such as the programming language. The initial release of CVEfixes contains information on 5,365 CVEs extracted from 5,495 vulnerability fix commits from 1,754 open-source projects.

Fan *et al.* [9] created a dataset called Big-Vul, containing 3,754 vulnerabilities published from 2002 to 2019 and spanning 348 GitHub projects writing in C/C++. The authors extracted descriptive information about the vulnerabilities from the CVEdetails, a free security vulnerability database/information source. They linked together this data, and the result is a dataset containing pieces of code before and after a patch is applied.

Jimenez *et al.* [10] introduced VulData7, a framework to collect security vulnerabilities mined from four open-source projects: Linux Kernel, WireShark, OpenSSL and SystemD. The dataset contains information about the vulnerabilities (e.g., CVE, description, severity, etc.), code (list of affected versions), and corresponding patches (commits before and after the fix). The authors were able to obtain information on 2,800 vulnerabilities and 1,600 fixes. Nonetheless, the framework performs the analysis on XML files that are no longer provided by the NVD.

Gkortzis *et al.* [11] presented VulinOSS, a dataset containing 17,738 vulnerabilities from 153 open-source projects. The data was collected by mapping project metadata (commits tags, branches, etc.) to CVE information obtained by crawling the NVD. The dataset provides information about the CVE, code metrics, and data related to modern development trends, such as continuous integration and testing. The authors only consider an older version of CVSS, i.e., CVSSv2.

Alves *et al.* [12] analyzed the bug-tracking systems of five open-source projects (Linux Kernel, Mozilla, Xen Hypervisor, httpd, and glibc) and looked for occurrences of CVE identifiers. They found 2,875 security patches and used the associated commit to create a manually labeled dataset of vulnerable and non-vulnerable code. Then, the authors investigated whether software metrics (e.g., number of lines of code, function nesting depth, etc.) correlated with the number of vulnerabilities.

Unlike previous works, Ponta *et al.* [13] manually curated a dataset comprising 624 publicly disclosed vulnerabilities affecting 205 open-source Java projects. The dataset presents information about the CVE, description, severity score, corresponding fixes, etc. The authors claimed that 29 vulnerabilities found do not have a CVE, and 46 had no additional information beyond the CVE identifier in the NVD.

The bibliographic research showed relevant works that seek to create security-related datasets. However, they have a limited scope, focusing their analysis on vulnerabilities in big open-source projects developed in a particular programming language. In addition, the datasets do not address hardware and operating system vulnerabilities (except those that affect the Linux Kernel). Moreover, few consider multiple security feeds, and none bring data about threat intelligence information, which is essential to assess the risk of vulnerabilities being exploited and the impact on the organization's infrastructure. Thus, our work proposes to overcome these limitations by creating a general-purpose security dataset with vulnerability characteristics and threat intelligence information collected from ten different sources, including vendor security advisories and social networks.

Table 1. Comparison between related work.

Reference	Vulnerability characteristics	Threat intelligence	# Sources	Program. language	Platforms	Size
[8]	×		2	Multiple	Software & OS (Linux Kernel)	5,365
[9]	×		2	C/C++	Software & OS (Linux Kernel)	3,754
[10]	×		2	C	Software & OS (Linux Kernel)	2,800
[11]	×		2	Multiple	Software	17,738
[12]	×		4	C/C++	Software & OS (Linux Kernel)	2,875
[13]	×		2	Java	Software	624
This work	×	×	10	Multiple	Hardware, Software & OS	200,000+

3   Dataset Construction

3.1   Dataset Background

Risk-Based Vulnerability Management (RBVM) is the ongoing process that seeks to identify, categorize, prioritize and remediate vulnerabilities in software, corporate applications, and operating systems [14]. In this discipline, vulnerabilities are prioritized based on their risks, which can be defined as the impact on the organization given the exploitation of a vulnerability after an adverse security event occurs [15]. Therefore, in addition to vulnerability characteristics, efficient RBVM strategies also consider threat intelligence and context information.

Threat intelligence is the discipline that collects and analyzes information in real-time to determine the global threat landscape and identify which networks, systems, and computers are most at risk of being exploited [16]. This information is collected from many sources, such as open-source feeds, social networks, data acquired through forensics, Internet traffic analysis, etc. Such information provides the necessary knowledge so that security analysts can identify and prioritize the vulnerabilities that are critical to remediate [17].

Context is any information used by analysts to categorize the criticality of assets within their network. Context is relevant as it allows the rank of precedence among the organization's assets. In other words, which network asset is more important and should be prioritized when fixing vulnerabilities, e.g., a server with a database that stores sensitive user information or a simple workstation?

The available dataset is for general use and has information about the intrinsic characteristics of the vulnerabilities and threat intelligence, as described in Subsect. 3.3. However, as context information plays a significant role in risk assessment, we have developed a script, described in Subsect. 3.4, to facilitate the incorporation of this data into the dataset.

3.2   Data Collection Methodology

The dataset was built by automatically mining public security feeds. We use the Python 3.8 programming language to collect, process, and analyze the gathered information. Altogether, we consulted ten different databases, listed as follows:

- NVD[1]: a repository maintained by the National Institute of Standards and Technology (NIST) that contains various information about security flaws, including the CVE, which is a unique identifier for vulnerabilities;
- Mitre[2] and OWASP[3]: not-for-profit organizations that enumerate dangerous software and hardware weaknesses, i.e., Common Weakness Enumeration (CWE). If these flaws were left unaddressed, systems, networks, or hardware could be vulnerable to attacks;
- ExploitDB[4]: a public archive for exploits and Proofs-of-Concept (POC), which can be used to explore vulnerable assets;
- EPSS[5]: an open scoring system that uses threat information and real-world exploit data to estimate the likelihood (probability) that a vulnerability will be exploited in the wild in the next 30 days;

---

[1] National vulnerability database website: https://nvd.nist.gov/.
[2] The Mitre corporation website: https://cwe.mitre.org.
[3] The Open Web Application Security Project website:https://owasp.org.
[4] An archive of vulnerable software and exploits: https://www.exploit-db.com/.
[5] Metric for estimating the probability of a vulnerability being exploited: https://www.first.org/epss/.

- Microsoft[6], Adobe[7], and Intel[8] security advisories: these companies provide fixes or workarounds for critical vulnerabilities identified in their products;
- And finally, vulnerabilities that received some attention on social networks, i.e., those cited by users in their publications, specifically on Twitter and Google Trends.

We construct the dataset in three steps:

1. First, we collect the information made available by these feeds through Web Scraping, API queries, or downloading files with vulnerability information in JSON format. That was possible thanks to libraries like urllib[9] and BeautifulSoup[10], which allow accessing and manipulating the content of websites from their URLs;
2. We then process the data collected about each vulnerability, deleting non-relevant information and mapping it to the CVE identifier;
3. Finally, we merge all the information into a single dataset, a process done using the CVE as the merge point. In all, we collected vulnerability information published between 2002 and 2022.

The scripts used in data mining and dataset creation, as well as a copy of the generated dataset, can be found on GitHub[11]. All vulnerability information is stored in a CSV file that can be consulted by analysts using, e.g., a spreadsheet tool or the Python pandas[12] library. Altogether, we collect information on more than 200,000 vulnerabilities. It is important to note that vulnerability information constantly changes over time. Thus, the users must run the script periodically to keep the dataset updated, as new vulnerabilities are published daily.

3.3   Attributes Description

The dataset has 34 attributes grouped into two sets: vulnerability characteristics and threat intelligence information. Below we describe some of these attributes, but a detailed description can be found in the project repository. The vulnerability characteristics group provides constant information that does not change over time. Such information helps analysts identify vulnerabilities and the impact of exploiting them on the organizations' systems. Some attributes that belong to this group are:

---

[6] Microsoft security advisory: https://msrc.microsoft.com/update-guide/en-us.

[7] Adobe security advisory: https://helpx.adobe.com/security.html.

[8] Intel security advisory: https://www.intel.com/content/www/us/en/security-center/default.html.

[9] Python package for working with URLs: https://docs.python.org/3/library/urllib.html.

[10] Python library for scraping information from web pages: https://pypi.org/project/beautifulsoup4/.

[11] Code developed to create the dataset: https://github.com/rodrigoparente/cvejoin-security-dataset.

[12] Python package for data analysis and manipulation: https://pypi.org/project/pandas/.

- **Platform, vendor, product:** these attributes allow the analyst to identify which platform, vendor, and product is affected by the vulnerability;
- **CVSS score, severity, and type:** indicates the CVSS base score (ranging from 0 to 10), the severity (LOW, MEDIUM, HIGH, and CRITICAL), and the type, which can be CVSSv2 or CVSSv3. These values allow the analyst to better rank the vulnerabilities;
- **Impact on Confidentiality, Integrity, and Availability (CIA):** indicates the impact on the confidentiality, integrity, and availability of the affected system, should the vulnerability be exploited;
- **Vulnerability published and modification date:** indicates how many days passed since the vulnerability was published or modified. Newer vulnerabilities are often associated with malware campaigns, as they are unknown and usually do not have security patches [18];
- **Update available:** indicates whether there is a patch available or not that can fix the vulnerability.

The information set of threat intelligence varies over time. This group is essential for the analyst to know the current cybersecurity scenario, providing the necessary information for its characterization. For example, the analyst can detect the existence of public exploits or ongoing malware campaigns that affects a given vulnerability. Below, we present some attributes that belong to this group and discuss why they are important:

- **Presence of public exploit:** indicates the existence of a public exploit, i.e., a program that allows malicious users to automatically exploit software flaws. Vulnerabilities with exploit are considered critical, since attackers can exploit them easily if they have access to the vulnerable machine;
- **Attack vector:** indicates the attack type that was used to exploit the vulnerability, e.g., remote code execution, denial-of-service, etc. With this information, analysts can identify the point of origin and prevent the same attack from happening again;
- **Feeds from Twitter and Google Trends:** identify how many users of these services are discussing a particular vulnerability. Such information may indicate the existence of a proof-of-concept (the basis for developing an exploit) or malware campaigns. It is common to see information of this nature appearing on these services before the specialized media;
- **Security Advisories:** some companies communicate the presence of critical vulnerabilities in their software and inform how to fix them. This type of communication usually occurs through a post on the company's blog. An example is a blog maintained by Microsoft, which has published a list of vulnerabilities and fixes every second Thursday of the month since 2003. Currently, several broad companies have a security advisory channel.

3.4   Incorporating Context Information into the Dataset

Context information is specific to each environment, so it is not initially present in our dataset. Automating the collection of context information is not a trivial task and requires knowledge of network inventory tools. However, since this information is crucial for the risk management of vulnerabilities, we developed a script that facilitates the incorporation of context data into our dataset.

For that, the user can execute a script, passing as input a CSV file containing the identifiers of the network assets (e.g., their IP addresses) with the associated context data and the CVEs of the affecting vulnerabilities. The output is a CSV file containing vulnerability characteristics, threat intelligence, and context information mapped by the asset identifier. The script can be found on the project's GitHub.

4   Dataset Exploration

The following analyzes are for vulnerabilities published between 2017 and 2021, i.e., the last five years. Analysts published more than 20,000 vulnerabilities in the NVD only in 2021. An increase of approximately 10% compared to 2020. The analysis of Fig. 1 shows that the number of unique vulnerabilities reported has been growing since 2017. Furthermore, if we consider the shortage of skilled workers in cybersecurity [2], it is clear the importance of using strategies for analyzing, prioritizing, and remediating these security flaws.

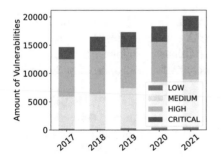

Fig. 1. Vulnerabilities per year.

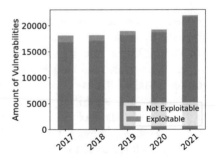

Fig. 2. # of exploits per year.

Some analysts use CVSS as a single metric in the prioritization process, as they believe it correctly indicates risk. In this way, efforts are focused on fixing vulnerabilities with CRITICAL severity. However, this strategy is inadequate, as CVSS is designed to categorize vulnerabilities by their severity, not their risk.

A study by RecordFuture shows that among the ten most exploited vulnerabilities of 2020, only 4 of them had CVSS CRITICAL [19]. Furthermore, we can see in Fig. 1 that the number of vulnerabilities that concentrate in the MEDIUM and HIGH classes is more than 50%. Also, the CRITICAL severity appears only for vulnerabilities published after 2016, when the *v3* version of CVSS was released.

Thus, besides CVSS, it is relevant to consider other information. One of the main issues to look at is the presence of public exploits, which is a threat intelligence information. Figure 2 shows the number of vulnerabilities that have exploits publicly available, which represents a small fraction of the total. Nonetheless, the analysis of the distribution of exploits by severity showed that the HIGH severity has the highest amount of exploits, with approximately 45% of the total value. Thus, by considering only the CVSS, we are underestimating the risk of vulnerabilities being exploited.

To help analysts understand the vulnerability under analysis, they must gather and analyze its characteristics. For example, Fig. 3 shows that vulnerabilities are classified into three groups: hardware, software, and operating systems. Note that 85% of all vulnerabilities concentrate on the software and operating system groups.

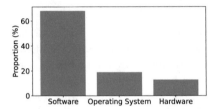

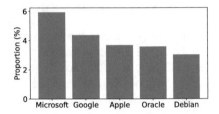

Fig. 3. # of vulnerabilities by platform.   **Fig. 4.** Top vulnerable vendors.

Another way to classify vulnerabilities is according to the vendor. Figure 4 shows the five most vulnerable vendors. Note that Microsoft is the manufacturer with the highest number of vulnerabilities published, followed by Google and Apple. Analysis showed that ten vendors account for approximately 30% of all reported vulnerabilities.

There are also security organizations, such as Mitre and OWASP, that publish in their sites lists of the most critical weaknesses found in software vulnerabilities. Such weaknesses, known as CWEs, are important indicators that help the analyst understand how vulnerability exploitation occurs and what can be done to protect against it. Furthermore, we have compiled a list of the top ten CWE that most affect software vulnerabilities: CWE-79, CWE-787, CWE-20, CWE-119, CWE-125, CWE-200, CWE-89, CWE-22, CWE-416, and CWE-352.

Finally, the analysis of what is being said about vulnerabilities on social networks is also of interest. By querying Twitter for the CVE of a vulnerability, we can account for the number of people talking about that flaw and its reach (i.e., views), identifying those that are critical and deserve attention. Table 2 shows the ten most mentioned vulnerabilities on Twitter on January 8, 2022. It is possible to notice the presence of the Apache Log4j vulnerability, which was a zero-day that affected millions of Java devices worldwide[13].

Furthermore, we can see in Table 2 some vulnerabilities that, despite not being officially published in the NVD, are known and considered critical by the

---

[13] News about Log4J vulnerability and how it was exploited: https://blog.qualys.com/vulnerabilities-threat-research/2021/12/10/apache-log4j2-zero-day-exploited-in-the-wild-log4shell.

security community. That is the case of CVE-2021-44142, which occupies the 7th position in the table. This vulnerability is a flaw in the Samba open-source implementation of the Server Message Block (SMB) protocol[14]. On January 21st, 13 days after our analysis, CVE-2021-44142 was published, receiving a base score of 8.8 (HIGH).

Table 2. Most mentioned vulnerabilities on Twitter on January 8, 2022.

Rank	CVE	Name	Publication date	CVSS Score	Views
1°	CVE-2021-44228	Apache Log4j	Dec. 2021	10	450K
2°	CVE-2022-21882	Windows Server	Jan. 2022	7.8	270K
3°	CVE-2021-4034	Polkit	Jan. 2022	7.8	110K
4°	CVE-2022-0509	–	Feb. 2022	–	110k
5°	CVE-2021-39137	Go Ethereum	Aug. 2021	7.5	100K
6°	CVE-2014-6271	GNU Bash	Sept. 2014	9.8	100K
7°	CVE-2021-44142	–	–	–	97K
8°	CVE-2022-23263	–	Feb. 2022	7.7	95K
9°	CVE-2022-23262	–	Feb. 2022	6.3	95K
10°	CVE-2020-24807	NodeJS	Oct. 2020	7.8	83K

5   Dataset Applications

To illustrate how our dataset can help analysts, consider a fictional scenario with two vulnerabilities whose identifiers are CVE-2020-8025[15] and CVE-2020-15849[16], respectively, as shown in Fig. 5. From a query to our dataset, we discover that CVE-2020-8025 is a vulnerability that affects Linux operating systems developed by the company SUSE. This vulnerability has a CVSS score of 9.3, its impact on the CIA is high, and it is exploitable only from the local network. CVE-2020-15849 is a software vulnerability that affects a customer service tool called Re:Desk. This vulnerability has a CVSS score of 7.2, has a high impact on the CIA, and can be exploited remotely.

Following the classic VM approach of patching vulnerabilities with the highest CVSS scores, analysts would prioritize CVE-2020-8025. However, as discussed in this paper, we need to analyze the context in which these vulnerabilities

---

[14] News about CVE-2021-44142: https://www.helpnetsecurity.com/2022/02/02/samba-bug-may-allow-code-execution-as-root-on-linux-machines-nas-devices-cve-2021-44142/.

[15] Vulnerability affecting SUSE OS: https://nvd.nist.gov/vuln/detail/CVE-2020-8025.

[16] Vulnerability affecting a help desk tool: https://nvd.nist.gov/vuln/detail/CVE-2020-15849.

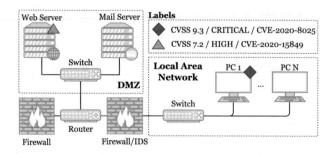

Fig. 5. Experimental environment.

are and the global threat landscape to come to a proper conclusion. CVE-2020-8025 affects a workstation, which receives security updates frequently and is in a network protected by a robust firewall and an Intrusion Detection System (IDS).

CVE-2020-15849 affects a web server in the Demilitarized Zone (DMZ), protected only by a basic firewall that performs packet filtering. Let's further suppose that an exploit for CVE-2020-15849, which would be noticeable in our dataset, emerges. In this scenario, CVE-2020-15849 has a higher risk to the organization than CVE-2020–8025, since it is facing the Internet and has a public exploit that can be used to exploit it remotely. Therefore, security analysts should fix it first, as it can be the gateway for the attacker to invade the company's network and compromise sensitive assets and data.

Additionally, this process of analyzing and classifying the risk of vulnerabilities can be automated using a Machine Learning (ML) technique. Security analysts can, for example, train an ML model using our dataset and their expertise as input. Thus, it would be possible to create a classifier capable of assessing the risk of exploiting a vulnerability similarly to the intuition and expertise of a cybersecurity expert.

6 Conclusion and Future Works

In the vulnerability management process, besides using VM tools, analysts often need to gather additional data, such as threat intelligence and context-specific information, to assess the risk of exploiting vulnerabilities. This task is complex and unreliable.

To help analysts in this activity, we created a dataset containing information about more than 200,000 distinct vulnerabilities, published between 2002 and 2022. The data was obtained by consulting several security feeds through API queries and web scraping and can help users in several security activities.

In the future, we intend to increase the number of security-related feeds consumed (including, e.g., GitHub repositories) and to develop a mechanism to automatically extract context information from users' networks to incorporate into the dataset.

Acknowledgment. The authors would like to thank CAPES for the financial support.

References

1. NVD, NIST. (2022, November 18). NIST National Vulnerability Database. Retrieved November 18 2022. https://nvd.nist.gov/
2. Furnell, S., Fischer, P., Finch, A.: Can't get the staff? The growing need for cybersecurity skills. Comput. Fraud Secur. **2017**(2), 5–10 (2017)
3. Forum of Incident Response and Security Teams (2019, June). CVSS v3.1 Specification Document [White paper]. Retrieved November 18 2022. https://www.first.org/cvss/v3.1/specification-document
4. Dey, D., Lahiri, A., Zhang, G.: Optimal policies for security patch management. INFORMS J. Comput. **27**(3), 462–477 (2015)
5. Spring, J., Hatleback, E., Householder, A., Manion, A., Shick, D.: Time to Change the CVSS? IEEE Security Privacy **19**(2), 74–78 (2021)
6. Trifonov, R., Nakov, O., Mladenov, V.: Artificial Intelligence in Cyber Threats Intelligence. In: 2018 International Conference on Intelligent and Innovative Computing Applications (ICONIC), (pp. 1–4). IEEE (2018)
7. Elbaz, C., Rilling, L., Morin, C.: Automated Risk Analysis of a Vulnerability Disclosure Using Active Learning. In: Proceedings of the 28th Computer and Electronics Security Application Rendezvous (2021)
8. Bhandari, G., Naseer, A., Moonen, L.: CVEfixes: Automated Collection of Vulnerabilities and Their Fixes from Open-Source Software. In: Proceedings of the 17th International Conference on Predictive Models and Data Analytics in Software Engineering , pp. 30–39. Association for Computing Machinery (2021)
9. Fan, J., Li, Y., Wang, S., Nguyen, T.: A C/C++ Code Vulnerability Dataset with Code Changes and CVE Summaries. In: Proceedings of the 17th International Conference on Mining Software Repositories, pp. 508–512 Association for Computing Machinery (2020)
10. Jimenez, M., Le Traon, Y., Papadakis, M.: [Engineering Paper] Enabling the Continuous Analysis of Security Vulnerabilities with VulData7. In: 2018 IEEE 18th International Working Conference on Source Code Analysis and Manipulation (SCAM), pp. 56–61. IEEE (2018)
11. Gkortzis, A., Mitropoulos, D., Spinellis, D.: VulinOSS: A Dataset of Security Vulnerabilities in Open-Source Systems. In: Proceedings of the 15th International Conference on Mining Software Repositories, pp. 18–21. Association for Computing Machinery (2018)
12. Alves, H., Fonseca, B., Antunes, N.: Software Metrics and Security Vulnerabilities: Dataset and Exploratory Study. In: 2016 12th European Dependable Computing Conference (EDCC), pp. 37–44. IEEE (2016)
13. Ponta, S., Plate, H., Sabetta, A., Bezzi, M.,Dangremont, C.: A manually-curated dataset of fixes to vulnerabilities of open-source software. In: 2019 IEEE/ACM 16th International Conference on Mining Software Repositories (MSR), pp. 383–387. IEEE (2019)
14. Foreman, P.: Vulnerability management. Auerbach Publications (2019)
15. Alexander, J.: Risk, threat, or vulnerability? what's the difference. Retrieved November 18, 2022 (2021). https://www.kennasecurity.com/blog/risk-vs-threat-vs-vulnerability/

16. Conti, M., Dargahi, T., Dehghantanha, A.: Cyber threat intelligence: challenges and opportunities. In: Dehghantanha, A., Conti, M., Dargahi, T. (eds.) Cyber Threat Intelligence. AIS, vol. 70, pp. 1–6. Springer, Cham (2018). https://doi.org/10.1007/978-3-319-73951-9_1

17. Bromander, S.: Understanding Cyber Threat Intelligence: Towards Automation [Doctoral's Thesis, University of Oslo] (2021). The University of Oslo Institutt for informatikk. https://www.duo.uio.no/handle/10852/84713

18. Suciu, O., Nelson, C., Lyu, Z., Bao, T., Dumitras, T.: Expected exploitability: Predicting the development of functional vulnerability exploits. In: 31st USENIX Security Symposium (USENIX Security 22), pp. 377–394 (2022)

19. RecordFuture, Inc. (2021, February). Top Exploited Vulnerabilities in 2020 Affect Citrix, Microsoft Products [White paper]. Retrieved November 18 2022. https://go.recordedfuture.com/hubfs/reports/cta-2021-0209.pdf

Big Data Management for Machine Learning from Big Data

Anifat M. Olawoyin[1], Carson K. Leung[1(✉)] (iD), Connor C. J. Hryhoruk[1], and Alfredo Cuzzocrea[2]

[1] University of Manitoba, Winnipeg, MB, Canada
Carson.Leung@UManitoba.ca
[2] University of Calabria, Rende, CS, Italy

Abstract. The world is dynamic, and so are big data. The evolving challenges of managing big data volume and velocity have resulted in several studies focusing on machine learning models. Despite the usefulness of these models, further explanation is often required to interpret, understand, and effectively use the outcome of machine learning models. In this paper, we examine challenges of machine learning models in processing big data. These include the inherent uncertainty in data collection and questionable validity of machine learning model outcome. Motivated by the challenges arising from complex varieties due to the rigid schema required by the prevalent relational database model and data warehouse, we present (a) an architectural design of a schema-less big data repository aiming at capturing all data type (e.g., structured, semi-structured, and unstructured data) and (b) a data-driven approach to metadata collection for managing the big data.

Keywords: Advanced information networking and applications · Big data · Big data challenges · Data management · Data models · Machine learning · Structured data · Semi-structured data · Unstructured data · Velocity · Volume

1 Introduction

Nowadays, big data [1,2] are everywhere. Examples include disease data [3–5], environmental data [6], music [7], news [8], social networks [9–11], sports statistics [12], and transportation data [13,14]. The volume of big data is growing exponentially resulting in adoption of machine and deep learning to process the data. More recently, the application of machine learning models has been extended to decision making such as credit ratings, loan processing, hiring decision, re-offence risk assessment in policing, as well as voice and image recognition. However, the output and decision from these models are sometimes questionable and difficult to be understood by human. A common example in research literature is a criminal risk assessment machine learning prediction model [15], which uses a statistical measure of false positive rates by the Correctional Management

Profiling for Alternative Sanctions (COMPAS). Kilbertus et al. [16] and Chiappa et al. [17] argued that using statistical fairness as a measure of discrimination may not be appropriate and proposed causal reasoning approach. Consistent with ongoing research on interpretability and fairness of machine learning models [18–22], *our key contributions* of the current paper include:

1. our investigation and review of challenges of big data management with machine learning models—such as complexity in handling input from disparate data types and questionable validity of machine learning models use in big data management tasks.
2. our design of a schema-less big data repository for capturing structured, semi-structured, and unstructured data types; and
3. our design of a framework for metadata collection using a data-driven approach.

The remainder of this paper is organized as follows: Sect. 2 discusses challenges of big data management. Sections 3 and 4 describe our big data repository conceptual model and metadata design. Section 5 shows the evaluation of our model on Amazon Web Services and Sect. 6 presents the conclusions.

2   Machine Learning and Big Data Processing Challenges

Challenge 1: Complex Varieties. A notable challenge of big data is complexity of capturing, storing and modeling evolving varieties of big data. The advancement in technology has resulted in collection of semi-structured, and unstructured big data in different domains—including telecommunication [23], social network [24] and biological data [25]—which are difficult to model using the predominant relational database. For instance, in relational databases, structured traditional data can be captured using tables to represent a data record in each row and its attribute in each column. Relationships among entities are defined using referencing such as one-to-one, one-to-many, or many-to-many. Tables can be queried using Structured Query Language (SQL). One-to-one and one-to-many relationships are simple to represent using a primary key as a foreign key in the other table. However, a more complex relationship (e.g., many-to-many) requires creation of a relation table to model the relationship. Thus, tabular representation poses a challenge when the relationship among various entities of the big data increases. The relationship representation becomes difficult to manage and may require schema restructuring. As a result, query processing may become time-consuming and expensive.

Over time, other data storage and models (e.g., document database, NoSQL database) have evolved to resolve diversity of big data type variety challenge. Recently, the need to adequately model and visualize dynamic and complex relationship among entities of a network-based data (e.g., telecommunication, biochemical interaction, social network) has led to graph databases. In general, it combines concept of graph theories [26] from the field of mathematics with database, data mining [27–29] and information retrieval to model complex relationship among various entities in the social network, telecommunication network, customer purchase pattern, information security network design, biochemical interaction of genes, human evolution, churn prediction [23], neighbourhood detection, crime, and other societal issues.

Challenge 2: Uncertain Veracity. Another challenge of big data is the variation in the degree of uncertainty inherent in most data collections. For instance, data such as hydrographic and bathymetric data collected during Arctic expedition, patient trajectories collected for COVID-19 contact tracing, sensor data, social network data, traffic flow data, smartphone data and continuous weather data are spatio-temporal in nature rooted in continuous space and time. Although storing spatio-temporal data in relational database is not an issue, the challenge is spatio-temporal data mining (i.e., a non-trivial task of extracting previously unknown pattern to prevent, solve or predict future societal issues such as spread of COVID-19 and other infectious disease, drug development and proactive patient testing, storm prediction, crime detection in police, signal loss and intrusion detection in communication and cyber attack monitoring). The traditional data mining methods are sometimes restricted to sampling methods with the assumption that samples are independent and stationary [30], whereas spatio-temporal are highly structural correlated in space and time. For instance, ocean and Arctic expedition data are collected using a *moving* research vessel supported with a rigid-hulled inflatable boat (RHIB), sound velocity profiler, multi-beam echo sounder and antennas. Additionally, the weather condition varies from one expedition to another. Hence, the data collection may not be replicable. Thus, domain knowledge and expertise are needed to pre-process, design, and transform the raw spatio-temporal data before applying the data mining techniques. For instance, the Society of Exploration Geophysicists (SEG) developed an open standard SEG-Y file format to store geophysical data; Triton Imaging Inc. developed an industrial standard known as Extended Triton Format (XTF) to record side scan and back-scatter data collected during Arctic expedition. The list of such standards keeps growing by the day. Hence, uncertainty data mining [31] research has emerged and remains an active research area.

Challenge 3: Privacy and Accessibility. Privacy and accessibility can be considered as double-sword edges for big data. Nowadays, government open data initiatives aiming at promoting transparency in governance and the accessibility of data through social network has created yet another challenge in the era of big data. Hornung et al. [32] defined the social notion of privacy as a dynamic social process of (a) negotiating personal boundaries and territories; (b) managing disclosure, concealment and relationship considering the context and situations. However, *the boundary of personal privacy in the public domain is currently undefined.*

Westin [33] defined privacy as the right of an individuals, institutions, or group to control when, how and to what extent information about them is

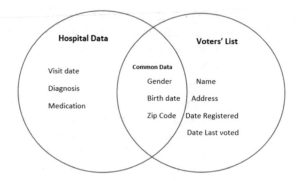

Fig. 1. Data linking

communicated to others. In other word, preserving privacy is a necessity in the entire life cycle of big data management starting from data collection, data storage, data processing, data analysis, data publishing to the point where data are eventually destroyed or archived. Challenges of data acquisition includes issues like terms of use, government legislation, copyright, disclaimers, and privacy. For instance, Statistics Canada[1] data collection and reuse is guided by government legislation such as the *Access to Information Act*[2] and the *Privacy Act*[3] while special permits are sometimes required to meet legal requirements in cases where the data are sensitive and proprietary. In addition, some publicly available data include copyright rules for data reuse, content ownership and disclaimers are often included on social media platform to isolate platform content from individual contributor's content. For instance, Twitter's Terms of Service[4] exclusively states:

> "You are responsible for your use of the Services and for any Content you provide, including compliance with applicable laws, rules, and regulations. You should only provide Content that you are comfortable sharing with others."

The term "public consent" has emerged as a justification for accessible open data and reusability. There is growing concern over what constitutes privacy of individuals within the publicly available data [34]. Though open data publishing involves anonymization of personal information (e.g., social insurance number, names and date of birth among others that could directly identify an individual), research has shown that despite the removal of sensitive attributes, re-identification is not impossible. For instance, Sweeney [35] demonstrated how individual could be identified from the US 1990 census data with *US 5-digit Zip code* using infrequent pattern modelling technique. In particular,

---

[1]https://www.statcan.gc.ca/.
[2]https://laws-lois.justice.gc.ca/eng/acts/a-1/.
[3]https://laws-lois.justice.gc.ca/eng/ACTS/P-21/index.html.
[4]https://twitter.com/en/tos.

Sweeney identified the governor of Massachusetts by linking common attributes as shown in Fig. 1 from anonymous voters' list and hospital level data collected by Massachusetts Group Insurance Commission (GIC) containing attributes such as gender, birth date, ethnicity and Zip code. Due to the possibility of re-identification of individual record. several studies have investigated and designed anonymization models including k-anonymity [36], l-diversity [37] and t-closeness [38]. A major drawback of these models is the inherent utility-privacy trade-off such that strong privacy often resulted in minimal utility. Then, differential privacy [39–41] emerged recommending data distortion through noise addition. However, adding noise to data may distort the data beyond its usefulness [42]. Thus, preserving privacy of individual record within big data remains an active research area.

Challenge 4: Questionable Validity. Objectivity and accuracy are two other challenges for big data. Boyd and Crawford [43] defined big data as a scholarly phenomenon resting on the interaction of technology, analysis and mythology of truth, objectivity, and accuracy. *However, we are far from the objectivity and accuracy of big data due to interpretability and fairness risks associated with machine learning (including deep learning) models used in processing big data.* In an attempt to solve big data processing time issues arising from the ever-increasing volume of big data that are infeasible to be processed by human, several machine learning models have been developed including artificial neural networks (ANN) [30,44,45], deep neural networks (DNN) [46,47], convolutional neural network (CNN) [48,49] and most recently graph neural network (GNN) [50]. These models are often subjective to the inherent pattern in the dataset or underlying nature of statistics measure. For instance, while some researchers [15,51] argued that the four statistics metrics—true positive (TP), true negative (TN), false positive (FP), and false negative (FN)—can be combined to measure fairness and bias of machine learning models, other researchers [16,17,20] suggested that counterfactual analysis approach [18,19] is more appropriate. Meanwhile, models are created and trained by human using past and present data to predict the future pattern. Hence, irrespective of the methods of explaining the output of machine models, big data collections are not error-free. Duplication, inconsistency, missing words and labels, natural disaster, change in weather condition among other errors are inherent in the heterogeneous sources of data collection. Further to the complexity of bias in machine output and inherent inaccuracy of data sources, human involvement in the process of data cleaning, feature selection and other pre-processing tasks is subjective in nature and may introduce bias to the input even before processing by the machine model. Moreover, interpretation of machine learning output by different domain remains a subjective endeavour relying on the purpose of use. Thus, bias, fairness, accuracy, and subjectivity are challenges that will most likely linger into the future.

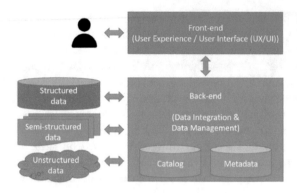

Fig. 2. Data REPO architecture

3   Our Schema-less Big Data Repository Conceptual Model

Motivated by the prevailing challenges of big data, several models and concepts have evolved over time aiming at capturing varieties of big data in different format including relational database models for storing, entities, entities relationships and defined constraints for ensuring integrity of relationship among entities and reducing redundancy of data as much as possible. Subsequently, document databases, extensible markup language (XML), NoSQL databases have emerged for storing and modelling semi-structured data. Yet, the growing varieties of big data in the world today are not static. Hence, *we design a conceptual model for capturing and storing different variety of big data* including structured, semi-structured and unstructured data along with metadata describing the data in support of open data FAIR principle of Findable, Accessibility, Interoperability and Re-usability. Our model is schema-less repository for variety of raw data along with metadata collection framework for managing the big data stored in the data repository. As presented in Fig. 2, our model is comprising of (a) data sources, (b) a front-end user interface, and (c) a back end with an integrated metadata.

There are **diverse heterogeneous data source** for big data collections. Usually, data collection, origination or acquisition has source and purpose. For instance, Statistics Canada collects data from various sources for further analysis to provide information such as population estimates, Consumer Price Index (CPI), unemployment rates and other key economic indicators to the policy makers and the general public[1]. IBM's The Weather Company (TWC)[5] continuously collects atmospheric data such as temperature, allergy tracker and air quality from diverse weather stations and satellites to provide weather forecast to various users of the weather app and researchers; arctic expedition data such as sub-bottom samples, salinity sample, bathymetric, side scan, sound velocity

---

[5]https://weather.com/.

profile are collected for specific project with a well-defined objectives. Other data may be collected at random or for security purpose. For instance, the prevalent of surveillance cameras in most houses today are intended to provide monitoring of activities around individual property, though the cameras are now helping the law enforcement agency in crime investigation. Data type can be structured, semi-structured or unstructured. Structured data are typically arranged in row and column for easy access, aggregation, filtering, and sorting. Semi-structured data type has a delineating delimiters or tags to identify each element of the data. Examples include comma-separated values (CSV), XML and JavaScript Object Notation (JSON). Unstructured data type has no pre-defined data model and is characterised mostly by uncertainly and irregularity. Examples include portable document format (PDF), video, audio, pictures, surveillance cameras data are particularly complex containing combination of audio, video, image and pictures.

Our **front-end user interface** consists of three web interfaces:

1. The search interface, which allows users to find data and metadata available in the data repository using different keywords and filters.
2. The visualization (Viz) interface, which aims for visualizing subsets of any dataset that can be retrieved from the data repository.
3. The application programming interface (API), which allows developers and other third-party interactions. Users can interact with the data repository pragmatically to obtain a dataset rather than using the search or Viz interface.

Our **back end** for the big data repository accepts structured, semi-structured, and unstructured data from heterogeneous sources and stored the data as-is. Each data source has rich metadata collected using a data driven approach and the data repository is surrounded by security layers to protect the ever growing big data in the repository. On top of the data repository are three additional layers:

1. Access control layer, which controls the users access to the data repository.
2. Query processing layer, which accepts tasks from users through the search user interface and fetch the requested data to the user.
3. Visualization tool/plugin layer, which provides flexible options for users to visualize a subset of a search results requested through Viz in user interface.

4   Our Metadata Framework Design

Regarding the sources of metadata, we consider the nature of data collection and acquisition from the perspective of WHAT, WHO, WHERE, WHEN and HOW and translate the resulting output into a baseline metadata conceptual design, so as to comply with the "reusable" principle in the FAIR Principles:

1. **WHAT:** Data collection and acquisition are done with specific goals. For instance, an expedition is usually project based. A project is initiated with clearly stated objectives and may have one or more events including equipment mobilization, testing and demobilization, sea trial, survey operation, data collection, and data analysis.

2. **WHO:** Who are data owner or data collector? In some cases, data collection and acquisition events are handled by multiple stakeholders. For instance, an expedition project team may includes principal investigator, research collaborators, research assistant, marine surveyors, data scientists, technicians and local guards.
3. **WHERE:** Data are collected from different locations (e.g., on-the-go through mobile devices, retail stores, social network platforms). For instance, for arctic expedition, events are undertaken at one or more sites within the Arctic region[6].
4. **WHEN:** The temporal metadata about a project include project initiation date, data collection period, data analysis, project closure and other project tasks period of execution. The temporal metadata may be simple, capturing only the timeline for the entire project or complex capturing individual task timelines, time lags and anticipated dependency of other projects over several years.
5. **HOW:** What are methodology and approaches for the project execution (e.g., data collection approach, method for data analysis and equipment deployment approach). For instance, recall from Sect. 2, a collaborative prototyping methodology and long-term participatory design workshops were utilized to study physical and digital privacy concerns of older adults adoption of technology.

Our data-driven approach is applicable to any project enabling extraction of metadata at every stage of data life cycle management by storing the metal data separately. The metadata are available beyond the project life cycle and when the data are restricted for publication (which may be due to legal or privacy issues) or outdated.

4.1   Baseline Metadata Conceptual Design

We also design the baseline metadata. The purpose of the metadata is to provide rich information about data. The metadata of a data will always be available even when the data is restricted for publication due to legal or privacy issue and when the data is outdated. We extract metadata at every stage of data life cycle management. Big data life cycle management is made up of one or more **tasks** including

- Planning, which defines the purpose and use of the data;
- Data collection;
- Data quality control, data cleaning, and data integration;
- Knowledge discovery through machine learning and data mining/analysis; and
- Data use, data sharing and reuse.

Data collection has both temporal and spatial components. Data are collected from heterogeneous location **site** and by multiple **stakeholders**. Data custodian has

---

[6]https://www.state.gov/key-topics-office-of-ocean-and-polar-affairs/arctic/.

contact information. **Data** types and formats depend on varieties of data. For instance, data collection during any Arctic or ocean expedition typically include hydrographic and bathymetric data, animal stock assessments, soil and salinity samples, snow samples, wind speed, air and water temperature, Data are collected during **events/tasks** using one or more **equipment resources** (e.g., research vessels, computer system, software, echo sounder, navigator, GPS receiver). The outcome of data analysis may be shared with the public through publications, which are associated with **publication** citation count. Moreover, **keywords**— when associated with the data collection event, purpose or use type—ensure that the data are **findable**, so as to comply with the "findable" principle in the FAIR Principles. Data licence is also an important metadata to collect because it gives the users information on any constraints on re-usability of the data, when and how to contact the data owner and if payment is required to reuse publicly available data. Common license includes government open licence[7] and Creative Common (CC) licenses[8] or government open licence.

5   Evaluation

We evaluated our repository conceptual model using combination of tools:

- Amazon Web Services (AWS) Simple Storage Service (S3), which provides scalable object storage infrastructure through a web service interface
- Amazon DynamoDB, which is NoSQL database service that supports key-value and document data structures for the catalogue and metadata storage

Our solution provides several benefits including (a) the solution accepts all forms of data type, structured, semi-structured and unstructured; (b) the metadata layer provides additional benefits. First, it supports machine learning to discover relationship between data and project as well as among project. Second, it aligns the data repository with the FAIR Principles of open data.

We used real-life disparate data collected from heterogeneous Arctic expeditions having diverse data types and format including:

- text file (*.txt);
- Excel file data (*.xls), say from mobile science labs and renewable-energy powered plant production pods;
- Portable Document Format (PDF) file;
- images (*.PNG, *.JPEG);
- side scan and back-scatter data in extended Triton format (*.xtf);
- geophysical data in open standard SEG-Y file format (*.segy);
- database file (*.db), which are QPS QINSy proprietary database files as maritime software solution for survey planning and real-time hydrographic data processing;

---

[7] http://open.canada.ca/en/open-government-licence-canada.
[8] https://creativecommons.org/licenses.

- Geo-database table files (*.GDBTable) and index files (*.ATX), both created with ArcGIS software;
- universal sandbox file (*.Horizon), which contains the positions and velocities of objects; and
- other semi-structured and unstructured data types such as audio, video, graph, charts, CSV, XML, JSON, hypertext markup language (HTML) and links to other web pages.

We stored the diverse big data using AWS S3 and used dynamoDB, a *NoSQL* pair-value storage database within the same web services for storing the metadata.

6   Conclusions

In this paper, we extensively discussed machine learning and big data processing challenges. First, we discussed the complexity of data type resulting in several research investigating options to overcome the rigid structure of relational database models in managing the ever growing variety of big data types. Second, we identified privacy and accessibility challenges as double edge sword research challenge of achieving balance between privacy and transparency concurrently with privacy and utility. In addition, we discussed the challenge relating to fairness and interpretability of machine learning models used in processing big data. Finally, motivated by the ever-growing big data types variety challenges, we designed a conceptual model for managing all types of big data including structure, semi-structured and unstructured data and discussed a data driven approach to collect metadata for managing the data about data beyond the data life cycle. An ongoing and future work, we explore other architectures for data repository and metadata management, as well as its applications to other real-life scenarios [52–54].

Acknowledgement. This work is partially supported by Arctic Research Foundation (ARF), Mitacs, NSERC (Canada), and University of Manitoba.

References

1. Dhaouadi, A., et al.: A multi-layer modeling for the generation of new architectures for big data warehousing. In: Barolli, L., Hussain, F., Enokido, T. (eds.) AINA, vol. 2. LNNS, vol. 450, pp. 204–218. Springer, Cham (2022). https://doi.org/10.1007/978-3-030-99587-4_18
2. Di Martino, B., et al.: Anomalous witnesses and registrations detection in the Italian justice system based on big data and machine learning techniques. In: Barolli, L., Hussain, F., Enokido, T. (eds.) AINA, vol. 3. LNNS, vol. 451, pp. 183–192. Springer, Cham (2022). https://doi.org/10.1007/978-3-030-99619-2_18
3. Fung, D.L.X., et al.: Self-supervised deep learning model for COVID-19 lung CT image segmentation highlighting putative causal relationship among age, underlying disease and COVID-19. J. Trans. Med. **19**(1), 1–18 (2021)
4. Liu, Q., et al.: A two-dimensional sparse matrix profile DenseNet for COVID-19 diagnosis using chest CT images. IEEE Access **8**, 213718–213728 (2020)

5. Souza, J., et al.: An innovative big data predictive analytics framework over hybrid big data sources with an application for disease analytics. In: Barolli, L., Amato, F., Moscato, F., Enokido, T., Takizawa, M. (eds.) AINA, AISC, vol. 1151, pp. 669–680. Springer, Cham (2020). https://doi.org/10.1007/978-3-030-44041-1_59

6. Anderson-Gregoire, I.M., et al.: A big data science solution for analytics on moving objects. In: Barolli, L., Woungang, I., Enokido, T. (eds.) AINA, vol. 2. LNNS, vol. 226, pp. 133–145. Springer, Cham (2021). https://doi.org/10.1007/978-3-030-75075-6_11

7. Barkwell, K.E., et al.: Big data visualisation and visual analytics for music data mining. In: IV, pp. 235–240 (2018)

8. Cabusas, R.M., et al.: Mining for fake news. In: Barolli, L., Hussain, F., Enokido, T. (eds.) AINA, Part II. LNNS, vol. 450, pp. 154–166. Springer, Cham (2022). https://doi.org/10.1007/978-3-030-99587-4_14

9. Cameron, J.J., et al.: Finding strong groups of friends among friends in social networks. In: IEEE DASC, pp. 824–831 (2011)

10. Leung, C.K., Jiang, F., Poon, T.W., Crevier, P.É.: Big data analytics of social network data: who cares most about you on facebook? In: Moshirpour, M., Far, B., Alhajj, R. (eds.) Highlighting the Importance of Big Data Management and Analysis for Various Applications. SBD, vol. 27, pp. 1–15. Springer, Cham (2018). https://doi.org/10.1007/978-3-319-60255-4_1

11. Leung, C.K., et al.: Personalized DeepInf: enhanced social influence prediction with deep learning and transfer learning. In: IEEE BigData, pp. 2871–2880 (2019)

12. Isichei, B.C., et al.: Sports data management, mining, and visualization. In: Barolli, L., Hussain, F., Enokido, T. (eds.) AINA, Part II. LNNS, vol. 450, pp. 141–153. Springer, Cham (2022). https://doi.org/10.1007/978-3-030-99587-4_13

13. Balbin, P.P.F., et al.: Predictive analytics on open big data for supporting smart transportation services. Procedia Comput. Sci. **176**, 3009–3018 (2020)

14. Leung, C.K., et al.: Urban analytics of big transportation data for supporting smart cities. In: Ordonez, C., Song, IY., Anderst-Kotsis, G., Tjoa, A., Khalil, I. (eds.) DaWaK. LNCS, vol. 11708, pp. 24–33. Springer, Cham (2019). https://doi.org/10.1007/978-3-030-27520-4_3

15. Angwin, J., et al.: Machine bias risk assessments in criminal sentencing. ProPublica, May 23 (2016)

16. Kilbertus, N., et al.: Avoiding discrimination through causal reasoning. In: NIPS, pp. 656–666 (2017)

17. Chiappa, S., Isaac, W.S.: A causal Bayesian networks viewpoint on fairness. In: Kosta, E., Pierson, J., Slamanig, D., Fischer-Hübner, S., Krenn, S. (eds.) Privacy and Identity. IFIP AICT, vol. 547, pp. 3–20. Springer, Cham (2018). https://doi.org/10.1007/978-3-030-16744-8_1

18. Mothilal, R.K., et al.: Explaining machine learning classifiers through diverse counterfactual explanations. In: FAT*, pp. 607–617 (2020)

19. Looveren, A.V., Klaise, J.: Interpretable counterfactual explanations guided by prototypes. In: Oliver, N., Pérez-Cruz, F., Kramer, S., Read, J., Lozano, J.A. (eds.) ECML-PKDD 2021. LNCS (LNAI), vol. 12976, pp. 650–665. Springer, Cham (2021). https://doi.org/10.1007/978-3-030-86520-7_40

20. Moraffah, R., et al.: Causal interpretability for machine learning-problems, methods and evaluation. ACM SIGKDD Explor. **22**(1), 18–33 (2020)

21. Leung, C.K., et al.: Explainable artificial intelligence for data science on customer churn. In: IEEE DSAA, pp. 235–244 (2021)

22. Leung, C.K., et al.: Explainable data analytics for disease and healthcare informatics. In: IDEAS, pp. 12:1-12:12 (2021)

23. Kostic, S.M., et al.: Social network analysis and churn prediction in telecommunications using graph theory. Entropy **22**(7), 753:1–753:23 (2020)
24. Leung, C.K., Jiang, F.: Big data analytics of social networks for the discovery of "following" patterns. In: Madria, S., Hara, T. (eds.) DaWaK, LNCS, vol. 9263, pp. 123–135. Springer, Cham (2015). https://doi.org/10.1007/978-3-319-22729-0_10
25. Yoon, B.H., et al.: Use of graph database for the integration of heterogeneous biological data. Genomics Inform. **15**(1), 19–27 (2017)
26. Bollobás, Béla.: Modern Graph Theor. GTM, vol. 184. Springer, New York (1998). https://doi.org/10.1007/978-1-4612-0619-4
27. Leung, C.K., et al.: Distributed uncertain data mining for frequent patterns satisfying anti-monotonic constraints. In: IEEE AINA Workshops, pp. 1–6 (2014)
28. Leung, C.K., Hayduk, Y.: Mining frequent patterns from uncertain data with MapReduce for big data analytics. In: Meng, W., Feng, L., Bressan, S., Winiwarter, W., Song, W. (eds.) DASFAA, Part I. LNCS, vol. 7825, pp. 440–455. Springer, Heidelberg (2013). https://doi.org/10.1007/978-3-642-37487-6_33
29. Rahman, M.M., et al.: Mining weighted frequent sequences in uncertain databases. Inform. Sci. **479**, 76–100 (2019)
30. Olawoyin, A.M., Chen, Y.: Predicting the future with artificial neural network. Procedia Comput. Sci. **140**, 383–392 (2018)
31. Leung, C.K., et al.: Fast algorithms for frequent itemset mining from uncertain data. In: IEEE ICDM, pp. 893–898 (2014)
32. Hornung, D., et al.: Navigating relationships and boundaries: Concerns around ICT-uptake for elderly people. In: CHI, pp. 7057–7069 (2017)
33. Westin, A.F.: Privacy and freedom. Washington Lee Law Rev. **25**(1), 166–170 (1968)
34. Olawoyin, A.M., et al.: Privacy-preserving spatio-temporal patient data publishing. In: Hartmann, S., Küng, J., Kotsis, G., Tjoa, A.M., Khalil, I. (eds.) DaWaK. LNCS, vol. 12392, pp. 407–416. Springer, Cham (2020). https://doi.org/10.1007/978-3-030-59051-2_28 DaWaK. LNCS, vol. 12392, pp. 407–416. Springer, Cham (2020). https://doi.org/10.1007/978-3-030-59051-2_28
35. Sweeney, L.: k-anonymity: a model for protecting privacy. Int. J. Uncertain. Fuzziness Knowl.-Based Syst. **10**, 557–570 (2002)
36. LeFevre, K., et al.: Incognito: efficient full-domain k-anonymity. In: ACM SIGMOD, pp. 44–60, (2005)
37. Li, N., et al.: Privacy beyond k-anonymity and l-diversity. In: IEEE ICDE, pp. 106–115 (2007)
38. Machanavajjhala, A., et al.: l-diversity: privacy beyond k-anonymity. ACM TKDD **1**(1), 3:1–3:52 (2007)
39. Cao, Y: Quantifying differential privacy under temporal correlations. In: IEEE ICDE, pp. 821–832 (2017)
40. Xiao, Y., Xiong, L.: Protecting locations with differential privacy under temporal correlations. In: ACM CCS, pp. 1298–1309 (2015)
41. Andres, M.E., et al.: Geo-indistinguishability: Differential privacy for location-based systems. In: ACM SIGSAC CCS , pp. 901–914 (2013)
42. Olawoyin, A.M., et al.: Privacy preservation of COVID-19 contact tracing data. In: IUCC-CIT-DSCI-SmartCNS, pp. 288–295 (2021)
43. Boyd, D., Crawford, K.: Critical questions for big data: provocations for a cultural, technological, and scholarly phenomenon. Inform. Commun. Society **15**(5), 662–679 (2012)
44. Leung, C.K., et al.: A machine learning approach for stock price prediction. In: IDEAS, pp. 274–277 (2014)

45. Leung, C.K., et al.: An innovative fuzzy logic-based machine learning algorithm for supporting predictive analytics on big transportation data. In: FUZZ-IEEE, 1905–1912 (2020)
46. Samek, W., et al.: Explaining deep neural networks and beyond: a review of methods and applications. Proc. IEEE **109**(3), 247–278 (2021)
47. Liu, C., et al.: Algorithms for verifying deep neural networks. Found. Trends Optim. **4**(3–4), 244–404 (2021)
48. Li, Z., et al.: A survey of convolutional neural networks: analysis, applications, and prospects. IEEE TNNLS **33**(12), 6999–7019 (2021)
49. Dhillon, A., Verma, G.K.: Convolutional neural network: a review of models, methodologies and applications to object detection. Progress Artif. Intell. **9**(2), 85–112 (2020)
50. Li, Y., et al.: Graph convolutional recurrent neural network: data-driven traffic forecasting. CoRR abs/1707.01926 (2017)
51. Larson, J., et al.: How we analyzed the COMPAS recidivism algorithm. ProPublica, May 23 (2016)
52. Camara, R.C., et al.: Fuzzy logic-based data analytics on predicting the effect of hurricanes on the stock market. In: FUZZ-IEEE, pp. 576–583 (2018)
53. Coronato, A., Cuzzocrea, A.: An innovative risk assessment methodology for medical information systems. IEEE TKDE **34**(7), 3095–3110 (2020)
54. Cuzzocrea, A., et al.: Tor traffic analysis and detection via machine learning techniques. In: IEEE BigData, pp. 4474–4480 (2017)

Assessment of the Use of Renewable Sources for Self-sustainability IoT Device Development

Luciana Pereira Oliveira[1]([⊠]), Edson Luís Vieira de Almeida[1],
Paulo José de Sousa Oliveira[2], and Luan Gomes de Carvalho[1]

[1] IFPB Campus João Pessoa, Av. Primeiro de Maio, 720,
Jaguaribe, João Pessoa, PB, Brazil
luciana.oliveira@ifpb.edu.br,
{edson.vieira,luan.gomes}@academico.ifpb.edu.br
[2] Banco do Brasil, João Pessoa, PB, Brazil

Abstract. Some IoT devices can be placed in hard-to-reach places and should require little maintenance. For example, avoiding replacing batteries and reducing maintenance in maritime containers far from the coast. The replacement of power supplies can have an impact on time and cost, which makes research relevant to prolong the life of these devices, providing greater use, and even self-sustainability. This work seeks to evaluate the source of solar energy and heat exchange as options that can be used to prolong the time of use of IoT devices. For this, evaluations were carried out with real devices in simulated environments containing the storage and generation of energy. Finally, it is hoped that the results of this research can be compared and used to assist in decision making for self-sustainability in IoT environments.

1 Introduction

There is a growing need to monitor different aspects: pollution level, water quality, monitoring of container loads and others using IoT (Internet of Things) mechanisms. In this context, there is a need to optimize the use of resources demanded by these devices, so it can be seen that the use of mechanisms that allow a longer life for these IoT devices is of paramount importance. That is, the optimization and balance between efficient energy consumption for transmitting data in a protocol with security.

Given these scenarios, focusing on the efficient consumption of devices and mechanisms that allow a longer useful life, it is important to reduce financial and ecological impacts. For this, the time of use of the devices must be extended and interventions avoided. Considering that the devices are possibly located in places that are difficult to access, solutions that provide a longer survival of the device can be extremely relevant. For example, a sensor can be on the high seas, collecting information and synchronizing this data with its gateway, this device is fully exposed to sunlight. This can make it possible to take advantage of this energy source, either through the sun's rays or the temperature exerted on it. These renewable sources allow you to have a self-sustaining device, using alternative energy sources during the day. So energy consumption in IoT devices can be considered one of the main points of relevance in research, given that this point can directly impact definitions of architectures and models. To study this, this article presents an investigation to answer the following questions:

- Which renewable resource can generate the highest level of energy?
- What is the additional energy consumption by the security mechanism?
- What is the additional communication time and electrical current by the security mechanism?

Although the answers to these questions can be expected to be additional costs generated by the security mechanism, it is important to perform measurements to obtain the real data. The values can be used in statistical models that will help in the study of the optimization behavior of the analyzed mechanisms. Measurement results in statistical models may assist in predicting and comparing solutions with a certain degree of safety.

Therefore, this work is based on the evaluation of different mechanisms aimed at harnessing energy from the environment and positively impacting the consumption and power of devices. Section 2 addresses a background in relation to sustainable scenario, security, models, architectures and protocols for IoT application. Section 3 discusses works related to the proposed theme. Section 4 presents the methodology, materials and methods applied in the work, while Sect. 5 shows the results of the experiments. Finally, Sect. 6 concludes the work by presenting a brief summary, applicability, problems and future work.

2 Background

2.1 Ocean Monitor as Sustainable Scenario

The implications of climate change include impacts to fisheries, aquaculture and transportation sectors [4]. For this reason, ocean monitoring systems and marine instruments have received increasing attentions and they are important to enable the definition of sustainable strategies and decision making [5]. Initially, in ocean observatories, the devices used protocols not standardized, serial communications interfaces that were not designed for network operation and many solutions were not economy viable for the most companies (generally small in oceanographic sector). For example, in 2011, work [6] proposed an ocean monitoring prototype consisting of a low-cost microcontroller running with an Ethernet interface. He considered different sleep modes to have very low power consumption, as battery replacement in various sensors may require expensive procedures: moving ships, vehicles and divers.

Recent works likes [5,7] considered wireless communication and maximum power efficiency as highly desirable for smart ocean systems, because energy is the scarce resource. In 2019, the paper [5] proposed MQTT (Message Queue Telemetry Transport) as communication protocol and microcontroller ESP32 with Wifi capacities instead of HTTP that was used by solution with Ethernet interface from [6] in 2011. The MQTT messages are smaller than HTTP messages and as a result they consume less power [8]. Moreover, the ESP32 is a widely used in wide scenario of Internet of Thing (ocean is one), because it is characterized by low cost, 802.11 and Bluetooth as transmission capabilities, a coprocessor and advanced power modes that allow it to be used with batteries.

However, since 2008 [9] and 2009 [10] the information captured by sensors (meteorological conditions) are considered as critical for effective decision making of users that use ocean observing systems in order to have a better understand the ecosystem. Moreover, the data captured and analyzed by devices are considered as critical for several other IoT scenarios. For this reason, several studies have been conducted pointing out the importance of using security mechanisms in the IoT environment, either in data privacy [11] or in the authentication of nodes that are part of the network [12]. Numerous approaches have been used in such research, such as using the MQTT protocol as a message transmission mechanism [13] and the use of Transport Layer Security (TLS) as a security mechanism [14]. A feature of some of these approaches is the use of tools that provide a virtual simulation environment [15].

2.2   Key Factors for the Implementation of the IoT Environment

The definition of a standardization model and a communication architecture between nodes are key factors for IoT implementation. This is something so important that the ITU has defined a document with the main indications and classifications for IoT solutions.

Different devices, whether 8, 16, 32 or even 24 bits, can be part of the same environment, as long as the adopted model provides for interoperability in their communication. The standardization in the communication between the nodes is important because it brings ease in the management of the communication between them. Architectures propose the use of brokers and middlewares that are capable of operating and converting different protocols, bringing interoperability in the communication between nodes.

So, IoT has specific characteristics and requirements, such as limited resources and available power, but the unified management of devices can become interesting as the IoT network expands. This requires greater prevention and detection of malicious or infected devices, the possibility of updating the device, controlling hardware resources, even monitoring its location.

2.3   Standardization Defined in the ITU Y.4460

The Architectural reference models of devices for Internet of things applications recommends numerous models and architectures varying according to mechanisms, solutions and protocols addressed, it is part of a series of recommendations defined in the ITU-T Y-SERIES RECOMMENDATIONS and is part of the INTERNET OF THINGS AND SMART CITIES AND COMMUNITIES recommendations section the ITU recommendations define the classification of devices by their processing power and communication capabilities, also demonstrating interactions and architectural references. the reference defines three types of devices:

1. low processing and low connectivity device - LPLC: an IoT device that acts only as an interface for collecting data on physical things or the surrounding environment, and/or performing operations on physical things or the surrounding environment. This device does not have enough processing resources to make decisions or execute complex algorithms; it also doesn't have enough connectivity features to connect directly to communication networks.
2. low processing, high connectivity device - LPHC: an IoT device that acts only as an interface for collecting data on physical things or the surrounding environment, and/or performing operations on physical things or the surrounding environment. This device has enough connectivity features to connect directly to communication networks, so it does not require an intermediary device (for example, the gateway).
3. high throughput and high connectivity device - HPHC: an IoT device that not only has high connectivity capabilities, allowing it to directly connect to cloud applications and services, and have high enough processing resources to make decisions and run complex algorithms (for example, artificial intelligence). Devices are standalone. They make decisions about their own functions and can coordinate other devices as well.

2.4   IoT Protocols and IoT Security

MQTT and CoAP are widely used protocols in IoT [18,30]. CoAP uses the UDP protocol to transmit messages and MQTT uses TCP. The following security mechanisms can be used: TLS and DTLS, respectively for MQTT and CoAP. However, MQTT is one of the most promising protocols for small devices in the IoT environment [19].

TLS is a protocol containing a handshaking phase in which it is used to authenticate client/server and to negotiate the type of cipher suites (encryption algorithm and keys that will be used in data transmission) [17]. For example, in ESP32 is possible to use TLS_ECDHE_RSA_WITH_AES_256_GCM_SHA384 as the cipher suite. It inform that ECDHE_RSA is a method to exchange secret keys over an insecure channel in order to start encrypted communication. AES_256_GCM is an encryption method, using the 256 bit key learned over the TLS connection using ECDHE RSA. GCM is a mode of AES. GCM is an improvement over the old CBC mode. SHA384 is a data integrity method to ensure the message has not been tampered with by using a calculated 384 bit hash quantity.

In [14], TLS is identified as a solution to ensure security. The paper presents the advantages of using the TLS handshake protocol as the core of TLS, because it is responsible for all the negotiation of the security parameters. The paper recognizes that current handshake mechanisms have high overhead and too much time to complete the handshake. Therefore, a new form of handshake is proposed to reduce overhead and long handshake latency. The paper proposes alternatives using IBC-ECDH and parity mechanisms. Finally, the performance is analyzed by comparing storage cost, latency and power consumption.

The paper [16] analyzes the energy demand and processing of the TLS protocol in IoT applications. They compare the performance of the TLS handshake with 3 different cipher suites. In the study, a Cypress CYW43907 IoT was used as a client and a Raspberry Pi as a server. The work uses a platform to measure the power consumption of IoT devices and the measurements taken prove the efficiency of ECDSA in IoT applications when compared to RSA.

3   Related Self-sustainable Works

There are papers that contribute with models and algorithms considering self-sustaining IoT devices and an paper with experimental values related to solar panels to be used by these models was found.

The paper [2] is a model defined to be used with multiple energy sources. This article presents probabilistic energy models for collecting hybrid energy for IoT nodes. The work aims to group the energy obtained from multiple sources to a single power supply, the model classifies each source into clusters and uses Gaussian models to calculate the possible combinations, showing a general expression and providing a probabilistic model for each cluster.

The work [26] defines a design for different power supplies for IoT devices, starting from the premise that the devices must operate and survive in unforeseen conditions. Additionally, it uses power conversion mechanisms that implement algorithms capable of estimating by prediction allowing the input of multiple power supplies, such as light, vibration and temperature.

It is possible to notice in [3] a study that shows the relevance of low power devices using solar panels, proposing a methodology that analyzes the behavior of solar systems by evaluating control algorithms to track the maximum power, aiming to improve the use of energy generated, the work gets from 87% to 97% efficiency.

Finally, in [1], the paper presents a self-sustainable IoT design and prototype based on capturing energy through a small solar panel. A proof of concept was made with the device in a controlled environment. Capacitors and cells were used with lithium battery to store energy. The study points out that the device can work for a year with transmission intervals of 55 s, with an average range of 1.8 km.

Therefore, this paper is similar to the last one that contributed with experimental values. The difference is that we provide values related to peltier source and compare with solar panel and consider energy consumption by security mechanisms.

4   Methodology

This work is classified, in terms of its approach, as a quantitative research, considering that its results and samples can be quantified numerically. In relation to its nature, this work is classified as an applied research, since it aims to generate knowledge for practical application. It is an explanatory work, when classified according to the objective, since it seeks to identify the factors that contribute to the occurrence of phenomena. Regarding the procedures used, this research can be classified as experimental, considering that it aims to analyze and evaluate different mechanisms that can bring benefits to the energy consumption of IoT devices. In addition, the present research was carried out with six stages that is explained as following:

In the first stage, two activities related to identify and design scenarios were defined to capture data for experiments with ESP32:

- Investigate two MQTT scenarios: the communication with TLS (specifically the version 1.2 with cipher suite TLS-ECDHE-RSA-WITH-AES-256-GCM-SHA384") and without TLS.
- Investigate a third scenario with peltier and another with solar panel in order to reduce the need to change batteries for monitoring systems.

In the second stage, this research defined metrics to investigate renewable resource and security mechanism in IoT device, considering the systematic reviews in [20, 28, 29]:

- The frame and payload size in bytes of the MQTT messages;
- The mean, the standard deviation, the median and the mode of communication time in ms between publish (ESP32) and subscribe device (desktop). The communication time can be seen in Fig. 1. This metric corresponds to "ttotal", where "tpub" is the time of publication by ESP32, "tp2b" is the time of message delivery from the publisher to the broker, "tbroker" is the time of message being processed by the broker, "tb2s" is the time of sending data from the broker to the subscriber and "tsub" is the confirmation that the subscriber has received data;
- The mean, the standard deviation, the median and the electric current in mA;
- Energy consumption estimation in mW, mWh and quantity of hours to use one battery with 3000 mAh, considering 5 V and the current measured during communication.
- Energy generated by peltier and solar panel in mW.

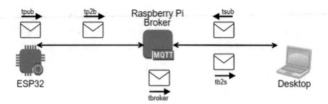

$$ttotal = tpub + tp2b + tbroker + tb2s + tsub$$

Fig. 1. Communication.

In the third stage, the environment (topology, protocols and traffic) was defined. The topology followed Fig. 2 was centralized in IP layer with four devices, but in MQTT layer only one publish node generated message that were forwarded by broker to subscribe node. There was a component to measure size of the frame and payload and "ttotal" metrics in device with broker functionality. Moreover, a component to measure the energy consumption was in publish node with battery. All these devices was performed on a bench with about 0.8 m² and the addressable devices used the IPv4 protocol at the network layer and 802.11n at the link layer.

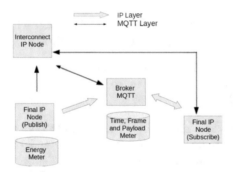

Fig. 2. Topology.

In terms of the total amount of frame size increase over MQTT without TLS was 29 bytes, such an increase was already expected given the increased information in the upper layer protocol headers. According to Fig. 3, it is noteworthy that using TLS added 17 bytes to the payload. The frame and payload size for both was unchanged during all publications, so the standard deviations was zero.

In terms of traffic, it was synthetic. The ESP32 has been configured to generate and to send one MQTT message per second to the "ESP32/PUB" topic where such a post used 0(zero) as QoS level and the information in message was a random number between 0 and 40.

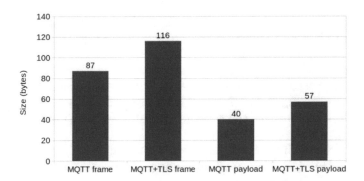

Fig. 3. Message size.

In the fourth stage, a study was carried out to understand the tools used in the environment defined in the previous stage. The components were defined to acquire the data from the experiments, considering the metrics in the second stage and the lower cost to extract the data.

In terms of software components, the tcpdump tool was located at the broker to collect the metrics according to Fig. 2. The mosquitto application was at the broker component to allow the MQTT communication. The operational system raspbian was at the broker in order to allow the installation of mosquitto and tcpdum tool at the broker component. However, the application to generate and to receive MQTT message, with and without TLS, needed to be elaborated in the five stage.

In terms of hardware, the definition of each one and the mapping between device and functionality are in Table 1. In addition, a tool developed in [27] was used to automate the capture of the electric current and to reduce the error when the multimeter was used to measure the electric current of the ESP32.

In the sixth step, the measurements were performed. In particular, the measurement of energy production for the ESP32 was carried out between 12 noon and 1 pm. The measurement of energy production considered the use of panel solar in the period from 1/1/2019 to 12/1/2019. In a second step, the value of the energy produced by a peltier coupled to esp32 from 13/1/2019 to 24/1/2019.

Table 1. List of devices

Device	Functionality
ESP32 [21]	Final node (publisher)
Raspberry pi 3 model B [22]	Broker MQTT
Notebook acer a515-51g-58vh [23]	Final node (subscriber)
Wireless router intelbras iwr 1000N [24]	Interconnect final nodes
Multimeter minipa ET-2231 [25]	Measure the eletric current

To answer the question about energy measurement, three applications were designed in fifth stage to interact with MQTT broker: one software for Desktop and two programs for ESP32.

The Paho-MQTT library was used to allow desktop and broker interaction without TLS (port 1883) and with TLS (port 8883 and containing one certificate). Considering Fig. 1, this application receives a message at "tb2s" time and immediately, after receiving the message, publishes, in "tsub" time, the response message to the broker with Mosquitto.

The WiFiClient library was used to build the application without TLS for ESP32. The WiFiClientSecure library was used in the implementation of the program with TLS for ESP32. These two programs have the PubSubClient as common library that allows ESP32 to connect to the MQTT broker. In addition, each one was responsible for posting random temperature values to the "ESP32/PUB" topic.

In the fifth stage, the total amount of measurement was 10 transmissions of the messages (each one for 100 s) by scenario type (with and witout TLS). Each one was repeated 100 times totaling 1000 execution per scenario that was evaluated by the metrics defined in the second stage. After having the collected data in the spreadsheet, graphs have been prepared for further analysis.

In the sixth stage two mechanisms to extend the time and the operation were evaluated: peltier and solar plate. The peltier allows the conversion of thermal energy into electricity. The solar plate converts the sun's energy into electrical energy. In both, it was sought to identify how many of these pieces of equipment would be necessary so that the device did not need a battery during sunny and hot days (Fig. 4).

(a)Peltier   (b)Solar panel

Fig. 4. Devices.

5   Results

5.1   Which Renewable Resource Can Generate the Highest Level of Energy?

The proposed solutions were evaluated regarding the gain in energy consumption, based on the specified metrics, it is possible to observe that the proposed solution regarding the use of solar energy proved to be significantly relevant, considering that it generated the highest level of energy, when compared with the peltier insert, providing longer life and faster charging for the battery that powers the device (Fig. 5).

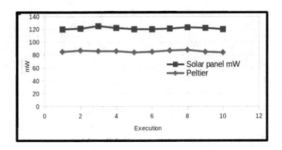

Fig. 5. Level of energy consumption.

The device lifetime considered the execution of the MQTT experiments that analyzed the additional cost of TLS. So, the results are based on the time it can remain active. That is, the lifetime using the battery plus the extra power supply analyzed, so that it covers the partial or total energy consumption generated by the device. Observing the lifetime in the results in Fig. 6, the use of the solar panel proved to be more efficient, providing the device with a longer lifetime.

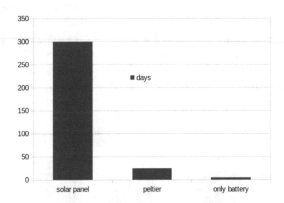

Fig. 6. Lifetime prediction for devices

5.2   What Is the Additional Energy Consumption by the Security Mechanism?

Considering the Power Dissipation as the result of voltage * electric current and the voltage as 5V, the Table 2 presents the estimation of the power dissipation mean, as well as the energy consumed and a predicted battery life based on the values obtained in the experiment.

Table 2. Estimation of energy consumption

Scenario	Power Dissipation mean	Energy consumption	Life (hs with 3000 mAh)
MQTT	313,5 (mW)	26,1 (mWh)	47,8
MQTT+ TLS	368,1 (mW)	32,5 (mWh)	40,7

The Fig. 6 shows the energy generated by solar panel and peltier. They can support the additional cost of using TLS. The solar panel offers benefit in terms of an energy level higher than that spent on TLS, when compared to the peltier that can offers to recharge the battery and also to cool the ESP32. This cooling can prevent packet loss due to high temperature in regions close to the Earth's equator.

5.3   What is the Additional Communication Time and Electrical Current by the Security Mechanism?

The Fig. 7 present energy consumption (eletric current) ant it was highest in MQTT + TLS. It shows 52.789 ms as the mean communication time in the sample of 1000 publications for MQTT, while MQTT with TLS was 52.0787 ms. The standard deviation was 13.897 for use without TLS and 8.812 for use with TLS. The median obtained was 51 ms without TLS and 54 ms with TLS. The mode was 58 without TLS and 54 ms with TLS. These values show that encryption has little influence on the communication time of the message. The explanation for this lower TLS value is that the measurements were performed by activating the hardware acceleration function for encryption in ESP32.

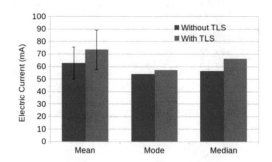

Fig. 7. The additional electrical current

The Fig. 8 presents the communication time in terms of mean, mode, median and standard deviation. The time in scenario without TLS was more variable (high standard deviation with 15.81) when compared 12.79 to time with TLS, although the standard deviation of communication time was only 8.812 for use with TLS.

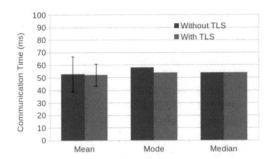

Fig. 8. The additional communication time

6   Conclusion

This work used a methodology with six stages related security and energy consumption of IoT devices in order to answer three research questions. The experiments used MQTT messages without and with security using real device (the ESP32). The security mechanism was TLS_ECDHE_RSA_WITH_AES_256_GCM_SHA38.

The results pointed out the solar panel as the renewable resource that generated the highest level of energy. The peltier had low energy output as it did not produce energy when it was in thermal equilibrium. Therefore, the device with solar panel and battery was able to remain in operation during the nocturnal period, and at dawn the recharge cycle restarts.

Others results identified that after the TLS negotiation phase, there was an increase in the value of ESP32 energy consumption corresponding to 14.83%. Additionally, an increase in secure messages sizes has been identified. The increase was 29,82% for payload size and 25% for frame size.

Although there was an increase in value of message size and energy consumption, this case study demonstrated that there was a reduction by 3.49% in the average communication time of secure MQTT messages, because the accelerator hardware from ESP32 was active for AES algorithm.

These results can encourage the design of new solutions with ESP32 using TLS, because the solution will have encrypted messages and also a better performance when there is no power restriction. For example, a sustainable scenario where ESP32 uses solar, peltier, wind and other energy.

The main contributions of this paper are these results of the evaluation and the methodology used to analyze the MQTT with and without TLS. The results presented in this work can be used in statistical models that will help in the study of new optimization solution. Moreover, the methodology adopted in this work can be replicated to evaluate the energy consumption and the researches will not need to acquire expensive devices such as oscilloscope.

For future works, it would be of vital importance to propose and to evaluate new compression mechanisms to reduce the payload size of TLS messages, because the smaller messages consume less energy. An another important work to be done is the investigation of the security mechanisms in terms of standby strategies from ESP32.

For instance, the authors intend to study, propose and evaluate an adaption for MQTT+TLS considering compression and some environment characteristics in order to control the protocol and device to dynamically to force standby mode or to turn off. One example is when ESP32 to use solar energy and the environment has clouds, in this situation, the work will define how to change ESP32 for standby mode and to keep MQTT+TLS connection.

Another future work, it is suggested the creation of a mechanism to avoid the loss of packages in moments of high temperature. The system can be composed of peltier to cool the ESP32 and the solar panel can power the battery. Moreover, other solutions can be evaluated to generate energy, such as energy and wind.

Therefore, the research can create and evaluate new solution or optimization mechanisms and identify the gap between energy reduction identified for new approach and the results of this study.

References

1. Kjellby, R.A., et al.: Self-powered IoT device based on energy harvesting for remote applications. In: 2018 IEEE International Conference on Advanced Networks and Telecommunications Systems (ANTS), Indore, India, pp. 1–4 (2018). https://doi.org/10.1109/ANTS.2018.8710171
2. Altinel, D., Karabulut Kurt, G.: Modeling of multiple energy sources for hybrid energy harvesting IoT systems. IEEE Internet Things J. 6(6), 10846–10854 (2019). https://doi.org/10.1109/JIOT.2019.2942071
3. Ram, S.K., Sahoo, S.R., Sudeendra, K., Mahapatra, K.: Energy efficient ultra low power solar harvesting system design with MPPT for IOT edge node devices. In: 2018 IEEE International Symposium on Smart Electronic Systems (iSES) (Formerly iNiS), Hyderabad, India, pp. 130–133 (2018). https://doi.org/10.1109/iSES.2018.00036.
4. Barange, M., Cochrane, K.: Chapter 28: impacts of climate change on fisheries and aquaculture: conclusions, vol. 2018, pp. 611–628, July 2018
5. Huang, A., et al.: A practical ma-rine wireless sensor network monitoring system based on lora and mqtt, vol. 2019, pp. 330–334, May 2019
6. Toma, D.M., et al.: Smart sensors for interoperable smart ocean environment. In: OCEANS 2011 IEEE - Spain (2011)
7. Orekan, T., Zhang, P., Shih, C.: Analysis, design, and maximum power-efficiency tracking for undersea wireless power transfer. J-ESTPE (2018). https://doi.org/10.1109/JESTPE.2017.2735964
8. Dizdarevi, J., et al.: A survey of communication protocols for internet of things and related challenges of fog and cloud computing integration. ACM Comput. Surv. 51(6), 1–29 (2019)

9. Cater, N.E.: Smart ocean sensors web enabled ocean sensors for aquaculture. OCEANS **2008**(2008), 1–9 (2008). https://doi.org/10.1109/OCEANS.2008.5151870

10. Cater, N.E., Eng, P.: O'Reilly, T.: Promoting interoperable ocean sensorsthe smart ocean sensors consortium. In: OCEANS 2009, vol. 2009, pp. 1–6 (2009)

11. Kozlov, D., Veijalainen, J., Ali, Y.: Security and privacy threats in iotarchitectures. In: Proceedings of the 7th International Conference onBody Area Networks, BodyNets, ICST (2012)

12. Chavan, A.A., Nighot, M.K.: Secure and cost-effective applicationlayer protocol with authentication interoperability for IoT. In: Procedia Computer Science (2016)

13. Giambona, A.E., Redondi, C., Cesana, M.: Mqtt+: Enhanced syntaxand broker functionalities for data filtering, processing and aggregation. In: Proceedings of the Q2SWinet (2018)

14. Mzid, R., et al.: Adapting tls handshakeprotocol for heterogenous ip-based wsn using identity based cryptog-raphy. In: International Conference on Wireless and UbiquitousSystems (2010)

15. Hasan, H., Alhusainy, B.: Evaluation of MQTT protocol for IoT basedindustrial automation **8**(2018), 19364–19369 (2018)

16. Gerez, A.H. et al.: Energy and processing demand analysis of TLS protocol in internet of things ap-plications. In: IEEE International Workshop on Signal ProcessingSystems, SiPS (2018)

17. Chalouf, M., Krief, F.: A secured service level negotiation in ubiquitous environments. IJCNIS **1**(2), 9–18 (2009)

18. MARTINS, ZEM, Estudo dos protocolos de comunicação mqtt e coap para aplicações machine-to-machine e internet das coisas (2015). https://pdfs.semanticscholar.org/7d08/7f8cbc95f8d73ba402a9e7ee515ba545e8a6.pdf

19. Al-Fuqaha, A., et al.: Internet of things: a survey on enabling technologies, protocols, and applications. In: IEEE Communications Surveys Tutorials, pp. 2347–2376 (2015)

20. Oliveira L.P., Vieira M.N., Leite G.B., de Almeida E.L.V.: evaluating energy efficiency and security for internet of things: a systematic review. In: Advanced Information Networking and Applications. AINA 2020 (2020). https://doi.org/10.1007/978-3-030-44041-1_20

21. https://www.espressif.com/sites/default/files/documentation/esp32-wroom-32datasheeten.pdf

22. https://docs-europe.electrocomponents.com/webdocs/14ba/0900766b814ba5fd.pdf

23. https://www.pauta.com.br/pdf/32539.pdf

24. http://hotsite.intelbras.com.br/sites/default/files/downloads/datasheetiwr1000nsite02-180.pdf

25. https://www.vectus.com.br/wp-content/uploads/2018/04/multimetro-digital-ET-2076-manual.pdf

26. Siskos, S.: Design of a flexible multi-source energy harvesting system for autonomously powered IoT:The PERPS project. In: 29th International Symposium on Power and Timing Modeling. Optimization and Simulation (PATMOS), vol. 2019, pp. 133–134 (2019). https://doi.org/10.1109/PATMOS.2019.8862078

27. Oliveira, L.P., et al.: Deep learning library performance analysis on raspberry (IoT Device). In: Barolli, L., Woungang, I., Enokido, T. (eds.) AINA 2021. LNNS, vol. 225, pp. 383–392. Springer, Cham (2021). https://doi.org/10.1007/978-3-030-75100-5_33

28. Filho, R.M.P.T., Oliveira, L.P., Carneiro, L.N.: Security, power consumption and simulations in IoT device networks: a systematic review. In: Barolli, L., Hussain, F., Enokido, T. (eds.) AINA 2022. LNNS, vol. 451, pp. 370–379. Springer, Cham (2022). https://doi.org/10.1007/978-3-030-99619-2_35
29. Oliveira, L.P., da Silva, A.W.N., de Azevedo, L.P., da Silva, M.V.L.: Formal methods to analyze energy efficiency and security for IoT: a systematic review. In: Barolli, L., Woungang, I., Enokido, T. (eds.) AINA 2021. LNNS, vol. 227, pp. 270–279. Springer, Cham (2021). https://doi.org/10.1007/978-3-030-75078-7_28
30. Menezes, A.H., Kelvin, R.D.O., Oliveira, L.P., Oliveira, P.J.D.S.: IoT environment to train service dogs. In: IEEE First Summer School on Smart Cities (S3C), Natal, Brazil, pp. 137–140 (2017). https://doi.org/10.1109/S3C.2017.8501386

BAIN: Bluetooth Adaption for IoT Device

Luciana Pereira Oliveira[1]([⊠]), Cinaglia Adagles Silva[1],
Djamel F. Hadj Sadok[3], Tarciana Dias Silva[2], Auristela Silva[3],
and Judith Kelner[3]

[1] IFPB campus João Pessoa, 720 Primeiro de Maio Ave.,
João Pessoa, PB 58015-435, Brazil
`luciana.oliveira@ifpb.edu.br`, `cinaglia.silva@academico.ifpb.edu.br`
[2] Rua Benfica, 455, Recife, PE 50720-001, Brazil
`tarciana.dias@ecomp.poli.br`
[3] UFPB, Av. Prof. Moraes Rego, 1235, Recife, PE 50670-901, Brazil
`{jamel,auristela,kelner}@gprt.ufpe.br`

Abstract. Many Internet of Things (IoT) distributed applications
adopt the Message Queuing Telemetry Transport (MQTT) architecture
based on publisher/subscriber model. However, it is considered a relatively heavy protocol for its handling by constrained devices and sensor
networks. As a result, the MQTT for Sensor Networks (MQTT-SN) is
seen as variant application protocol targeting the design of an energy
efficient solution that can be deployed using wireless technologies such
as Bluetooth. However, both architectures fail to attend the challenges
faced by the publish message for more than eight IoT devices (large networks with low energy). In this paper, we adapt the Bluetooth protocol
to work in a network based on topics as MQTT-SN and MQTT, which
we call BAIN (Bluetooth Adaption for IoT Networks). This new protocol
performs routing considering topics information and manages to build a
topology with a larger number of nodes. As with the original Bluetooth,
BAIN applies to short-range networks. We used the PROMELA language
and SPIN tool for its formal verification.

1   Introduction

The Internet of Things (IoT) provides brings new benefits to diverse domains
such as agriculture, healthcare, homes, transport logistics, smart cities, climate
and ocean monitoring, manufacturing among many others. The Message Queuing
Telemetry Transport (MQTT) architecture is often the design of choice adopted
by IoT applications. This paradigm is based on a publisher/subscriber model
and has gained a great deal of interest. The MQTT architecture organizes client
server communications according to topics where servers publish their data.
Clients are request to subscribe to a given topic in order to gain access to its
content. In actual implementation terms, the word topic refers to an UTF-8
coded string that is used to forward messages for other devices. Overall, MQTT

is considered a heavy protocol when used by constrained devices and sensor networks.

Consequently, MQTT for Sensor Networks (MQTT-SN) [17] was adapted from MQTT, especially designed as an energy-efficient solution that can be used by non-TCP/IP networks, such as Bluetooth. However, MQTT-SN is not projected to solve how to route messages in a network with more eight Bluetooth devices.

Bluetooth is a popular technology for short range communications. It provides some useful features for IoT routing such as device discovery (Bluetooth inquiry mode allows devices to be discovered) and consequently their relatively easy adaptation for handling topological changes, low processing and communication costs, robustness and reliability and low user maintenance.

On the other hand, Bluetooth does not support native mechanisms to disseminate information based on the publisher/subscriber model and was not designed to handle a high number of participating nodes. These features will be supported by BAIN (Bluetooth Adaption for IoT Networks) that is presented in this paper. The BAIN includes the design of a routing protocol based on topics and support for a larger topology. In this context, topic related information has been included into routing messages of Blueline, an algorithm used to create and maintain a large number of Bluetooth nodes. BAIN manages the topology on demand.

To summarize, this paper presents BAIN, a protocol created to deal with the challenge imposed by low energy constrained devices that collect and transmit data, satisfying topic requests. BAIN will be applied as an intra-domain protocol across short-range networks - typically wireless networks with stringent power control. The routes will be associated not only to target addresses but also to the topic and data they offer. This article seeks to answer the following issues:

- How could more than eight Bluetooth devices be integrated to MQTT-SN?
- How to verify the routing based on topics of a Bluetooth network?

This article is organized as follows: Sect. 2 provides an overview of Bluetooth technology, MQTT-SN and IoT, Sect. 3 presents the BAIN protocol. Section 4 details the formal verification of the BAIN using SPIN and PROMELA. Finally Sect. 5 describes the conclusion and future works.

2 Background

2.1 IoT Protocols

The paper [4] cited IoT scenarios considering HTTP, CoAP, MTTQ and AMQP as protocols. The systematic review considering power constrained devices in [18] identified 11 papers with CoAP, 3 with HTTP, 1 with MQTT and 0 with AMPQ. The major number of papers with CoAP can be explained that device's HTTP, MTTQ and AMQP are always actives, impacting in higher energy consumption due to processing retransmissions and maintaining open TCP connections.

Although MQTT and CoAP are lightweight architecture, MQTT is often not used by sensors.

Consequently, MQTT-SN was designed and tailored for use by low-energy devices and microcontrollers with low consumption of resources. The MQTT-SN architecture is based on topics to publish or subscribe data and consists of three main MQTT-SN functionalities in Fig. 1: MQTT-SN clients subscribed on topics, MQTT-SN gateways (GW), and MQTT-SN forwarders (FW). MQTT-SN clients connect to an MQTT broker via an MQTT-SN GW using the MQTT-SN protocol. An MQTT-SN GW can be an in or out type of MQTT broker. In case of a GW out broker, the MQTT protocol is used between the MQTT broker and the MQTT-SN GW. Its main function is the translation between MQTT and MQTT-SN. For example, the MQTT broker can have two interfaces: Wi-Fi and Bluetooth. MQTT messages may be exchanged over Wi-FI using TCP, while MQTT-SN messages could be transported using Bluetooth. MQTT-SN clients can also access a GW via an FW that encapsulates the MQTT-SN frames received on the wireless link and forwards them unchanged to the GW.

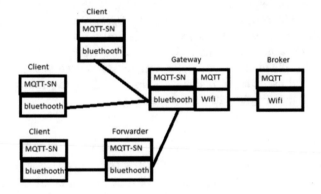

Fig. 1. MQTT-SN components.

The MQTT-SN architecture reduces bandwidth and memory consumption using topic ID instead of topic name. This can be configured in two ways: a client sends a registration request with topic ID and topic name to a broker in order to use only ID after the registration is accepted; or topic ID can be configured in MQTT-SN gateway and topic registration message can be skipped before publishing messages.

2.2   Bluetooth

Bluetooth is the industry standard for Wireless Personal Area Network (WPAN) communication and was developed to address short-ranged environments characterized by low overhead, flexibility, topology adaptation and low power consumption. The Bluetooth nodes form a network topology called piconet that is composed at most 8 nodes - 1 master and 7 slaves. This organization results in a star formation of up to 8 time-slot and frequency synchronized nodes. A set of piconets may form a larger network called scatternet, since the designation of nodes that may synchronize more than one piconets, and therefore may receive and send data between them. However, actual scatternet support remains missing as the standards only describe simple forms of scatternet formation such as the connection between two piconets. This connection can be made by any node that may become a bridge if it senses the presence of a foreign master node, through the INQUIRY/PAGING procedure [1]. The bridge is a node that has two roles, but not within the same piconet, since it may be slave in one piconet and master in the another one.

Given that Bluetooth does not directly support scatternet formation, strategies have been designed to create, maintain and route data across a scatternet topology. Existing protocols can be classified into four classes based on the used topology: star [6], tree [7], ring [8] and graph [11]. Star, Tree and Ring topologies are the simplest ones since all devices or nodes are located within the same transmission range or even in a scatternet hierarchy, where there are components connected through their respective leaders ([9] and [10]). Tree and Star topologies create bottlenecks at the root nodes because they provide only a single path to reach any given node, hence offering limited robustness and efficiency. Besides, the nodes need to remain within connectivity range. Differently from other protocols, Blueline [11] builds and maintains scatternets in a distributed way, without the need for all nodes to be in the same transmission range, providing as a result multiple paths among nodes.

2.3   Bluetooth and IoT

Considering a piconet as a short-range with MQTT-SN clients and MQTT-SN FW or GW, these collection can be considered as a larger MQTT-SN that consists of several Bluetooth networks linked together in order to reach or share data in an efficient way. Hence, the aggregation of piconets can be seen as a type of composition in MQTT-SN with the goal of providing a specific service to the user. Recall that Bluetooth does not directly support scatternet formation. As a result, it provides no path redundancy considered an important requirement to achieve network robustness, fault tolerance and high availability.

MQTT-SN GW can create a Wi-Fi network where each one also has a Bluetooth interface associating it to up to 7 other Bluetooth devices (as limited by Piconet membership).

To address the above limitation, a new routing protocol is required for use by IoT devices. The dynamic routing disseminates topic ID from MQTT-SN, instead of working only with end-points addresses, as currently is the case with Bluetooth. Therefore routes must be associated with the topic ID publish or subscribe message sent by each node in a network. Moreover, routing protocols in this new approach need to have a high level of adaptability to topological changes.

3   The Routing Proposal - BAIN (Bluetooth Adaption for IoT Device)

BAIN introduces a number of changes to the Blueline algorithm. It is unique in that it offers scatternet formation due to its support of multipath, an important design requirement for example, for networks with only a single MQTT-SN GW. The following sections will detail the BAIN.

3.1   Services in Packets and Structure of BAIN

The services are functionalities from MQTT-SN described on Sect. 2.1. The BAIN protocol route services request with following messages:

- Request Routing Packet (RRP)- This is sent when a node wants a service (MQTT-SN functionality). This packet contains: *a)* Packet-type. For instance, the type of a RRP is set to 1; *b)* ID: a unique identifier of the packet; *c)* Services: this is a set of the service ID field; *d)* Hop to live: a field that is decremented whenever a required service is not encountered in the node.
- Response Routing Packet - In response to request related MQTT-SN funcionality, the returned packet consists of: *a)* Packet-type: type of this packet, set to number 2 type; *b)* ID: this is a unique identifier of the packet as also its sequence number; *c)* BD_ADDR: the physical address of the node with MQTT-SN functionalities or knows where they are located; *d)* Hop number: the number of hops to reach the service (MQTT-SN functionality); *e)* Services: those that were found inside the network. For example, the path to find the nodes subscribe on topic that starts at the MQTT-SN gateway until the set of BD_ADDR of nodes indicated on the packet.

The protocol has also two storage structures (query buffer and cache). They have the following respective entry formats:

- Query Buffer Entry - Whenever a node does not have the service in its cache, a query buffer entry is stored and then the node has to send a request routing packet to its neighbors in order to ask for the service(s), as was described in the routing protocol state machine. The entry contains: *a)* the ID. The value of this field is a unique identifier of the entry, similar to a sequence number. This ID is also like the request routing packet id that did not receive a response yet. *b)* BD_ADDR: physical address of node requiring service(s). *c)* Services: the service(s) required.
- Cache Entry - used to store which service(s) were already discovered and which node owns them. The entry contains the additional fields *a)* BD_ADDR: physical address owning the service(s). *b)* Services: service that the node knows where are located. *c)* Hop number: the hop number needed to reach a service.

Therefore, BAIN introduces new features to MQTT-SN. It associates services to routes and floods the data. For BAIN to work with a service, the request message from Blueline was modified to attend the node that seeks to service (MQTT-SN functionalities based on topics). The response message was modified to discover the devices for a given service.

Each node establishes a service route to several destination on demand. A node stores the information about the location of a service and according to it receives a service request. Based on this new cache and buffer structures, considering host and services, each node builds on demand a routing table, where the service name is the key. A node stores and floods service requests and keeps in its cache the service and its location in order to reach subscriber nodes for a given specific service. Additionally, as BAIN has to be compatible with the legacy systems, cache and buffer structures of the new algorithm also contain service and address information.

Therefore, the service is an option field that was added to all Blueline messages and structures, because it adapts Bluetooth for IoT networks (BAIN). Moreover, if BAIN is used to request the functionality of MQTT-SN client subscribed to a topic, the routing table can grow according to the number of topics. So, if the network has many topics and devices, the administrator network should use request/respond messages for MQTT-SN gateway, MQTT-SN forward and legacy routing (based on address).

3.2 Example of Scenario for BAIN Nodes

A scenario example is the phase "subscribe on topic /home/room by MQTT-SN gateway". Each MQTT-SN client subscribed on topic /home/room has the service B and the MQTT-SN gateway has the service A. The design of the intra domain routing protocol for Bluetooth has as main requirement adding service information to the routing protocol. This new information has to be disseminated to the other nodes on demand. The request and response messages of Blueline execute the dissemination and discovery of neighbor location. The cache and buffer of Blueline store the routing information.

Figure 2 depicts an example of a wireless network using the BAIN Protocol. In this example, Node 5 offers the Service_A whereas Node 9 hosts Service_B. After all nodes build the scatternet formation, Node 5 requests the Service B from its neighbors (node 4, when it is master). Initially Node 4 has no information regarding Service B in its cache. As a result, it forwards the request to its slaves. When Node 7, when acting as a slave node, receives the request, it verifies that its cache does not have the targeted service. Therefore Node 7, when acting as a master, forwards next the request to its slaves. Among these, only Node 9 responds to Node 7 giving the the location of Service B. This is because the cache of Node 9 already has information regarding Service B as it hosts this service. Recursively each one of the nodes 7, 0, 4 and 5 receive the location of service B, and store the service name and its location into their cache.

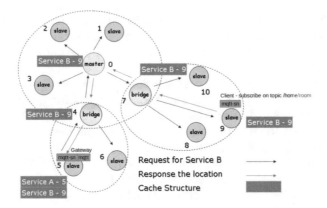

Fig. 2. Topology.

After these routing tables (cache of nodes 5, 4, 0, 7 and 9) are built on demand through the request and response messages, a set of nodes now have a topological view of the nodes and services of their adjacent nodes. Then when a node that is adjacent to them, sends a request related to service B, it quickly finds the location of service, because the adjacent node searches its cache for this information, and discovers if the service is in its network or if the requested service must start the discovery mechanism to update the caches. When a node does not locate a service after some given period of time, it considers that the service does not exist in network and that it is not supported by any node. Note that the mechanisms for aggregating services and routes are out of the scope this paper.

3.3 BAIN Algorithm

The BAIN algorithm has two phases. In the first phase, the nodes create and maintain the scatternet formation that will depend on the dynamicity and mobility of the nodes. At some given time, the scatternet topology can be represented as shown in Fig. 2, but few seconds later the topology may change. This mechanism was implemented from Blueline described in [11]. Therefore, the result represents the connections among nodes without involving services.

In the second phase, nodes create/update the routing table (cache) and route based on service at same time. A request routing packet is used when the node wants to find Client subscribed on topic or other MQTT-SN functionality, while a response routing packet is sent when the node owns the MQTT-SN functionality or knows where it is located. When a node that receives the request does not know how to respond directly, the result is the creation/updating of routing table. When the node receives a request routing packet, it can either have the service or know where it is (by querying its cache).

When a node holds information regarding the location of service within its cache, the node builds a response routing packet and sets its hop number to 0 (if the node is the owner of the service) or sets the hop number to the number stored in its cache (signaling that it does not have the service, but knows where it is located). Then the node sends this routing response packet, and goes back to listen to other routing packets.

When a node has no information regarding the location of a service, it has to lower the hop_to_live variable of the received request routing packet and forward it to its neighbors. At the same time, it is necessary to write in the query buffer of the node which receives the packet: ID, physical address of the node which makes the request and its requested services.

When a node receives a response routing packet, it has to increment the hop number. It next stores the node's physical address that sends the packet, the services contained in the response (not all requested services), and the hop number to reach them. Upon receiving a response routing packet, the node has to compare the packet response ID with all the identifiers stored in its query buffer, which is a structure containing all pending request routing packets. If any of the identifiers in the query buffer coincides with the packet response ID, the respective entry is erased. Then the node sends the response to the requester and returns to listen to other packets.

4   Formal Verification of the BAIN Using SPIN

The formal methods allow the description of the requirements of algorithms at a high level [5] and model checking is one method to verify specifications.

For example, Process Meta Language (PROMELA) [2] allows the modeling of new algorithms. The verification can be performed using SPIN [3], a model checking tool in order to formally verify and evaluate BAIN. During the verification stage, SPIN allows to check for some important safety and liveness properties that must be satisfied in order to prove the correctness of the protocol. Safety properties can be checked to avoid deadlock, unreachable code (or "dead code") and unspecified receptions (receiving unexpected information). Liveness properties can be used to check if the data was accepted by the receiver, if there is a match between what was sent and received.

PROMELA was used to build BAIN abstractions and SPIN is a tool for analyzing the logical consistency of concurrent systems, specifically for protocols. SPIN and PROMELA have been used for modelling and verification of a number of protocols and software modules [12–14].

4.1   Adaptations

The verification of BAIN behavior in SPIN requires some features that are not supported nativity in SPIN. Hence, we defined the properties of typical wireless environments such as reachable nodes, type of communication, resource limitations, in order to ensure that the SPIN model simulates a real-world deployment scenario. Firstly, we have to model a wireless environment where not all nodes are in the same range. In order to calculate distance among nodes, and for allowing a direct communication among nodes in the same range, each node in our model has a position (X,Y). Assuming that all nodes are in a two-dimensional space, we calculate the distance between nodes as in (1) and compare it to see if it is less than or equal to the radio range established. This process is performed during the scatternet formation, when the nodes are searching for each others, and this comparison is pre-requisite in order for nodes to establish the communication among them.

$$d = \sqrt{(x1 - x2)^2 - (y1 - y2)^2} \tag{1}$$

Multicast transmission in BAIN is also required to simulate the transmission between one node and its direct neighbors, since a node is only able to send and receive messages to/from its neighbors. The neighbors of a node were defined depending on the state (master node can have at most 7 slaves and a slave node can have at most 2 masters). Additionally, the PROMELA language does not provide a direct support for multicast or even broadcast communication. Thus, we had to simulate multicast communication, creating a process based on [15], while also adapting it according to our SPIN model. Hence, during the node process, each node uses the to_mcast channel to send packets to the multicast process. Nodes receive the packet through from_mcast channel. The multicast process is described in the following section.

4.2 Processes

The BAIN model was implemented in three processes: Init, Node and Communication (multicast). The Init process initiates the model in parallel with the communication process. Each node receives the following parameters:

- (X,Y) - position in network;
- State - randomly a node can be slave or master;
- Service - service that each node offers.

In this initial step the nodes do not have a valid state, but we know the position and the service of each one. For example, node with position X=2 and Y=4 has 6 as service identifier (Fig. 3).

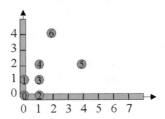

Fig. 3. Node formation.

The Node process implements the logic for a BAIN node. This process receives the parameters described above and initiates itself the Scatternet Formation and Routing Protocol by sending and receiving packets through the Communication process. This process receives the parameters described above and initiates itself by sending the modified packets through the Communication process. In scatternet formation step, the node calculates the distance between the node and the source node that sent the packet, verifying whether this distance is valid and whether they have opposite states. If this verification are valid, the node builds a link with this new neighbor and ads it to its neighbor list.

In the routing step, a node can receive/send three types of packets. When an invalid packet is received (type 0), then it is discarded. If a node receives a request packet (type 1), then it creates a response packet to source node (which originated the request), indicating who knows the location of the requested service. This answer could be found within the node profile or could come from another reachable node indicated into the cache of the node. However, if the node does not have any information about the service, then it forwards the request packet to its masters or slave neighbors, and stores the request into the buffer.

If a node receives a response packet (type 2), then it gets a request packet into the buffer and forwards a response packet to the requesting source. It also puts this service and registers who knows it into its cache. After executing any of these conditions (packet with type 0, 1 or 2), a node returns to the initial state.

Moreover, each node randomly searches by services that were to put into a request packet. A node has to have one valid state (master or slave) and it will forward the search to neighbors with state different from its state.

The Communication Process sends and receives packets forming the communication structure of BAIN model and it is started in parallel with all instances of Node Processes. Each node writes packets in a multicast channel. These packets have all the information necessary to maintain the communication between routers, including location (X and Y) of the packet's destination and the location of source node that sent the packet. A neighbor node receives such packet in the other multicast channel, which integrates the communication system.

4.3   Verification Results

The SPIN offers two different modes to process verification and validation: an exhaustive state space search (for small to medium size models) and a supertrace search (for larger models). The latter mode was chosen, because the verification can be performed using much smaller amounts of memory while still retaining excellent coverage of the state space.

The specification of the BAIN (extracted from the Bluetooh and Blueline) written in the PROMELA language included service information. After the code was compiled, the tool executed a supertrace search for all reachable states.

In PROMELA, valid end-states are those system states which mean that either the process reached the end of its body or it became stuck at an expected statement - identified by an 'end' label in its beginning. All other states are invalid end-states. As shown in Fig. 4, we found one invalid end-state in the communication process. But this is understandable because it is the end of the code and since the communication process was written to be an infinite loop, the state would be unreachable. During the verification, no invalid end-states inside the init and node processes were found. Thus, it can be concluded that there is no dead code in our spin model and this correctness property was accomplished.

```
Bit statespace search for:
        never claim             - (not selected)
        assertion violations    - (disabled by -A flag)
        cycle checks            - (disabled by -DSAFETY)
        invalid end states      +

State-vector 5296 byte, depth reached 41727, errors: 0
   159750 states, stored
   894813 states, matched
1.05456e+06 transitions (= stored+matched)
   797689 atomic steps

hash factor: 1.64096 (best if > 100.)

bits set per state: 3 (-k3)

Stats on memory usage (in Megabytes):
846.675 equivalent memory usage for states (stored*(State-vector + overhead))
0.033    memory used for hash array (-w18)
28.000   memory used for DFS stack (-m1000000)
0.991    other (proc and chan stacks)
29.024   total actual memory usage

unreached in proctype Node
        (0 of 258 states)
unreached in proctype Multicast
        line 426, "Bluetooth-AN", state 18, "-end-"
        (1 of 18 states)
unreached in proctype :init:
        (0 of 9 states)
```

Fig. 4. Verification results with SPIN.

Moreover, BAIN presented a relatively low complexity considering the number of states and transitions (see Table 1) in accordance with the criteria used in [16].

Table 1. BAIN: Algorithm verification results.

Search Completed	State Vector Size	Depth reached	Number of states	Number of transitions
Yes	5296	40598	804187	963492

5   Conclusion

In this paper, BAIN was an adaptation of the Blueline and MQTT-SN. In this context, BAIN eliminated some disadvantages of Bluetooth or bottlenecks. Features such as scatternet formation, routing among nodes in different ranges, routing based on services, multipath and path adaptation, simple and clear messages were covered.

The routing mechanism associates services to routes. This makes sense as in the context of IoT the requests are made for services and not anymore based on names or an address. Services can be topics in MQTT-SN, subscribe to a topic by a gateway, publish data on topic by a gateway and so on.

In order to make BAIN work with service and support larger number of nodes, the Blueline messages were modified. In BAIN, each node knows about its services as well as those of its neighbours. On demand services are used to initiate route search instead of destination addresses. Hence, it allows an intra domain protocol that can operate in IoT scenarios and, together with other protocols of other types of networks, can attend to user service requests. Additionally, BAIN is able to interact with legacy systems using Blueline.

Moreover, this protocol was evaluated formally where the PROMELA language was used to develop a BAIN formal model. In addition, the verification process conducted using a well known model checker - SPIN.

As part of future work, there is the intention to simulate the BAIN protocol, comparing its performance to that of Blueline, and evaluating the consumed link bandwidth and energy. Different scenarios will be used to validate BAIN, including its integration with an inter domain protocol for IoT in order to evaluate its interoperability with different intra and inter domain protocols.

References

1. Bluetooth Specification. http://www.bluetooth.com
2. Holzmann, G.J.: The model checker spin. IEEE Trans. Software Eng. **23**(5), 279–295 (1997)
3. Spin Overview Document. http://spinroot.com/spin/whatisspin.html
4. Menezes, A.H.S., de O, K.R.M., Oliveira, L.P., de S. Oliveira, P.J.: IoT environment to train service dogs. In: 2017 IEEE First Summer School on Smart Cities (S3C) (2017). https://doi.org/10.1109/s3c.2017.8501386
5. Oliveira, L.P., da Silva, A.W.N., de Azevedo, L.P., da Silva, M.V.L.: Formal methods to analyze energy efficiency and security for IoT: a systematic review. In: Barolli, L., Woungang, I., Enokido, T. (eds.) AINA 2021. LNNS, vol. 227, pp. 270–279. Springer, Cham (2021). https://doi.org/10.1007/978-3-030-75078-7_28
6. Petrioli, C., Basagni, S., Chlamtac, M.: Configuring BlueStars: multihop scatternet formation for bluetooth networks. IEEE Trans. Comput. **52**, 779–790 (2003)
7. Zaruba, G.V., Basagni, S., Chlamtac, I.: Bluetrees-scatternet formation to enable Bluetooth-based ad hoc networks. In: Proceedings of IEEE International Conference on Communications, vol. 1, pp. 273–277 (2001)
8. Lin, T.-Y., Tseng, Y.-C., Chang, K.-M., Tu, C.-L.: Formation, routing, and maintenance protocols for the bluering scatternet of bluetooths. In: Proceedings of the 36th Hawaii International Conference on System Sciences (HICSS 2003) (2003)

9. Salonidis, T., Bhagwat, P., Tassiulas, L., LaMaire, R.: Distributed topology construction of bluetooth personal area networks. In: Salonidis, T., Bhagwat, P., Tassiulas, L., LaMaire, R.: Proceeding of 20th Annual Joint Conference of the IEEE Computer and Communications Societies, INFOCOM 2001, vol. 3 (2001)
10. Law, C., Mehta, A.K., Siu, K.Y.: Bluetooth: performance of a new Bluetooth scatternet formation protocol. In: Proceedings of ACM International Symposium on Mobile Ad Hoc Networking and Computing, pp. 183–192 (2001)
11. Chang, R.-S., Chou, M.-T.: Blueline: a distributed bluetooth scatternet formation and routing algorithm. J. Inf. Sci. Eng. **21**, 479–494 (2005)
12. Gluck, P.R., Holzmann, G.J.: Using SPIN model checking for flight software verification. In: Proceedings of IEE Aerospace Conference, vol. 1, pp. 1–105–1–113 (2002)
13. Khan, S., Waheed, A.: Modeling and Formal Verification of IMPP. In: SERP 2003, Las Vegas, Nevada, USA June 23–26 (2003)
14. Yongjian, L., Rui, X.: Using SPIN to model cryptographic protocols. In: Proceedings of International Conference on Information Technology: Coding and Computing, ITCC 2004, vol. 2, pp. 741–745 (2004)
15. Ruys, T.: Towards Effective Model Checking, Phd Thesis (2001)
16. Renesse, R., Aghvami, A.H.: Formal verification of Ad-hoc routing protocols using SPIN model checker. In: IEEE MELECON 2004, Dubrovnik, Croatia, May 12–15 (2004)
17. Clark, A.S., Truong, H.L.: MQTT for sensor networks (MQTT-SN) protocol specification. (2013). http://mqtt.org
18. Oliveira, L.P., Vieira, M.N., Leite, G.B., de Almeida, E.L.V.: Evaluating energy efficiency and security for internet of things: a systematic review. In: Barolli, L., Amato, F., Moscato, F., Enokido, T., Takizawa, M. (eds.) AINA 2020. AISC, vol. 1151, pp. 217–228. Springer, Cham (2020). https://doi.org/10.1007/978-3-030-44041-1_20

An Authentication Protocol for Healthcare Application: A Case Study of a Diabetic Patient

Neila Mekki[1]([envelope]), Mohamed Hamdi[2], Taoufik Aguili[1], and Tai-hoon Kim[3]

[1] Communications Systems Laboratory, (SysCom), National Engineering School of Tunis, (ENIT), University of Tunis El Manar, (UTM), Tunis, Tunisia
{neila.mekki,taoufik.aguili}@enit.rnu.tn

[2] Higher School of Communication of Tunis, Sup'Com, University of Carthage, Carthage, Tunisia
mmh@supcom.rnu.tn

[3] School of Information Science, University of Tasmania, Hobart, Australia
taihoonn@empal.com

Abstract. We provide in this paper, a secure authentication protocol based on Restful approach for monitoring the diabetic patient. We combine information theory and restful API in the security of smart healthcare monitoring application to address the issue of emergency situation. On the one hand, there is not an unified basic theory which completely covers all aspects of IoT security. Inspired by Shannon's information theory, we attempt to construct a general theory to solve security problem which is suitable to IoT security. On other hand, a medical IoT device publishes the physical condition (such as blood glucose) of a diabetic patient to a remote healthcare center periodically, with limited processing and memory capabilities. So, Restful web service has been employed to make data accessible by both local and remote user to monitor physiological parameters. In line with, the restful (Request/ Response) paradigm on HTTP messages has been exploited in WBAN design, by leveraging our authentication protocol.

Keywords: Internet of Things · Healthcare · Monitoring · Patient · Doctor · Diabetic · Security · Authentication

1 Introduction

With the rapid growth of digitized content and wireless network, security has become the most important concern to protect content from unauthorized access. To respond to this challenge, many researchers have investigated using cryptographic approaches to avoid unauthorized access and eavesdropping. However, it is proved that most presently-used cryptosystems schemes are not highly secured. A significant challenge is to search schemes with unconditional security. The corresponding proofs should be based on theoretical security. One reason to approve this, is that information quantification efficiency establishes security solution according to Shannon's [1] definition.

© The Author(s), under exclusive license to Springer Nature Switzerland AG 2023
L. Barolli (Ed.): AINA 2023, LNNS 661, pp. 434–445, 2023.
https://doi.org/10.1007/978-3-031-29056-5_38

Let's consider our IoT-based healthcare scenario [2], physiological signal-based schemes typically require sensor nodes that can monitor certain unique physiological signals, such as electrocardiograms.

As IoT- healthcare offers continuous monitoring for individuals, it is a highly sensitive against insecure communication.

For this purpose, an emergency framework for a diabetic patient should be developed to ensure the security and privacy of user data.

In order to overcome these issues our contribution can be formalized as the following:

- We propose to design and implement a Secure Authentication Architecture (SAA). In the SSA, the authentication protocol has been designed, for secure communication, based on the Representation State Transfer (REST), and web architecture. It provides an authorized layer (HTTPs) between the perception layer and application layer over the network layer (A Smart Health Gateway (SHG)).
- We propose to exploit a smart health gateway to perform an authentication protocol based Restful approach.

The rest of this paper is organized as follows. First, in Sect. 2, we begin by analysing the literature review, based on light-weight protocols. Second, in Sect. 3, we formulate our research problem by designing our novel Secure Authentication Architecture (SAA). The concrete construction of the proposed SAA is described in Sect. 4. Section 5 presents the scheme's security model, the validity proof of our scheme by BAN logic in detail, and security analysis of defense variety attacks. Concluding and future trends are drawn in Sect. 6.

2   Related Work

In 2004, Malan et al. [3], have been developed CodeBlue, who that is a most popular healthcare project at the Harvard sensor network Lab. In this approach, several medical sensors have been placed in the patients' body. However, security aspects are still pending as future work, in which author CodeBlue admit the necessity for secured medical application.

Subsequently, Lorincz et al. [4], contributed an Elliptic Curve Cryptography (ECC) [5], and TinySec [6], to CodeBlue project, to assure an efficient solution. Unfortunately, their proposed approach has not been implemented yet.

Whilst, Kambourakis et al. [7], discuss and describe several attack models concerning a CodeBlue project such as snooping attack, denial-of-service attack, grey-hole attack, and masquerading attacks.

In the last years, light-weight protocols are needed to minimize a power consumption of electronic mobile devices [8,9], and specifically to serve constraints imposed IoT [10–12]. Consequently, several lightweight protocols have been proposed to cope with this issue in recent years.

In 2015, He et al. [13], presented an analysis survey of IoT healthcare authentication system using Radio Frequency Identification (RFID). They discussed and analyzed some recently proposed Elliptic Curve Cryptography (ECC) based RFID authentication schemes. However, the author admits that most of them are still vulnerable to malicious attacks. Therefore, they are proposed, the necessity to construct a suitable model ECC based RFID.

In 2016, Gope et al. [14], proposed a first authentication mechanism for IoT-based healthcare system, using a body sensor network (BSN). They presented a light-weight protocol which consists on one way hash function and a bitwise exclusive OR operation. Therefore, their proposal satisfies some important security properties e.g. mutual authentication, scalability, and resistance to impersonation attack.

In 2017, Yeh et al. [15], proposed a secure light-weight protocol authentication scheme for IoT healthcare application. They exploited two authentication process: (i) phase 1: between the local processing and BSN server, and (ii) phase 2: between biosensor, local processing, and BSN server; to satisfy an efficient analysis and implementation. To evaluate and prove the proposed scheme, the author implemented and developed a platform based Raspberry Pi II, with adopting SHA-3 (512 bits) as one-way-hash, ECC via open JDK and eclipse 3.8. In the same year, Roy et al. [16] proposed an anonymous user authentication for IoT healthcare system. They investigated three factors: (1) smart card, (2) password, and (3) personal biometrics. The authors prove that the proposed scheme is light-weight. They have adopted an efficient chaotic map for the authentication scheme. Therefore, the proposal is successfully satisfied by mutual authentication.

Also, in 2017, Wazid et al. [17] proposed a secure light-weight protocol for IoT healthcare system. Their proposal uses three factors: (1) smart card, (2) password and (3) personal biometrics. Therefore, they exploited a hash function based on symmetric encryption/ decryption. To evaluate the proposed scheme, the authors prove resiliency against attacks by using a VISPA tool simulation.

In 2018, Mughal et al. [18], proposed a light-weight protocol for IoT healthcare system. They presented a shortened Complex Digital Signature to assure secure communication, which resists against attack. Their proposal satisfies less computational time and communication overhead.

In the same year, Uddin et al. [19] proposed an authentication mechanism for healthcare application. They presented a secure light-weight authentication protocol between BSN, patient's smartphone and the patient Centric Agent (PCA). Their proposal includes a symmetric key encryption/ decryption algorithm.

To the best of our knowledge, most researches (e.g. [13, 14, 16–18] and [19]) propose an authentication lightweight protocols. There is no major study, except [15], which has been addressed the issue of implementation IoT healthcare. Different from all the previous work, our work combines information theory and restful approach to monitor the diabetic patient, in which we summarize problems into two major parts: a theoretical and a practical part as follows.

- The theoretical one was based on efficient IoT-healthcare architecture of a diabetic patient.
- The practical part was to propose and implement a strong light-weight authentication protocol which guarantees privacy users. Moreover, our research partial part was to familiarize with the development tools of e.g. Restful and android.

3   A Secure Authentication Architecture

In this section, we aim to design and implement a Secure Authentication Architecture (SAA).

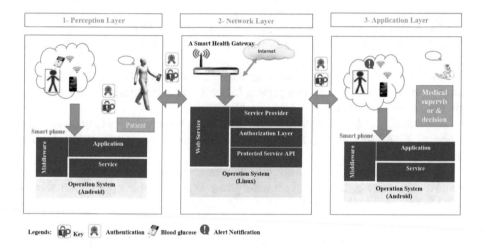

Fig. 1. Secure authentication of IoT healthcare application

As shown in Fig. 1, we consider three main components in our conceived SAA: **(i) the wearable body bio-sensors** (i.e. the smart objects blood glucose), **(ii) The Smart Healthcare Gateway, (iii) the user interface visualization**. Implementation details of SAA are provided in the following subsections

3.1   Wearable Body Sensors (i.e. Smart Objects: WBAN, E.g. Blood Glucose)

It is adopted by the patient for collecting and monitoring physiological parameters. In order to achieve interoperability with internet, the restful (Request/ Response) paradigm on HTTP messages has been exploited in WBAN design. Note that, HTTP is one of communication protocol on Representational State Transfer (REST), to provide an authentication lightweight access.

In our proposed SAA, the WBAN embeds sensor to maintain patients with a good health condition. However, if their values blood glucose exceed a specific range, the blood glucose threshold an alert event. In addition, HTTP provides a request model mapped to the GET method interaction, which allows notification mechanism every change in the state of a resource. The use of conditional observation allows doctors to be notified when critical emergency situation values change considerably.

3.2   Smart Healthcare Gateway

It represents the core of our proposed SAA. Hence, restful API has been adopted to enable a secure authentication scheme between WBAN nodes and remote users. The Restful API (Request/ Response) paradigm has been exploited, to provide lightweight access to physical resources.

The smart gateway takes two different tasks: (1) it allows secure communication, (2) it is responsible for monitoring the patients' health status and alerting doctors in case of critical situations.

For this purpose, we develop a management application (MA), which is a standalone PHP and Nginx web server. MA is easily configurable and accessible via a web browser, which provides low latency.

Specifically, MA is able to receive and reply to any request, coming from users, in JSON format. An MYSQL database stores the information retrieved from WBAN.

So, doctors do not need to directly integrate WBAN during normal mode of operation. However, in case of an emergency situation, our application should be able to send an alert notification to the medical staff.

3.3   Users Interface Visualization

It represents the interface allows users (patients or doctors) with the specific privilege to access. For those reasons, users need to be authenticated before the access to the platform, accessible via smartphone. Specifically, these interfaces implement restful services, which allow the user to communicate with WBAN through the Smart Health Gateway (SHG). It offers two main functionalities depending on two possible users (patient or doctors)

Patient interface: it allows the patient to register to SAA. Furthermore, the patient can be equipped with a smartphone and running application, named MobiDiabetic, a customized Android Application with specific privilege access.

Doctor interface: it allows medical staff to register to SAA. Furthermore, doctors can be equipped with a smartphone and running the same healthcare Application, to check and interact with patient vital signs. Since the system collects sensitive and confidential data to ensure the adequate security level.

In our proposed SAA, users need to perform registration with smart health gateway in advance, to achieve our goal authentication protocol. There are four phases: Initialization phase, User registration phase, Sensor node registration phase, and Authentication phase. These phases will be explained in Sect. 4.

4   Protocol Description

We exploit a smart health gateway to perform an authentication protocol. The idea of our proposed scheme contains, four phases: Initialization phase, User registration phase, Sensor node registration phase, and Authentication phase. The specification of this step is detailed in the next Sub-Sections.

4.1   Initialization Phase

In order to ensure the proposed operation of designed SAA, an initialization phase is required. During this phase, important parameters have to be set up for both sensor nodes and gateways. In particular, we envision the patient wearing a sensor node which is capable to detect blood glucose. To make the collected data easily accessible, the restful API has been adopted. Since the Restful API interface allows the patient to manage and detect important physiological parameters.

First, the sensor node SN submits its identity ID_{SN} to the gateway GW via a secure channel. After the gateway receives the request from a sensor node, the gateway generates a secret key SK_{SN} with the length of at least 256 bits, a random number $x \in (-\infty, +\infty)$, and a one way hash function $h(.)$. Next, the GW securely sends the secret value SK_{SN} back to the sensor node via a secure channel. As soon as the sensor node obtains the secret value SK_{SN} from the gateway, it stores the secret value SK_{SN} in its secure storage.

Therefore, the gateway stores each managed sensor node's important information in the binding table which field contains the identity ID_{SN}, the corresponding secret value SK_{SN}, authentication period T.

Specifically, the notation used in our proposed mechanism is specified in Table 1.

4.2   User Registration Phase

In order, to be known within the infrastructure, and to allow her or him to be authenticated, user needs to be enrolled into the system: this step is called registration.

The registration phase is divided into three parts users, (i) the patient registration and (ii) medical staff registration, and (iii) sensor node registration.

Patient Registration. If the patient (P) wants to be a legal user of secure authentication IoT healthcare application, the following steps must be executed between P and smart health gateway SHG shown in the algorithm 1:

- **Step 1:** P chooses ID_{pa}, and PW_{pa} and send ID_{pa}, and $h(PW_{pa})$ to GW through a secure channel.
- **Step 2:** Upon receiving ID_{pa}, and $h(PW_{pa})$, GW select a random number x and computes registration message $MID_{pa} = h(ID_{pa} \parallel sk_{pa})$, and $M_{pa} = MID_{pa} \oplus h(PW_{pa}) + x$ and return it to P via a secure channel.
- **Step 3:** Patient P computes $M_{pa}^* = M_{pa} \oplus x$ and deletes M_{pa}

Table 1. Notation of proposed mechanism

Notation	Definition (s)
$T_n(x)$	Chebyshev map defined by equation $2xT_{n-1}(x) - T_{n-2}(x) \mod p$
P	Patient
ID_{pa}	The identity of a patient
PW_{pa}	The password of a patient
SK_{pa}	The secret key of a patient
D	Doctor
ID_d	The identity of a doctor
PW_{do}	The password of a doctor
SK_d	The secret key of a doctor
SN	A sensor node
ID_{SN}	The identity of a sensor node SN
SK_{SN}	The secret key of a sensor node SN
GW	Gateway
r	Random number
T	The authentication period
$H(.)$	A one-way hash function
$\|$	A concatenation operation
$\oplus$	The bit wise XOR operation
Δt	Maximum transmission delay
ACK	Acknowledge message

Input : A login request from patient
Output: user logged in
for *For each period T: Smart health gateway Side* **do**
   | **1** wait to receive a ID_{pa}, and PW_{pa} request from the patient
   | **2** $Setup_{Connection}$ the P calculates the registration message message
   | $MID_{pa} = h(ID_{pa} \| sk_{pa})$, and $M_{pa} = MID_{pa} \oplus h(PW_{pa}) + x$ and return
   | it to P via a secure channel.
   | **3** Patient P computes $M_{pa}^* = M_{pa} \oplus x$ and deletes M_{pa}
   | **4** ACK login successes
end

Algorithm 1: User Login Patient

Medical Staff Registration. If the doctor (D) wants to be a legal user of secure authentication IoT healthcare application, the following steps must be executed between D and smart health gateway SHG shown in algorithm 2:

– **Step 1:** D chooses ID_{do}, and PW_{do} and send ID_{do}, and $h(PW_{do})$ to GW through a secure channel.

- **Step 2:** Upon receiving ID_{do}, and $h(PW_{do})$, GW select a random number x and computes registration message $MID_{do} = h(ID_{do} \parallel sk_{do})$, and $M_{do} = MID_{do} \bigoplus h(PW_{do}) + x$ and return it to P via a secure channel.
- **Step 3:** Patient P computes $M_{do}^* = M_{do} \bigoplus x$ and deletes M_{do}

Input   : A login request from Doctor
Output: user logged in
for *For each period T: Smart health gateway Side* **do**
> **1** wait to receive a ID_{do}, and PW_{do} request from the doctor
> **2** $Setup_{Connection}$ the P calculates the registration message message $MID_{do} = h(ID_{do} \parallel sk_{do})$, and $M_{do} = MID_{do} \bigoplus h(PW_{do}) + x$ and return it to D via a secure channel.
> **3** Doctor D computes $M_{pa}^* = M_{pa} \bigoplus x$ and deletes M_{pa}
> **4** ACK login successes

end

Algorithm 2: User Login Doctor

4.3 Sensor Node Registration Phase

SN proceeds registration with the help of GW as below:

- **Step 1:** SN generates its identity ID_{sn}, and computes the corresponding secret key SK_{sn} and $h(ID_{sn} \parallel SK_{sn})$. Then, SN sends $(ID_{sn}, h(ID_{sn} \parallel SK_{sn}))$ to GW via a secure channel.
- **Step 2:** After receiving the message $(ID_{sn}, h(ID_{sn} \parallel SK_{sn}))$ from SN, GW select a random number x and computes $R_{sn} = h(ID_{sn} \parallel SK_{sn}) \bigoplus h(ID_{sn}) + x$, and send it to SN.
- **Step 3:** Upon receiving the message R_{sn} from GW stores it into its memory.

4.4 Authentication Phase

When the user U (patient or doctor) wants to access the sensor node SN, he or she initiates this phase by issuing a request via GW. This phase enables GW, U and SN to effectively authenticate each other and then establish a session key between U and SN. If a session key is negotiated successfully by U and S, then they can exchange private messages with each other via a public channel. A detailed description of the steps of this phase are as follows:

- **Step 1:** U selects his/her PW, and Gateway GW generates a current timestamp T_1 and computes $C_1 \equiv T_u(x) \mod p$, $KA \equiv T_u(T_{mk}(x)) \mod p$, $X_u = Y \bigoplus h(PW)$ then creates the message $M_{us} = h(ID \parallel DID \parallel X_u \parallel C_1 \parallel KA)$, and send the login message $M_1 \equiv \{C_1, DID, M_{us} \parallel T_1\}$ to GW through a public channel, where T_1 is the current timestamp.

- **Step2:** Upon receiving the request message, GW checks whether $T_2 - T_1 \leqslant \Delta T$ holds, where T_2 is the current timestamp. If it does not hold, GW terminates the session; otherwise GW computes $KA' \equiv T_{mk}(T_u(x)) \mod p$, $ID' = DID \oplus h(KA')$, and $X'_u = h(ID \parallel mk)$. Then GW checks whether $h(ID' \parallel DID)$ $X'_u \parallel C_1 \parallel KA' = M_{us}$; if not, GW also terminates the session; otherwise GW generates a random number r and computes $C_1 \equiv T_r(x) \mod p$ and the session key $sk \equiv T_r(T_u(x) \mod p)$. Finally GW computes $M_{su} = h(ID' \parallel C_2 \parallel KA' \parallel sk)$ and sends the response message $M_2 = \{C_2, M_{su} \parallel T_3\}$ to the user U where T_3 is the current timestamp.
- **Step 3:** After receiving the response message M_2, GW verifies whether $T_4 - T_3 \leqslant \Delta T$ holds, where T_4 is the current timestamp. If not, GW terminates the session; otherwise GW computes $sk' \equiv T_u(T_r(x)) \mod p$; Then GW checks whether $h(ID \parallel C_2 \parallel KA \parallel K') = M_{su}$. If not, GW terminates the session; otherwise U computes $M_{sk} = h(KA \parallel sk')$ and sends $M_3 = \{M_{sk}\}$ to GW.
- **Step 4:** Upon receiving the message M_3, GW checks whether $h(KA \parallel sk') = M_{sk}$ holds; if it is true, the verification between U and GW succeeds and mutual authentication is accomplished. The session key is correct and both U and GW can use sk to communicate with each other in safety. Otherwise, this connection will be stopped.

5   Performance Analysis: Security Model

In this section, we will analyze the validity, and security of our protocol. First, we demonstrate the security model and then use Burrows-Abadi-Needham (BAN) logic to confirm the correctness of the proposed protocol. Second, we will explain that our protocol can withstand various attacks. Our scheme should resist the known attacks in the authentication protocol, by using provable security. The proof of security is based on the random oracle model and the model proposed by Abdalla and Pointcheval.

Participants. Each participant of an authentication protocol is either a patient $P = \{P_1, P_2, P_3..P_i...P_n\}$ or a doctor. We refer to the *ith* instance of P_i in a session as $\prod_p^i$, and the instance of the Gateway is denoted by $\prod_G$.

Adversary Model. The communication network is assumed to be potentially controlled by an adversary A, who has the ability to intercept, block, inject, remove, or modify any messages transmitted over the public network. The adversary A is allowed to have access to the following queries in any order.

- $Execute(\prod_p^i, \prod_g)$ This query models passive attacks. It outputs the messages that were exchanged during the honest execution of the patient instance $\prod_p^i$ and gateway instance $\prod_g$
- $send(\prod_c^k, M)$ This query models active attacks. Adversary A can send a message through this oracle to $\prod_c^k$, where $c \in (U, S)$. Then $\prod_c^k$ returns some messages, which are computed by $\prod_c^k$ based on the proposed scheme, to A.

- **Reveal** $\left(\prod_{c}^{k}\right)$ This query models the misuse of session key. A can obtain a session key from the oracle $\prod_{c}^{k}$. If the oracle $\prod_{c}^{k}$ has accepted, then it returns the session key to A. Otherwise, $\prod_{c}^{k}$ returns a null value to A.
- **Corrupt** (P) This query models the adversary A to corrupt a protocol participant P; that is, A can get the secret information about P.
- **Test** $\left(\prod_{c}^{k}\right)$ This query measures the semantic security of the session key sk. To respond to this query, the oracle $\prod_{c}^{k}$ chooses a random bit $b \in \{a, b\}$. if $b = 1$, then $\prod_{c}^{k}$ returns the session key sk. Otherwise, it returns a random value. Adversary A can send only a single query of this form to $\prod_{c}^{k}$.
- $(h m_i)$. In this query, when an adversary A does this hash query with message m_i, $\prod_{c}^{k}$ returns a random number r_i and adds (m_i, r_i) to a list L_h From every beginning, the list is empty.

Security Proof. Here we show that the proposed scheme can provide the secure authentication and key agreement under the assumption of CMDHP.

Theorem supposes that A can violate the proposed protocol with a non negligible probability. A makes q_p query to the oracle of the patient $\prod_{p}^{i}$, q_s query to the oracle of the gateway $\prod_{g}$, and q_h query to $h(.)$. Then we can design an algorithm C to solve the chaotic maps-based Diffie-Hellman Problem (CMDHP) with a non negligible probability.

Proof we assume the type of attack which forges the user (patient or doctor) to communicate with gateway.

Then we can construct algorithm C to solve the CMDHP; that is, C returns $T_{ur}(x) \mod p$ from an instance of $\{x, T_u(x) \mod p, T_r(x) \mod p\}$ by CMDHP, where $u, r \in Z_P^*$

For instance, CMDLP is $\{x, T_{mk}(x), mk\}$. B simulates the system initializing algorithm and registration phase to generate the parameters $\{x, T_{mk}(x), h(.)\}$ to A. B interacts with A as follows.

$h(.)Query$ B holds a list L_h of tuples (str_i, h_i). When A queries the oracle $h(.)$ on (str_i, h_i), B responds as follows.

If str_i is on L_h, B return h_i to A. Otherwise, B randomly chooses an integer h_i which is the only one in L_h and adds (str_i, h_i) to L_h and then responds with h_i.

$Reveal(.)Query$ When the adversary A makes a $Reveal(\prod_{c}^{k})$ query, B responds as follows.

If $\prod_{c}^{u}$ is not accepted, B returns a null value to A. Otherwise, B examines the list L_h and responds with the corresponding h_i

$send(.)Query$ When the adversary A makes a query $send(\prod_{c}^{u}, "start)$, B responds as follows.

If $\prod_{c}^{u} = \prod_{u}^{u}$, B follows the proposed steps. Otherwise, B generates a random number mk^* computes $T_{mk*}(x)$ and replaces $T_{mk}(x)$ with $T_{mk*}(x)$. B responds with $\{C_1, DID^*, M_{us*}\}$. The simulation works successfully since A cannot distinguish whether $\{C_1, DID^*, M_{us*}\}$ is correct or not only when A knows ID and PW.

When the adversary A makes query $send(\prod_c^u, (C_1, DID^*, M_{us*}))$ query, B responds as follows.

$\prod_c^u = \prod_u^u$, B cancels the game. otherwise, B computes KA', ID', and X_u' with mk'. B ckecks whether $h\left(ID' \parallel DID^* \parallel C_1 \parallel KA'\right) = M_{us*}$ hold or not. If it holds, B computes $C_2 \equiv T_r(x) \mod p$ and $sk \equiv T_r(T_u(x)) \mod p$ and responds message $\{C_2, M_{su*}\}$ according to the proposed protocol.

When the adversary A makes a query $send(\prod_c^u, (C_2, M_{su*}))$, B responds as follows.

if $\prod_c^u = \prod_u^u$, B cancels the game. Otherwise, B computes $sk \equiv T_r(T_u(x)) \mod p$ from $x, T_u(x) \mod p, T_r(x) \mod p$ and get $h(ID \parallel PW)$ from the list l_h. Therefore, if A can violate a user to the gateway authentication, B must solve the CMDHP problem with a non negligible probability. This is contradiction to the computation infeasible to the CMDHP problem.

Through analyzing, we can conclude that it is almost impossible for A to violate the user to the gateway authentication.

6   Conclusions

In this paper, we have introduced a new lightweight authentication protocol for healthcare application. The idea of our proposed scheme contains four phases: Initialization phase, User registration phase, Sensor node registration phase, and Authentication phase

Therefore, the IoT diabetic healthcare application appliance should utilize an effective authentication mechanism to correctly identify and authorize the communication requests from legitimate mobile apps. Besides the security of mobile apps, the security of mobile operating platforms, namely Android, and iOS, plays a vital role in the smart healthcare appliance protection.

References

1. Shannon, C.: Bell Syst. Techn. J. **28**(4), 656 (1949). https://doi.org/10.1002/j.1538-7305.1949.tb00928.x
2. Mekki, N., Hamdi, M., Aguili, T., Kim, T.H.: pp. 554–559 (2017). http://www.scitepress.org/PublicationsDetail.aspx?ID=Rw/BhvXf+cw=&t=1
3. Malan, D., Fulford-Jones, T.R.F., Welsh, M., Moulton, S.: CodeBlue: An Ad Hoc Sensor Network Infrastructure for Emergency Medical Care (2004)
4. Lorincz, K., et al.: IEEE Pervasive Comput. **3**(4), 16 (2004). https://doi.org/10.1109/MPRV.2004.18
5. Koblitz, N.: Math. Compu. **48**(177), 203 (1987). https://doi.org/10.1090/S0025-5718-1987-0866109-5. https://www.ams.org/home/page/
6. Karlof, C., Sastry, N., Wagner, D.: In: Proceedings of the 2nd International Conference on Embedded Networked Sensor Systems, SenSys 2004, pp. 162–175. ACM, New York (2004)
7. Kambourakis, G., Klaoudatou, E., Gritzalis, S.: In: The Second International Conference on Availability, Reliability and Security (ARES 2007) (2007). https://doi.org/10.1109/ARES.2007.135

8. Patton, E.W., McGuinness, D.L.: A power consumption benchmark for reasoners on mobile devices. In: Mika, P., et al. (eds.) ISWC 2014. LNCS, vol. 8796, pp. 409–424. Springer, Cham (2014). https://doi.org/10.1007/978-3-319-11964-9_26

9. Chabarek, J., Sommers, J., Barford, P., Estan, C., Tsiang, D., Wright, S.: In: IEEE INFOCOM 2008 - The 27th Conference on Computer Communications (2008), pp. 457–465 (2008). https://doi.org/10.1109/INFOCOM.2008.93

10. Gubbi, J., Buyya, R., Marusic, S., Palaniswami, M.: Futur. Gener. Comput. Syst. **29**(7), 1645 (2013)

11. Raza, S., Shafagh, H., Hewage, K., Hummen, R., Voigt, T.: IEEE Sens. J. **13**(10), 3711 (2013). https://doi.org/10.1109/JSEN.2013.2277656

12. Sehgal, A., Perelman, V., Kuryla, S., Schonwalder, J.: IEEE Commun. Mag. **50**(12), 144 (2012). https://doi.org/10.1109/MCOM.2012.6384464

13. He, D., Zeadally, S.: IEEE Internet Things J. **2**(1), 72 (2015). https://doi.org/10.1109/JIOT.2014.2360121

14. Gope, P., Hwang, T.: IEEE Sens. J. **16**(5), 1368 (2016). https://doi.org/10.1109/JSEN.2015.2502401

15. Yeh, K.: IEEE Access **4**, 10288 (2016). https://doi.org/10.1109/ACCESS.2016.2638038

16. Roy, S., Chatterjee, S., Das, A.K., Chattopadhyay, S., Kumari, S., Jo, M.: IEEE Internet Things J. **5**(4), 2884 (2018). https://doi.org/10.1109/JIOT.2017.2714179

17. Wazid, M., Das, A.K., Odelu, V., Kumar, N., Conti, M., Jo, M.: IEEE Internet Things J. **5**(1), 269 (2018). https://doi.org/10.1109/JIOT.2017.2780232

18. Mughal, M.A., Luo, X., Ullah, A., Ullah, S., Mahmood, Z.: IEEE Access **6**, 31630 (2018). https://doi.org/10.1109/ACCESS.2018.2844406

19. Uddin, M.A., Stranieri, A., Gondal, I., Balasubramanian, V.: IEEE Access **6**, 32700 (2018). https://doi.org/10.1109/ACCESS.2018.2846779

Micro IDS: On-Line Recognition of Denial-of-Service Attacks on IoT Networks

Henrique Fell Lautert[1]([⊠]), Douglas D. J. de Macedo[2], and Laércio Pioli[1]

[1] Graduate Program in Computer Science (PPGCC) - Federal University of Santa Catarina, Florianópolis, Santa Catarina, Brazil
{henrique.lautert,laercio.pioli}@posgrad.ufsc.br
[2] Department of Information Science (CIN) - Federal University of Santa Catarina, Florianópolis, Santa Catarina, Brazil
douglas.macedo@ufsc.br

Abstract. The growth in the number of Internet of Things (IoT) devices and applications, as well as their heterogeneity and hardware limitations, make it difficult to apply traditional security mechanisms. Thus, the IoT layer has become a highly vulnerable part of the network. In this article, a low computational complexity intrusion detection system is proposed for online recognition of denial-of-service attacks. A common feature of denial-of-service attacks is the sudden surge of a certain type of packet or request. To track this sudden spike, network traffic is reduced to the number of packets per minute, segmented by protocol. On these data, we applied sliding window and moving average comparison techniques to identify anomalies. After identification, a selective search is performed only in the anomalous protocol, to isolate the target and neutralize the attack. Tests performed on data extracted from *pcap* file, containing attacks carried out on real devices, demonstrate the accuracy in recognizing attacks. In addition, the tools and techniques for implementing the proposed model in a realistic environment are described.

1 Introduction

The Internet of Things is already a reality in our daily lives. Clocks, televisions, security cameras, speakers with virtual assistants, are some of the devices connected to the internet that are increasingly common in our routines. In addition, there are more and more connected devices helping industrial automation, collecting data for smart city management and improving agricultural production. Whether for ease, comfort, entertainment, technological or productive differential, IoT (Internet of Things) devices are on the rise. A report prepared by IoT Analytics predicts about 27 million connected devices by 2025 [1].

The rapid growth combined with the heterogeneity and hardware limitation of IoT devices brings with it security issues for the networks where they are connected. It is challenging to think of solutions for secure networks, since most

L. Barolli (Ed.): AINA 2023, LNNS 661, pp. 446–459, 2023.
https://doi.org/10.1007/978-3-031-29056-5_39

of its components do not support traditional security mechanisms [2]. In view of this, we can observe several researches on this topic, but only a few of them are close to being implemented in real environments [3].

Meanwhile, we are faced with devastating impact attacks, Denial of Service (DoS) attacks and their variation of Distributed Denial of Service (DDoS). This type of attack takes advantage of the large number of devices in IoT networks, using these devices as robots to attack a given victim. In October 2016, a terabit of traffic per second was registered against Dyn, an important DNS (Domain Name Service), directly impacting the availability of Twitter, Netflix, Spotify, Airbnb, Reddit, Etsy, SoundCloud and The New York Times services [4]. In 2020, another DDOS attack reached a 2.3 Tbps [5] against Amazon Web Services. In the same year, the New Zealand stock exchange was the target of a DDoS attack, which resulted in the interruption of service for 4 consecutive days.

DoS attacks are difficult to prevent due to the fact that they use packets that, at first, are just a part of the natural connection establishment of protocols, so they cannot be simply blocked [6]. The TCP SYN attack, for example, uses the connection establishment mechanism to overload the target. Blocking this type of traffic would prevent TCP connections from being established. Setting a fixed threshold can also drop benign traffic and harm the network [7].

There are several studies using Machine Learning (ML) techniques on datasets [8,9] that demonstrate high accuracy in recognizing denial of service attacks. However, few consider using a time window and online learning in reading dataset or network traffic. In the real world, using an entire dataset to perform training and only after that starting the recognition of attacks opens up a space of vulnerability and also makes it difficult to generalize for application in different IoT networks [3].

The number of features extracted and used to carry out the training is also worth mentioning. In addition to the high computational cost involved [10], gathering all the features and simply delegating them to the Principal Component Analysis (PCA) process for the selection of the main features can compromise the accuracy in attack recognition [8]. When evaluating the feature selection process, it is possible to see that there are a small number of features, such as destination address, source address and protocol that are similar in different types of DoS attacks, where there is greater variation and therefore a potential gain in recognition of attacks [11].

In this work we explore the association of 2 features for the anomaly recognition step: The number of packets in relation to time segmented by network protocols. Based on this association, we calculated the moving average and performed a comparison of the last 3 averages using the sliding window technique, so that the variations of each protocol are highlighted, allowing the online recognition of attacks. In the attack counter-action stage, we use 3 more features: destination ip, port and protocol. Enough to identify the target of the attack and neutralize its effect through firewall rules at the edge of the network.

The strategy of recognizing anomalies at the edge of the network with lower computational cost becomes possible due to more regular transmission patterns in IoT networks [12]. In a specific research on the recognition of anomalies in IoT networks, Cook et al. [3] points out challenges such as: data reduction for analysis and recognition of attacks with online learning, using only a window of time to identify the attack.

In this context, the main contributions of this article are: First, analysis of changes in network protocols in the face of denial of service attacks, opposing IoT networks in normal operation and in the face of these attacks. Next, the proposition of network traffic data reduction to relevant information in the identification of denial of service attacks. Furthermore, development of an Intrusion Detection System (IDS), capable of recognizing DoS attacks online in IoT networks with low computational cost, describing the techniques and tools for implementation in real environments.

This section introduces the article. In Sect. 2, the background is presented. Section 3, discusses related works. In Sect. 4, details of the proposal validation are given. Section 5 describes the prototype. Section 6 presents the results and limitations of the proposal. Finally, in Sect. 7, the final considerations are outlined.

2   Background

Technically, the objective of a denial of service attack is to overload the target, making it unavailable, thus characterizing a denial of service. For this, the attacker fires a very high number of requests making the target respond slowly, until it becomes totally unable to respond to legitimate requests. A variation of this technique is distributed denial of service (DDoS) attacks, where multiple devices are used to amplify the effect of the attack.

The rapid growth in the number of devices connected via the Internet of Things has been an enabling environment for massive DDoS attacks. IoT devices often use default passwords and lack traditional security mechanisms, making them vulnerable to exploitation. Infection of IoT devices often goes unnoticed by users, and an attacker can easily compromise hundreds of thousands of these devices to carry out a large-scale attack without the knowledge of the device owners [13].

Using small devices in a distributed way, in addition to enhancing the attack, increases the difficulty of recognition, as the true source of the attack becomes more difficult to identify. In Fig. 1(a), we can visualize a DoS denial-of-service attack, while in Fig. 1(b) this same technique is performed in a distributed manner, generating an even greater burden on the attack target.

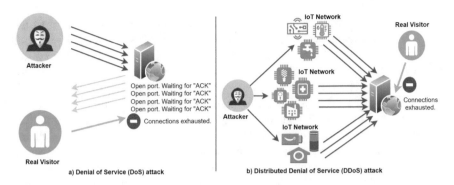

Fig. 1. A Denial of service (DoS) attack and a Distributed Denial of Service (DDoS) attack.

3   Related Works

In order to face DoS-type threats in IoT, Sousa et al. [14] proposes an IDS (Intrusion Detection System) for IoT, aiming at the detection of some DoS-type attacks. The proposal was validated using resources from the *iptables* tool. The proposed mechanism uses the *iptables* statistics, dispensing the integral analysis of the packages. In this way, reconnaissance and attack contraction are applied at the edge of the network, in this experiment through the Raspbian operating system and Raspberry Pi 3 hardware. Despite the efficiency, the results are limited to just 2 attacks, TCP SYN and UDP Flood. The use of a fixed limiter *max_packets* to classify as an attack, in addition to causing an increase in false positive cases, also makes it difficult to generalize the solution in different scenarios.

Sharma et al. [9] makes a remarkable contribution on how to define the line that separates benign traffic from a DoS attack. For this, the PCA technique is first applied to unify the data. Then the data is processed by the Continuous Ranked Probability Score (CRPS) type algorithm. Defining the threshold between benign traffic from an attack based on dataset statistics is a great tactic to parameterize the recognition algorithm. However, in real environments we do not have long datasets available for this parameterization. Problematizing the application of research in real environments, Cook et al. [3] suggests the use of the time window technique to enable the identification of anomalies at the edge of the network.

Some proposals already partially adopt the time window for training Deep Learning (DL) algorithms and anomaly recognition [15]. However, they still use data from the entire dataset for threshold selection, a parameter that distinguishes benign traffic from an attack. Salahuddin et al. [16] points out incremental learning and online recognition of DDoS attacks as future work.

Despite the high accuracy obtained through DL techniques, reaching more than 99% with different methods [17], the application at the edge of the network is challenging due to hardware limitations. Many IoT devices fail due to

overheating, accelerated battery drain, and runtime stalling. The main reason for
such a failure causing problems is the exhaustion of device memory (especially
SRAM) [13]. Within the scope of this research, Sudharsan el al. [18] is the only
one involving DL to present the computational costs involved in the operation in
a detailed way, where the proposed algorithm reaches the complexity of $O(2^O)$.

In this context, the use of mathematical models or simplified ML models
to recognize the anomalies generated by denial of service attacks, in addition
to having a lower computational cost than solutions that use DL [10], have
achieved satisfactory accuracy rates. Alzahrani et al. [19] observes the variations
in processing, memory and energy of IoT devices, for this data is compared
through the Euclidean distance formula, reaching an accuracy of 93% to 100%.
Santoyo et al. [20] uses the *BPFabric* architecture [21] to apply exponential
averaging and Shannon entropy in the recognition of anomalies from network
traffic statistics, with an accuracy of 93 to 95%, limited to the TCP SYN flood
attack. Related works are summarized and compared in Table 1.

Table 1. Comparison of Related Works

Autor	Year	F1	F2	F3	F4
Sousa et al. [14]	2017	Yes (ND)	No	ND	No
Sharma et al. [9]	2021	Yes (ND)	No	ND	Yes
Doshi et al. [15]	2018	No	No	99%	Yes
Salahuddin et al. [16]	2021	No	Yes	96%	Yes
Tann et al. [19]	2021	No	Yes	99%	Yes
Sudharsan et al. [18]	2021	Yes. $O(2^O)$	Yes	ND	Yes
Alzahrani et al. [19]	2022	No	Yes	80%–100%	Yes
Santoyo et al. [20,22]	2019	Yes (ND)	Yes	95%	No
Current Work	**2022**	**Yes.** $O(n)$	**Yes**	**87%–100%**	**Yes**

F1: Online Recognition and (Computacional Complexity). F2: Dynamic
Threshold. F3: Precision. F4: Multi Protocol. ND: Not declared

Most of the works do not describe how to perform the online recognition of
threats, nor the computational complexity involved, which is a differential due
to the limitation of resources at the edge of the network. Despite the high level
of accuracy achieved by DL works, in real environments we cannot read an entire
dataset to differentiate what is an attack from what is benign traffic. This would
be like accepting a series of attacks until the defense system starts operating,
which could cause the network to be unavailable. Within these findings, we
propose an intrusion detection system with low computational complexity, with
feasibility of online operation and dynamic threshold through the history of
moving averages.

4   Micro IDS

Our proposal presents the use of sliding window and moving averages to recognize anomalies with $O(n)$ complexity, the smallest among the works found in the literature. In addition, online recognition and dynamic threshold are prioritized to detect denial-of-service attacks. The model was validated in the datasets of MQTTset-2020 [22] and UNSW-2019 [23], being tested in TCP SYN flood, Smurf (ICMP), SSDP reflection, SNMP reflection, ARP spoof and MQTT-Malaria attacks.

In the validation phase, the reading of active network protocols was performed using the *Wireshark* tool [24]. It also extracted minutes and the number of packets per segment segmented by protocol through Berkeley Packet Filters (BPF). This data was saved in csv, and then processed through the *python* language, supported by the *numpy*[1] . Figure 2 shows the steps taken to validate the proposal.

A common feature of denial-of-service attacks is the sudden increase of a certain type of packet or request. Therefore, by isolating the protocols, we can visualize the variations in the number of packets in relation to the time of each protocol, being generalizable to new types of attacks and protocols. After extracting the number of packets per minute segmented by protocol, we start processing the data and recognizing the attacks.

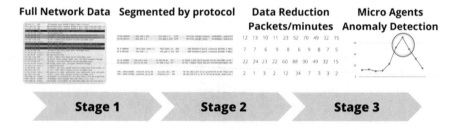

Fig. 2. Micro Intrusion Detection System

First, the data series is transformed into a matrix, where each line contains the last *(k)* 10 measurements (packets per minute segmented by protocol). This algorithm is known as sliding window [26], and it can be executed through a function of the *numpy* routine[2], being the first step to obtain the moving average (Eq. 1).

---

[1] https://numpy.org/.

[2] https://numpy.org/devdocs/reference/generated/numpy.lib.stride_tricks.sliding_window_view.html.

$$MA_k = \frac{p_{n-k+1} + p_{n-k+2} \cdots + p_n}{k}$$

$$= \frac{1}{k} \sum_{i=n-k+1}^{n} p_i \tag{1}$$

Threshold reached IF

$$MA(n) > MA(n-1) + 10\% \text{ AND } MA(n-1) > MA(n-2) + 10\% \tag{2}$$

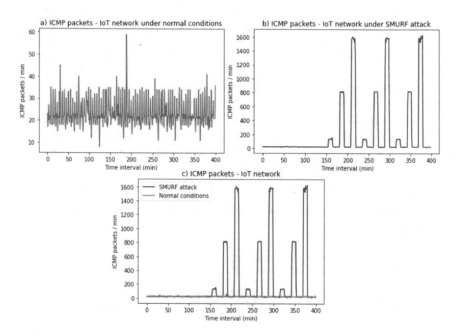

Fig. 3. ICMP packets under normal conditions vs. Smurf attack - IoT network

After obtaining the matrix where each row contains the last 10 measurements, just get the average of the last 3 rows. Finally, it compares whether the current line is 10% greater than the penultimate one, and the penultimate one is 10% greater than the antepenultimate one (Eq. 2). If so, the event is classified as an anomaly, an abnormal growth of a type of protocol in the network is recognized.

For better understanding, we use the case of the DoS Smurf attack as an example, which has as a symptom the increased number of ICMP packets. In Fig. 3(a) it is possible to observe the variations of benign traffic, in Fig. 3(b) the variations of malignant traffic. By joining the lines of benign and malignant traffic in Fig. 3(c) the difference becomes clear. We noticed that the variations of the benign traffic are small, and it maintains a constant line in relation to the malignant traffic, which rises radically in the moments of attack. In addition, there are 3 levels of red line elevation, which reflects the frequency parameter at which the SMURF attack was performed, 1, 10, and 100 packets per second.

By zooming in and applying the last 3 moving averages we can see how false positives are eliminated and the attack identified. In Fig. 4(a), the moving averages are observed during normal traffic. As the variations in ICMP packet traffic are punctual, the averages do not deviate. In Fig. 4(b), the moving averages move away, so that the current moving average is 10% greater than the penultimate one, and the penultimate one 10% greater than the antepenultimate (Eq. 2), which activates the limit, and then, the attack is recognized.

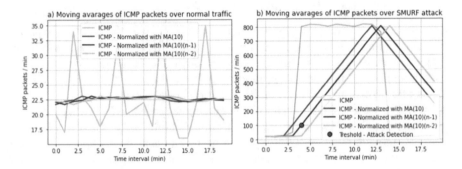

Fig. 4. Closer view of Moving Avarages Lines with ICMP packets over normal conditions vs. Smurf attack - IoT network

This method was performed on 6 different attack types and protocols: TCP SYN flood, Smurf (ICMP), UDP reflection, SSDP reflection, ARP spoof and MQTT-Malaria, all available on the GitHub platform, in Jupyter Notebook [27], to facilitate later research. In Table 2, it is observed that most of the results tend to 100%, in the TCP, UDP, SSDP, ICMP and MQTT protocols. The method of comparing moving averages, shown in Fig. 4, is able to differentiate benign traffic from an attack with very high precision. In the ARP protocol, we have a particular situation generated by the ARP spoofing technique. In addition to the increase in ARP packets at the time of the attack, the gateway itself on the network ends up echoing alert messages that there is a repeated IP on the network. In this way, it generates additional traffic that impairs the identification of anomalies based on the comparison of moving averages.

Another finding was that the change of state from 0 (zero) packets to some value was read as a false positive. Some protocols are intermittent, they transmit for a while, then end the transmission, and when reactivating the transmission, the moving averages change from 0 to some value, surpassing the growth rate established by Eq. 2. To work around this problem, it was decided not to count 0's (zeros), this way the anomaly is verified only in the transmission period, when there are packets being transmitted. The policy of not accounting or storing zeros, in addition to eliminating false positives, reduces the space needed to store statistics, making the system even more efficient.

Table 2. Datasets, files and results

Dataset	Files	Attack Type	Launched	Detected	Precision
UNSW-2019 [23]	18-05-31.pcap	TCP SYN	30	30	100%
	18-06-01.pcap	UDP reflection	9	9	100%
	18-06-02.pcap	SSDP reflection	4	4	100%
	18-10-19.pcap	ARP spoof	15	13	86, 67%
	18-10-23.pcap	Smurf (ICMP)	22	22	100%
MQTTset-2020 [22]	1w_malDoS.pcap	MQTT-malaria	4	4	100%
IDS	Overall	Precision	84	82	97, 62%

4.1 Computational Complexity Analysis

As we have limited computational resources at the edge of the network, computational complexity is an important metric to watch. It is necessary to reflect on how many resources are needed to achieve the objective. In addition to the aspect of the feasibility of implementing models in IoT networks, the improvement of models and algorithms brings operational and energy efficiency gains [28].

The computational complexity of the moving average is $O(w)$, where w is the size of the data window, which in the proposed model is 10. Online operation requires 3 moving averages $O(3w)$. Considering the continuous flow of network n, or in the case of the dataset, the entire series of n data, the micro IDS continuously obtains the moving average, thus raising the complexity of operation to $O(n)$.

5   Prototype

To simulate the operation in real environments, a prototype was implemented with open source tools. The implementation allowed the improvement of the model, with emphasis on data reduction, which is being recommended for data from IoT devices [29]. The tools are described in Table 3 and the implementation steps explained throughout this subsection.

Table 3. Tools used in the prototype

Software	Version	Size
OpenWrt	21.02.2	30 MB
iptables	1.8.7	included (*OpenWrt*)
python3	3.9.13	8, 6 MB
scapy	2.4.4-1	1, 5 MB
daemonlogger	1.2.1	146.3 KB
mailsend	39.5	39.5 KB

The first step is reading the network traffic, segmenting the protocols, and converting it into statistics. As mentioned before, in the validation phase this task was performed manually using the *Wireshark* tool [24]. In a production environment, the *deamonlogger* tool was adopted, a fast packet logger specifically designed for use in network and security management environments [25] and compatible with *OpenWrt*. Through it, the initial step is carried out, which is the reading of the protocols on the network: *daemonlogger eth0 -t 60 -M 75*. The parameter -t 60 was used to divide the capture every minute. The second parameter "–M 75" writes files to the disk until it is 75% utilized and then rolls over and deletes the oldest file in the logging directory.

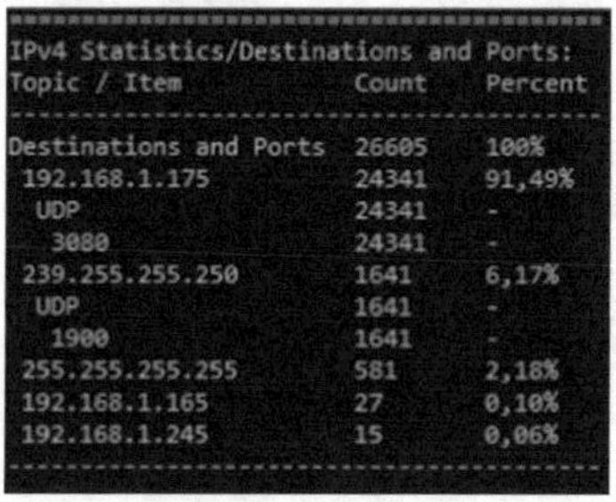
IPv4 Statistics/Destinations and Ports:

Topic / Item	Count	Percent
Destinations and Ports	26605	100%
192.168.1.175	24341	91,49%
UDP	24341	-
3080	24341	-
239.255.255.250	1641	6,17%
UDP	1641	-
1900	1641	-
255.255.255.255	581	2,18%
192.168.1.165	27	0,10%
192.168.1.245	15	0,06%

Fig. 5. Target of attack, destination IP 192.168.1.175, destination port 3080

Following that, the protocol segmentation is performed with the packet manipulation tool *scapy*[3]. The contents of the file created by the *daemonlogger* are read by the *scapy*, and then segmented by protocol, using the script from Di Vita [30]. This completes the statistics storage part. Through this step, the data required for network recognition is reduced by approximately 99.9%. As our model is based on statistics, initially it is only storing the number of packets per minute, segmented by the protocol, which is 0.1% in relation to the total network traffic. In the tests performed, the total network traffic amounted to a total of 11.2 GB, the saved statistics amounted to only 230 KB.

5.1 Neutralization of the Attack

The statistics are read and processed by micro agents [27] with the help of the *numpy* library, a comparison of the percentage growth of the moving averages is performed, in the same way that it was presented in the validation step (Eq. 1 and 2). If an anomaly is recognized in any protocol, a targeted search is carried out in the files generated by the *daemonlogger* to discover the target of the attack.

The protocol with anomaly is used as a filter, ex: SSDP (SSPD Reflection Attack). In Fig. 5 we present the filter of all SSDP application protocol packets. It is observed that 91.49% of the packets are destined to IP 192.168.1.175, which in this case is the target of the attack. Below the IP, the transport protocol is

[3] https://scapy.net/.

visualized, in this case it is the UDP and finally the destination port 3080. In this way, we obtain the necessary information to neutralize the effects of the attack.

The strategy of looking for the attack target after recognizing the anomaly reduces the volume of data to be processed, which contributes to the efficiency of the IDS. In the case of the SSDP reflection attack, we consider the volume of partial network traffic to be 1.36 GB, when reduced to the period in which the attack took place, and for the SSDP protocol the volume drops to 14.6 MB. The target search is performed by reading approximately 1% of the total traffic. As a countermeasure, in addition to creating a firewall rule, an email is sent to inform the responsible specialist through the *OpenWrt* commands listed below.

```
uci add firewall rule
uci set firewall.@rule[-1].name='DROP SSDP attack - IDS rule'
uci set firewall.@rule[-1].src='lan'
uci set firewall.@rule[-1].dest_ip='192.168.1.175'
uci set firewall.@rule[-1].proto='UDP'
uci set firewall.@rule[-1].dest_port='3080'
uci set firewall.@rule[-1].target='DROP'
uci commit firewall
mailsend -f microids@openwrt -t sysadmin@example.com -smtp smtp.example.com -sub "DOS attack blocked" -msg-body "DOS SSDP attack blocked, verify your network ASAP"
```

6   Results and Discussion

After studying the related works, it was possible to propose an efficient IDS to recognize denial of service attacks online, compatible with the computational resources available at the edge of the network. The ability to contain attacks at the edge of the network has been proven, preventing DoS and DDoS attacks from taking massive proportions.

Using moving averages with a sliding window contributes to the ability to generalize, allowing the identification of different types of DoS attacks. Because the limit is adaptive, it triggers only when there is a spike in growth, not when it reaches a certain number or rate, as with a fixed limit. The engine contains a preloaded dynamic recognition model, detailed on Sect. 4, which looks for the most common characteristic of DoS attacks and not a specific pattern of variation. This way you don't need hundreds of epochs of learning. Its reference is constructed from the current average (n) and the previous sliding windows $(n-1)$ and $(n-2)$.

With this technique the complexity was kept at $O(n)$, lower than the work of [18], which after optimization of the deep learning technique reaches the complexity of $O(2^O)$. The accuracy achieved was 100% on most attacks, staying at 86.67% on ARP Spoof. Overall, the accuracy was 97.62%, the highest among jobs with online recognition.

It is also worth mentioning the data reduction in the extraction of statistics from network traffic, which was performed with a proportion rate of 99.9%. Radically reducing the cost of processing to recognize the threat. Finally, we added disk space occupied by the tools used in the prototype implementation, reaching a total of 40.65 MB, previously described on Table 3. This configuration allows detection of DoS attacks to be implemented at the edge of the network, in resource-restricted routers.

This IDS proposal was conceived for IoT networks of sensors, which have regular transmission patterns [12]. In traditional networks with computers and human interactions, or even in mixed networks, with IoT devices, cell phones and laptops, the proposal will not work properly, the irregularity and randomness of the traffic will cause many false positives.

7   Final Considerations

Analyzing how denial-of-service attacks happen and their symptoms through network traffic statistics opens up a range of models for their recognition. Unlike traditional networks, the profile of IoT devices has its behavior on the network better defined, therefore more viable to be mapped through mathematical models, as in this proposal, where learning takes place online, without the need to read the entire dataset for the model to be trained and finally become able to recognize the attacks.

The micro-IDS experiment was successful in online recognizing denial-of-service attacks by prioritizing implementation efficiency and feasibility. Comparison between moving averages is a metric that has proven to be highly generalizable, with the potential to recognize new types of denial-of-service attacks, regardless of protocol. With $O(n)$ complexity, it was possible to identify TCP SYN flood, Smurf (ICMP), UDP reflection, SSDP reflection, ARP spoof and MQTT-Malaria attacks in 2 different datasets with an overall accuracy of 97.62%.

The neutralization of the attack at the edge of the network, prevents DoS and DDoS attacks from taking massive proportions. Being an important part of the damage reduction of this type of attack. As future works, we intend to add a primary line of defense against floods with a response close to real time, since this first defense will be implemented even before the attack takes place, applying a dynamic limit rate based on historical moving average data.

References

1. Hasan, M.: State of IoT 2022: Number of connected IoT devices growing 18% to 14.4 billion globally (2022). https://iot-analytics.com/number-connected-iot-devices/. (Accessed 13 Sep 2022)
2. Al-Fuqaha, A., Guizani, M., Mohammadi, M., Aledhari, M., Ayyash, M.: Internet of things: A survey on enabling technologies, protocols, and applications. IEEE Commun. Surv. Tutorials **17**(4), 2347–2376 (2015)

3. Cook, A.A., Mısırlı, G., Fan, Z.: Anomaly detection for IoT time-series data: A survey. IEEE Internet Things J. **7**(7), 6481–6494 (2019)
4. Perlroth, N.: Hackers Used New Weapons to Disrupt Major Websites Across U.S (2016). https://www.nytimes.com/2016/10/22/business/internet-problems-attack.html. (Accessed 13 Sep 2022)
5. Porter, J.: Amazon says it mitigated the largest DDoS attack ever recorded. https://www.theverge.com/2020/6/18/21295337/amazon-aws-biggest-ddos-attack-ever-2-3-tbps-shield-github-netscout-arbor. (Accessed 13 Sep 2022)
6. Balaban, I.: Denial-of-service attack. Int'l J. Info. Sec. Cybercrime **10**, 59 (2021)
7. Mergendahl, S., Li, J.: Rapid: Robust and adaptive detection of distributed denial-of-service traffic from the internet of things. In 2020 IEEE Conference on Communications and Network Security (CNS), pp. 1–9. IEEE (June 2020)
8. Saghezchi, F.B., Mantas, G., Violas, M.A., de Oliveira Duarte, A.M., Rodriguez, J.: Machine learning for DDoS attack detection in industry 4.0 CPPSs. Electronics **11**(4), 602 (2022)
9. Sharma, D.K., et al.: Anomaly detection framework to prevent DDoS attack in fog empowered IoT networks. Ad Hoc Netw. **121**, 102603 (2021)
10. Li, F., Shinde, A., Shi, Y., Ye, J., Li, X.Y., Song, W.: System statistics learning-based IoT security: Feasibility and suitability. IEEE Internet of Things J. **6**(4), 6396–6403 (2019)
11. Sahoo, K.S., Puthal, D.: SDN-assisted DDoS defense framework for the internet of multimedia things. ACM Trans. Multimedia Comput. Commun. Appli. (TOMM) **16**(3s), 1–18 (2020)
12. Wan, Y., Xu, K., Wang, F., Xue, G.: Characterizing and mining traffic patterns of IoT devices in edge networks. IEEE Trans. Netw. Sci. Eng. **8**(1), 89–101 (2020)
13. Cisa. Understanding Denial-of-Service Attacks (2022). https://www.cisa.gov/uscert/ncas/tips/ST04-015. (Accessed 13 Sep 2022)
14. Sousa, B.F.L.M., Abdelouahab, Z., Lopes, D.C.P., Soeiro, N.C., Ribeiro, W.F.: An intrusion detection system for denial of service attack detection in internet of things. In: Proceedings of the Second International Conference on Internet of things, Data and Cloud Computing, pp. 1–8 (March 2017)
15. Doshi, R., Apthorpe, N., Feamster, N.: Machine learning ddos detection for consumer internet of things devices. In: 2018 IEEE Security and Privacy Workshops (SPW), pp. 29–35. IEEE (May 2018)
16. Salahuddin, M.A., Pourahmadi, V., Alameddine, H.A., Bari, M.F., Boutaba, R.: Chronos: Ddos attack detection using time-based autoencoder. IEEE Trans. Netw. Serv. Manage. **19**(1), 627–641 (2021)
17. Tann, W.J.W., Tan, J.J.W., Purba, J., Chang, E.C.: Filtering DDoS attacks from unlabeled network traffic data using online deep learning. In Proceedings of the 2021 ACM Asia Conference on Computer and Communications Security, pp. 432–446 (May 2021)
18. Sudharsan, B., Patel, P., Breslin, J.G., Ali, M.I.: Enabling machine learning on the edge using sram conserving efficient neural networks execution approach. In: Dong, Y., Kourtellis, N., Hammer, B., Lozano, J.A. (eds.) ECML PKDD 2021. LNCS (LNAI), vol. 12979, pp. 20–35. Springer, Cham (2021). https://doi.org/10.1007/978-3-030-86517-7_2
19. Alzahrani, M.A., Alzahrani, A.M., Siddiqui, M.S.: Detecting DDoS attacks in iot-based networks using matrix profile. Appl. Sci. **12**(16), 8294 (2022)
20. Santoyo-González, A., Cervelló-Pastor, C., Pezaros, D.P.: High-performance, platform-independent DDoS detection for IoT ecosystems. In: 2019 IEEE 44th Conference on Local Computer Networks (LCN), pp. 69–75. IEEE (October 2019)

21. Jouet, S., Pezaros, D.P.: Bpfabric: Data plane programmability for software defined networks. In: 2017 ACM/IEEE Symposium on Architectures for Networking and Communications Systems (ANCS), pp. 38–48. IEEE (May 2017)
22. Vaccari, I., Chiola, G., Aiello, M., Mongelli, M., Cambiaso, E.: MQTTset, a new dataset for machine learning techniques on MQTT. Sensors **20**(22), 6578 (2020)
23. Hamza, A., Gharakheili, H.H., Benson, T.A., Sivaraman, V.: Detecting volumetric attacks on lot devices via sdn-based monitoring of mud activity. In: Proceedings of the 2019 ACM Symposium on SDN Research, pp. 36–48 (April 2019)
24. WireShark (2022). https://www.wireshark.org/index.html#download. (Accessed 29 Oct 2022)
25. Daemonlogger (2022). https://talosintelligence.com/daemon. (Accessed 29 Oct 2022)
26. Lee, C.H., Lin, C.R., Chen, M.S.: Sliding-window filtering: an efficient algorithm for incremental mining. In Proceedings of The Tenth International Conference On Information And Knowledge Management, pp. 263–270 (October 2001)
27. Lautert, H.: Multiple Moving Avarages to Anomaly Detection in IoT networks (2022). https://github.com/hflautert/AnomalyDetection. (Accessed 29 Oct 2022)
28. Dean, W.: Computational Complexity Theory, The Stanford Encyclopedia of Philosophy (Fall 2021 Edition), Edward N. Zalta (ed.). https://plato.stanford.edu/archives/fall2021/entries/computational-complexity/. (Accessed 29 Oct 2022)
29. Pioli, L., Dorneles, C.F., de Macedo, D.D., Dantas, M.A.: An overview of data reduction solutions at the edge of IoT systems: a systematic mapping of the literature. Computing, 1–23 (2022)
30. Di Vita, L.: Protocols Counter (2019). https://github.com/lucadivit/Protocols_Counter. (Accessed 31 Oct 2022)

Using Hidden Markov Chain for Improving the Dependability of Safety-Critical WSNs

Issam Alnader[(✉)], Aboubaker Lasebae, and R. Raheem

Middlesex University, The Burroughs, London, Hendon NW4 4BT, UK
{I.Al-Nader,A.Lasebae,R.H.Raheem}@mdx.ac.uk

Abstract. Wireless Sensor Networks (WSNs) are distributed network systems used in a wide range of applications, including safety-critical systems. The latter provide critical services, often concerned with human life or assets. Therefore, ensuring the dependability requirements of Safety critical systems is of paramount importance. The purpose of this paper is to utilize the Hidden Markov Model (HMM) to elongate the service availability of WSNs by increasing the time it takes a node to become obsolete, via optimal load balancing. We propose an HMM-algorithm that, given a WSN, analyses and predicts undesirable situations, notably, nodes dying unexpectedly or prematurely. We apply this technique to improve on the Randomized coverage-based scheduling algorithm (RCS) by C. Lius, a scheduling-based algorithm that has served to improve the lifetime of WSNs. Our experiments show that our HMM technique improves the lifetime of the network, achieved by detecting nodes that die early and rebalancing their load. Our technique can also be used for diagnosis and provide maintenance warnings to WSN system administrators. Finally, our technique can be used to improve algorithms other than the RCS.

1 Introduction

Thanks to their easy deployment, WSNs have been used in a wide range of application systems, from simple detection systems such as humidity detection [1] to safety critical systems such as intrusion detection systems [2–5]. Our research findings revealed weaknesses in the Randomised Coverage Based Scheduling (RCS) algorithm [6] and bridged the gap in increasing the WSN's lifetime, service continuity, reliability, and availability. Through our analysis, we captured scenarios whereby some nodes are over-worked, while others are under-worked, hence compromising the availability and lifetime of the network. The reason for this is the randomized nature of the RCS algorithm, which makes optimal load balancing for future performance optimization a challenging task. For example, towards the end of the network life-cycle, some nodes are found to be already obsolete ("dead"), which can lead to breaking the WSN into different segregations. This situation can lead to data loss due to the inability to transmit messages across the WSN.

Our main objective here is to work on multi-objective optimisation, i.e., trade-offs between network lifetime (service availability) and network coverage & connectivity. Therefore, the aim is to address the problem of improving the system dependability of

© The Author(s), under exclusive license to Springer Nature Switzerland AG 2023
L. Barolli (Ed.): AINA 2023, LNNS 661, pp. 460–472, 2023.
https://doi.org/10.1007/978-3-031-29056-5_40

safety-critical WSNs, thus achieving their quality of service (QoS) requirements [7, 19]. The RCS algorithm in [6] already works towards optimising network coverage. In this work, we extend network lifetime optimisation by using a Hidden Markov Model or HMM [8] to reason about WSNs. As we recreated the RCS algorithm in our simulation environment, we were able to achieve two goals: (1) avoid duplicating efforts in designing a scheduling algorithm for WSN while RCS already exists, (2) improve on RCS optimisation algorithm by increasing the network's lifetime via optimising the nodes' ON states. Thus, this work acknowledges reaching a state of obsoleteness is an inevitable state for nodes. However, delaying reaching this state is useful and has arguably economic benefits. The work utilised HMM to intelligently extend the usability of nodes to collectively increase the service availability & reliability of the WSN [9].

The authors of this paper acknowledge to the best of their ability that the literature lacks focus on extending the WSN lifetime while maintaining coverage and connectivity in a significant manner [10, 11, 12]. Related work in the literature has focused on extending the last remaining nodes alive, while we aim for extending the collective lifetime of the WSN. Consequently, another objective in this work is to guarantee minimum QoS levels e.g. connected coverage sensor networks, even in undesired situations such as example, when the WSN becomes segregated due to nodes' premature death. The paper is structured as follows; Sect. 2 reviews related work in the literature, while the proposed new HMM scheduling algorithm with its mathematical representation is introduced in Sect. 3. Section 4 covers the results and compares the proposed HMM algorithm to the RCS algorithm. Finally, Sect. 5 is the conclusion drawn from this work.

2 Literature Review

HMM is one of the most common methods used in unsupervised learning. HMM is a hidden state machine in which each state has a certain transition probability to another state. Each transition state generates an observation state, which follows a law of probability associated with the current state. The observations can be discrete or continuous. In the case where they are discrete, each state will be associated with the probability of making the observation of each of the possible discrete symbols. In the case where they are continuous, each state will be associated with a density function (often a Gaussian mixing model) as depicted from the definition of [13]. There are several papers in the literature that use the HMM method to address multiple problems that are related to energy consumption in WSNs. Therefore, the use of the HMM method in WSNs is not new [8, 14]. What follows shed the light on some related work that utilizes HMM methods in WSNs.

The use of HMM methods in WSNs was presented by [15], where a scheduling algorithm that utilized HMM method to observe states for a resource-constrained WSN is proposed. Krishnamurthy, 2002 [15] addresses the problem of data reliability, of obtaining certain physical data, e.g. measurements of a certain process known as signal processing application. The basic principle is to utilize the HMM finite states to locate noisy sensor nodes in the WSN, then select these nodes to send the required data. The solution utilized dynamic stochastic programming which is achieved in two folds: (1) By finding the optimal channel allocation for various components of a measurement vector, for example, when a noise sensor must transmit over a time-shared communication channel with limited bandwidth; (2) By finding the optimal time of measurements of the sensor when the number of possible measurements is limited because of energy constraints [15].

Goudarzi et al., 2010 [16] propose the use of HMM method in combination with particle swarm optimization (PSO) to predict the energy level in WSN. The algorithm uses PSO to select the cluster head Nodes. The proposed method reduces the cost of clustering and improves the performance of the WSN. The optimization problem defined for the best possible energy values can be obtained by solving it with PSO to find the best position that is evaluated from the fitness function. To initialize the PSO all the defined variables are assigned random values in the search space. In each run, each particle generates the best individual position (associated with energy) and global best position to obtain the best possible solution in search space. The communication overhead is reduced due to the use of PSO. However, there is always a trade-off between the cost of finding the energy versus the accuracy [16]. If the cost is higher the accuracy is better and a lower cost leads to lower accuracy. Hence, we have to work between these extremes. This solution introduces an overhead in the WSN as a result of the complexity involved in using the PSO algorithm at the expense of obtaining a good level of the accurate energy solution.

Qihua et al., 2015, [17] propose a scheduling algorithm to extend the network lifetime. The HMM is utilized to model sensor node states in the WSN where a node can be in two states: 0 or asleep and 1 or active. The work in [17] bases its HMM method on two factors that are the energy cost and the errors of the sensor nodes reading. In addition, they consider the use of an actuator that takes the reading of the node's energy and sends that data to the central controller, which manages the entire WSN. Qihua et al.' algorithm [17] seems to improve the lifetime of the WSN, however, the introduction of a coordinator leads to an overhead control message on the WSN.

In summary, to our best knowledge, there are not many works that utilize HMM to address multi-objective optimization problems involving energy consumption, network lifetime, coverage, and connectivity, in the context of the WSNs scheduling approach. In comparison, our work utilizes a very simple HMM algorithm to optimize the energy as well as the position of nodes for effective scheduling and energy management. Hence, our approach utilizes the HMM method to extend the network lifetime and introduces no communication overhead.

3 Problem Formulation

A WSN network is composed of sensors connected through radio links in a given target area, where the function of the WSN is embedded in the sensors [18]. In an HMM, a node/sensor is modeled as a state-transition diagram, indicating the possible states of the node and the transitions from one state to the other. Each transition is assigned a probability for its occurrence. The state of the WSN is then given by the states of all its nodes. The possible states of a sensor vary depending on the application. Typical states include a Transmission state and a Receiving state. Other possible states include Active, Sleep, Relay, Idle, and Fail, amongst others. Different states consume different levels of energy.

Here, we distinguish the ON state, when the sensor is performing some action, from the OFF state, where no energy is consumed. Problem: we are trying to optimize energy consumption in the ON state, by switching the sensor from the (ON, Tx) transmission state to the (ON, Rx) receiving state.

The following are conditions to be considered for switching:

1. The energy level (threshold energy level) of nodes
2. The probability assignment (of transition between states), based on the distance of a node from the sink

   1. The initial probability assignment is proportional to the threshold energy levels and the distance from the sink, for send and receive states.
   2. We call hidden states those that lie between (ON, Tx) and (ON, Rx), e.g., Sleep. The probability of their occurrence is randomly derived from the immediate probability assigned to the parent node such that the sum of the distributed probabilities is equal to the probability assigned to the parent node.

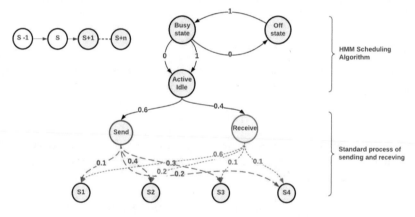

Fig. 1. Balancing Tx (Send) and Rx (Receive) states using HMM

Figure 1 illustrates the state-transition diagram of a simple WSN, with two states, Send and Receive, and four hidden states (chosen for the purpose of illustration), named S1 to S4.

A sensor node initially has two states, but after using HMM a sensor node will be assigned four hidden states (S1, S2, S3, S4) as illustrated in Fig. 1 where the greatest probability amongst S1, S2, S3, S4 will be assigned to the sensor node instantaneously. A sensor node starts its operation with an active idle state. This means, as the network progresses in time, e.g. nodes executing their task routine; Tx (Send) and Rx (Receive), the energy of each node reduces, hence probability using HMM is a good fit to keep up with which nodes to transmit and which nodes to receive. As time evolves, the energy of the sensor nodes depletes as result probability of receiving is higher than transmitting because the energy for transmitting is always higher.

The following represents the pseudo code of the proposed HMM Algorithm.

1. *At an instant time (t).*
2. *For each (N) set of nodes starting from node*
3. *Assign two states Tx and Rx states for a node*
4. *Do the granularity of each states be (m) states with observations*
5. *Assign the probability for transmit > receive*
6. *Update transition_probability and emission_probability (w.r.t the energy levels)*
7. *Until probability for transmit < receive*
8. *Set the final probability transmit and receive to the maximum of Emission_probability*
9. *Next node*

Let N be the set of the Sensor Nodes over the area to be monitored, j is the minimum number of nodes, jmax the maximum number of nodes, and λ is the binding variable between j and jmax that represent the minimum and the maximum number of Sensor Nodes respectively. In our scenario, j is a none zero variable. Let #N be the size of set N. Then:

$$j \leq \#N \leq jmax \tag{1}$$

Similarly, let CHs be the set of the Cluster Head Nodes that are chosen randomly amongst the Sensor Nodes and CHmax is the maximum number of CHs.

$$1 \leq \#CH \leq CHmax \tag{2}$$

Upon the deployment of the nodes in the network, each node is able to set up a link to the other nodes or to the CH accordingly. Through the above discussion, we have found this connectivity parameter:

$$P(a, b) = \begin{cases} 1 \ If \ a \ establishes \ a \ connection \ with \ b \\ 0 \ otherwise \end{cases} \tag{3}$$

The a,b $\in$ N, and a $\neq$ b, where b is either a member node or a CH in the network. If b is a CH, then a is a member node. The quality of links is directly related to the received signal strength and the distance. Suppose that c(t) denotes a stochastic process to signify the selected number of CHs at a specific time instant t. As the process of CHs selection starts from the beginning of each round, an integer scale t and discrete-time t + 1 instant are selected at the beginning of two successive rounds. We express that r(t) is the round at a time instant t, and x(t) is a stochastic procedure that signifies the period of a scheme at a time instant t, which is x(t) = r(t)mod(1/P) (this equation guides the selection round). We also assume another integer 1/P, which denotes n = 1/P. The state space of this model is:

$$\{0, N\} \cup \{(i, x) : i \in [0, N], x \in [1, n-1]\} \tag{4}$$

The state space model represent only those nodes which are connected. This is because the connected nodes form the network, which is our point of interest. Where i and x are integers; transition state (active nodes) and observation(connected nodes) state respectively, as this process $\{x(t), c(t)\}$ holds the Markov property and $n - 1$ is number of connections link between the nodes.

We use the bi-directional Hidden Markov Chain model stationary distribution and one-step transition probabilities from [14] to estimate the Probability Mass Function (PMF) for the nodes and CHs to switch its states as follows:

$$P(CH = L) = \pi(0, N).\left[P(0, N) \rightarrow (1, N - L) + f(n - 1, L) + \sum x \right] \quad (5)$$

where π denotes the initial probability distribution, P is one step transition probability matrix, and f signifies a factor matrix (is the matrix which is assigned to 1 for active state, 0 otherwise), $\sum x$, is included due to stochastic effect of the WSN, fxi, $x \in [1, n - 1]$, and $i \in [0, N]$ are elements of the factor matrix. The sensor nodes can be switching from the (ON, Rx) transmission state to the (ON, Tx) receiving state using previously set threshold probabilities.

The HMM optimization method assigns random transition probabilities for switching into these four different states (S1, S2, S3, and S4). As the network progresses in time, the energy of each node reduces, and the HMM algorithm adjusts the probability of reaching the state. As time evolves, the probability of receiving is higher than transmitting and nodes that deplete their energy at a higher rate eventually enter into the Rx state only. Nonetheless, there will remain TX nodes too, depending on their current predicted energy levels.

4   Results and Discussion

The main objective of this experiment is to analyse the performance of the new proposed HMM algorithms with the original RCS scheduling algorithm. The proposed solution can be applied to any scheduling solution since it alters the state of the ON node at a very particular point in time at a very particular event, e.g., when a node's level of energy reaches a certain level.

In the experiment of the new solution using the MATLAB simulation environment, we set up our experiments in a 100 m^2 area to be monitored. The simulation consists of 100 stationary sensor nodes that are randomly deployed. The base station is located on the far-left edge of the network. All nodes are homogeneous which means they have the same sensing and communication capabilities. Figure 2 illustrates the simulated environment as explained above.

The simulations were performed over 2000 rounds as the upper bound of our simulation experiment. The breaking point of our simulation experiment (when the simulation will completely stop) is when 95% of the sensor nodes have depleted their energy. In each simulation round, an event is detected and reported to the base station. In this experiment, we analyse and compare the performance of the original RCS algorithm against the improved version using the HMM process, in terms of the following metrics: Energy Used, WSN Lifetime, Throughput, Connectivity, Coverage, and Coverability. Below are

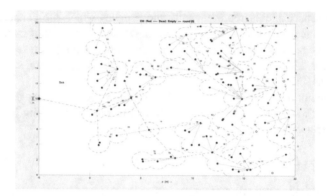

Fig. 2. Simulation Set-up of HMM in Matlab environment

the diagrams (the output of our experiments) obtained per metric, in each diagram, there are two simulation factors to evaluate the achieved results:

1. RCS, referred to as "scheduling_o";
2. RCS with HMM, referred to as "scheduling_m".

All Used Energy

Figure 3 represents the total energy used during the rounds in the WSN. This metric is concerned with the sum of consumed energy in the WSN during its lifetime of operability (simulation time for the purpose of this work). The X-axis is the number of simulation rounds/times, while the Y-axis represents the energy consumption unit in Joules.

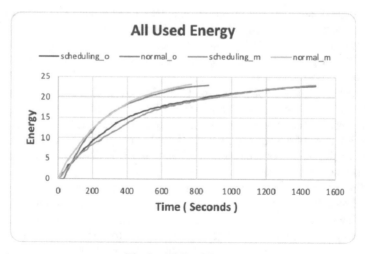

Fig. 3. All Used Energy

The performance of the curves in Fig. 3 can be interpreted as a metric performance evaluation whereby the less the curvature with respect to Energy levels the better because this shows the WSN achieves the same number of rounds (Time) with less energy. As can be seen in Fig. 3, at round 168 the energy consumed by the Original RCS is 7.867 J, but the energy consumed by our improved algorithm (using HMM) is 7.133 J. Although this seems an insignificant improvement due to the parameters set up in our simulation, if we tuned the energy parameter in our experimentations to a higher value, we would have certainly observed a significant improvement with our proposed HMM-assisted algorithm. Hence, the HMM-assisted algorithm shows better performance with respect to energy consumption, than its RCS peer. Furthermore, we also notice that as time progresses, energy will eventually deplete, then the difference in energy consumption levels between the two algorithms is almost non-existent. This is because there is a positive correlation between the granularity value (of the parameter) and the distance between the curves.

Lifetime of Sensor Nodes

Figure 4 shows the number of live sensor nodes in the WSN. The 'Lifetime of Sensor Nodes' metric refers to how long the sensor nodes will last before their energy deplete. The X-axis is the time referred to the number of Rounds, and the Y-axis is the energy of each sensor node.

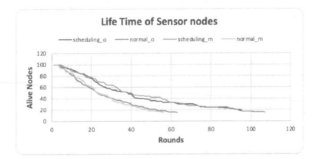

Fig. 4. Lifetime of sensor nodes

We notice from the Y-axis that the remaining number of nodes at the end of the simulation run-time is 15 for both HMM and RCS. Yet, the X-axis shows with HMM the WSN lifetime reaches 107 rounds before the end of the simulation, but RCS lasted 102 rounds only. Despite the small improvement made with the HMM (of 5 rounds only), this can be explained due to the small value allocated to the sensor parameters (number of nodes, energy, communication & sensing ranges, etc.). Note, assigning large parameter values is an unrealistic experimentation practice (Xianglin et al., 2012).

Throughput

Figure 5 represents the throughput utilized by HMM assisted algorithm and its RCS peer in the WSN. The throughput is the number of data packets sent and received in the network. The X-axis is the time which is in rounds, and the Y-axis is the number of data packets transmitted in the network.

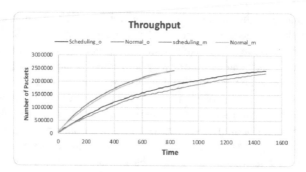

Fig. 5. Throughput

We can notice the distance between the two curves of HMM and RCS algorithms. There are lots of packets generated in the network which indicates that a sizable number of packets are discarded due to collision. The excess generation of packets leads to the wastage of energy and hence, the resultant data transfer is low. With HMM, we managed to transfer fewer packets which results in better performance with respect to throughput. For example, the number of packets generated with HMM at the 1414th round is 2275957, meanwhile, with RCS, the number of generated packets is 2379214. Hence, RCS generated 10,000 more packets than HMM.

Connectivity

Figure 6 reflects the achieved results with respect to connectivity in the WSN. Connectivity is the number of nodes connected at any instance in time. The X-axis is the time in rounds, and the Y-axis is the fraction or percentage of nodes being connected.

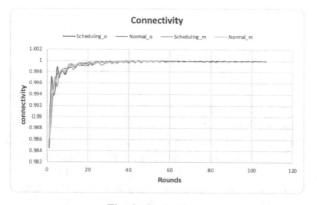

Fig. 6. Connectivity.

The HMM's connectivity performance is similar to the RCS' performance with respect to the fractions of the nodes that are connected. Both algorithms scored equal connectivity values of 99.99%, which indicates that there is almost no scope for improvement over the RCS as far as this metric is concerned. The fluctuations can be attributed to the stochastic nature of the HMM algorithm over the randomization nature of the RCS algorithm.

Coverage
Figure 7 represents the coverage results in the WSN. The coverage metric is defined as the percentage of the area covered by the WSN. The X-axis is the number of rounds in time, and the Y-axis is the ratio of the coverage for the WSN.

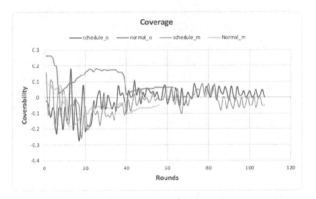

Fig. 7. Coverage

If we take the 65th round, for instance, the HMM performance is 0.005, while the RCS algorithm scores 0.05. This indicates that network coverage has been increasedtenfold using the HMM algorithm.

Coverability
Figure 8 represents the achieved Coverability results in the WSN. The X-axis represents the nodes' sectional analysis number, while the Y-axis is the coverage percentage.

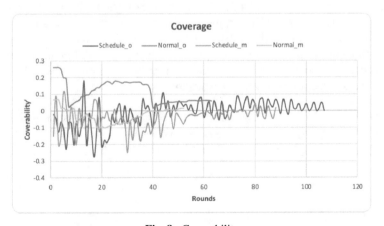

Fig. 8. Coverability

The objective here is to study the network's coverage per section of nodes. For 70 nodes out of a total of 150 nodes, the HMM normal coverage (0.31) > RCS normal coverage (0.27869) and HMM scheduling coverage (0.231885) is also greater than RCS scheduling (0.15254). This trend follows as the number of nodes increases. It is worth noting that the term 'Coverability', while sounding synonymous with 'Coverage', actually refers to the percentage of the coverage per sensor node. Figure 8 explains the result of the experiment. The RCS algorithm was originally tested on the basis of coverability, not coverage, in C Liu's work (Liu et al., 2006). This is understandable because the algorithm is not compared or evaluated with other algorithms with respect to (1) the coverage metric of the extra ON nodes, and (2) network lifetime improvement. In our HMM improved algorithm, we are concerned with improving network lifetime so as to contribute to the safety critical system requirements, and further contribute to the dependability of such systems. The Coverage metric is concerned with all of the network's nodes, while the Coverability metric is concerned with a certain number of nodes and can be used to analyse single nodes only. Therefore, we choose to focus on the Coverage metric in our experimentations.

5   Conclusion

In this work, we introduced a novel improvement to the RCS algorithm, by proposing an HMM algorithm that is based on a probability distribution, unlike the original RCS which is based on random scheduling. Our HMM-based algorithm has increased the network lifetime and also improved the coverage and connectivity. The transition states of the nodes are restricted to four states in our algorithm to avoid computational overheads. The main goal was to predetermine the receive and transmit states' as defined by the algorithm of each node in the WSN in the design time. In this process, we adjust the probability of transition between our receive states and transmit states. As a result, each node knows its receive states' as defined by the algorithm and operates accordingly in the runtime. In network connectivity and path optimality, this prediction is important considering the broadcast collision and channel errors as metrics for quality of service. The RCS algorithm has had several errors due to high throughput values, but our proposed HMM algorithm has mitigated this by the use of path optimality (optimization) to reduce traffic overheads. This is achieved by ensuring both receiving and transmitting nodes at every round of simulation time to ensure the network's operability. In the future, we will be addressing the limitation of the RCS and the HMM algorithms using bioinspired computational and artificial intelligence methods.

Acknowledgement. The authors would like to thank all those who contributed to the completion and success of this work. Also, the Faculty of Science and Technology at Middlesex University played a significant role in backing this work at all stages of the study.

References

1. Ambrose, A.I.: Scheduling for Composite Event Detection in Wireless Sensor', Master Thesis Fla. Atl. Univ. (2008). http://fau.digital.flvc.org/islandora/object/fau%3A2881/datast ream/OBJ/view/Scheduling_for_composite_event_detection_in_wireless_sensor_networks. pdf (Accessed Aug. 07 2015)
2. García-Hernández, C.F., Ibarguengoytia-Gonzalez, P.H., García-Hernández, J., Pérez-Díaz, J.A.: Wireless sensor networks and applications: a survey. IJCSNS Int. J. Comput. Sci. Netw. Secur. **7**(3), 264–273 (2007)
3. Knight, J.C.: Safety critical systems: challenges and directions. In: .Proceedings of the 24rd International Conference on Software Engineering, 2002. ICSE 2002, pp. 547–550 (2002). http://ieeexplore.ieee.org/xpls/abs_all.jsp?arnumber=1007998 (Accessed Oct. 18 2016)
4. He, T., et al.: Energy-efficient surveillance system using wireless sensor networks. In: Proceedings of the 2nd international Conference on Mobile Systems, Applications, and Services, pp. 270–283 (2004). http://dl.acm.org/citation.cfm?id=990096 (Accessed Jul. 15 2016)
5. Aslan, Y.E., Korpeoglu, I., Ulusoy, Ö.: A framework for use of wireless sensor networks in forest fire detection and monitoring. Comput. Environ. Urban Syst. 36(6), 614–625 (2012). http://www.sciencedirect.com/science/article/pii/S0198971512000300 (Accessed Nov. 05, 2015)
6. Wang, L., Wei, R., Lin, Y., Wang, B.: A clique base node scheduling method for wireless sensor networks. J. Netw. Comput. Appl. **33**(4), 383–396 (2010)
7. Avizienis, A., Laprie, J.-C., Randell, B., et al.: Fundamental concepts of dependability. University of Newcastle upon Tyne, Computing Science (2001). http://edi-info.ir/files/Fundam ental-Concepts-of-Dependability.pdf (Accessed Nov. 13 2015)
8. Hu, P., Zhou, Z., Liu, Q., Li, F.: The HMM-based modeling for the energy level prediction in wireless sensor networks. In: 2007 2nd IEEE Conference on Industrial Electronics and Applications, pp. 2253–2258 (2007)
9. Sailhan, F., Delot, T., Pathak, A., Puech, A., Roy, M.: Dependable wireless sensor networks In: 3th workshop Gestion des Données dans les Systèmes d'Information Pervasifs (GEDSIP) in cunjunction with INFORSID, pp. 1–16 (2009). http://cedric.cnam.fr/~sailhanf/publicati ons/gedsip.pdf (Accessed Jun. 21 2015)
10. Cerpa, A., Estrin, 'D.: SCENT: Adaptive self-configuring sensor networks topologies. IEEE Trans. Mob. Comput. **3**(3), 272–285 (2004). http://ieeexplore.ieee.org/xpls/abs_all.jsp?arn umber=1318596 (Accessed Oct. 10 2015)
11. Cardei, M., Thai, M.T., Li, Y., Wu, W.: Energy-efficient target coverage in wireless sensor networks. In: INFOCOM 2005. 24th Annual Joint Conference of the IEEE Computer and Communications Societies. Proceedings IEEE 2005, vol. 3, pp. 1976–1984 (2005). http://iee explore.ieee.org/xpls/abs_all.jsp?arnumber=1498475 (Accessed Feb. 09 2016)
12. Tian, D., Georganas, N.D.: A Coverage-preserving node scheduling scheme for large wireless sensor networks. In: Proceedings of the 1st ACM International Workshop on Wireless Sensor Networks and Applications, New York, USA, pp. 32–41 (2002): https://doi.org/10.1145/570 738.570744
13. Castellani, U., Cristani, M., Fantoni, S., Murino, V.: Sparse points matching by combining 3D mesh saliency with statistical descriptors. Comput. Graph. Forum **27**(2), 643–652 (2008)
14. Patra, C., Mullick, S.: Calculation of the Duty Cycle for BECA. Int. J. Comput. Appl. **121**(14) (2015)
15. Krishnamurthy, V.: Algorithms for optimal scheduling and management of hidden Markov model sensors. IEEE Trans. Signal Process. **50**(6), 1382–1397 (2002)
16. Goudarzi, R., Jedari, B., Sabaei, M.: An efficient clustering algorithm using evolutionary hmm in wireless sensor networks. In: IEEE/IFIP International Conference on Embedded and Ubiquitous Computing, vol. 2010, pp. 403–409 (2010)

17. Qihua, W., Ge, G., Lijie, C., Xufeng, X.: Scheduling strategy for Hidden Markov Model in wireless sensor network. In: 2015 34th Chinese Control Conference (CCC), pp. 7806–7810 (2015)
18. Hill, J.L.: System architecture for wireless sensor networks' University of California, Berkeley (2003). http://www.eps2009.dj-inod.com/docs/09-02-01/system_architecture_for_wireless_sensor_networks.pdf (Accessed Oct. 24 2015)
19. Avižienis, A., Laprie, J.-C., Randell, B., Landwehr, C.: Basic concepts and taxonomy of dependable and secure computing. IEEE Trans. Dependable Sec. Comput. 1(1), 11–33, (2004). http://ieeexplore.ieee.org/xpls/abs_all.jsp?arnumber=1335465 (Accessed Nov. 18 2015)

A Fault-Tolerant IoT Solution for Solid Waste Collection

Raimir Holanda Filho[1]([✉]) [iD], Wellington Alves de Brito[1],
Debora Carla Barboza de Sousa[1], Victor Pasknel de Alencar[2],
Joan Lucas Marques de Sousa Chaves[1], and Emanuel Leão Sá[1]

[1] University of Fortaleza, 1321, Washington Soares Avenue, Fortaleza, Brazil
raimir@unifor.br
[2] Morphus Labs., Fortaleza, Brazil
http://www.unifor.br , http://www.morphuslabs.com

Abstract. The Internet of Things (IoT) is a revolutionary paradigm which aims to provide ubiquitous connectivity between smart objects and humans. In this sense, an interesting use of the IoT technology is on Smart Cities, as it enables the collection and analyses of data from sensors spread across many points. In the context of smart cities, solid waste management is one of the problems that has gained greater prominence in recent years, discussed with greater emphasis in conjunction with environmental issues. This paper implements a solution based on Low Power Wide Area Network (LPWAN) and blockchain technologies to provide a fault-tolerant solution for collecting data from the position (latitude and longitude) and the amount of garbage into trash bins. These data are used to calculate a daily optimal route for trucks that collect solid waste. Additionally, the performance of the blockchain network was analyzed in a cloud environment under multiple workloads and tests were performed to evaluate the impact of network failures.

Keywords: IoT · LoRaWAN · Blockchain · Solid waste management

1 Introduction

Waste management is a name given to the waste collection system that includes transportation, disposal or recycling. This term is assigned to waste produced by human activities that must be handled to avoid its adverse effects on health and environment. One of the goals of controlling these processes is the reuse of available resources. Waste management methods may differ between developed countries, between an urban and rural environment, or between an industrial and residential area. According to brazilian legislation, waste management in metropolitan and rural areas is the responsibility of a municipality, while the waste produced by industries is their own responsibility and managed by themselves.

Supported by University of Fortaleza.

Solid waste management affects everyone; however, those most affected by the negative impacts of poorly managed waste are largely's most vulnerable-losing their lives and homes from landslides of waste dumps, working in unsafe waste-picking conditions, and suffering profound health repercussions. Too often, the environment also pays a high price. Also, solid waste management is a critical-yet often overlooked-piece for planning, sustainable, healthy, and including cities and communities for all. However, waste management can be the single highest budget item for many local administrations. Municipalities in low-income countries are spending about 20% of their budgets on waste management, on average-yet over 90% of waste in low-income countries is still openly dumped or burned. As these cities and countries grow rapidly, they desperately need systems to manage their growing waste and mechanisms to pay for the essential services that keep their citizens healthy and their communities clean.

The "UN World Urbanization Prospects" provide estimates of urban shares across the world through to 2050. Across all countries, urban shares are projected to increase in the coming decades, although at varied rates. By 2050, it's projected that 68% of the world's population will live in urban areas, an increase from 54% in 2016, which corresponds to almost 7 million people. According to the most recent survey publishing by the "International Solid Waste Association", developed in partnership with the Brazilian Association of Public Cleaning and Special Waste Companies (ABRELPE), the costs of poor solid waste management total more than R$ 5.5 billions per year in Brazil. The balance concerns expenses with environmental recovery and health treatments related to marine pollution.

Among all, waste collection and transportation are one of the costliest stages in solid waste management [1], as the truck driver must go to each bin every single day and check whether the bin is full or not. The most used service model in Brazilian municipalities that have this type of service working, pays service providers through a fixed monthly amount and pre-determined daily routes, in which, the solid waste collection process is carried out, usually three days a week, when the truck passes by the residences to collect the garbage and send it to the sanitary landfills. This process works based on established fixed routes and the company responsible for this activity is payed for this service based on the distance traveled. If the bin is not full, it is not only a waste of time but also a waste of fuel used by the truck. It increases pollution due to smoke released from trucks and it is not economic because more men are needed for checking all the bins on different routes.

To face the problem described above, the use of technologies based on Internet of Things (IoT) becomes essential for the gradual transformation of large urban centers around the world into smart cities [2,3]. In this work, we are proposing the use of sensors to measure the amount of waste in the bin, and LoRaWAN technology to send the data for further information processing. When the trash inside the bin reaches a certain threshold level, its location and the current position of the truck driver are detected. Both pieces of information are sent to be processed and the shortest path between them is shown. Using this approach, the garbage bins can be emptied before the dustbin overflow.

Despite the proposal to use smart bins is not new, our research is focused on providing a high degree of data availability of the information sent by the sensors, improving the LoRaWAN architecture. The current specification of LoRaWAN points to some security vulnerabilities as, a centralized authentication process, which under a denial-of-service (DoS) attack, for example, all services may become unavailable. This work proposes a distributed authentication process, based on blockchain, to provide a fault-tolerant IoT platform.

2 Related Work

Fault tolerance can be defined as the capacity of a system to continue operating properly despite the failure of one or multiple components. Within the scope of LPWAN technologies, fault tolerance means that the whole network is not stopped due to network failures. Therefore, the LPWAN infrastructure must contain redundancy mechanisms to ensure a series of replicated components, such as network servers and gateways.

Since LPWAN technologies can be divided in two categories (proprietary and open standards), fault-tolerance must be handled by the network owner. For example, replicated components must be managed by the network operator in a NB-IoT environment. On the other hand, the user is fully responsible for managing redundancy mechanisms in LoRaWAN environments. Some examples of research aimed at improving fault-tolerance in LoRa environments can be found in the literature [4,5].

Most of these works use approaches based on network design to enhance fault-tolerance against interference and gateway failures. An algorithm for planning large LoRa networks based on gateway placement and end-device configuration is presented in [4]. The authors formulate the problem of planning LoRa networks as a Mixed-Integer Non-Linear Program.

In [5], authors introduce m-gateway connectivity to ensure fault-tolerance in LoRa networks. M-gateway connectivity leverages the ability of end-devices to switch to higher SFs in order to transmit over greater distances and communicate with more distant gateways. Additionally, the authors also formulate a Integer Nonlinear Program to minimize the number of gateways. A LoRaWAN gateway placement model, named DPLACE, is proposed in [6]. DPLACE calculates the number of gateways based on the Gap statistics method and their locations using K-Means and Fuzzy C-means algorithms.

3 LPWAN and Blockchain Background

3.1 LPWAN

Low Power Wide Area Network (LPWAN) is one of the enabling technologies of the Internet of Things (IoT). An LPWAN is a network designed to allow long-range communications among smart devices, at low bit-rates [7]. This type of network applies to IoT scenarios where low-cost devices need to transmit small messages over long distances (a few kilometers) and transmission delay is tolerable by the application. In LPWANs, end-devices are connected to a gateway node in a direct manner, thus forming a star topology. The simplified design of LPWANs and duty cycling provide a significant reduction in power consumption. Moreover, the great majority of communications in an LPWAN occur from the end-node to the gateway (uplink). This enables end-devices to enter in sleep mode after sending a message, and remain sleeping for long time intervals. Currently, several LPWAN solutions are available, among them LoRa [8] stands out as one of the most promising LPWAN technologies.

In LPWANs, especially for LoRa, security is a major concern. In the literature, several works have exposed the susceptibility of LoRa to attacks, in the phases of key management, network connection, and communications [9–11]. Although many security improvements have been added to the architecture of LoRa in more recent specifications, much work still needs to be done in this regard.

LoRa is a wireless modulation technique, designed by Semtech, that provides lowpower and long-range communication for small low-cost devices [12]. The LoRa stack can be divided in two layers: physical (PHY) and media access control (MAC). The LoRa PHY layer is a proprietary spread spectrum modulation technique designed by Semtech and it is based on a variation of the Chirp Spread Spectrum (CSS) modulation [13]. LoRaWAN is an open MAC layer protocol developed and maintained by the LoRa Alliance [8]. LoRaWAN is based on a star-of-stars topology, as presented in Fig. 1. End-devices are connected to the LoRaWAN network through wireless gateways. The communication between end-devices and gateways is performed using LoRa PHY modulation. Gateways are responsible for relaying uplink messages to the network server. The communication between gateways and the remaining servers is performed over traditional IP networks.

The Network Server (NS) is one of the main components of a LoRaWAN network (center of the star topology) and it has many features, such as packet routing from end-devices; frame authentication; data rate adaptation; relaying authentication messages between the end-devices and the join-server; and roaming. The Application Server (AS) handles all the messages sent by the end devices and provides high-level data management to the end-user. The AS may also send downlink messages to the associated end-devices. The Join Server (JS) is responsible for root keys storage, the Over-the-Air Activation (OTAA) procedure and derivation of session keys. JS is a new element in the LoRaWAN network architecture since it was only included in the latest release of the LoRaWAN specification.

3.2   Blockchain

Blockchain is a revolutionary technology built on top of a peer-to-peer (P2P) network infrastructure that implements a public distributed ledger where trans-

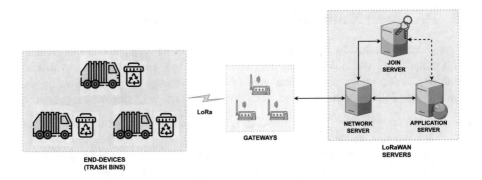

Fig. 1. Illustration of a LoRaWAN network architecture.

actions are recorded as an immutable chain of blocks [14]. In other words, a blockchain can be described as a distributed database that stores all digital operations that participating entities execute in a given system. The need for a third party to intermediate a transaction between two participating entities is eliminated with the use of a blockchain due to its decentralized design and the strong cryptographic primitives it is based on. Furthermore, a consensus algorithm is used by the participants to verify each transaction registered in the system. This ensures that the public ledger is tamper-proof. Blockchain also provides auditability for systems, since the information added to the chain can never be altered or erased [14].

There are two types of blockchain: public and permissioned. They basically differ from each other in terms of performance, consensus algorithm, and read permission. In public blockchains, the elevated amount of participating nodes and validations result in decreased throughput and increased latency [15]. Also, any node in the world can make part of the consensus process in public blockchains and all transactions are visible to anyone else. On the other hand, in permissioned blockchains, the smaller number of validations and the limited amount of nodes participating in the consensus process result in much faster transactions.

Hyperledger Fabric is an open-source permissioned blockchain platform maintained by the Linux Foundation [16]. Hyperledger Fabric is highly modular as it allows pluggable components, such as consensus algorithms and membership services. Hyperledger Fabric supports smart contracts which are self-executing code used to digitally enforce agreements and other rules in a blockchain environment. As opposed to public blockchains, such as Bitcoin and Ethereum, Hyperledger Fabric does not rely on a cryptocurrency which results in no transaction costs. Smart contracts are referred to as chaincode in Hyperledger Fabric and can be written with general purpose languages, such as Golang and Java. Chaincode stores and retrieves data from the blockchain as key-value pairs.

4   Proposed Solution

The proposed solution includes the creation of smart bins, using the Internet of Things (IoT) technology, that give the current volume and position (latitude and longitude), which will be scattered in several strategic points. Based on this system, the following steps are carried out for the route optimization process:

- Get the coordinates - latitude and longitude - of the bins, taking their volume as a collection criterion;
- Generate a graph between the coordinates of the bins, which need to be collected, including the starting and ending point coordinates;
- Obtain the displacement time from a given coordinate to any other coordinate (taking into account all the coordinates included in the generated graph);
- Generate an optimized route, using the traveling salesman problem algorithm;
- Finally, send the optimized route - alternative to the fixed route - to the driver which will collect solid waste.

Based on these steps, it is necessary, to access all smart bins to obtain the coordinates and volume. Based on the volume threshold, the bins that will be part of the next collection are selected and they will compound the route. Using the basic concept of a graph - the coordinates represent the vertices (following the concept of origin and destination) and the displacement time between them represents the weight of the edge, that is, the one responsible for the connection of the vertices in both directions; To obtain the travel times between all coordinates, we use the API (Application Programming Interface) for the "Open Route Service" - in which, the parameter is lowest time, resulting in a multidimensional matrix, referring to time which, in turn, takes traffic into account when calculating the displacement time between all given coordinates. Finally, with time and coordinates available, the traveling salesman algorithm is applied, which will return an ordered list based on their visit priorities, containing all the coordinates, including origin and destination.

We apply the traveling salesman algorithm to optimize it considering the same points. The route obtained is shown in Fig. 2. To obtain this route map, in addition to the distance, the time spent on the route was taken into account. The result was that the distance covered was 46.9 km in a time of 2 h and 11 min, implying a reduction of 30% in the total time of the journey and 40% in the distance covered, showing the potential for cost reduction and improvement the efficiency of the waste collection system that this method can bring. It should be added that the potential for cost reduction is even greater when compared to the current system that works based on fixed routes, regardless of demand.

This proposed IoT application is based on LoRaWAN networks. However, in the current LoRaWAN architecture, the Join Server (JS) is a Single Point of Failure (SPOF), from a security point of view. This is due to the fact that the JS is responsible for handling the authentication procedure and storing copies of all encryption keys. Therefore, it becomes an important target for attackers. If, for example, the JS suffers a Denial-of-Service attack, the entire LoRaWAN network can be significantly impacted and stop working.

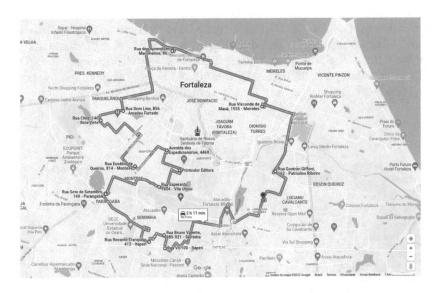

Fig. 2. The optimized route graphic.

To handle this problem, this paper proposes to modify the LoRaWAN architecture by adding a secure and distributed storage functionality based on blockchain to support the key management procedure. In this new architecture, the JS is replaced by a smart contract and a permissioned blockchain infrastructure. Key update and key storage are among the main security challenges of key management schemes. Key update ensures that keys are only updated by secure entities, in a secure fashion. Key storage ensures that keys are held in a secure location that guarantees their confidentiality, integrity and availability. While the JS satisfies the key update feature in the current architecture of LoRaWAN, it falls short with respect to the key storage feature, especially regarding availability. On the other hand, availability and fault tolerance are built-in features provided by the distributed design of blockchain.

Specifically, in permissioned blockchain networks, a group of authenticated peers is responsible for maintaining the distributed ledger, which is replicated in each of these peers. Therefore, even if some peers become unavailable, the data can still be obtained from the remaining peers.

The architecture of the proposed solution is presented in Fig. 3. In the proposed architecture, the JS is replaced by a smart contract, which is responsible for performing the main management tasks and authentication. Hyperledger Fabric is used to create the permissioned blockchain, in which NS and the smart contract share a common distributed ledger (defined as channel in Hyperledger). In the permissioned blockchain, the smart contract and NS are divided in two different groups (GRP1 and GRP2). Private Data Collection (PDC) is a security feature, implemented by Hyperledger Fabric, used for scenarios where access to sensitive data in a channel must be restricted to a specific group. The PDC is used in the proposed architecture to prevent NS from having access to the root encryption keys (which are only visible to members of GRP1).

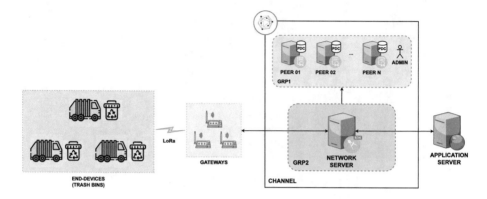

Fig. 3. Modified LoRaWAN network architecture.

To demonstrate the feasibility of the proposed architecture, a working prototype has been created using open-source tools and commodity hardware to enhance security and availability in LoRaWAN networks. Additionally, a cloud environment under multiple workloads and fault-tolerance tests were performed to evaluate the impact of network failures.

5   Performance Analysis

The goal of the fault-tolerance tests is to evaluate the impact of network failures in the environment. Figure 4 shows the network topology used during the tests. A new virtual machine was added in order to generate failure scenarios within the working prototype. Pumba[1] is a chaos testing tool that is able to create network failures in Docker containers. Pumba was used in order to kill random Docker containers running endorsing peers.

A Golang script was developed to simulate 10 simultaneous clients (each client sends continuous requests for the chaincode as fast as possible). The Golang script also monitors the number of requests sent by the clients to the endorsing peers during 10 s intervals. Therefore, it is possible to analyze how the Hyperledger network behaves during the entire course of the faul-tolerance test (before and after Pumba is activated).

The fault-tolerance test consists of running the Golang script for 10 min in two different scenarios: 8 and 16 endorsing peers. During each experiment, two

---

[1] https://github.com/alexei-led/pumba.

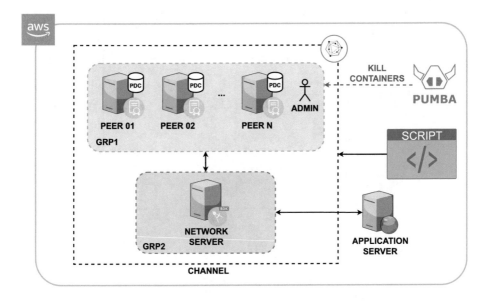

Fig. 4. Fault-tolerance test topology

endorsing peers were killed at 300th second. Each scenario was executed 15 times. Figure 5 shows that once two endorsing peers are killed, the Hyperledger network is momentarily impacted. After a short period of time, the remaining endorsing peers are able to resynchronize and handle new incoming requests. Additionally, an increase in the number of requests processed per second can be seen in the network after Pumba is activated. This behavior can be explained by the fact that a smaller number of endorsing peers tends to result in higher throughput values.

5.1 Comparison with Traditional Setup

The performance results obtained from the working prototype were compared with a traditional ChirpStack setup. The performance test with the traditional ChirpStack setup was executed a total of 15 times. A comparison for the throughput values between the working prototype and the original ChirpStack setup is presented in Fig. 6. Initial results show that the traditional ChirpStack setup has greater throughput values for a small number of simultaneous requests (up to 250 requests). However, the performance of the ChirpStack setup starts to degrade after 500 requests. This situation occurs because the JS in the ChirpStack setup fails to process a large number of simultaneous Join-Request messages and starts

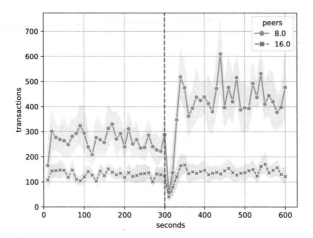

Fig. 5. Fault-tolerance test (killing two peers at 300th second)

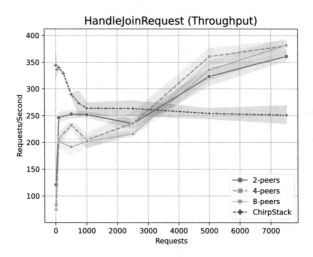

Fig. 6. Performance comparison of the OTAA procedure with traditional ChirpStack setup and Hyperledger Fabric

to drop OTAA requests. A performance transition happens around 3000 requests. At this point, the working prototype shows better throughput values than the traditional ChirpStack setup (for any number of endorsing peers).

Additionally, the Fig. 7 presents an analysis of the number of error messages seen in the ChirpStack setup during the performance tests. Results show that after 500 requests, the JS starts to drop Join-Request messages. The number of dropped messages quickly increases from 2500 messages. Based on values obtained, it can be concluded that the traditional ChirpStack setup is not suited for large LoRaWAN networks since it fails to process a large number of simultaneous OTAA requests.

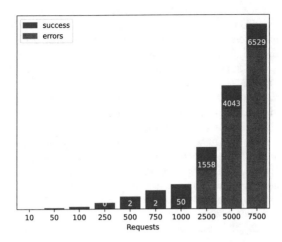

Fig. 7. Analysis of error messages during OTAA procedure for the traditional Chirp-Stack setup

6 Conclusions

The Join Server can be seen as a Single Point of Failure in LoRaWAN networks due to its centralized design. The availability of the entire network may be affected if malicious users are able to attack this server. This paper proposes a novel architecture to improve the levels of security and fault-tolerance in LoRaWAN networks by replacing Join Server with a permissioned blockchain. The use of smart contracts allowed authentication and key management to be done in a secure and decentralized manner. The results show that there is a trade-off between performance and availability when choosing the number of endorsing peers in small scenarios. A reduced set of endorsing peers demonstrates better throughput values in situations with a few simultaneous requests (up to 2500). However, this behavior is reversed in scenarios with a large number of transactions, as the performance of larger sets of endorsing peers stands out. Therefore, we conclude that using multiple endorsing peers is best suited for realistic LPWAN scenarios.

References

1. Torkashvand, J., Emamjomeh, M.M., Gholami, M., Farzadkia, M.: Analysis of cost-benefit in life-cycle of plastic solid waste: combining waste flow analysis and life cycle cost as a decision support tool to the selection of optimum scenario. Environm. Developm. Sustainability **23**(9), 13242–13260 (2021). https://doi.org/10.1007/s10668-020-01208-9
2. Mdukaza, S., Isong, B., Dladlu, N., Abu-Mahfouz, A.M.: Analysis of IoT-enabled solutions in smart waste management. In: Proceedings of the 44th Annual Conference of the IEEE Industrial Electronics Society, pp. 4639–4644 (2018)

3. Kokila, J., Devi, K.G., Dhivya, M., Jose, C.H.: Design and implementation of IoT based waste management system. Middle-East J. Sci. Res. **25**(5), 995–1000 (2017)

4. Ousat, B., Ghaderi, M.: Lora network planning: Gateway placement and device configuration. In: 2019 IEEE International Congress on Internet of Things (ICIOT), pp. 25–32. IEEE (2019)

5. Yu, X., Xu, W., Cherkasova, L., et al.: Automating reliable and faulttolerant design of LoRa-based IoT networks. In: 2021 17th International Conference on Network and Service Management (CNSM), pp. 455–463. IEEE (2021)

6. Matni, N., Moraes, J., Oliveira, H., et al.: Lorawan gateway placement model for dynamic internet of things scenarios. Sensors **20**(15), 4336 (2020)

7. Raza, U., Kulkarni, P., Sooriyabandara, M.: Low power wide area networks: An overview. IEEE Commun. Surv. Tutorials **19**(2), 855–873 (2017)

8. Alliance, L.: lora-alliance (2018). Disponível em: <https://www.lora-alliance.org>. (Accessed October 2 2022)

9. Butun, I., Pereira, N., Gidlund, M.: Security Risk Analysis of LoRaWAN and Future Directions. Future Internet **11**(1) (2018)

10. Yang, X., Karampatzakis, E., Doerr, C., et al.: Security vulnerabilities in LoRaWAN. In: 2018 IEEE/ACM Third International Conference on Internet-of-Things Design and Implementation (IoTDI), pp. 129–140. IEEE (2018)

11. Tomasin, S., Zulian, S., Vangelista, L.: Security analysis of LoRaWAN join procedure for Internet of Things Networks. In: Wireless Communications and Networking Conference Workshops (WCNCW 2017), pp. 1–6. IEEE (2017)

12. De Carvalho Silva, J., Rodrigues, J.J., Alberti, A.M., et al.: LoRaWAN-A low power WAN protocol for Internet of Things: A review and opportunities. In: 2017 2nd International Multidisciplinary Conference on Computer and Energy Science (SpliTech), pp. 1–6. IEEE (2017)

13. Semtech, A.: 120022, LoRa Modulation Basics, May 2015 (2018). Disponível em: <https://www.semtech.com/uploads/documents/an1200.22.pdf>. (Accessed January 25 2018)

14. Zheng, Z., Xie, S., Dai, H.-N., et al.: Blockchain challenges and opportunities: A survey. Int. J. Web Grid Serv. **14**, 352 (2018)

15. Bragagnolo, S., Marra, M., Polito, G., et al.: Towards scalable blockchain analysis. In: 2019 IEEE/ACM 2nd International Workshop on Emerging Trends in Software Engineering for Blockchain (WETSEB), pp. 1–7. IEEE (2019)

16. Hyperledger Fabric. https://www.hyperledger.org/projects/fabric (Accessed 2 Oct 2022)

Management of Power Supply for Wi-Fi Module Based on IoT Device Connections

Kazuhiko Sugimoto[✉] and Takayuki Kushida

Graduate School of Computer Science, Tokyo University of Technology, Hachioji, Tokyo, Japan
g212102894@edu.teu.ac.jp, kushida@acm.org

Abstract. The development of IoT has increased the demand to use IoT devices for indoors and outdoors. IoT devices need to reduce power consumption because batteries power them. Conventional methods are that IoT devices split data and transfer it to neighboring devices to reduce power consumption. One of the reasons for increased power consumption in IoT devices is to activate the Wi-Fi module. This paper reduces the number of IoT devices that activate the Wi-Fi module and prolongs the total operating time of IoT devices. The proposed method determines the IoT device that activates the Wi-Fi module by Clustering and Weighting. Clustering means that the server creates groups for IoT devices. The element of the cluster is the IoT device to which the neighbor IoT device one hop forward is connected. Weighting means that the IoT device sets its weight as a unique indicator. The IoT device with an activated Wi-Fi module has the most remaining battery power and the highest weight value. The experiment compares the operation time of IoT devices with the proposed method, which distributes the data transmitted by IoT devices to neighboring devices. The experimental results were 521 [min] for the proposed method and 456 [min] for the existing method, which is prolonged 14.2 [%].

1 Introduction

Growing demand for IoT has led to the development of software and hardware for sensor nodes that sort and collect a variety of information [1–3]. The multi-hop network consists of geographically distributed nodes. The purpose of the node is to send sensor data to the server. IoT devices are generally called a node. Nodes send power generation data to the server via a gateway installed on the server side. Sensor data can be sent to the server via the gateway when a node is placed within the radio range. Nodes send data to the gateway via another node when it cannot connect to the gateway. The node through which the data is transmitted relays the received data when it forwards data to the gateway through another node. Node power consumption increases, and operating time decreases when nodes relay data. One of the solutions to solve energy efficiency in IoT is clustering sensor nodes [4]. The nodes create clusters and elect one cluster head from the cluster. The cluster head communicates with each node and relays data to the gateway. This paper defines a cluster head as a node that activates Wi-Fi access point mode (Wi-Fi module).

ⓒ The Author(s), under exclusive license to Springer Nature Switzerland AG 2023
L. Barolli (Ed.): AINA 2023, LNNS 661, pp. 485–496, 2023.
https://doi.org/10.1007/978-3-031-29056-5_42

2   Issue

This paper addresses the issue of generating the isolated CH without considering the cluster heads and node connection when the server selects it. The power consumption of each node will increase, and the operation time of nodes will decrease when the number of cluster heads is increased to maintain the robustness of the connection. To Prolong the operation time of nodes required to reduce the number of CH. The reason is that the CH consumes more power consumption than other nodes because it relays data. Power consumption of IoT devices increases as the amount of data forwarded increases [5].

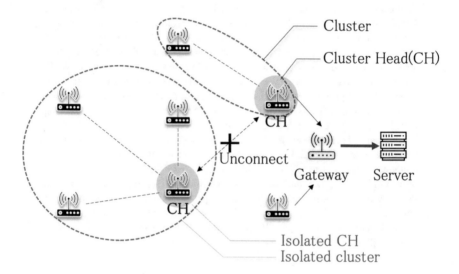

Fig. 1. There will be IoT devices that can deactivate the Wi-Fi module

Figure 1 shows seven nodes, one gateway, one server, an isolated cluster, CH, and a node. Each node generates data and sends it to the server through the gateway. A node cannot send data to the server if it cannot connect to other CHs or the gateway, as shown in Fig. 1. Because the nodes are distributed, the topology of the nodes is unknown to the server. Each node must exchange battery capacity and candidate nodes, perform clustering individually, and select cluster heads. On the other hand, conveying information from one node to all nodes can strain bandwidth and increase node power consumption. The optimal processing between each node while limiting the range of data propagation is required to optimize the entire process.

3   Related Studies

John et al. have intensively studied the LEACH protocol to improve network performance. the LEACH protocol is a vigorous contributor in multi-hop networks and

WSNs to improve the performance of IoT systems [6,7]. S. H. Kang et al. propose a clustering method by distance per node. This reduces unnecessary power consumption as the distance increases [8]. Xiangning et al. propose the Energy-LEACH protocol, which improves the existing LEACH protocol by selecting cluster heads based on their remaining battery capacity. The nodes with more battery power become cluster heads to increase the network lifetime in this paper. The node with the least amount of battery power becomes the common node [9]. Behera et al. proposed a method to dynamically select cluster heads by considering the remaining battery level. After a cluster head is selected once, the cluster head is reselected when the ranking of the remaining battery level changes. This method solves the problem of cluster heads stopping early [10].

4   Proposed Method

The proposed method reduces the number of Wi-Fi module-activated IoT devices to prolong their operation time. The purpose of the proposed method is to reduce the total power consumption of entire IoT devices. The sensor data are sent to the server via IoT devices. The central server determines whether the IoT device's Wi-Fi module is activated. This determination is based on Clustering and Weighting. Clustering is that the server creates groups for IoT devices. The item of the cluster is the IoT device to which the IoT device one hop forward is connected. Weighting sets the weight of an IoT device as a unique metric for itself. The weight is the number that indicates how many groups an IoT device belongs to it. The weight is an index of how many IoT devices can connect to other IoT devices. A high-weight value for an IoT device can connect more to other IoT devices. IoT devices need to have more remaining battery power to be operated time by battery longer.

The IoT device that activates the Wi-Fi module has the highest battery level and the highest weight. Only one Wi-Fi module is activated among the IoT devices in the same hop. The server needs to decide on one or more IoT devices to activate the Wi-Fi module in all groups. The server performs the process mentioned above and then decides to activate the Wi-Fi module of the IoT device with the lowest weight. Activating the Wi-Fi module of the IoT device with the lowest weight creates an IoT device that activates only one Wi-Fi module in every group. The topology of the multi-hop network changes depending on IoT devices' installation and connection conditions.

Figure 2 shows the proposed method. The proposed method consists of three processes. These summaries are shown below.

1. Get the shortest hop counts
   Obtain the IoT device topology for the server to determine the activation of the IoT device's Wi-Fi module.
2. Weighting and grouping
   This item shows concrete methods of weighting and grouping with diagrams and pseudo code.
3. Wi-Fi module of IoT device activation determination method
   This item shows a concrete explanation of how the server decides to activate the Wi-Fi module by grouping and waiting.

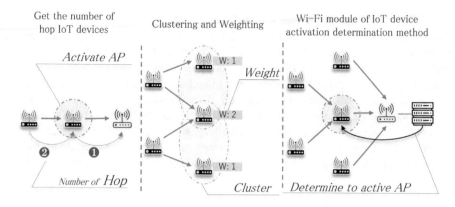

Fig. 2. Overview of proposed method

4.1   Get the Shortest Hop Counts

The proposed method requires the server to get the shortest number of hops for the IoT device. This requires knowing the number of hops beforehand to perform clustering according to the number of hops. Entire sensor data is sent from the IoT device to the server. On the other hand, the IoT device does not know which route the data will be sent when the IoT device's link is not built. This proposed method is also necessary for IoT devices to know which route to take to send sensor data to the server. The shortest hop or route is constructed starting from IoT devices close to the gateway. IoT devices that are connected to the gateway will unconditionally connect and send hop count data to the server with a hop count of 1. IoT devices that have secured a route to the server will then initiate Wi-Fi access point mode. IoT devices that cannot connect to the gateway will catch the SSID of other IoT devices' Wi-Fi and connect, adding 1 to the connected IoT device's hop count and setting it as their hop count.

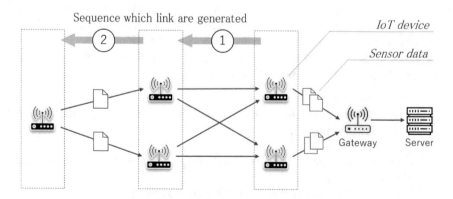

Fig. 3. Sequence which links are generated

4.2   Clutering and Weighting

The server performs clustering and weighting to determine which IoT device's Wi-Fi module to activate. The Clustering method is that the server groups IoT devices based on the shortest hop count received from IoT devices and the connectable IoT devices. The weighting method is that the IoT device sets the unique indicator called weight and lamination weight for itself.

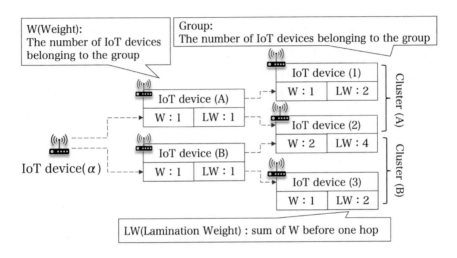

Fig. 4. Clustering and Weighting

Figure 4 is an example of clustering and weighting for IoT devices. The multi-hop network in Fig. 4 consists of 6 IoT devices. The connection relationships of the IoT devices are indicated by dashed lines in Fig. 4. The groups are shown in Fig. 4 as Group (A) and Group (B). Group (A) belongs to IoT devices (1) and (2), which are IoT devices to which IoT device (A) connects. All other groups are also connected to the IoT device one hop later, as in group (A). The weight defines how many IoT devices belong to the group. The weight is two because IoT device (2) belongs to groups (A) and (B) in the example of IoT device (2) in Fig. 4. The weight W_r is $1 \leq W_r \leq E_n$ because IoT devices connect to two or more IoT devices. E_n is the number of IoT devices sending data. The lamination weight is the sum of the lamination weights of connectable IoT devices one hop before and the weight itself.

$$W_l = \sum_{k=1}^{n} W_{ln} + W_r \qquad (1)$$

Activating the Wi-Fi module of an IoT device with a large lamination weight has much data that can be forwarded across multiple hops. W_l in formula (1) is the lamination weight. W_{lb} in formula (1) is the lamination weight of connectable IoT devices one hop back. n is the number of connectable IoT devices one hop back. W_r in formula (1) is the weight.

4.3 Activate and Deactivate the Determination of the Wi-Fi of IoT Devices

The server determines the lamination weight and remaining battery power to decide whether to activate or deactivate the IoT device's Wi-Fi module. The purpose of the proposed method is to reduce the number of IoT devices that activate the Wi-Fi module, and the IoT devices send data to the server. At least one IoT device in the group must activate the Wi-Fi module. Algorithm 1 shows the judgment algorithm for activating and deactivating the Wi-Fi module.

The IoT device that activates the Wi-Fi module is the IoT device that has the largest lamination weight among all IoT devices. The server activates the Wi-Fi module of the IoT device with the highest remaining battery power when there are multiple IoT devices with the highest lamination weight. The server checks whether there is a group in which the Wi-Fi module of all IoT devices is stopped when the server determines one IoT device to activate the Wi-Fi module. The server activates the Wi-Fi module of one IoT device in that group when there is a group in which the Wi-Fi modules of all IoT devices are stopped.

Figure 5 shows an example of starting the Wi-Fi module in the remaining groups. Figure 5 shows four IoT devices, denoted by (1) to (4). Since the proposed method

---

Algorithm 1. Activate/deactivate Wi-Fi module determination

Input:

LWD: The dictionary that stores the IDs and lamination weights of IoT devices in the identical hop

BD: The dictionary that stores the ID and remaining battery power of IoT devices in the identical hop

Output:

RA:The ID of the IoT device that activates the Wi-Fi module

```
1:  function GET_WI-FI_ACTIVE(LWD,BD)
2:      max_LW = max(LWD.values())
3:      max_LWD = [key for key in LWD if LWD[key] == max_LW]
4:      if len(max_LWD) > 1 then
5:          RA = max_LWD[0]
6:          return RA
7:      else
8:          LWDB = dict()
9:          for all ID, battery ← BD.items() do
10:             if ID in max_LWD then
11:                 LWDB.update ({(ID, battery)})
12:             end if
13:         end for
14:         max_LWDB = max(LWDB.values())
15:         RA = max_LWDB.key()
16:         return RA
17:     end if
18: end function
```

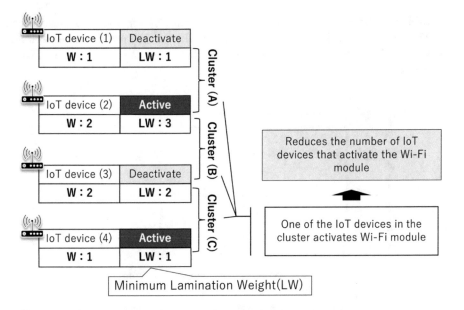

Fig. 5. Determining whether to start the Wi-Fi in an undetermined group

reduces the number of IoT devices that activate the Wi-Fi module, the server must have only one IoT device that activates the Wi-Fi module in the group. The second and subsequent determinations activate the Wi-Fi module of the IoT device with the minimum lamination weight after the server determines the activation of the IoT device of the Wi-Fi module for the first time.

5 Experiments

This section presents the implementation for the evaluation of the proposed method. The preliminary experiment compares the power consumption when the Wi-Fi module is activated and deactivated and the power consumption when forwarding data.

Preliminary Experiment

This subsection shows the preliminary experimental results to confirm that the activation of the Wi-Fi module results in higher current consumption. The preliminary experiment uses ESP32 as the IoT device and INA219 as an ammeter sensor. It set that measurement time is 30 [s] and measurement interval is 0.5 [s]. The average power consumption of ESP32 is 128.1 [mA] when the Wi-Fi module is activated and 50.5 [mA] when it is deactivated. Current consumption when the Wi-Fi module is activated is about 2.5 times higher than when it is deactivated. Deactivating the Wi-Fi module is an effective way to reduce the power consumption of IoT devices. The analysis shows that SSID broadcast is a factor that causes the current consumption to be higher when the Wi-Fi module is activated than when it is deactivated. The preliminary experiment confirms how the current consumption of IoT devices increases with the number of photovoltaic power data.

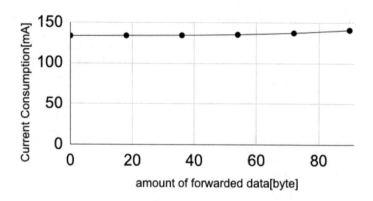

Fig. 6. Comparing the current consumption and rate of current increase by the amount of data

Figure 6 shows that the current consumption of ESP32 increases with the number of photovoltaic power data. The preliminary experiment measures the power consumption of ESP32 to change the amount of transmission photovoltaic power data 0, 18, 36, 54, 72, and 90 [byte]. The transmission data is one 18-character string. The current consumption of ESP32 increased by about 2.7 [%] when the amount of transmission data increased to 90 [byte]. Thus, it is clear that decreased number of deactivated IoT devices and the Wi-Fi module is insignificant to increased power consumption due to the number of relay photovoltaic power data. The implementation mainly uses ESP32, a server, and an IoT device.

Implementation

The multi-hop network is constructed using nine ESP32 units in the evaluation experiment. Figure 7 shows how to connect ESP32 in the evaluation experiment. Figure 7 shows nine ESP32s as ESP-1 to ESP-6 and ESP-A to ESP-C. The three units from ESP-A to ESP-C do not measure their current consumption.

The Wi-Fi module is activated and deactivated to balance the power consumption of all IoT devices in the conventional methods. Activating and deactivating with the conventional method activates the Wi-Fi of the IoT device with the maximum battery level in the group and stops the others.

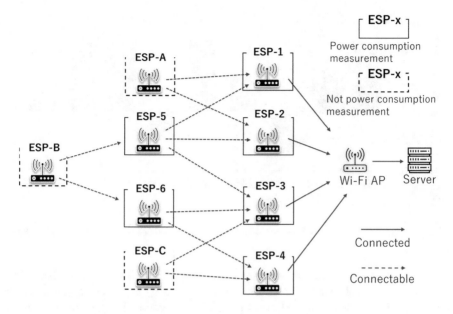

Fig. 7. IoT device topology in evaluation experiment

6 Result

This paper compares the existing and proposed methods and measures the changes in remaining battery power and operating time for each IoT device. The results show in Fig. 8 to Fig. 11.

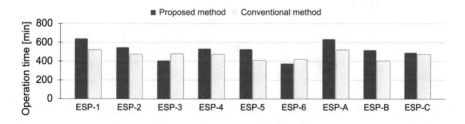

Fig. 8. Comparing the operating time of the proposed method to the conventional method

Figure 8 compares the operation time of IoT devices between the conventional method and the proposed method for each IoT device. Figure 8 is a bar graph showing the operation time of 9 units from ESP-1 to ESP-6 and ESP-A to ESP-C. The average operating time of all ESP32 was about 456 [min] for the conventional method and about 521 [min] for the proposed method. It is approximately 14.2 [%] of the operating time of ESP32 compared to the conventional method.

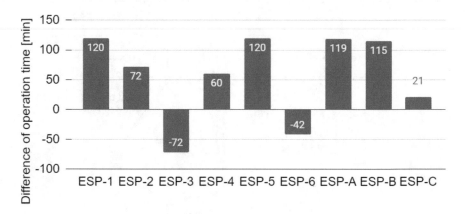

Fig. 9. Comparing the proposed method's operation time difference to the conventional method

Figure 10 shows the difference in operation time of ESP32 in Fig. 8 as a bar graph. The formula for calculating the difference in Fig. 9 is $T_{Dif} = T_{pro} - T_{exi}$. T_{Dif} is the difference, T_{pro} is the operating time of ESP32 by the proposed method, and T_{exi} is the operating time of it by the existing method. The total prolonged operation time is 513 [min] in Fig. 9.

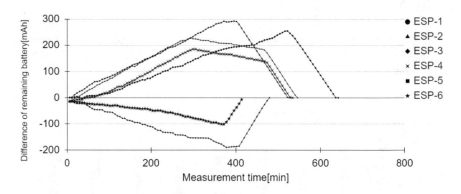

Fig. 10. Comparing the proposed method's operation time transition to the conventional method

Figure 10 shows the difference in the remaining battery of IoT devices between the conventional and the proposed method. The formula for calculating the difference in Fig. 10 is $T_{Dif} = W_{pro} - W_{exi}$. T_{Dif} is the difference in remaining battery capacity, T_{pro} is the operating time of ESP32 by the proposed method, and T_{exi} is the operating time of it by the conventional method.

Figure 11 shows the transition of the number of operating ESP32 during the experiment period. Figure 11 compares the conventional and proposed methods. The number of ESP32 in operation decreases when the ESP32 stops due to battery exhaustion. The evaluation result is better when one ESP32 runs for a longer time and more ESP32s run.

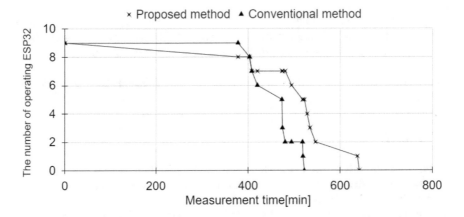

Fig. 11. Comparing the number of ESP32 in operation with conventional and proposed methods

The evaluation results showed that before 400 [min], the conventional method operated more ESP32s than the proposed method. However, the evaluation results showed that after 417 [min], the proposed method operated more ESP32s for longer than the conventional method. The proposed method extends the operation time of the last ESP32 unit by 120 [min] compared to the conventional method.

7   Discussion

The proposed method is that the server activates or deactivates the Wi-Fi module when the IoT device with the Wi-Fi module activated stops. This method reduces the power consumption of IoT devices when switching Wi-Fi modules. There is an issue that the IoT device that receives the decision to activate the Wi-Fi first stops earlier than the conventional method. An improved method smoothes the remaining battery level of IoT devices to solve this issue. The algorithm for finding the switching timing is formula (2).

$$T_c = \frac{R_c}{W_{active} W_{passive}} \tag{2}$$

T_c is the switching timing, R_c is the remaining battery level, W_{active} is the power consumption when the Wi-Fi module is activated, $W_{passive}$ is the power consumption when the Wi-Fi module is deactivated. This algorithm further prolongs the operating time of IoT devices.

8   Conclusion

This paper proposed a method to reduce power consumption by reducing IoT device activation of the Wi-Fi module. The server determines whether the Wi-Fi module is activated by weighting and clustering processes. The evaluation result is the average operation time of nine IoT devices, 456 [min] for the conventional method and 521

[min] for the proposed method, which means that the data reception time is approximately 14.2 [%] longer than the conventional method. This paper contributes to using IoT devices in outdoor applications such as photovoltaic power monitoring.

Acknowledgements. This work was supported by JSPS KAKENHI Grant Number JP20K11776.

References

1. Pham, Q.-V., Mirjalili, S., Kumar, N., Alazab, M., Hwang, W.-J.: Whale optimization algorithm with applications to resource allocation in wireless networks. IEEE Trans. Veh. Technol. **69**(4), 4285–4297 (2020)
2. Minoli, D., Sohraby, K., Occhiogrosso, B.; Iot considerations, requirements, and architectures for smart buildings-energy optimization and next-generation building management systems. IEEE Internet Things J. **4**(1), 269–283 (2017)
3. León, O., Hernández-Serrano, J., Soriano, M.: An efficient energy-aware predictive clustering approach for vehicular ad hoc networks rasmeet. Int. J. Commun Syst **23**(5), 633–652 (2015)
4. Nayak, P., Devulapalli, A.: A fuzzy logic-based clustering algorithm for wsn to extend the network lifetime. IEEE Sens. J. **16**(1), 137–144 (2015)
5. Neto, J.H.B., Rego, A., Cardoso, A.R., Celestino, J.: Mh-leach: A distributed algorithm for multi-hop communication in wireless sensor networks. ICN **2014**, 55–61 (2014)
6. Stankovic, J.A.: Research directions for the internet of things. IEEE Internet Things J. **1**(1), 3–9 (2014)
7. Duan, J., Gao, D., Yang, D., Foh, C.H., Chen, H.-H.: An energy-aware trust derivation scheme with game theoretic approach in wireless sensor networks for iot applications. IEEE Internet Things J. **1**(1), 58–69 (2014)
8. Kang, S.H., Nguyen, T.: Distance based thresholds for cluster head selection in wireless sensor networks. IEEE Commun. Lett. **16**(9), 1396–1399 (2012)
9. Xiangning, F., Yulin, S: Improvement on leach protocol of wireless sensor network, In: International Conference On Sensor Technologies And Applications (SENSORCOMM 2007). vol. 2007, pp. 260–264. IEEE (2007)
10. Behera, T.M., Mohapatra, S.K., Samal, U.C., Khan, M.S., Daneshmand, M., Gandomi, A.H.: Residual energy-based cluster-head selection in wsns for iot application. IEEE Internet Things J. **6**(3), 5132–5139 (2019)

Time-Based Ray Tracing Forwarding in Dense Nanonetworks

Eugen Dedu[1(✉)] and Masoud Asghari[2]

[1] FEMTO-ST Institute, Univ. Franche-Comté, CNRS Numérica, cours Leprince-Ringuet, 25200 Montbéliard, France
eugen.dedu@univ-fcomte.fr

[2] Department of Computer Engineering, Faculty of Engineering, University of Maragheh, P.O. Box 55136-553, Maragheh, Iran
mas.asghari@maragheh.ac.ir

Abstract. Wireless nanonetworks, consisting of nodes of nanometric size, are an emerging technology with various applications, such as in medicine, metamaterial and agriculture. As in any multi-hop network, routing is an important primitive. Due to their extremely small sizes, nanonodes' energy budget is rather limited, hence any method to reduce the energy used is of paramount importance. The energy is influenced, among others, by the number of packets sent and received, and forwarding nodes. This paper proposes ray tracing forwarding, where only nodes having received duplicate packets in the same time slot can forward them. Simulations in a dense nanonetwork show that the forwarders form a quasistraight line and can deviate at borders. In long nanonetwork environments such as blood vessels and branches, the ray tracing forwarding outperforms the coordinate-free and coordinate-based routing methods found in the literature.

1 Introduction

Nanotechnology allows the design of integrated nanodevices at the nano scales, opening a wide range of novel applications that could not be imagined in macro scales. Due to the extremely limited size, resources and energy of nanodevices, one cannot expect very much from an individual nanodevice. Thus, for a viable solution, the nanonodes should collaborate in a distributed nanonetwork structure to perform complex tasks in a considerable range. Electromagnetic nanonetworks consist of nodes of nanometric size (i.e. their components are less than $1\,\mu m$, to make nodes with sides less than around $10\,\mu m$) communicating in the terahertz band. They usually contain a nanoprocessor, a nanomemory, a nanosensor and a nanodevice to communicate with the other nanonodes. Nanonetworks can be dense, making them a very challenging environment for traditional routing protocols. Moreover, due to their tiny size, classical network techniques are unsuitable. The modulation proposed, Time Spread On-Off Keying (TS-OOK) [7], does not use signal carriers, but pulses. A bit 1 is transmitted by generating a power pulse, and a bit 0 is "transmitted" as silence. The time length of a pulse is $T_p = 100\,fs$ [7]. Due to hardware constraints, a nanonode cannot send bits one right after the other; the time between two consecutive bits is $T_s = \beta \times T_p$, and $\beta = 1000$ is given as an example of time spreading ratio [7].

L. Barolli (Ed.): AINA 2023, LNNS 661, pp. 497–508, 2023.
https://doi.org/10.1007/978-3-031-29056-5_43

Fig. 1. In TS-OOK, bits are sent with T_s interval, and are received (with a bit of delay) with the same T_s interval.

Given that bits are sent with T_s interval between them, a receiver can match the bits from one packet by reading bits at T_s interval too, i.e. at time x, $x+T_s$, $x+2T_s$, etc. Bits received at different times belong to another packet. An example of sending bits and matching them is given in Fig. 1. For simplicity of understanding, we assume that any T_s interval is divided into β time slots, and all bits from one packet are received at the same slot.

Nanonetworks can be connected to the macro world by gateway nodes, enabling the Internet of Nano-Things (IoNT). The communication of nanodevices in the sense of intrabody nanonetworks [3] opens the door to innovative medical applications inside the human body such as molecular-level detection and reporting of pathogens including viruses and bacteria, high-precision drug delivery, targeted monitoring, and neurosurgery. The combination of nanonetworks and metamaterials (artificial structures with unnatural properties) in the form of Software-Defined Metamaterials (SDMs) allows performing geometrically-altering actions on the metamaterial and tuning of its electromagnetic behavior by sending commands to nanodevices. Other applications include wireless robotic materials, military and agriculture.

The communication range of the nanodevices is usually less than 1 mm, because the path loss of the THz band in environments with high water concentration such as blood is very high, and nanodevices do not have enough power density (due to their limited sizes) to overcome the path loss in longer distances. Therefore, a multi-hop routing is needed to send data packets over a considerable distance in the nanonetworks. The traditional routing approaches proposed in the literature are not suitable to nanonetworks (e.g. backoff flooding is not destination-oriented, but does flooding). Due to nanodevices' limited resources and energy, using specialized routing techniques based on the nanonetwork operating conditions (e.g., perimeter dimensions and shape, nodes placement and mobility, nodes energy source and limitations, desired applications) to reduce packet dissemination in unnecessary directions and reduce the number of forwarders can improve nanonetwork performance and lifetime.

In this paper, we focus on long nanonetwork perimeters such as pipes and blood vessels, where previously proposed nanonetwork routing techniques cannot work properly. To this end, we propose ray tracing forwarding for nanonetworks by using TS-OOK characteristics to forward packets in quasistraight lines. The ray tracing forwarding creates a narrow packet propagation path from a source to the gateway across a long nanonetwork perimeter without flooding the packet to all the nodes in the network. This reduces the dissemination of the packets to unnecessary nodes, leading to fewer forwarders and less network energy and resource usage.

In the ray tracing forwarding, only nanodevices having received duplicate packets in the same time slot will forward them. With the proposed algorithm, the forwarders form a quasistraight line and can deviate at the borders of the long perimeter, moving the packet across the perimeter from the source to the gateway at the end of the perimeter. The contributions of this paper are as follows:

- For the first time in literature, we utilize the time slots of TS-OOK and signal propagation duration as means of constructing a quasistraight path from the source across the nanonetwork.
- We propose the ray tracing forwarding algorithm to utilize these quasistraight paths for routing in long perimeters and reduce the number of forwarders.
- We implemented and evaluated the ray tracing forwarding in a simulator, comparing its results with the related forwarding methods.

2   Related Work

The routing protocols either deliver the data packets to a destination node through the nodes located between the source and destination (unicast or merely zone-cast routing), or flood the whole network so that every node receives a certain message (flooding). Due to the shape of the perimeter, mobility of nanonodes and very high number of homogeneous nanonodes in the dense nanonetworks, traditional addressing and routing protocols cannot be applied in nanonetworks, making the flooding-based forwarding the only viable option in most cases. In *pure flooding*, every node in the network forwards every unique packet that it receives. This flooding results in resource wastage and broadcast storms in dense environments, leading to poor network performance. In *probabilistic flooding*, nodes forward packet with a certain probability [11]. The probability needs to be tuned to prevent broadcast storms and guarantee message delivery. *Backoff flooding* is a highly efficient flooding scheme using a counter-based mechanism to count copies received, and adds a random waiting time (backoff) to packet forwarding [2]. The backoff window depends on the neighbor density. Any node forwards the packet to all nanonodes within its wireless transmission range only if it did not receive r (redundancy factor) copies of that packet during the backoff period. As a general fact, in flooding (including probabilistic and backoff), a packet is broadcasted in the whole network instead of moving toward the destination, leading to redundant packet reception and wastage of nodes resources (e.g. bandwidth and energy). Therefore, any attempt to reduce packet dissemination in unnecessary directions and reduce the number of forwarders can improve network utilization.

Geoforwarding protocols rely on geographic information about node positions to take the forwarding decision [1]. Node position can be obtained either directly through a GPS receiver in each node, or by triangulation techniques using relative positions of the nodes obtained by anchors and beacons [4]. Due to the small size, limited resources, low energy and very high density of the nanonodes, neither GPS or triangulation-based nanonode positioning, nor any protocol that relies on detailed neighborhood knowledge or routing table can be employed in the nanonetworks [2]. In *RADAR* routing, the nanonetwork environment is a circular area and a central node emits a directional signal at an angle [10]. Nanonodes receiving the signal inside the angle of radiation are in the ON state, and all the other nodes are in the OFF state. RADAR requires a convex region and a symmetric perimeter with a special node in the center. Similar to RADAR, *DEROUS* also requires a beacon node set at the center of a 2D circular area sending packets to classify nanonodes [8]. In *SLR*, the network is in a 3D cubic space and a few anchor nodes are placed at corners. During the setup phase, anchors transmit their packets in sequence, allowing nanonodes to set their coordinates as hop counts from the anchors [13].

Straight-line routing protocol is proposed for wireless sensor networks (WSNs) in [5,9]. It is a form of random-walk routing protocol which constructs a straight path between the source and destination using two-hop information without using geographic information. The method constructs the path in a hop-by-hop manner, where after each hop a node that lies on the extended line of the path is selected as the next hop. It assumes that after receiving a packet, each node is able to determine the distance from its transmitter according to the signal strength [5,9]. The distances between nodes are much shorter in nanonetworks than in WSNs. Unlike WSNs, nanonodes usually use TS-OOK modulation in the THz band where bits 1 are transmitted by ultra-short pulses of duration T_p, and bits 0 are implied by the absence of the pulse (as described in the Introduction). The receiver on the nanonodes is very basic the can only detect THz band pulses. Due to the short distances, TS-OOK modulation and simplification of the radio receivers, nanonodes cannot determine accurate distances based on the received pulse strength, rendering the protocol inappropriate for nanonetworks. However, TS-OOK employs temporal multiplexing by using very short pulses with an accurate pulse duration to detect consecutive bits of each packet. In this paper, we employ these TS-OOK characteristics (pulse propagation and arrival times) in a ray tracing fashion to forward packets in quasistraight lines and reduce the number of forwards.

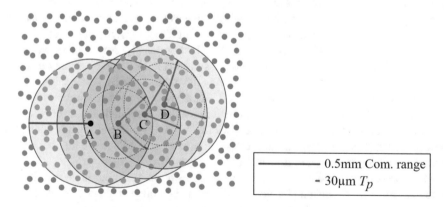

Fig. 2. Sketch of ray tracing forwarding: potential forwarders are in green (they are "collinear" with the previous two transmitters), and effective forwarders (*B*, *C* and *D*) are in blue.

3   Ray Tracing Forwarding

The network model is a long strip such as a pipe or blood vessel. The network is dense, i.e. any node has at least 50–100 neighbors.

The ray tracing method *does not make any assumption* by itself. As presented below, it needs nodes to be able to listen at T_s interval, and to send bits at a given time (i.e. there is no random/nondeterministic backoff added at the device level), but these characteristics are already fulfilled in nanonetworks.

The ray tracing aims to choose forwarders in a straight line. For that, it uses the *timing of packet receptions*: when two 1 bit pulses (note: 0 bits do not emit pulses and cannot be used) from two copies of the same packet sent by two transmitters arrive in the same time slot (explained in the Introduction), the receiver considers itself to be collinear with the two transmitters. Which 1 bits are taken from the packets does not matter, since consecutive bits of one packet are spaced by $1\,T_s$, hence are in the same slot.

The ray tracing forwarding is shown in Fig. 2 and is explained in the following.

All the nodes execute the same procedure, in a loop. Thus, we describe the procedure recursively: the base case (initialization), and the recursive step (general case).

Initialization step: A source A, called the *sender*, sends a packet. All the nodes inside the communication range receive it, and one of them, B, called the *steering node* because it produces the forwarding direction, forwards it.

Recursive step: We are at the time where node A has already sent a message, and node B is currently sending a message. This message will be received in all its communication range. All the nodes in the network are in one of the following cases:

- It is the first packet it receives: node just memorizes the time slot.
- It is the second packet it receives: if the current packet is *not* in the same time slot as the first one, then it discards the packet and forgets about this communication; otherwise (i.e. the two slots are the same), it uses a backoff: if it receives any copy before the backoff, discard it, elsewhere send it at the same time slot.
  The window from which the backoff is chosen is large, hence only one of them will forward, whereas all the others will see the copy of this packet and as such will discard their packet.
  Geometrically, all the nodes in the intersection of the discs (communication range) of the first and the second packet receive both packets. Some of them receive the two packets in the same time slot, depicted as green nodes in Fig. 2, and consider themselves as collinear with A and B. The other nodes receive them in different slots, and they discard the packet and forget about the communication.
- It is the third or ensuing packet it receives: discard it (because its forwarding time has already passed, and the propagation is already in front of it).

Once node B sent the packet, the procedure is repeated (recursive step) with node B becoming sender A, and node C becoming steering node B.

In a multi-flow case, we suppose each packet generated by a source has a different id (made for example from the concatenation of three strings: source id, flow/port id, packet sequence number). The algorithm executed by each node upon reception of a packet p[id] is shown in Fig. 3.

Input:
  timefirst[] = null

if timefirst[id] == null **then**   // packet id is seen for the first time
    timefirst[id] = (crttime % T_s) / T_p // % is modulo operation
    second[id] = false
elseif second[id] == false **then**   // 2nd packet id received
    second[id] = true
    if (crttime − timefirst[id])%T_s < T_p **or** (crttime − timefirst[id])%T_s > $T_s − T_p$ **then**
      // same time frame, so collinearity
      backoff = pktsize ×T_s× random int number in [0,1000)
      schedule event e[id] to forward p[id] at crttime+backoff
      scheduled[id] = true
    end
else // 3rd or subsequent packet
    if scheduled[id] == true **then**
      unschedule event e[id]
      scheduled[id] = false
    end
end

Fig. 3. Ray tracing forwarding algorithm.

Finally, after the multi-hop forwarding, the packet reaches the destination zone. Here, a special procedure is executed instead of the ray tracing. We consider that the destination node is a gateway with large resources, and especially with a high receiving sensitivity, which can capture the packet sent by the nodes in its vicinity (i.e., the strip width). Upon reception of a packet, it sends a high power packet, received by all the nodes in its vicinity, which informs nodes to stop further propagation. The ray tracing forwarding ends as the gateway has received the packet and the propagation stops.

4 Evaluation

4.1 Nanonetwork Simulator Used

Given the high number of nodes involved, experiments are not possible, hence we use simulations. Among the nanonetwork simulators, only BitSimulator [6] allows to simulate more than around one thousand nodes. It is much more scalable compared to the other nanonetwork simulators [12]. BitSimulator targets routing and transport protocols, and pay attention to some low-level bit transmission peculiarities, such as transmission time and bit-dependent collision at the receiver. It uses TS-OOK modulation. It also provides a visualization program, very useful to *see* what happens in the network. It is free software and has been used to validate the results of several articles[1].

We implemented the ray tracing forwarding in BitSimulator. The other routing protocols were already included in the simulator.

We provide a Web page[2] allowing to reproduce all the results of this article.

[1] http://eugen.dedu.free.fr/bitsimulator.
[2] http://eugen.dedu.free.fr/bitsimulator/aina23.

4.2 Base Scenario

The simulation parameters are shown in Table 1. The network is a rectangular strip (6 mm × 3 mm) and has 5000 nanonodes placed randomly using a uniform distribution, plus 2 nanonodes placed manually, as shown below. Node antennas are omnidirectional, and the communication range is CR = 0.5 mm, i.e. the network is $3/0.5 = 6$ CR high and $6/0.5 = 12$ CR wide, enough to test the linear forwarding. Neighbour density (also called node or network density) is $5002 * \frac{\pi*0.5^2}{6*3} = 218$ neighbors. The TS-OOK low-level parameters (described in the Introduction) are the ones proposed in the literature, i.e. $T_p = 100$ fs and $\beta = 1000$.

The sender node (of id 0), on the left of the network, and the steering node (of id 1), a bit on its right, are manually placed to start the algorithm. The sender node generates one packet (the packet has 80 bytes, a random number, and a random payload). The steering node retransmits it. Afterward, all the nodes execute the ray tracing algorithm.

Table 1. Simulation parameters.

Parameter	Value
Number of nodes	5002
Network size	6 mm × 3 mm
Communication range	0.5 mm
⇒ Neighbour density	218

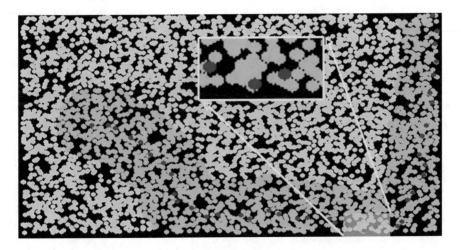

Fig. 4. VisualTracer capture, showing forwarders (in blue), receivers (in green) and the other nodes (in grey) during the ray tracing forwarding in the 5002 node scenario; the small region shows the auto-deviation capability of the ray tracing forwarding.

4.3 Ray Tracing Propagation Features

Figure 4 presents the forwarders, the receivers and the remaining nodes for the whole simulation in one representative case. It can be seen that the forwarding nodes are quasilinear, and that the propagation auto-deviates when reaching borders. Both features are explained in the following.

4.3.1 Quasilinear Forwarding

The goal of ray tracing is to reduce the number of forwarders when transmitting a packet from a source to a destination. Ideally, the propagation would go straight from the source to destination. However, in reality the line is *quasi*straight. The nonlinearity is due to time resolution error and to the next node placement gaps in space.

The time resolution error (T_p) occurs because the locus of points having received two packets in the same time slot (of length T_p) resembles to a triangle, as shown in Fig. 2. Any of these points can be the forwarder.

The node gaps are due to the impossibility of having nodes *everywhere* in the network. Obviously, there are spaces, and no forwarder can be there.

To conclude, the propagation line is not straight, but *quasi*straight. An example is given in Fig. 4, the left part, from the left source until reaching the bottom border, where the propagation deviates.

4.3.2 Auto-deviation at Borders

Our ray tracing technique has auto-deviation capabilities when reaching network borders or walls, i.e. it deviates the forwarding path pushing further the packet propagation. This section explains the reason of this desired and unexpected feature.

Let us look a zoom of some part at the bottom border of Fig. 4, shown at its top. At this time, the sender is the left blue node and the steering node is the middle one.

The potential forwarders, i.e. having received the packet in the same time slot, are located at the right of the middle node, in a triangle. The bottom 3/4 of the triangle is black, i.e. no node is there, because it is outside the network. Hence, the potential forwarders are found only on the top quarter of the triangle, and one of them will become a forwarder (the right blue node in our case). This turns counterclockwise the line between the sender and steering nodes, making it deviate from the bottom border.

To conclude, the deviation happens because:

- The potential forwarders are inside the network, not outside.
- There *are* nodes inside the network that are potential forwarders.

It should be noted that the deviation is not always successful, as can be seen at the top-right of Fig. 4, where the propagation stops. Some specific mechanism, outside of the scope of this article, is needed for that.

4.4 Comparison with Related Forwarding Methods

The ray tracing method does not use any coordinate. We therefore compare it with coordinate-free protocols, and afterward with coordinate-based protocols.

4.4.1 Comparison with Coordinate-Free Methods

The only coordinate-free methods we found appropriate to nanonetworks are flooding protocols: pure flooding, probabilistic flooding (with probability of 4 %, the minimum value which still allows propagation in the whole network), and the highly optimized auto-adaptive backoff flooding, all of them described in Related work (Sect. 2). We compare the ray tracing method with them.

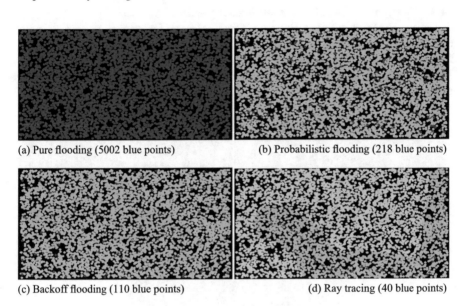

(a) Pure flooding (5002 blue points)　　　　　(b) Probabilistic flooding (218 blue points)

(c) Backoff flooding (110 blue points)　　　　　(d) Ray tracing (40 blue points)

Fig. 5. VisualTracer capture of the four compared methods, showing forwarders (blue nodes) and receivers (green nodes) for the 5002 node scenario.

Table 2. Comparison of the ray tracing with related flooding (left) and destination-oriented routing (right) methods.

Method	Forwarding nodes	Packet receptions	Method	Forwarding nodes	Packet receptions
Pure flooding	5002 (100 %)	980 146	SLR	1654 (33.0 %)	335 423
Probabilistic flooding	218 (4.3 %)	42 804	Counter-based SLR	33 (0.66 %)	6987
Backoff flooding	110 (2.2 %)	20 326	Ray tracing	29 (0.58 %)	5 259
Ray tracing	40 (0.8 %)	6 825			

Figure 5 shows the propagation of the compared protocols. It shows the placement of senders in the whole network for flooding, and the directed routing for ray tracing. It also shows that the ray tracing has the fewest number of senders and receptions.

Table 2 shows the number of forwarding nodes and of *packet receptions* (a node receiving two packets is counted as two packet receptions). It can be seen that ray tracing reduces considerably the number of forwarders (5002 to 40) and receptions. It outperforms the highly optimized backoff flooding, given that it has a quasistraight forwarding, unlike backoff flooding which floods the network.

4.4.2 Comparison with Coordinate-Based Methods

SLR is a destination-oriented routing protocol proposed for nanonetworks [13], which uses a setup phase where nodes assign coordinates. We note that it can be improved by combining it with backoff flooding (called "counter-based SLR" in the following). Both SLR and backoff flooding are described in Related work (Sect. 2). We compare ray tracing method with them.

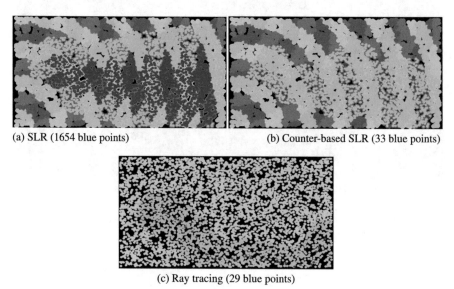

(a) SLR (1654 blue points)   (b) Counter-based SLR (33 blue points)

(c) Ray tracing (29 blue points)

Fig. 6. VisualTracer capture of the three compared destination-oriented methods, showing forwarders (blue nodes) and receivers (green nodes) for the 5002 node scenario.

In the scenario, one of the nodes found on the bottom right of the network is selected as the destination. Also, for a fairer comparison, for the two SLR-based methods, the packet exchanges during the SLR setup phase is not counted, given that it is done only once, at the network deployment.

The propagation is shown in Fig. 6 (showing the SLR zones too), and the results are given in Table 2. Ray tracing (29 forwarders) outperforms both SLR (1654 forwarders) and counter-based SLR (33 forwarders). The reason is that ray tracing is more linear than counter-based SLR. Thus, even if the latter is quite optimized, ray tracing reduces even more the number of forwarders and receptions.

To conclude, ray tracing outperforms both related coordinate-free and coordinate-based methods. Besides, the latter methods need an additional step at network deployment (to create the coordinate system), which might prevent node mobility and might create problems in nonconvex or very long environments such as blood vessels, giving the only coordinate-free methods comparable to the ray tracing. Instead, ray tracing does not need any coordinate system, but only a direction.

5 Conclusion and Perspectives

This paper presented a ray tracing method to forward packets in a quasistraight line in a multi-hop dense nanonetwork. It allows to reduce the number of forwarders and the number of receptions through a quasistraight line, and has deviating capabilities on network borders. Simulations show that it outperforms, in terms of number of forwarders and receptions, all the related methods, either coordinate-free or coordinate-based, without relying on a coordinate system. Thus, our method proves that the packet reception times can indeed be used to do a linear routing.

This paper presented preliminary results of the ray tracing forwarding. We think that this method has a big potential, and several improvements can be done on it. Future work includes analyzing the algorithm in less dense scenarios, where sometimes there is no node in the forwarding direction, in even denser networks, where forwarding can take several directions if the backoff window is not tweaked, and in various networks (with walls or curved). Also, how to choose the steering node automatically, based on the desired direction, and how to improve the deviation.

Acknowledgements. This work has been funded by Pays de Montbéliard Agglomération (France).

References

1. Abu-Ghazaleh, N., Kang, K.D., Liu, K.: Towards resilient geographic routing in wsns. In: International Workshop on Quality of Service and Security in Wireless and Mobile Networks, pp. 71–78 (2005)
2. Arrabal, T., Dhoutaut, D., Dedu, E.: Efficient density estimation algorithm for ultra dense wireless networks. In: International Conference on Computing and Communication Networks (ICCCN), pp. 1–9 (2018)
3. Asghari, M.: Intrabody hybrid perpetual nanonetworks based on simultaneous wired and wireless nano communications. Nano Commun. Netw. **32**, 100–406 (2022)
4. Bulusu, N., Heidemann, J., Estrin, D.: Gps-less low-cost outdoor localization for very small devices. IEEE Pers. Commun. **7**(5), 28–34 (2000)
5. Chou, C.F., Su, J.J., Chen, C.Y.: Straight line routing for wireless sensor networks. In: 10th IEEE Symposium on Computers and Communications (ISCC 2005), pp. 110–115. IEEE (2005)
6. Dhoutaut, D., Arrabal, T., Dedu, E.: BitSimulator, an electromagnetic nanonetworks simulator. In: 5th ACM/IEEE International Conference on Nanoscale Computing and Communication (NanoCom), pp. 1–6. ACM/IEEE, Reykjavik, Iceland (2018)
7. Jornet, J.M., Akyildiz, I.F.: Femtosecond-long pulse-based modulation for terahertz band communication in nanonetworks. IEEE Trans. on Commun. **62**(5), 1742–1753 (2014)
8. Liaskos, C., Tsioliaridou, A., Ioannidis, S., Kantartzis, N., Pitsillides, A.: A deployable routing system for nanonetworks. In: IEEE International Conference on Communications (ICC), pp. 1–6 (2016)

9. Liu, H.H., Su, J.J., Chou, C.F.: On energy-efficient straight-line routing protocol for wireless sensor networks. IEEE Syst. J. **11**(4), 2374–2382 (2017)

10. Neupane, S.R.: Routing in resource constrained sensor nanonetworks. Master's thesis, Tampere University of Technology, Finland (2014)

11. Reina, D.G., Toral, S., Johnson, P., Barrero, F.: A survey on probabilistic broadcast schemes for wireless ad hoc networks. Ad Hoc Netw. **25**, 263–292 (2015)

12. Sahin, E., Dagdeviren, O., Akkas, M.A.: An evaluation of internet of nano-things simulators. In: 6th International Conference on Computer Science and Engineering, Ankara, Turkey, pp. 670–675 (2021)

13. Tsioliaridou, A., Liaskos, C., Dedu, E., Ioannidis, S.: Packet routing in 3d nanonetworks: A lightweight, linear-path scheme. Nano Commun. Networks **12**, 63–71 (2017)

MoON: Flow-Based Programming with OpenAPI in the Web of Things

Vasileios Papadopoulos, Aimilios Tzavaras, and Euripides G.M. Petrakis[(✉)]

School of Electrical and Computer Engineering, Technical University of Crete (TUC),
Chania, Greece
{vpapadopoulos1,atzavaras}@tuc.gr,petrakis@intelligence.tuc.gr

Abstract. The Web of Things (WoT) Architecture recommendation of
W3C suggests a model for handling Things (i.e. IoT devices) as ordinary Web pages. Thing descriptions are a central building block of the
recommendation. In previous work, we proposed that Things should be
described similarly to Web services using OpenAPI. The uniformity of
the representation allows Things and Web services to interact with each
other in an application. MoON builds upon this idea and shows how
WoT applications can be created using mashups and Flow-based programming. MoON can compose new applications at a larger scale by
reusing existing devices, Web services, or applications. The application
generation process is applied to create a smart home from smart devices
and a smart city from smart homes.

1 Introduction

The Web of Things (WoT) initiative [3] aims to unify the world of interconnected
devices over the Internet. A Thing may refer to any device: a temperature or
pressure sensor, a window or door actuator, a smart coffee machine, or a smart
car. In response to this requirement, the Web of Things (WoT) architecture
model of W3C [7] defines a framework for integrating devices into the Web.
Thing Description (TD) [6] is a central building block of the WoT Architecture.
TDs are used to expose Thing metadata (in JSON) on the Web so that other
Things or clients (i.e. services or users) can interact with them. Things and Web
services coexist in applications and in the Web of Things. In our recent work [13]
Things (e.g. devices) are described similarly to Web services using OpenAPI [8].
OpenAPI provides information about service endpoints, message formats, and
the conditions for invoking the service.

Leveraging principles of OpenAPI and WoT, this work proposes a Mashup
of OpenAPI Nodes (MoON) using Flow-Based programming [4]. Mashup is the
process of creating new services by combining existing services (e.g. Things in
this work). Flow-Based Programming (FBP) facilitates the composition of new

© The Author(s), under exclusive license to Springer Nature Switzerland AG 2023
L. Barolli (Ed.): AINA 2023, LNNS 661, pp. 509–520, 2023.
https://doi.org/10.1007/978-3-031-29056-5_44

applications on visual interfaces (i.e. the coding takes place in the background). Node-RED[1] is a flow-programming tool that brings the two technologies together. MoON forwards the concept of WoT closer to reality and streamlines the process of creating new applications from existing resources (i.e. Things, Web services, or applications). Things (e.g. devices), Web services, and applications can become components of more complex applications at (any) larger scale. As a use case and starting from home applications implementing simple human-to-device or device-to-device interaction (e.g. lamps, air conditioners, motion sensors), MoON was applied to implement more complex scenarios for smart homes and a smart city. For example, a smart home energy consumption scenario can be part of an innovative smart city application.

MoON makes the assumption that all resources are described using OpenAPI. The widespread use of OpenAPI during the last few years makes this a realistic assumption. SwaggerHub [12] is an online platform with more than 100,000 OpenAPI documents of REST APIs. In addition, OpenAPI is supported by a complete tools palette[2]. In our previous work [13] we proposed a mechanism for the generation of OpenAPI descriptions of devices from minimum user input. MoON is realized by means of Node-RED and its openapi-red[3] extension node that enables interacting with APIs that have an OpenAPI description. The output of MoON is the code of the new application with its OpenAPI description.

Infrastructure, tools, and related work on WoT mashups are discussed in Sect. 2. The architecture of MoON is presented in Sect. 3. Example applications are discussed in Sect. 4 followed by conclusions and issues for future work in Sect. 5.

2   Background and Related Work

The idea of using mashups to create IoT applications is not new. Most methods deal with the heterogeneity of resources by making the assumption that the Things expose their functionality as Web services. They rely on a model-based solution and mashups for their orchestration and composition. Most methods resort to FBP for creating applications with flows (i.e. networks of *black box* processes with well-defined service endpoints). By connecting the output of a component with the input of another, data is exchanged and processed. A black box may represent a device or a service.

Node-RED is a popular representative of FBP. Figure 1 displays a sample of a Node-RED flow. In this example a DHT22 sensor[4] provides temperature measurements and an Air Conditioner is switched on or off based on whether the temperature surpasses a threshold value.

---

[1] https://nodered.org.

[2] https://openapi.tools/.

[3] https://flows.nodered.org/node/openapi-red.

[4] https://www.adafruit.com/product/385.

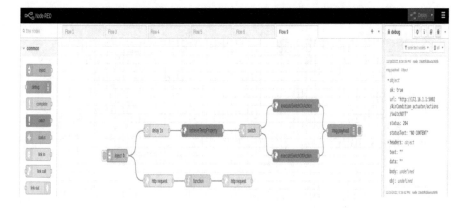

Fig. 1. Example of a data flow in Node-RED.

Additional Node-RED nodes can be imported. Openapi-red is an extra Node-RED node that enables interaction with REST APIs that have an OpenAPI description. Things are handled as REST APIs and their OpenAPI description is the means for interacting with other nodes (e.g. other Things, timestamps, HTTP requests). Using this node, a user can get all the Thing operations based on tags, set various parameters (e.g. JSON request bodies) and handle the outputs by connecting them to other nodes for further process. Another example of an extra Node-RED node, the MongoDB node allows the connection to a running MongoDB server through Node-RED's graphical environment.

WoTKit [1] is an IoT mashup platform based on Node-RED. The data model consists of sensors with fields describing a sensor or an actuator. The communication between components is implemented with Active MQ[5], a standard Java Messaging Service that enables data aggregation from sensors and passing simple control messages to actuators. IoTMaaS [5] is a mashup service model for connecting Things in applications. To counter the IoT heterogeneity, they assume that the Thing manufacturers provide a Thing Service Driver (TSD) with the programming interface for each Thing. To create an application, a Service Planner (SP) creates templates, which can be customized (i.e. select Things and configure their parameters) by the user. The WoTDL2API (Web of Things Description Language to API) tool [10] uses the WoTDL ontology [9] to generate and deploy RESTful APIs for devices as instances of this ontology using OpenAPI. The process begins by identifying the available devices and their capabilities for an IoT scenario. A toolchain creates an API for the devices, by transforming their WoTDL model to an OpenAPI document. The OpenAPI Generator[6] is used to produce the REST API code.

---

[5] https://activemq.apache.org.

[6] https://openapi-generator.tech/.

Table 1. Comparison of IoT mashup solutions.

System	WoTKit	IoTMaaS	WoTDL2API	MoON
Representation	No	TSD	OpenAPI	OpenAPI
Uniformity	No	Template	Ontology	OpenAPI
Communication	Active MQ	Custom	HTTP-REST	HTTP-REST
Planning	Node-RED	Template	Template	Node-RED
FBP	Yes	No	No	Yes
Web of Things	Yes	No	Yes	Yes

WoTKit is a full-featured IoT platform with an intuitive visual FBP tool. It lacks a uniform representation of the devices. IoTMaas comes with the most complex design. It provides uniformity of representation and allows the creation of custom applications. A notable drawback of this system is that Things are not described in a standard way. Instead, it is assumed that the manufacturers provide a Thing Service Driver (TSD) for each device. The process of combining devices to form applications seems convoluted, as users that have the Service Planner (SP) role need to create templates that contain the application's configuration (i.e. devices, parameters, and resources to allocate). WoTDL2API uses OpenAPI to represent things and applications using HTTP REST. A platform to create applications by combining existing services is missing, as developers create applications by manually choosing the HTTP endpoints of the devices. MoON is a solution that combines all desirable features and in addition, it is extensible, and easy to use. Table 1 summarizes the results of this comparison.

3   Design and Architecture

MoON comprises three major components (subsystems) [11]. The user is opted to select one from the user interface. Figure 2 illustrates the components that MoON comprises. It is a merge of the three subsystems, enriched with the Context Broker service [2], a MongoDB, and the REST API provided by the Web Thing Model service (WTMs) [13]. WTMs implements a REST API that allows clients (i.e. users or Things) to interact with all the Things of a system, as long as they follow the OpenAPI Thing Template. It is designed to support operations for retrieving and updating abstract description payloads of Things (i.e. not TDs) and their properties. It also implements functions that support actions (i.e. through an actuator), retrieval of action executions and functions related to subscriptions on Web Thing resources.

Describing applications with OpenAPI offers scalability and reusability, as these descriptions can be reused and be part of more complex applications at a larger scale. For example, applications of smart cars can be combined and comprise a new service to monitor the traffic, or a smart home's application can be reused in a smart city context. Node-RED exports a JSON file that contains the configuration settings, such as name, wiring, and functionality, from each node used in the flow. The URL of the Things is an input to the openapi-red nodes and therefore it is included in the file as a configuration setting. When a new application description is generated, it is stored in the *Applications* collection of MongoDB.

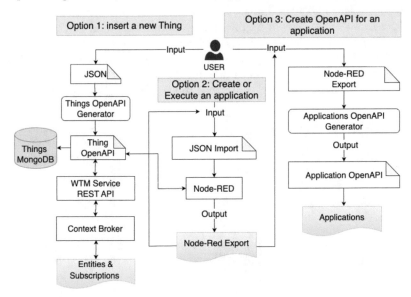

Fig. 2. MoON's subsystems.

The openapi-red extra node provides an interface for OpenAPI documents in Node-RED, thus enabling the interaction with Things as RESTful services. Operations, payload schemas and endpoints are conveniently visible, grouped by tags and described with textual metadata explaining each operation. Things must have an OpenAPI TD to be integrated in application flows and thus allow other components of the application to communicate with them. OpenAPI allows Things to be used without knowledge requirements about source code or documentation, and also be reused to compose new applications. OpenAPI provides documentation for the system's infrastructure components (e.g. MongoDB, Context Broker) and indicates how they can be used in applications by exposing the available HTTP endpoints.

MoON utilizes existing technologies such as a Context Broker to support publication and subscription functionality. This is of critical importance to event-driven architectures (such as the IoT) and is a requirement for improved responsiveness by allowing a system to react in real-time to events (e.g. changes to data). The MongoDB database system comprises four collections (i.e. *Things, Applications, Entities* and *Subscriptions*). Each collection document corresponds to an entity, which can be properties (e.g. humidity, pressure), device's internal state, action executions, or a list of available actions. The *Entities* and *Subscriptions* collections are components of the Context Broker to store measurements, subscriptions, device state, action executions, etc. When a new Thing is registered, its OpenAPI TD is stored in the *Things* collection of MongoDB, which automatically creates a unique identifier (*id* field). A URL, whose last part is the identifier (e.g. http://172.16.1.1:5001/things/6238e9b5e3447e4e4d13006e) is created; The page referenced by this URL exposes the document content, which in this case is the TD of the Thing it refers to. The *Applications* collection stores the OpenAPI descriptions of applications generated. Similar to the TDs, MongoDB automatically creates a unique *id* field for each new application stored. A unique URL is created for every document and the page referenced by this URL exposes the application's OpenAPI document and it will be used as an input to the openapi-red node. Figure 3 illustrates the structure of the database.

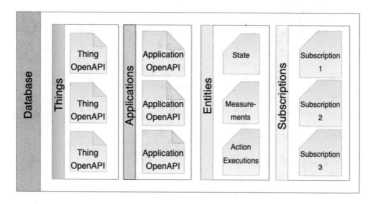

Fig. 3. The structure of MongoDB Database of MoON.

3.1   Component 1: Insert a New Thing

Its purpose is to add new Thing Descriptions into the database. The input can be either an OpenAPI TD to store in the Things collection, or a JSON file that follows the OpenAPI Generator format [13] to generate the Thing's OpenAPI. This JSON file contains Thing characteristics such as external documentation, type, supported actions, and security settings. If the user's input is a TD, the OpenAPI Generator validates and stores it in MongoDB. Otherwise, it is parsed to produce its OpenAPI description. Each new OpenAPI TD is assigned a unique URL and is stored in the Things collection.

3.2   Component 2: Application Development and Execution

This subsystem creates an application from the OpenAPIs of its resources (i.e. Web services, Things, applications). Node-RED is the core of the application development process subsystem. The openapi-red nodes get TDs as an input and connect Things in applications. Being combined with other Node-RED nodes (e.g. HTTP request nodes), they are able to form applications. To create an application, the user defines a flow of connected items (e.g. sensors, relational operators, computing functions). The application is exported as a Node-RED compliant JSON file. It can be given as input to Node-RED to execute the application.

An application is the outcome of the combination (i.e. a flow) of nodes in Node-RED's visual programming interface. Node-RED nodes implement devices, relational operators, computing functions, provide timestamps or time intervals, and output messages for debugging. There are *function* nodes that allow JavaScript code to run and provide functionality that is not included in other basic nodes (e.g. a random number generator). For applications that use Things such as sensors and actuators, the minimum requirement is that Node-RED communicates with the *Things collection* of the database to get the Thing's OpenAPI descriptions, as well as with the *Entities* collection in order to execute actions or retrieve properties (i.e. states, measurements) during run-time. For instance, an application with a sensor needs this collection to store temperature measurements, as well as read measurements to decide actions.

To interact with a Thing, the TD URL is given as an input to the openapi-red node. In the Node-RED Graphical User Interface, a user can choose from a list of Thing operations based on a selected tag (e.g. if the Tag *Actions* is chosen, a list of available Thing actions is shown). To perform actions that require data stored in the database (e.g. a device's state), the MongoDB extra node is used. The database's port and name and the collection's name are required in the input, while the queries are passed from other function nodes (JavaScript). For example, a node can be connected to the *Entities* collection of MongoDB to retrieve temperature measurements.

3.3   Component 3: Generation of OpenAPI of Applications

The purpose of the third component is to create the OpenAPI description of an application from a Node-RED JSON-compliant description. The user provides a title and a short description of the application. The OpenAPI of the application is stored in MongoDB and can be reused (by the second component) to create new applications. In addition to the title and description, the user provides the JSON file that was exported from Node-RED. An advantage of this mechanism is that no technical knowledge is required by the user, as the input, aside from the textual metadata, is provided directly from Node-RED.

Figure 4 is the flow chart of this process. The process begins by parsing the exported JSON file to find the URLs of all used Things (step 1). Each URL is split into parts and only the last part is kept; it is the Thing identifier (i.e. the *id* field) in MongoDB. The OpenAPI TD of the Thing is searched in the MongoDB *Things* collection using the identifier (step 2). To create the various OpenAPI objects, Python's dictionaries are utilized, ensuring that no duplicates exist in case a Thing is used more than once in the same application. A dictionary is created for each JSON object of each Thing (step 3). The *info* object is appended to an *info* dictionary along with the title and the description that the user initially provided for the final OpenAPI document. The *tags* array object is appended to a *tags* dictionary, the *paths* object in a *paths* dictionary, the components object in a *components* dictionary, and the *externalDocs* object in an *externalDocs* dictionary. Each dictionary contains all the information regarding the corresponding OpenAPI object of all devices (e.g. the *info* object contains the

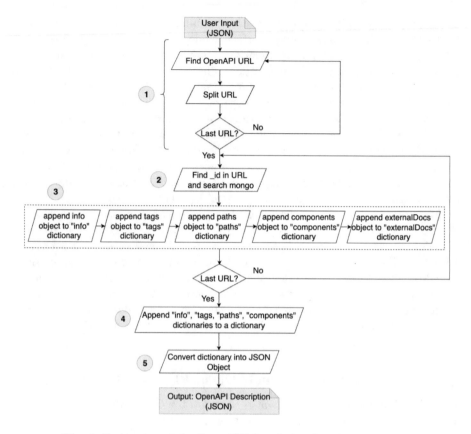

Fig. 4. Generation of the OpenAPI description for an application.

info objects of all devices). The dictionaries are appended to a final dictionary (i.e. dictionary of dictionaries) containing all the OpenAPI objects and fields for each separate Thing (step 4). The process ends (step 5) by converting the final dictionary into a JSON file, which is the algorithm output (i.e. the OpenAPI description of the application).

4   Use Cases and Evaluation

A Smart Lamp Actuator, an Air Conditioner (AC), a Motion Sensor, and a DHT22 (i.e. temperature and humidity) sensor are used to compose applications. Their OpenAPI descriptions are created with the aid of the Thing Generator service [13] (also on Github[7]). The following applications are composed:

Smart Home A: Switches the Smart Lamp on at 20:00 and switches it off at 02:00. Figure 5 illustrates the flow of the application. The time range is defined

---
[7] https://github.com/Emiltzav/wot_openapi_generator.

by the user. The application starts with an *inject* node (in light blue color) that triggers the flow. *Time Range* (in orange color) is the second node. The *executeLampOnAction* and *executeLampOffAction* (in purple) are nodes to execute actions on the device using its OpenAPI TD. The Smart Lamp will be switched on or off depending on the previous step's time range. The *msg.payload* node (in green)) outputs a message to the user. The application has been extended to work with the addition of a Smart Door Actuator: an action to lock or unlock the Smart Door is executed in a given time range (e.g. 24:00–09:00).

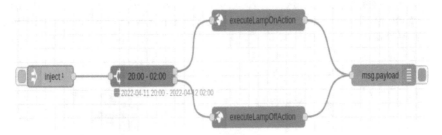

Fig. 5. Application to switch on or off the lights.

Smart Home B: Switches the AC on when the temperature surpasses a certain threshold (set by the user). Figure 6 illustrates the flow of the application. An *inject* node (in light blue color) triggers the flow to start the application. Two *HTTP requests* nodes (in yellow) and a *function* node (in orange) simulate the functionality of the sensor. The bottom row which comprises two *HTTP requests* nodes (in yellow) and a *function* node (in orange), simulates the functionality of the sensor. The *function* node generates a random number as a temperature measurement and then issues an HTTP request to the Context Broker service. The *delay* node (in pink) ensures that the sensor's simulation process will be completed before observing the new temperature. Following this delay, the *retrieveTempProperty* node (in purple) retrieves the temperature value using the DHT22 Sensor's OpenAPI TD. Using a *switch* node, a temperature threshold is set by the user and, with a relational operator ($\geq$ or $<$), the next node

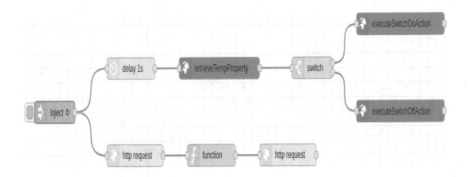

Fig. 6. Application to switch on or of the AC depending on temperature.

will either switch the AC off with the *executeSwitchOffAction* node or on with the *executeSwitchOnAction node* (purple nodes).

Smart City: The OpenAPI descriptions of smart homes and DHT22 sensors are used to calculate the average energy consumption in a city. A city comprises 4 neighborhoods with 4 houses each. Figure 7 illustrates the flow of the application for one neighborhood. The application starts with an *inject* node (in light blue) that triggers the flow every 10 s. The first *HTTP request* node (yellow) initializes the state of the devices, followed by a *function* node (in orange) that initializes the counters. The counter variables hold the number of repetitions and how many ACs are switched on at any given time.

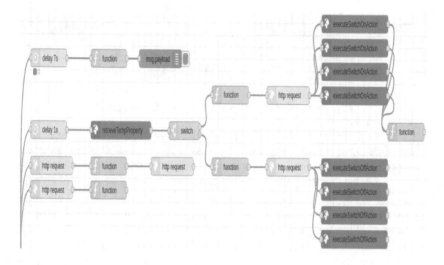

Fig. 7. One neighborhood of the Smart City application.

The second row contains two *HTTP requests* nodes (in yellow) and a function node (in orange), simulating the functionality of the sensor and generating a random number as a temperature. Following the *delay* node (in pink) that issues a delay of one second, the temperature is retrieved using the *retrieveProperty* node (in purple). Using a *switch* node (in gold) the threshold is set by the user and with a relational operator ($\geq$ or $<$) the next node (i.e. yellow *HTTP request* node) will change the AC's state to on or off by issuing an HTTP request to the Orion Context Broker, depending on the result of the *switch* node. If the AC gets switched on, the function node increments a global counter of switched-on devices. The next step will either switch off the AC by interacting with the device through the *executeSwitchOffAction* node or switch on the AC through the *executeSwitchOnAction* node (purple nodes) using their OpenAPIs. This process is repeated for each neighborhood of the Smart City and finally, after a delay of seven seconds (pink node) a *function* node calculates the average number of devices switched on by dividing the counter by the number of repetitions. The flow ends with a *message* node (in green) that outputs the average number of ACs that are switched on.

The following results demonstrate that MoON can be used to generate running applications that respond in a reasonable time. Table 2 presents (indicative) response times of the above applications. The first application flow involves only one device and has the lowest response time (i.e. 18 ms). The next three flows issue requests to the API of two devices and they have similar response times. The fifth flow involves the devices of 16 smart homes and which leads to a significantly longer response time.

Table 2. Response time of applications in Node-RED.

Action	Time (ms)
Turn on the lights at the designated time	18 ms
Turn on the lights and lock the door at the designated time	32 ms
Turn on the AC when the temperature surpasses the threshold	29 ms
Turn on the lights when motion is detected	30 ms
Calculate the average AC power consumption in the city	298 ms

5 Conclusions and Future Work

MoON is a Mashup system for composing applications at any scale in the Web of Things. Things are handled similarly to Web services and must be described using OpenAPI so that they can be ported to the openapi-red extension node of Node-RED. MoON enables seamless integration of even complex applications as long as they are also described by OpenAPI. MoOn is currently extended for incorporating security mechanisms (e.g. OAuth2.0) for enabling authentication and authorization in applications. Also incorporating a query language for OpenAPI descriptions would facilitate the search services and Things to use in applications. Replacing the simulated devices with real ones is also an interesting issue for future work.

Acknowledgment. We are grateful to Google for the Google Cloud Platform Education Grants program. The work has received funding from the European Union's Horizon 2020 - Research and Innovation Framework Programme H2020-SU-SEC-2019, under Grant Agreement No 883272- BorderUAS.

References

1. Blackstock, M., Lea, R.: IoT mashups with the WoTKit. In: Proceedings of 2012 International Conference on the Internet of Things, pp. 159–166 (2012). https://doi.org/10.1109/IOT.2012.6402318
2. Fiware: ORION Context Broker (2022). https://fiware-orion.readthedocs.io/en/master/

3. Guinard, D., Trifa, V.: Building the Web of Things. Manning Publications Co., Greenwich, CT, USA (2016). https://webofthings.org/book/
4. Harwood, T.: Visual Programming Guide (2019). https://www.postscapes.com/iot-visual-programming-tools/. Postscapes
5. Im, J., Kim, S., Kim, D.: IoT Mashup as a service: cloud-based mashup service for the Internet of Things. In: IEEE International Conference on Services Computing, pp. 462–469 (2013). https://doi.org/10.1109/SCC.2013.68
6. Kaebisch, S., Kamiya, T., McCool, M., Charpenay, V., Kovatsch, M.: Web of Things (WoT) Thing Description (2020). https://www.w3.org/TR/wot-thing-description/. W3C Recommendation
7. Kovatsch, M., Matsukura, R., Lagally, M., Kawaguchi, T., Toumura, K., Kajimoto, K.: Web of Things (WoT) Architecture (2020). https://www.w3.org/TR/wot-architecture/. W3C Recommendation
8. Miller, D., Whitlocak, J., Gartiner, M., Ralphson, M., Ratovsky, R., Sarid, U.: OpenAPI Specification v3.1.0 (2021). https://spec.openapis.org/oas/latest.html. OpenAPI Initiative, The Linux Foundation
9. Noura, M., Gaedke, M.: WoTDL: Web of Things description language for automatic composition. In: IEEE/WIC/ACM Intern. Conference on Web Intelligence (WI 2019), pp. 413–417 (2019). https://dl.acm.org/citation.cfm?id=3352558
10. Noura, M., Heil, S., Gaedke, M.: Webifying heterogenous Internet of Things devices. In: International Conference on Web Engineering (ICWE 2019), pp. 509–513. Daejeon, Korea (2019). https://link.springer.com/chapter/10.1007/978-3-030-19274-7_36
11. Papadopoulos, V.: Flow-based programming support with OpenAPI in the Web of Things. Diploma thesis, Technical University of Crete (TUC), Chania, Crete, Greece (2022). https://dias.library.tuc.gr/view/93710
12. SmartBear, S.: SwaggerHub (2022). https://swagger.io/tools/swaggerhub/. (Latest editor's draft)
13. Tzavaras, A., Mainas, N., Petrakis, E.G.: OpenAPI framework for the Web of Things. Internet Things **21**, 100675 (2023). https://www.sciencedirect.com/science/article/pii/S2542660522001561

iBot: Secure and Trusted Access to IoT Data with Blockchain

Anastasios Pateritsas and Euripides G.M. Petrakis[✉]

School of Electrical and Computer Engineering, Technical University of Crete (TUC),
Chania, Greece
apateritsas@tuc.gr, petrakis@intelligence.tuc.gr

Abstract. iBoT brings together all desirable features of current blockchain-backed IoT systems that use the blockchain to implement authorized access and control policies to data. Compared to other approaches, iBoT foresees the same high level of security and trust for all system actors including devices, users, and applications and applies multi-level authentication by means of Decentralized Identifiers (DIDs), Verifiable Credentials (VCs), and Smart Contracts. It uses Hyperledger Fabric to implement a private blockchain for all system entities (i.e. users, devices, applications and IoT data). iBoT conforms to the principles of the Web of Things (WoT) architecture model of W3C that defines a framework for integrating devices into the Web. iBoT has been deployed on the Google Cloud Platform (GCP). The experiments demonstrated that iBoT is capable of responding in real-time under heavy workloads.

1 Introduction

The idea of using blockchain to secure data in the IoT backend is not new [13, 19]. In addition to its capacity to offer distributed and safer data storage, blockchain guarantees that only validated data are stored (i.e. data must satisfy a set of rules to be valid) by applying *smart contracts*. Smart contracts are programs stored on the blockchain. They are executed autonomously (i.e. without the involvement of any central authority or user) to ensure that only trusted transactions are allowed.

iBot features all desirable characteristics of existing blockchain IoT architectures (e.g. smart contracts, safe storage) and in addition, applies a series of technologies to ensure that only trusted and validated actors (i.e. users, devices, applications) interact with the system. This is realized by means of *Distributed Identifiers* [16] and *Verified Credentials* [15]. iBoT incorporates event notification mechanisms [10] that track system transactions. Similar to smart contracts, they are activated autonomously to notify the subscribed entities (i.e. users or services) on the availability of information from devices or when operations of interest to them are executed (e.g. deletion of a user causes applications created by that user to become inactive).

iBoT is a Service Oriented Architecture (SOA) and complies with the principles of the Web of Things (WoT) Architecture model of W3C [8]. The Web of Things adds a new level of abstraction over IoT that hides network, architecture, and hardware detail, and enables communication and use of Things (i.e. devices) using HTTP REST.

L. Barolli (Ed.): AINA 2023, LNNS 661, pp. 521–533, 2023.
https://doi.org/10.1007/978-3-031-29056-5_45

Thing Descriptions (TDs) [7] are used to expose device metadata on the Web so that other devices or clients (i.e. services or users) can interact with them. In a recent work [18] we show that devices can be described similarly to Web services using OpenAPI [12]. OpenAPI is a mature framework for human and machine-readable descriptions of Web services. This enables the uniform handling of both devices and services. iBoT is deployed on the Google Cloud Platform (GCP) and evaluated by measuring its response times to common operations. The experimental results are a good support to the claims of real-time efficiency of iBoT.

The rest of this paper is structured as follows: existing solutions for Blockchain-based IoT architecture and their comparison with iBot are discussed in Sect. 2. iBoT design and functionality are discussed in Sect. 3. The evaluation testbed and experimental results are presented in Sect. 4 followed by conclusions and issues for future research in Sect. 5.

2   Related Work and Background

Hyperledger Fabric [1] is an open-source, modular, and extensible blockchain. Similarly to other architectures [5, 17], iBot uses Hyperledger Fabric to store data. A Hyperledger Fabric (henceforth Fabric) maintains two types of nodes on the blockchain, the *peers* and the *orderers*. The peers store the blockchain and are responsible for executing the smart contracts. The orderers receive transactions from peers, sort the transactions within a block (in a first-come-first-serve order) and build the block that is sent to the peers. There can be one orderer for each cluster of peers. Figure 1 shows a blockchain network with 4 peers and 1 orderer. The blockchain holds the history of all valid and invalid transactions. Each peer maintains a copy of the ledger that repeats the history of all transactions. A ledger consists of two distinct, though related components, a *world state* and a blockchain. The world state is a database that holds the current values of a set of ledger states (e.g. the current value of the connected sensors). The stored data (values) and the information for identifying the data (key) are stored on the blockchain as key-value pairs.

Li et al. [11] mitigate the problems of traditional cloud-based IoT architectures that impose high computation and storage demands on the cloud servers. Its distributed data storage relies on blockchain and Distributed Hash Tables (DHTs). The blockchain keeps only a pointer to the DHT storage address. The Sash framework [17] builds upon Fiware and Hyperledger Fabric. It stores IoT data off the blockchain but offloads the access control functionality to the blockchain where it is handled by the *IoT broker*. Data owners push data to the off-chain storage and advertise it to the smart contracts as an *offer*. Islam and Madria [5] use Hyperledger Fabric too. The IoT network comprises IoT devices, a sink node (i.e. the network coordinator) and a gateway. It proposes a distributed access control method where decisions are taken based on the consensus of the stakeholders. IoTeX [4] is a full-stack platform that enables trusted data from trusted devices for use in trusted applications. The platform uses Decentralized Identities that enable users or devices to own their data, identity, and credentials. It is compatible with 2 devices (i.e. Pebble Tracker, and Ucam).

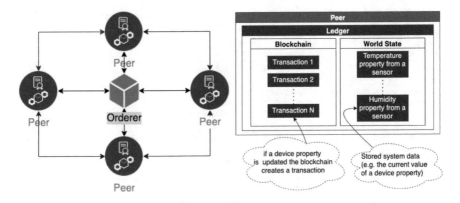

Fig. 1. Hyperledger fabric network with 4 peers and 1 orderer (left) and a ledger that is stored on a peer (right).

Table 1 summarizes the comparison with iBoT. iBoT is the only system to comply with the principles of Web of Things Architecture (WoT) of W3C [8], to handle devices similarly to REST services, and to support subscriptions. Subscription functionality [10] is of critical importance to event-driven architectures (such as the IoT) and is a requirement for improved responsiveness by allowing a system to react in real-time to events (e.g. changes to data). It is the only system to handle aggregate data.

iBoT applies the same method to describe and protect all system entities (i.e. users, devices, and applications). All entities need to register to create a public and private key pair. An authorization service uses their private keys to create their DIDs [16] and VCs [15] which are stored locally in *wallets*. For devices and applications they are created by the device and application owners, respectively. Identity information is also stored on the blockchain. IoTex [4] and Li et al. [11] store DIDs on the blockchain. They apply encryption and rely on protection performed by the blockchain. Truong et al. [17], and Islam and Madria [5] use wallets but do not describe the method that protects identity information.

Protecting IoT data is of critical importance. iBot uses the blockchain and smart contracts to protect raw or aggregated data and different operations. IoTeX [4] applies encryption. All other works use smart contracts to decide whether the requester has access to the data. Truong et al. [17] use data encryption as an additional security layer. Where data is stored is also important. iBot stores IoT data on the blockchain. IoTex [4] applies a custom solution, Islam and Madria [5] and Li et al. [11] store data in DHTs. iBot uses Hyperledger Fabric for indexing which, in turn, uses CouchDB to store and index data; Islam and Madria [5] uses SQLite.

Table 1. IoT Blockchain systems: comparison based on system features.

System Properties	iBoT	[4]	[11]	[17]	[5]
Identity protection	DID, VC (Blockchain, Wallet)	DID (Blockchain)	Blockchain	N/A	N/A
IoT data protection	Smart Contract	Encryption	Smart Contract (if requester is authorized)	Smart Contract (if requester is authorized) - Encryption	Smart Contract (if requester is authorized)
Usability	OpenAPI (device as REST)	Custom acess	Custom access	Custom access	Custom access
IoT data storage -	Blockchain	Blockchain	DHT	Cloud	SQLite
Indexing	Yes	Yes (Graph protocol)	N/A	Yes	Yes
User information storage	Wallet (VC, DID) Blockchain (DID)	Blockchain (DID)	Wallet (key pair)	Wallet (key pair)	Wallet (key pair)
Subscriptions	Yes	No	No	No	No
Services protection	OAuth2.0, Guards,	No	No	No	No
Web of things	Yes (W3C)	No	No	No	No
Aggregate data	Yes	No	No	No	No

3   iBoT Design and Architecture

iBoT design complies with the principles of the Web of Things (WoT) Architecture model of W3C [8] and enables communication and use of Things (i.e. devices) using REST HTTP. Thing Descriptions are device metadata that clients (i.e. services or users) can use to discover the devices and interact with them. In iBoT, each entity (i.e. user, device, or application) has a DID, a private, and a public key that is stored in the private *Wallet* of the entity. DID document (with the DID and public key of the entity) is stored on the blockchain. Anyone who wants to verify or authenticate the DID can check the blockchain to verify if the corresponding DID exists. The following user groups with corresponding use cases are defined [14].

System administrators: they configure, maintain, and monitor system operations (e.g. users' activities). They perform Create, Read, Update, Delete (CRUD) operations on users (e.g. they can register new users to the system and define their roles), devices, and applications. Similarly, they apply CRUD operations on identifier and authorization information (e.g. DIDs, VCs) of users, devices, and applications. They can remove DID documents from the blockchain, causing users, devices, or applications to become unauthenticated (i.e. they cannot log into the system). Only system administrators are entitled to install or update smart contracts. This is a risky operation, and the system could be compromised if smart contracts are not properly defined. When a DID Document is deleted, an event is generated from the blockchain. The *Publication and Subscription* service [10] listens to this event and to notify users or other services subscribed to it.

Infrastructure owners: they subscribe to iBoT for a fee and their subscriptions are stored on the blockchain. They are granted permission to register, configure, monitor, or remove devices. iBot provides functionality for connecting and controlling devices using IDAS service [2]. Infrastructure owners are responsible for creating VCs and DIDs for devices that are stored in the Wallet of the owner. As a result, the devices can connect to the system. New devices together with their OpenAPI and DID documents are stored on the blockchain. Infrastructure owners can delete a device (e.g. by deleting its DID document on the blockchain) if they are the owners of the device. If a device is deleted, the event is captured by the Publication and Subscription service and an e-mail is sent to users subscribed to this device. If a device used by applications is deleted, the developers of the affected applications are notified by e-mail. Infrastructure owners are entitled to create or update the OpenAPI of the devices they own.

Application developers: they subscribe to iBoT and to a set of devices for a fee. Their subscriptions are stored on the blockchain. Once subscribed to devices they can create applications. They can search for devices based on geographic location or by properties, purpose, or actions. The queries address the blockchain where the OpenAPI descriptions of the devices are stored. Alternatively, they can browse the OpenAPI documents of the devices. They create VCs and DIDs for their applications and store them in Wallets.

Customers: they are authorized to search for applications, view their OpenAPI metadata, and subscribe to applications for a fee. Once subscribed they are granted access to the application over the Web. iBot provides query mechanisms for selecting applications for subscriptions based on criteria such as location, name, functionality, etc. Their subscriptions are stored on the blockchain. Customers are granted only access rights to applications.

3.1   iBot Architecture

iBot is a Service Oriented Architecture (SOA) comprising communicating and independent RESTful microservices [14]. Figure 2 is an abstract view of iBoT. The *Web Application* implements a Graphical User Interface (GUI). A user logs into iBoT with a passphrase encrypted with the user's private key, DID, role, and proof (from the user's VC). For new users, a registration process is initiated to assign them a role. If authorized to access the system, a login session is initiated with their credentials and an access token encoding their access rights (i.e. authorization to access iBot services). This is a responsibility of the *Authorization service*. All requests are forwarded to *Application logic* and (from there) to the appropriate service. *Application Logic* runs the application. It orchestrates all other services. It determines the order in which they are run and how they are synchronized. When a request is received (from a user or service) it is sent to the appropriate service.

The *blockchain* is a NoSQL database that stores raw (i.e. unprocessed values) and aggregate historical data (i.e. average minimum, and maximum values per hour or per day etc.). The blockchain stores transactions, DID Documents, OpenAPI descriptions, subscriptions, and smart contracts. Smart contracts are services that can retrieve or store different types of information on the blockchain. The *Aggregate data* service is the RESTful interface for accessing historical data on the blockchain. It is connected to the blockchain and provides RESTful methods for retrieving raw and aggregated historical

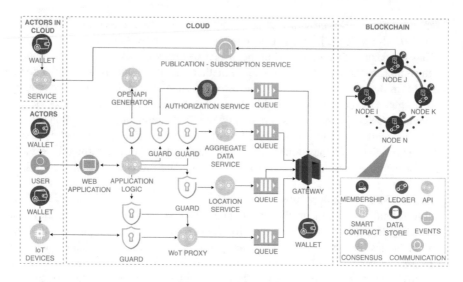

Fig. 2. iBot architecture.

data. The blockchain can be searched by device or application DID, or by aggregate data values (i.e. maximum, minimum and average values per hour, day, month, etc.). The *Location service* provides a RESTful interface to the blockchain for retrieving the location of a sensor or for retrieving sensors in a geographic zone (i.e. in a specific latitude and longitude range).

Data is stored as key-value pairs and can be any of the following types: (a) DID Documents with context, DID, controller, verification method and public key, (b) OpenAPI descriptions of devices, (c) user's billing transactions each one representing an association between users with devices (i.e. how many times the user used a device, and the type of transaction that took place), (d) current data that shows the current-time value of a device, (e) actions the type of actions that took place on devices, (f) aggregate data showing maximum, minimum, average values of devices over a month, a day, or an hour, (g) subscriptions to devices or applications.

3.1.1 Web of Things Interface

The devices are considered parts of the Web and are accessed using HTTP REST. However, physical devices may use one of a wide range of IoT or application-specific protocols (e.g. Bluetooth, MQTT, ZigBee, LoRa, etc.) and cannot be always accessed directly using REST. A solution is to deploy a *Web proxy* [18] that runs on a *IoT network gateway* (external to Fig. 2). Essential parts of the WoT proxy are the (a) *IoT management service* that maps device DIDs to system URLs and translates each IoT-specific protocol to REST. This is the responsibility of the IDAS backend IoT management service [2] of the proxy, (b) a RESTful API service (in Python Flask) that implements operations on IoT devices (e.g. for retrieving and updating device properties and their descriptions) and (c) *OpenAPI generator* [18] facilitates the creating of OpenAPI descriptions of devices from user input in YAML (or JSON).

3.1.2   Blockchain Interface

Various iBoT services are entitled to send data to the blockchain. To prevent data loss, *Queues* are installed in between these services (the producers) and the *Gateway* service (the consumer) that communicates with the blockchain. The requests from the producer services to the blockchain are stored in a queue and are consumed by the blockchain at the rate it can process them without missing any. Each queue is implemented as an AMQP message broker of RabbitMQ service [6] and is protected with username and password. The Gateway service is implemented in NestJS[1]. The Gateway translates the REST requests it receives to a smart contract method calls (i.e. it decides which smart contract and method to call). The Gateway has a Wallet with public and private keys to connect to the blockchain (a gRPC connection). After a smart contract method call, the service forwards the response from the blockchain to the producer.

3.1.3   Authentication and Authorization

The *Authorization service* is responsible for registering and connecting users to iBoT. It grants access to registered users based on roles (i.e. clients, infrastructure owners, application developers, and administrators). At login, the user provides a passphrase encrypted with the user's private key, DID, role, and proof (from the user's VC). The service searches the blockchain for the DID document and public key. With the user's public key, the authorization service can decrypt the encrypted passphrase. Using the administrator's public key and JWS[2] (i.e. a JSON Web Signature that is used to validate the DID document and encodes the user's role), the service can verify that user data is valid. After data validation, the service creates an *access token* (i.e. JSON Web Token) that includes the user's role and DID. The access token is a unique-dynamic token with limited time validity, which acts as the user's identity for accessing iBot's services. This sequence of steps in described in [14].

New users register to create a public and private key pair and a profile. iBoT issues the VC, DID, and DID documents as a response. Without DID and VC, the actor can not log into the system or generate an access token. The Authorization service creates an *access token* (a JSON Web Token) that encodes the user's role and DID. The access token is a unique-dynamic token with limited time validity, it acts as the user's identity within the system and is included in all user requests. The *Wallet* is an application (i.e. the personal data store) that stores the DID, the VC, and the public and private keys of their owner. The DID document with the DID, VC, and the public key is also stored on the blockchain.

Infrastructure owners and application developers have to apply to create DIDs and VCs for devices and applications they own. Similar to users, devices and applications have to log into connect to iBoT. This is related to the system's protection as a whole and prevents access to services and data over the network by non-authorized entities. With the addition of Smart Contracts, iBoT is protected against malicious actions or incorrect data of devices. The system executes automated checks every time a device sends data to the system via different types of smart contracts (depending on the device type). iBot ensures that all (user or service) requests have the appropriate authorization.

---

[1] https://nestjs.com.

[2] https://openid.net/specs/draft-jones-json-web-signature-04.html.

3.1.4   Distributed Identifiers and Virtual Credentials

A DID is a URI that identifies a user, device, or application. The DID document is stored on the blockchain. The DID document contains additional information that is used by the Authentication and Authorization service (the controller) to certify the existence of the user, device, or application. The DID document expresses cryptographic material, verification methods, or service endpoints that provide a set of mechanisms enabling a controller to prove the ownership of the DID. To verify or authenticate a DID, the controller can check the blockchain to see if the corresponding DID document exists. The DID document is a JSON file with fields *id* (i.e.the user's DID), and VerificationMethod with fields *Controller* (i.e. the DID of the person who has the right to change the DID Document with the string *#keys-1* at the end), *type* (e.g. *RsaVerificationKey2018*[3]), *publicKeyMultibase* (i.e. the public key is used to verify information signed with the private key of each user).

Once the DID document is saved on the blockchain, the Authorization service creates the user's *Virtual Credential* (VC). A Verifiable Credential (VC) is the equivalent of a physical credential (e.g. a driver's license). A VC is used to prove who issued the credential and who is the holder. A VC includes the identifier (i.e. DID) and metadata that describe the issuer, the holder, the expiry date and time, and a public key. The issuer and holder are declared by their DIDs. In iBoT, VCs declare the role of the holder (i.e. of a user, a device, or an application). The VC is also a JSON file with fields *id* (i.e. the user's DID), *type* (i.e.), *issuer* (e.g. *iBotLoginCredentials* that it is a credential for connecting to the system), *issuance date* (i.e. the date the VC is created), *credentialSubject* (with fields the user's DID and type which can be any of developer, user, infrastructure owner, administrator, IoT, or application) and *proof* (i.e. JWS, DID, and passphrase).

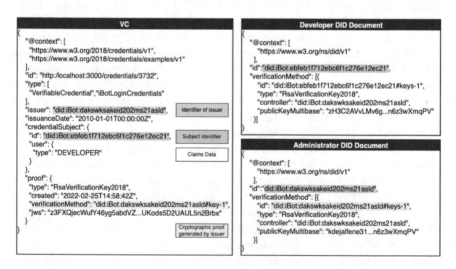

Fig. 3. VC for an application developer (left) and DID document for an administrator and a developer (right).

---

[3] https://w3c-ccg.github.io/ld-cryptosuite-registry/#rsasignature2018.

Figure 3 illustrates the VC of DID documents for an application developer and the DID document for an administrator and an application developer. The field *issuer* of the VC is the same as the DID of the administrator, and similarly for the *id* of *credential-Subject* field, and for the *Controller* of the VerificationMethod field. Applications and devices are registered similarly. The only difference is that the VCs for applications and devices are issued by application developers and infrastructure owners, respectively.

The DIDs and VCs are returned to the user to store in a *Wallet*. For the devices and applications they are kept in their own wallets. The advantage is that the data is not exposed to the database or the Web, and each user controls her/his own data. The user's role and private key are kept only in the user's personal wallet. At the same time, the blockchain only stores the DID Document that declares the existence of each user. The blockchain is searchable by *id* (i.e. the user's DID). The sequence of steps of creating a user's VC and DID document is described in [14]. Depending on their role, the users can then register applications and devices.

3.1.5 Service Guards

iBot protects devices and applications from unauthorized access. This mechanism is implemented by means of *Guards*. They implement a *Policy Enforcement Point Proxy* (PEP) based on Role Based Access Control[4] (RBAC) and the Guard class[5] of NestJS framework. Guards protect services against unauthorized access. They act as proxies that stand in front of the protected services. The header of each user request includes an access token. The proxy forwards the access token to the Authorization service that responds with the user's DID and role. If the token is valid and the user is entitled to access the service (depending on the role) the request is approved and the PEP forwards the request to the protected service. If the token is invalid, the request is rejected. There are also services which are accessible by other services only. Such requests include a *Master key* in their header. This secret code can be different for each protected service and PEP and is specified during the initialization of the PEP service. If the request carries the correct Master key, the PEP delivers the request to the protected service. Figure 4 illustrates the process of protecting services with an access token (left) and a Master key (right).

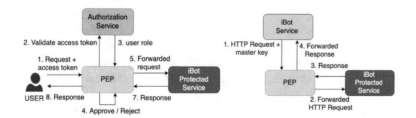

Fig. 4. Protecting a service with access token (left) and master key (right).

---

[4] https://auth0.com/docs/manage-users/access-control/rbac.
[5] https://docs.nestjs.com/guards.

3.1.6 Smart Contracts

Smart Contracts [3] add yet another level of security. They protect iBoT from risks due to malicious or abnormal operations, power failure, fatigue of material, or compromised devices. These are programs installed on blockchain nodes. Every time a device sends data to the blockchain, prior to storage, the corresponding smart contract (of the node) is invoked. By nature, the blockchain stores device measurements while maintaining an unchanging time series history. Similarly, aggregate data is stored and validated by smart contracts. IoT data is stored either as *raw* (i.e. unprocessed values), or *aggregated* (i.e. statistical). Maximum, minimum, and average values over predefined time intervals (e.g. every hour, day, week, etc.) are stored.

Smart contracts work autonomously or in collaboration with each other to ensure the validity of data and operations. *WoT proxy smart contracts* implement methods for retrieving or updating IoT devices, device properties, actions, or subscriptions on devices. They are triggered by calls to the *WoT proxy* service. The *transaction Log* smart contract, monitors user activity, including both valid and invalid transactions (i.e. unauthorized access attempts). Valid accesses to devices (e.g. to retrieve subscriptions or to receive measurements) and the time and duration of connections are stored on the blockchain so that users can be billed based on their activity. *Data validity* smart contracts are triggered to check the validity of measurements (e.g. if their values are in the correct range). There are three types of validity smart contracts: (a) *value range control*, (b) *amount of information*, and (c) *frequency of transmission control*. Malicious sensors might violate any of these smart contracts in cases they transmit a large volume of information or at very high frequencies (e.g. an attempt for a denial of service attack). *Raw and Aggregated* smart contracts undertake the retrieval of device data from the blockchain. They are triggered by calls to the *Aggregate data service*. It is connected to the blockchain and provides a RESTful interface for retrieving raw and aggregated historical device data.

3.1.7 Publication and Subscription

The *Publication and Subscription* service (the producer) listens to changes made on the blockchain and sends this information to a queue (i.e. a message broker). A user or another service gets notified when new information becomes available in the queue and forwards it to users or other services subscribed to it (the subscriber). In iBoT, publication and subscription information is stored on the blockchain in JSON. For example, the JSON of a subscription to a device may refer to an application developer and contain the list of devices that the developer has added to applications. The Publication and Subscription service listens to these events to send notifications to the subscribed entities. RabbitMQ [6] implements the Publication and Subscription service in iBoT.

Deletion of iBoT entities causes other entities to become unauthenticated or cause other events (e.g. e-mails to be sent to users affected by this action). This is also realized with the aid of the Publication and Subscription service. For example, the deletion of an entity (i.e. user, device, application) is caused by deleting its DID document on the blockchain (i.e. the entity can no longer log into the system). In that case, an event is generated from the blockchain. The Publication and Subscription service listens to this event and triggers a sequence of operations to maintain system integrity. For example, the deletion of a user causes the simultaneous deletion of all user entities (i.e. devices, applications, or subscriptions depending on user type). Deleting the DID document of the device or application triggers an e-mail to users subscribed to it. If a device is deleted, an e-mail is sent to the infrastructure owner of the device. If an infrastructure owner is removed, the devices belonging to that user are also deleted. If an application developer is deleted, her (his) applications will be deleted too. This, in turn, will trigger deletions of all her (his) subscriptions to devices.

4   Performance Evaluation

iBoT is deployed on Google Cloud Platform (GCP) on a Virtual Machine (VM) with x86_64@3.8 Ghz CPU, 8 GB of Memory, 256 GB HDD, and OS Ubuntu 18.04. Hyper-ledger Fabric deployed a blockchain with 5 nodes (i.e. 4 peers and 1 orderer). The fol-lowing experiments aim to study the performance iBoT (i.e. end-to-end latency) under stress for common user requests. Jmeter[6] is used to create various load conditions by specifying the number of requests and how many of them will be executed concurrently simulating the affect of many users working in parallel. In each experiment, Jmeter generated 2,000 requests with concurrency 1, 50, 100, 200, and 500. All times below are averages over 2,000 requests. The blockchain is initialized with 10,000 (simulated) connected sensors at random locations and 1,000 users.

Figure 5 (left) reports the average latency of requests that *write* a single record on the blockchain. For 1 user issuing a single request at a time, iBoT is fast responding in 143 ms. For parallel requests, the blockchain fails to respond in real-time. These results are in line with the official Website of Hyperledger Fabric [9] reporting a typical speed of 3,000 transactions per second (tps). The speed also depends on many other factors, such as infrastructure setup, the logic of operations, the type and complexity of the policies that are applied (e.g. consensus, smart contracts) and the ratio of concurrent requests.

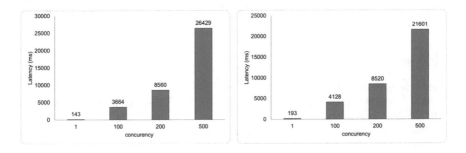

Fig. 5. Average latency (ms) for writing (left) and for reading (right) a single record.

---

[6] https://jmeter.apache.org/.

Figure 5 (right) reports the average latency of requests that *read* a single record from the blockchain. An interesting observation is that read operations perform about the same as write operations. Read operations differ from transactions in that there is no state change. The explanation for this behavior (i.e. read operations are expected to be much faster) is that regardless of read or write operation, latency accounts mainly for the delay in executing validity and security tests (e.g. smart contracts, user validation, and authorization), and the involvement of extra services (e.g. accessing security proxies, reading and writing from queues and subscription checking).

5   Conclusions and Future Work

iBot is a blockchain-backed Internet of Things (IoT) platform for the cloud. All system actors are identified and protected using distributed identifier technologies. The support for Subscriptions also distinguishes iBoT from other IoT blockchain systems. Access to the blockchain is protected by a Gateway service that apart from securing the blockchain, listens to events and notifies users subscribed to certain events. Even more important, the service listens to transactions (i.e. events that change the state of the blockchain such as registration, updates, or deletions of entities) and triggers a sequence of operations to maintain system integrity. This mechanism transforms Hyperledger Fabric into an event-based IoT architecture. iBoT is Web of Things (WoT) architecture and treats devices as REST.

Beyond its positive offers and integration of new technologies, iBoT can be improved in certain ways. Storing credentials in wallets can be risky. An obvious improvement would be to keep wallets in separate hardware or use Web browser extensions that keep information encrypted and difficult to recover from a third party. Also, HTTPS must be used to secure the connections of the users or services in different servers or cloud regions. Improving latency (i.e. response times) to meet the real-time requirements of modern IoT architectures is the biggest challenge for all blockchain IoT architectures.

Acknowledgment. We are grateful to Google for the Google Cloud Platform Education Grants program. The work has received funding from the European Union's Horizon 2020 - Research and Innovation Framework Programme H2020-SU-SEC-2019, under Grant Agreement No 883272- BorderUAS.

References

1. Hyperledger Fabric. Linux Foundation (2022)
2. Fiware: interface to internet of things (2022). https://fiwaretourguide.readthedocs.io/en/latest/iot-agents/introduction/
3. IBM: What are smart contracts on blockchain? (2022). https://www.ibm.com/topics/smart-contracts
4. IoTex: Connecting the Real World to Web3 (2022). https://iotex.io. The Platform for MachineFi Builders
5. Islam, M., Madria, S.: A permissioned blockchain based access control system for IOT. In: IEEE International Conference on Blockchain (Blockchain), pp. 469–476. Atlanta, GA, USA (2019). https://ieeexplore.ieee.org/document/8946172

6. Johansson, L., Vinka, E., Abrahamsson, S.: The Optimal RabbitMQ Guide, 3 edn. CloudAMQP (2022). https://www.cloudamqp.com/docs/index.html
7. Kaebisch, S., Kamiya, T., McCool, M., Charpenay, V., Kovatsch, M.: Web of things (WoT) thing description (2020). https://www.w3.org/TR/wot-thing-description/. W3C Recommendation
8. Kovatsch, M., Matsukura, R., Lagally, M., Kawaguchi, T., Toumura, K., Kajimoto, K.: Web of things (WoT) architecture (2020). https://www.w3.org/TR/wot-architecture/. W3C Recommendation
9. Krisha, C.: Understanding FastFabric: how to scale Hyperledger transactions per second (2022). https://blog.accubits.com/understanding-fastfabric-how-to-scale-hyperledger-transactions-per-second/. White paper
10. Lazidis, A., Tsakos, K., Petrakis, E.G.M.: Publish-subscribe approaches for the IoT and the cloud: functional and performance evaluation of open-source systems. Internet Things **19**(100), 538 (2022). https://doi.org/10.1016/j.iot.2022.100538
11. Li, R., Song, T., Mei, B., Li, H., Cheng, X., Sun, L.: Blockchain for large-scale internet of things data storage and protection. IEEE Trans. Serv. Comput. **12**(5), 762–771 (2019). https://ieeexplore.ieee.org/document/8404099
12. Miller, D., Whitlocak, J., Gartiner, M., Ralphson, M., Ratovsky, R., Sarid, U.: OpenAPI specification v3.1.0 (2021). https://spec.openapis.org/oas/latest.html. OpenAPI Initiative, The Linux Foundation
13. Mohanda, B., Jena, D., Panda, S., Sobhanayak, S.: Blockchain technology: a survey on applications and security privacy challenges. Internet of Things **8**(100), 107 (2019). https://www.sciencedirect.com/science/article/abs/pii/S2542660518300702
14. Pateritsas, A.: Enhancing data security in the internet of things with blockchain. Diploma thesis, Technical University of Crete (TUC), Chania, Crete, Greece (2022). https://dias.library.tuc.gr/view/93707
15. Sporny, M., Longley, D., Chadwick, D.: Verifiable credentials data model v1.1 (2022). https://www.w3.org/TR/vc-data-model/. W3C Recommendation
16. Sporny, M., Sabadello, D.L.M., Reed, D., Steele, O., Allen, C.: Decentralized identifiers (DIDs) v1.0 (2022). https://www.w3.org/TR/did-core/. W3C Recommendation
17. Truong, H., Almeid, M., Karame, G., Soriente, C.: Towards secure and decentralized sharing of IoT data. In: IEEE International Conference on Blockchain (Blockchain), pp. 176–183. Atlanta, GA, USA (2019). https://ieeexplore.ieee.org/document/8946129
18. Tzavaras, A., Mainas, N., Petrakis, E.G.: OpenAPI framework for the web of things. Internet of Things **21**(100), 675 (2023). https://www.sciencedirect.com/science/article/pii/S2542660522001561
19. Wang, Q., Zhu, X., Ni, Y., Gu, L., Zhu, H.: Blockchain for the IoT and industrial IoT: a review. Internet of Things **10**(100), 081 (2020). https://www.sciencedirect.com/science/article/abs/pii/S254266051930085X

Runtime Model-Based Assurance of Open and Adaptive Cyber-Physical Systems

Luis Nascimento[1], André L. de Oliveira[1(✉)], Regina Villela[1], Ran Wei[2], Richard Hawkins[3], and Tim Kelly[3]

[1] Universidade Federal de Juiz de Fora, Juiz de Fora, Brazil
{luis.felipe.almeida,andre.oliveira,regina.braga}@ice.ufjf.br
[2] Dalian University of Technology, Dalian, China
ranwei@dlut.edu.cn
[3] The University of York, York, UK
{richard.hawkins,tim.kelly}@york.ac.uk

Abstract. Cyber-Physical Systems (CPSs) in domains such as automotive and autonomous vehicles that perform safety-critical functions require the justification and demonstration of system dependability. Assurance cases provide an explicit means for assessing confidence in system safety, security, and other properties of interest. The Structured Assurance Case Metamodel (SACM) issued by the Object Management Group (OMG) defines a standardized metamodel for representing structured assurance cases. SACM provides the foundations for model-based system assurance with great potential to be applied in emergent open and adaptive CPS domains. Thus, assurance cases are expected to be exchanged, integrated, and verified at runtime to ensure the dependability of CPSs. However, existing design-time system assurance activities are insufficient to enable dynamic safety and security assurance of CPSs at runtime. In this paper, we introduce extensions to SACM to support the specification and synthesis of executable assurance cases from design, analysis, and process models to demonstrate CPS safety and security at runtime. We evaluate the feasibility of our approach in an illustrative study in the automotive domain.

1 Introduction

Cyber-Physical Systems (CPS) integrate physical processes and computer systems that contain sensors to observe the environment and actuators to influence physical processes [1]. CPSs harbor vast economic potential and societal impact, enabling new types of promising applications in different embedded system domains such as automotive[1] [2], avionics[2] [3], railway, healthcare, and home automation [4]. CPSs in such domains are highly open (systems have the ability to connect to each other at runtime), and adaptive (they are capable of adapting to changing contexts). These systems perform safety-critical functions and if they fail may harm people or lead to the collapse of important infrastructures

---

[1] Autonomous cars.

[2] Unmanned aerial vehicles.

© The Author(s), under exclusive license to Springer Nature Switzerland AG 2023
L. Barolli (Ed.): AINA 2023, LNNS 661, pp. 534–546, 2023.
https://doi.org/10.1007/978-3-031-29056-5_46

with catastrophic consequences to industry, and/or society [5]. Safety-critical CPS domains such as automotive require the justification and demonstration of system dependability. Assurance cases provide an explicit means for justifying and assessing confidence in safety, security, and other dependability properties of interest. An assurance case is a structured argument formed by a set of claims arguing and justifying the assurance of system safety and security properties in a particular operating environment based on compelling evidence (e.g., design, analysis, and process models) and the relationships between these elements [6]. An assurance case supported by evidence is the key artefact for safety/security acceptance of systems before their release for operation.

Since most open and adaptive CPSs domains are safety-critical, it is imperative to assure the safety, security, and other dependability properties of a CPS/CPS component face unknowns and uncertainties, introduced by Artificial Intelligence, at runtime [7,8]. Thus, assurance cases are expected to be exchanged, integrated, and verified at runtime to ensure the dependability of CPSs. However, established assurance approaches, e.g., Goal Structuring Notation (GSN) [9], Claims-Arguments-Evidence (CAE) [10], and standards designed to address standalone systems, building a complete understanding of the system and its environment at design time, are insufficient to assure the dependability of CPSs. CPSs are loosely connected and come together as temporary configurations of smaller systems that may dissolve and give place to other configurations. Thus, existing assurance approaches can limit runtime flexibility, and cannot cope with the complexity of CPSs [4].

The assurance of CPS and/or CPS component dependability properties demands a paradigm shift from moving part of the system assurance process activities at design time to runtime, where uncertainties can be resolved dynamically citech46Wei2018. With this shift, we'll have a transition from the current design time assurance cases produced from manually created artefacts to assurance case models that can be automatically synthesized and evaluated at runtime [8]. To achieve this goal, it is needed to equip a CPS or a CPS component with all the information that uniquely describes its dependability characteristics (design, analysis, and process models) within a Digital Dependability Identity (DDI) [4]. Thus, DDIs produced at design time provide the basis for automated integration of components into systems at development time, and dynamic integration of independent systems into systems of systems at runtime.

Model-based system assurance is particularly needed to assure both safety and security properties of open and adaptive CPSs. The Structured Assurance Case Metamodel (SACM) [11] issued by the Object Management Group (OMG) defines a standardized metamodel and visual notation for representing structured assurance cases. SACM was developed to support interoperability between different assurance case approaches (e.g., GSN, CAE). SACM provides the foundations for model-based assurance of CPSs at runtime. SACM was used in the DEIS project [4] as a backbone for its Open Dependability Exchange (ODE) metamodel, which defines the appropriate format of a DDI, in the first step towards runtime assurance of CPSs. Although SACM and ODE metamodels provide an ideal basis for the assurance of CPS at runtime, support for both specification and automated synthesis of executable SACM assurance case patterns (templates) from ODE design and analysis is still needed. Assurance case patterns are useful in capturing good practice in system argumentation for re-use, by defining the required system information to instantiate abstract assurance claims and evidence to support those claims [12]. Assurance case patterns and their traceability to evidence (ODE models), to reason on the validity of safety/security assurance claims, is still challenging. Achieving this goal demands: the provision of language support to add semantics to SACM Implementation Constraints to support the specification of executable assurance case templates linking evidence to external artefacts (ODE models) (**CH1**), and automated synthesis of assurance cases from design and analysis models (**CH2**).

The contributions of this paper are twofold: **i)** the proposal of extensions to SACM, built upon GSN patterns extensions [9], to support the specification of executable assurance case templates, and **ii)** a certification algorithm to automate the synthesis of these templates from design and analysis models to enable the assurance of CPSs at runtime. We evaluate the feasibility of our solution in an illustrative study in the automotive domain. This paper is organized as follows. In Sect.2, we present the SACM metamodel and its visual notation for representing assurance cases. Section 3 introduces the SACM patterns extensions for specifying executable assurance case templates. In Sect. 4 we describe our approach for automatic synthesis of executable SACM assurance case templates. Section 5 illustrates our approach in the automotive domain. We compare our approach with related work in Sect. 6. Section 7 presents conclusions and perspectives of future work.

2   Structured Assurance Case Metamodel

SACM defines a metamodel and a visual notation for representing assurance cases. It was developed to support interoperability with well-established assurance case frameworks such as GSN and CAE. SACM metamodel comprises the following packages [11]:

- **Structured Assurance Case Base Classes** captures the foundational concepts and relationships between SACM base elements used, through inheritance, by the rest of the metamodel;
- **Structured Assurance Case Terminology** package defines the concepts of Term, Expression, and external interface to capture terminology information used in the Assurance Case. Those concepts support the definition of a controlled and reusable vocabulary often referred to in the argumentation of safety and/or security of a system. Such vocabularies can relate to external heterogeneous (design, analysis, process) models that define the semantics for the terms;
- **Assurance Case Packages** define the concept of modularity in assurance cases. An AssuranceCasePackage is an exchangeable element that may contain arguments (ArgumentPackages), evidence descriptions (ArtefactPackages), and terminology definitions (Terminology Packages);

- **Argumentation Metamodel** defines the concepts for representing structured arguments;
- **Artefact Metamodel** specifies the concepts for structuring evidence referenced in assurance cases.

SACM Base Classes: SACMElement contains the basic properties shared among all elements of a structured assurance case. Each SACMElement has a $+gid$, a unique identifier within the scope of a model instance, an $+isCitation$ Boolean flag indicating whether it cites another element, and an $+isAbstract$ Boolean flag to indicate whether the SACMElement is considered to be abstract. The $+isAbstract$ flag is used to indicate whether an element is part of an assurance case pattern or template. ModelElement refines SACMElement with a $+name$ and references to UtilityElements such as Description and ImplementationConstraint. LangString is the SACM format for description. It has the same purpose of a String with the additional specification of the language ($+lang$) used for its $+content$. A ModelElement can contain a Description that provides its content, e.g., a Description that provides the text of a Claim. ExpressionLangString extends LangString to denote a structured expression, which can be (optionally) used to refer to an ExpressionElement (Term or Expression) in the TerminologyPackage. ModelElement extends SACMElement and it is the base element for the majority of SACM modeling elements. A ModelElement can contain zero or more ImplementationConstaints. An ImplementationConstraint specifies the details of a constraint that must be satisfied to convert a referencing ModelElement from $isAbstract = true$ to $isAbstract = false$. The language used in the specification of an ImplementationConstraint is limited to computer languages (e.g., OCL, EOL, SQL). ImplementationConstraints described in natural language cannot be used in the automatic instantiation of assurance case patterns.

SACM Argumentation Metamodel: Structured arguments are represented in SACM through Claims, citations of artefacts or ArtefactReferences (e.g., Evidence and Context for Claims), and the relationships between these elements expressed via AssertedRelationships. A detailed description of the visual notation (concrete syntax) for SACM Argumentation elements is available elsewhere [11, 13].

Although SACM supports the specification of ImplementationConstraints to abstract elements of an assurance case pattern, it is not possible yet to specify executable SACM assurance case templates with explicit links between evidence referenced within abstract Terms (placeholders) and external artefacts, i.e., ODE Design, Hazard Analysis and Risk Assessment (HARA), Failure Logic, Fault Tree Analysis, and Security Analysis models, which constitute the Digital Dependability Identity [8] of a CPS or CPS component, necessary to enable the automated reasoning on the validity of safety/security assurance claims of open and adaptive CPSs at runtime. Moreover, there is still a need to add semantics to the concept of ImplementationConstraints to support management and the systematic reuse of SACM assurance cases in the same way as pattern extensions [12] in GSN assurance cases. Thus, it is needed to provide semantic support for representing generalized *n-any*, *optional*, and *alternative* (choice) structural relationships between abstract SACM argumentation elements in an assurance case template, and generalization/specialization relationships between abstract argumentation elements, their links to abstract terminology elements, and information from external artefacts to enable the automated synthesis of executable assurance case templates. Thus, providing a solution for the realization of the concept of Executable Digital Dependability Identity [8,14].

Hence, it is needed to support the specification of complex dependence relationships between abstract Terms of a SACM assurance case template, e.g., the instantiation of HazardX abstract Term, from Hazard Avoidance [12] argument pattern, is dependent upon the *System.name* information retrieved from the design model required to convert SystemX Term from *isAbstract = true* to *isAbstract = false*. The potential of computer languages such as OCL and EOL in providing semantic support for ImplementationConstraints to connect abstract terminology and argumentation elements from SACM assurance case patterns to external artefacts was not fully exploited yet to enable the automated reasoning of safety and security assurance of CPSs at runtime.

3   SACM Patterns Extensions

In this section, we introduce a novel approach to support the specification of executable assurance case patterns (templates) using SACM metamodel [11] and its visual notation [13] as another step towards the realization of the concept of Executable DDI [4,14], and automated synthesis of safety/security cases of CPSs at runtime. Our SACM argument pattern specification approach comprises two levels of abstract structures: argumentation elements at the first level, and abstract expression elements (terms and expressions). We present patterns extensions to add semantics to SACM ImplementationConstraints to enable the specification of structural relationships between SACM argumentation elements and traceability links between terminology elements and external artefacts in assurance case patterns.

SACM pattern extensions add semantics to ImplementationConstraints to support the specification of executable argument patterns. Those extensions comprise *Multiplicity (m)*, *Optional (o)*, *Choice (c)*, and *Mapping (p)* constraints. *Multiplicity*, *Optional*, and *Choice* have the same semantics of GSN *Multiplicity* and *Optionality* patterns extensions [15] respectively. *Mapping* constraints are used to relate abstract *Terms* to elements from external artefacts, e.g., design, analysis, or process models, via model queries to obtain model elements [7].

Users can associate the ImplementationConstraint sub-types to any SACM element with no restriction in the used language. The language used in ImplementationConstraints are not limited to computer languages. Natural languages can also be used to describe instantiation rules of ImplementationConstraints, except the instantiation procedure is limited to manual. The automation of assurance pattern instantiation needs tool support and a model management engine to execute the ImplementationConstraints. Hence, we consider computer languages such as Object Constraint Language (OCL) [16] and Epsilon Object Language (EOL) [17] to specify model queries for *Mapping* constraints to link abstract *Terms* from SACM argument patterns to external artefacts (design or analysis models). The specification of ImplementationConstraints to elements of SACM argument patterns plays the role of a *weaving model* [18] linking abstract *Terms* to *elements* from external artefacts. Each ImplementationConstraint sub-type is detailed in the following.

Mapping (p): can be attached to an abstract *Term t* to impose restrictions on the *+value* slots of its instances. **Semantics**: a mapping (*p*) constraint is used to link the *+value* slot of an abstract *Term t* to information from an external artefact. The reference to an external artefact is stored in the *+externalReference* slot of an abstract *Term t*. *p* denotes *t* will be instantiated with one or a set of concatenated values retrieved from an external artefact, specified in the *+externalReference* slot from *t*, by executing a model query in a computer language, e.g., EOL.

Multiplicity (m): can be attached to SACM *ModelElements*, e.g., *argumentation* and *terminology expression* elements. **Semantics**: *m* denotes *zero-or-more n-ary* cardinality of an abstract *ModelElement*, e.g., *Claim*, *AssertedRelationship*, *Term*, in an argument pattern. In the same way as GSN [9], *multiplicity* indicates zero or more instances of an abstract *ModelElement* (*Claim*, *Term*) relate to *element property values* retrieved from an *external artefact*. A *m* constraint attached to a *ModelElement* denotes that zero or multiple instances of such an element should be created during argument pattern instantiation.

Optional (o): can be attached to any SACM *ModelElement*, except *Terms* and *Expressions*. **Semantics**: *o* is used to denote *zero-or-one n-ary* cardinality of an abstract *ModelElement*, e.g., *Claim*, *AssertedRelationship*, in an argument pattern. The semantics of an *o ImplementationConstraint* is similar to GSN *Multiplicity* pattern extensions attached to a *GSN SupportedBy* element. *Optional* are used to assign Boolean conditions under *element property* (*p*) values (*v*), retrieved from an *external artefact*, that should be satisfied for instantiating an abstract *ModelElement*.

Choice (c): constraint can be attached to *ArtefactAssetRelationship* and *AssertedRelationship* elements. **Semantics**: it can be used to denote possible alternatives (choices) in satisfying a relationship analogous to *GSN Optionality* extension [9]. A *choice* constraint can be used to represent *1-of-n* or *m-of-n n-ary* selection of *source nodes* of an abstract SACM *AssertedRelationship* or *ArtefactAssetRelationship*. The *source* nodes of an *AssertedRelationship* can be *Claim* or *ArtefactReference* elements. *Artefact* elements (*Artefact*, *Activity*, *Event*, *Participant*, *Technique*, *Resource*) can be the *source* nodes of an *ArtefactAssetRelationship*. Choice *ImplementationConstraint* supports the specification of *lower* and *upper bounds* for the selection of *source* nodes of an AssertedRelationship or ArtefactAssetRelationship.

4   SACM Argument Pattern Instantiation

This section presents our model-driven approach for automated synthesis of executable SACM argument patterns from system design models. We provide an overview of the process to interpret and execute EOL [17] model queries stored into *ImplementationConstraints* elements attached to abstract SACM terminology elements from executable argument pattern specifications.

The certification algorithm was implemented in Epsilon EOL and Java languages, and it executes on the Eclipse Modeling Framework [19] platform. It requires the following input artefacts: An executable SACM-compliant assurance case pattern model enriched with *ImplementationConstraints*, and MOF-compliant (e.g., EMF, UML, Simulink, XML) design, analysis, and/or process models stated as *+externalReferences* of abstract *Terms* of the pattern. External artefacts provide the required information to the pattern instantiation process. The instantiation program:

1. identifies abstract *Terms* in argument pattern models;
2. retrieves the required information from external artefacts to instantiate each abstract *Term* depending on its possible *+origin* slot, by executing OCL/EOL model queries stored into *mapping (p) constraints* attached to each abstract *Term*;
3. instantiates abstract *Terms* based on *multiplicity (m* constraint);
4. instantiates abstract *Expressions* based on instances of abstract *Terms* referenced by their *+element* slot, and *multiplicity constraints* attached to the them;
5. instantiates abstract *ModelElements* (e.g., *Claims, ArtefactReferences* argumentation elements) based on instances of abstract *ExpressionElements* referenced by their *Description* element, and attached *multiplicity* and/or *optional* constraints;
6. instantiates abstract SACM *Package, PackageInterface,* and *PackageBinding* elements based on attached *multiplicity (m)* referencing an abstract *Term* and/or *optional (o) ImplementationConstraints*;
7. instantiates abstract *AssertedRelationships* based on instances of its *+target ModelElement* and attached *m, o,* and/or *c ImplementationConstraints*; and
8. creates instances of *artefactAssetRelationships* based on attached *m* referencing an abstract *Term, o,* and/or *c ImplementationConstraints*.

Abstract *Terms* guide the instantiation of abstract SACM *ModelElements* (e.g., argumentation) that reference them directly, via *ExpressionElement* property, or indirectly, via *ImplementationConstraint, Description,* or another *ModelElement*. The execution of our Java/EOL certification algorithm outputs a SACM argument pattern instance in the form of Graphical Modeling Framework (GMF) diagram and .xmi model files compliant with SACM metamodel version 2.2 [11] and its visual notation [13]. Users can edit the instantiated argument using a SACM-compliant EMF modeling tool, e.g., ACME [7], SACM ACEditor [20]. Changes in the external artefacts modify the assurance case, requiring the re-execution of our pattern instantiation algorithm to keep argument-evidence traceability.

5   Illustrative Study

We demonstrate the feasibility of our approach in supporting specification/instantiation of SACM arguments from design models for an automotive braking system. HBS is a hybrid brake-by-wire system (Fig. 1) for electric vehicles propelled by four in-wheel motors (IWMs) taken from [21]. Hybrid means that braking is achieved through combined action of electrical IWMs, and frictional Electromechanical Brakes (EMBs). While braking, IWMs transform the vehicle's kinetic energy into electricity, which charges the power train battery, increasing the vehicle's range. The HBS architecture comprises 30 components: four wheel-brake modules, one AuxiliaryBattery, one PowertrainBattery, one MechanicalPedal, one ElectronicPedal, and two communication buses, with 74 ports, and 69 connections. Each wheel-brake module comprises six components: a Wheel Node Controller (WNC), an IWM, an IWM Power Converter, an EMB, an EMB Power Converter, and an Add Braking. The system is activated when the driver presses the mechanical pedal. The Electronic Pedal senses the driver's action, and it sends the braking forces, via a duplex bus system, to WNCs of each wheel brake module. Each WNC generates commands to the power converters to activate EMB and IWM braking actuators. While braking, the power flows from auxiliary battery to EMB, and from IWM to the powertrain battery. Different hazards with different criticality (i.e., ISO 26262 Automotive Safety Integrity Level - ASILs), and causes can raise from the interaction between wheel braking system components. We considered four hazards: No braking after request from the driver (**H1**: No Braking Four Wheels, ASIL D), No front wheels braking after request (**H2**: No Braking Front, ASIL C), No rear wheels braking after request (**H3**: No Braking Rear, ASIL B), and Braking more or less than requested (**H4**: Value Braking, ASIL D).

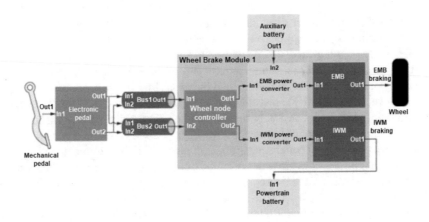

Fig. 1. Hybrid braking system architecture.

5.1 SACM Argument Pattern

We need to create a convincing argument based on evidence (design and safety analysis artefacts) that justifies and demonstrates the safe operation of a brake-by-wire system deployed into an electrical vehicle in a high speed road. We used a customized version of the Hazard Avoidance [12] argument pattern to structure the argument over the risks posed by hybrid brake-by-wire system hazards.

542   L. Nascimento et al.

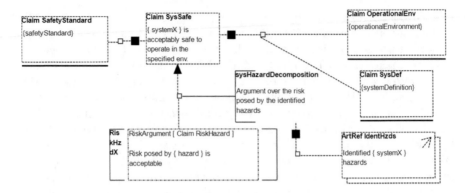

Fig. 2. The Hazard Avoidance Pattern in SACM visual notation.

Figure 2 shows the specification of this pattern in SACM visual notation [13] and its implementation constraints. We specified the pattern using the SACM Assurance Case Editor [20] built upon the EMF platform. In this pattern, the implicit definition of 'safe' is 'hazard avoidance'. (SysSafe) is an abstract claim arguing systemX (abstract *Term*) is acceptably safe to operate in the specified environment. This claim is stated in the context of SysDef, OperationalEnv, and SafetyStandard axiomatic abstract claims related to system properties, connected to SysSafe claim via AssertedContext relationships. SysSafe claim is addressed by arguing that the risk posed by each identified system hazard is acceptable (sysHazardDecomposition *ArgumentReasoning*). sysHazardDecomposition argument reasoning is only applicable in the context of some knowledge of plausible hazard, e.g., identified during Hazard Analysis, represented by IdentHzds abstract *ArtefactReference* connected to the *ArgumentReasoning* element via AssertedContext. Based on the identified hazards (IdentHzds *artefactReference*), we can have zero or multiple instances of RiskHzdX abstract citation claim arguing the risk posed by hazard is acceptable. The argument progresses from these 'hazard acceptability' goals. Instances of RiskHzdX abstract claim cite instances of RiskHazard claims from concrete instances of abstract RiskArgumentPackage element. RiskArgumentPackage is not detailed due to space limitations. The *p* implementation constraint attached to *hazard* abstract *Term* has the EOL model query *'ExternalReference!t_hazard.all.collect(h|h.a_name)'*, stored into its *+content* slot. RiskHzdX abstract cited claim also has a *m'* multiplicity *Implementation-Constraint*. *p* and *m ImplementationConstraints* attached to *hazard* abstract Term denote that one instance of hazard should be created for each element retrieved from *HBS artefact*, referenced in its *+externalReference* slot, by executing the EOL model query stored into *p* constraint. Since *hazard* abstract *Term* is referenced in the description of RiskHzdX abstract cited claim, one instance of RiskHzdX should be created for each concrete instance of *hazard*.

5.2   Design and Analysis Models

The HBS design and hazard analysis models are stored in an XML file (Listing 1) for illustrative purposes. HBS XML file stores system, subsystem and port properties, and hazard analysis information (e.g., NoBrakingFourWheels) required to instantiate the Hazard Avoidance argument pattern.

Listing 1. HBS design and safety analysis XML artefact.

```
<system name= ''HBS" enviroment= ''high speed roads">
 <standard name= ''ISO 26262"> </standard>
 <hazard name= ''NoBrakingFourWheels"></hazard>
 <hazard name= ''NoBrakingFront"></hazard>
 <hazard name= ''NoBrakingRear"></hazard>
 <hazard name= ''ValueBraking"></hazard>
 <subsystem name= ''MechanicalPedal">
    <Port type= ''out" name= ''Out1"/>
 </subsystem>
 <...>//others subsystems
</system>
```

5.3   SACM Argument Pattern Instantiation

The SACM Hazard Avoidance executable argument pattern enriched with *ImplementationConstraints*, and the HBS system design and analysis XML file were input to the instantiation program. The execution of the instantiation program generated a complete instance of the Hazard Avoidance argument pattern (see Fig. 3) arguing the risk posed by *no braking four wheels*, *no braking front*, *no braking rear*, and *value braking* system hazards are acceptable with traceability links between claims and fragment of HBS design and safety analysis models (i.e., the XML file). Changes in the XML file are automatically propagated throughout the structure of the assurance case, via re-execution of the instantiation program. The instantiated argument is stored into a .xmi file compliant with SACM metamodel version 2.2. The instantiation program also accepts pattern models compliant with SACM version 2.0, and ODE[3]. The instantiation program, the SACM argument pattern, and design models used in this illustrative study are available on GitHub[4]. The complete argument pattern comprises modules arguing the risk posed by hazards and failure modes are acceptable. The instantiated argument contains 83 claims, organized into 13 argument packages.

6   Related Work

A model-driven approach for automatic generation of GSN assurance cases from design models has been proposed [18]. In this work, the concept of model-based assurance cases was introduced, that used a weaving model to explicitly link entities in the assurance case to the external models that represent them. The work presented in this paper is built on the approach to automatic generation of

---

[3] https://github.com/DEIS-Project-EU/DDI-Scripting-Tools.

[4] https://github.com/LuisFelipeAN/sacm-mbac-updatesite.

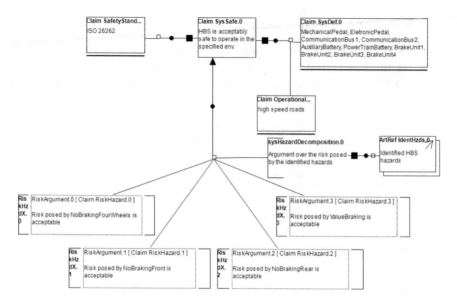

Fig. 3. Instantiated hazard avoidance argument pattern.

assurance arguments from design models first proposed by Hawkins et al. [18]. This paper extends this work by utilising SACM models and the ImplementationConstraint element to provide an implementation of a 'model weaving' approach to explicitly link placeholders (abstract *Terms*) to information from external artefacts inside the SACM argument patterns. Wei et al. [7] propose ACME tool to support model-based assurance cases developed in SACM. This tool supports model transformations and other model management capabilities, however a mechanism to support automatic instantiation of assurance cases is not provided. Another tool that has been developed is AdvoCATE [22]. AdvoCATE supports the automatic assembly of safety arguments and the instantiation of argument patterns. However, this tool does not enable the creation of assurance arguments directly from models, but instead requires the creation of a table with data entries needed to populate the arguments.

7   Conclusion

In this paper, we presented a model-driven approach to support the specification and synthesis of executable SACM argument patterns with traceability links between claims and external artefacts to enable the realization of the concept of Executable Digital Dependability Identity as another step towards system assurance at runtime. We added the semantics of GSN Multiplicity and Optionality extensions [15], and the concept of model weaving [18] to SACM ImplementationConstraints. We also provide a certification algorithm to support the automatic synthesis of SACM argument patterns from external artefacts, as another step towards the demonstration of CPSs assurance at runtime. Thus, changes in the system design or runtime information are propagated throughout the structure of the assurance case. We evaluated the feasibility of our solution in a medium-sized illustrative automotive brake-by-wire system. Our approach may contribute to improving argument-evidence traceability, and reduce the complexity of maintaining SACM argument structures. The medium size and complexity of the models used in the evaluation are threats to the validity of our study, not supporting generalizing the results. Case studies using systems and CPSs from other safety and security-critical domains need to be conducted to properly evaluate the feasibility, scalability, and performance of our SACM argument pattern specification/instantiation approach. An experimental evaluation of the effectiveness of our approach against related work in industry settings is still needed, demanding extending it to support specification/instantiation of executable argument patterns in GSN. Moreover, experimental studies in the industry still need to be conducted to evaluate the usability of both SACM visual notation and assurance case modeling tools in supporting the specification of executable argument patterns enriched with ImplementationConstraints. The lack of support for simplifying the specification of complex model queries involving references to elements from different external artefacts is another limitation of our work. In future work, we also intend to exploit ontology to simplify the specification of complex model queries.

Acknowledgments. This work was supported by CAPES, finance code 001, FAPEMIG under Grant APQ-00743-22, and CNPq Brazilian research funding agencies, by the Secure and Safe Multi-Robot Systems (SESAME) H2020 Project under Grant Agreement 101017258, and by the Assuring Autonomy International Programme (https://www.york.ac.uk/assuring-autonomy).

References

1. Kopetz, H., Bondavalli, A., Brancati, F., Frömel, B., Höftberger, O., Iacob, S.: Emergence in cyber-physical systems-of-systems (CPSoSs). In: Bondavalli, A., Bouchenak, S., Kopetz, H. (eds.) Cyber-Physical Systems of Systems. LNCS, vol. 10099, pp. 73–96. Springer, Cham (2016). https://doi.org/10.1007/978-3-319-47590-5_3
2. Dajsuren, Y., van den Brand, M.: Automotive Systems and Software Engineering: State of the Art and Future Trends, 1st edn. Springer Publishing Company, Incorporated (2019)
3. Aslansefat, K., et al.: Safedrones: Real-time reliability evaluation of uavs using executable digital dependable identities, In: IMBSA 2022. LNCS, vol. 13525. Springer, Cham. Proceedings, pp. 252–266. Springer, Cham (2022). https://doi.org/10.1007/978-3-031-15842-1_18
4. Wei, R., Kelly, T.P., Hawkins, R., Armengaud, E.: DEIS: dependability engineering innovation for cyber-physical systems. In: Seidl, M., Zschaler, S. (eds.) STAF 2017. LNCS, vol. 10748, pp. 409–416. Springer, Cham (2018). https://doi.org/10.1007/978-3-319-74730-9_37

5. Trapp, M., Schneider, D., Liggesmeyer, P.: A safety roadmap to cyber-physical systems. In: Perspectives on the Future of Software Engineering, pp. 81–94. Springer (2013). https://doi.org/10.1007/978-3-642-37395-4_6

6. Hawkins, R., Kelly, T.: A systematic approach for developing software safety arguments. In: 27th International System Safety Conference, pp. 25–33 (July 2010)

7. Wei, R., Kelly, T.P., Dai, X., Zhao, S., Hawkins, R.: Model based system assurance using the structured assurance case metamodel. J. of Syst. and Soft. **154**, 211–233 (2019)

8. Wei, R., Kelly, T., Reich, J., Gerasimou, S.: On the transition from design time to runtime model-based assurance cases. In: MoDELS (Workshops), pp. 56–61 (2018)

9. GSN, GSN Community Standard Version 3 (2022). https://scsc.uk/r141B:1?t=1. (Accessed January 4 2023)

10. Bloomfield, R., Bishop, P.: Safety and assurance cases: Past, present and possible future - an adelard perspective, London, pp. 51–67. Springer, London (2010). https://doi.org/10.1007/978-1-84996-086-1_4

11. OMG, Structured Assurance Case Metamodel (sacm) Version 2.2. https://www.omg.org/spec/SACM/2.2/About-SACM/. (Accessed January 12 2023)

12. Kelly, T.P., McDermid, J.A.: Safety case construction and reuse using patterns. In: Safe Comp 97, pp. 55–69, Springer (1997). https://doi.org/10.1007/978-1-4471-0997-6_5

13. Selviandro, N., Hawkins, R., Habli, I.: A visual notation for the representation of assurance cases using SACM. In: Zeller, M., Höfig, K. (eds.) IMBSA 2020. LNCS, vol. 12297, pp. 3–18. Springer, Cham (2020). https://doi.org/10.1007/978-3-030-58920-2_1

14. D. Consortium, D3.1: Digital dependability identities and the ode meta-model (2020). https://deis-project.eu/dissemination/. (Accessed January 12 2023)

15. Habli, I., Kelly, T.: A safety case approach to assuring configurable architectures of safety-critical product lines. In: Giese, H. (ed.) ISARCS 2010. LNCS, vol. 6150, pp. 142–160. Springer, Heidelberg (2010). https://doi.org/10.1007/978-3-642-13556-9_9

16. O.M.G., O.M.G: OCL Version 2.4 (2014). https://www.omg.org/spec/OCL/2.4/PDF.

17. Kolovos, D.S., Paige, R.F., Polack, F.A.C.: The epsilon object language (EOL). In: Rensink, A., Warmer, J. (eds.) ECMDA-FA 2006. LNCS, vol. 4066, pp. 128–142. Springer, Heidelberg (2006). https://doi.org/10.1007/11787044_11

18. Hawkins, R., Habli, I., Kolovos, D., Paige, R., Kelly, T.: Weaving an assurance case from design: a model-based approach. In: 2015 IEEE 16th International Symposium on High Assurance Systems Engineering (HASE), pp. 110–117. IEEE (2015)

19. Eclipse, Eclipse modeling framework (emf). http://www.eclipse.org/modeling/emf/. (Accessed January 4 2023)

20. Nascimento, L.F.A.: SACM: Editor: an OMG standard compliant model-based tool for specification of Assurance Cases for Safety-Critical Systems (2020). http://monografias.nrc.ice.ufjf.br/tcc-web/exibePdf?id=468.

21. de Castro, R.A.R., Freitas, D.: Hybrid abs with electric motor and friction brakes. In: Proceedings of 22nd International Symposium on Dynamic of Vehicle on Roads and Tracks (IAVSD11), pp. 1–7 (2011)

22. Denney, E., Pai, G., Pohl, J.: AdvoCATE: an assurance case automation toolset. In: Ortmeier, F., Daniel, P. (eds.) SAFECOMP 2012. LNCS, vol. 7613, pp. 8–21. Springer, Heidelberg (2012). https://doi.org/10.1007/978-3-642-33675-1_2

Data Logging and Non-invasive IoMT Approach for Rats Monitoring in Laboratory Experiments

Bryan C. M. Barbosa[1($\boxtimes$)], Steven A. Garan[2], Barbara Melo Quintela[3], and Mario R. Dantas[4]

[1] Universidade Federal de Juiz de Fora, Juiz de Fora, Brazil
bryan.barbosa@ice.ufjf.br

[2] Center for Research and Education in Aging, University of California, Berkeley and Lawrence Berkeley National Laboratory, Berkeley, CA, USA
sgaran@arclab.org

[3] Computational Physiology and High Performance Computing Laboratory and Computer Science Department at Universidade Federal de Juiz de Fora, Juiz de Fora, Brazil
barbara.quintela@ufjf.br

[4] Computer Science Graduate Program at Universidade Federal de Juiz de Fora, Juiz de Fora, Brazil
mario.dantas@ufjf.br

Abstract. It is being observed that the use of Internet of Medical Things (IoMT) in health sciences research grows as the technology and miniaturization of devices occur. Those devices often times suffer from several issues such as: being invasive, lacking the ability to collect many signals concurrently and communication. The present work proposes an approach to the use of an ultra-small IoMT device to fulfill the functions of a bio-signals data collector, a mathematical models processor, a data transmitter-receptor through the network, and a mechanical actuators controller. Our test environment was conceived to be used in animal models. This is an ongoing project from which partial results indicate that, when functioning, the device will use real-time data for monitoring bio-signals and helping with decision-making as it interacts with the organism. The data collection is going to be made through low-cost non-invasive sensors. Therefore, it can favor installation, maintenance, access, and promote a better quality of life for the user.

1 Introduction

The use of Internet of Things (IoT) has seen exponential growth since its first mention [5]. Through the years of advances in the Internet of Things field and medical devices, several new solutions for biosignals monitoring and for drug delivery have been presented [12]. The biosignals monitoring solutions mostly cover uniquely one or a specific group of biosignals (e.g. heart rate, oxygen levels, etc.). Furthermore, most sensors are very invasive either for biosignals monitoring and for drug delivery, resorting to subcutaneous solutions that decrease the life quality of the individual and are harder to access for maintenance or replenishment [6]. Typically those devices have no embedded decision-making solutions and lack interfaces for connection, data collecting, or remote control.

L. Barolli (Ed.): AINA 2023, LNNS 661, pp. 547–557, 2023.
https://doi.org/10.1007/978-3-031-29056-5_47

Adding to the issues mentioned, the history of experimental research is permeated by the use of animal models [13] e.g. rats. However, in animal model experiments, there is always a risk of being invasive and causing discomfort to the experimented subject. In this research, we aim to provide a platform to manage sensors with the minimum level of stress and discomfort to the rats involved as possible, which can be perceived through the choice of developing a small external device that will be placed at the rat's back as a vest, containing sensors that include infrared and temperature reading.

By using the Internet of Things (IoT) concept, for medical devices, we can develop the Internet of Medical Things (IoMT) [22] and add many types of medical sensors and devices to a system that can be integrated into a single package that uses a Single Board Computer (SBC), to collect data and control other devices. A SBC fits the needs of the project due its characteristic lightness and small size but still keeping enough processing power to perform a relatively large number of complex computation operations. It also usually has GPIO (General Purpose Input/Output) ports, as well as serial communication buses and both Wifi and Bluetooth connection [10].

The architecture presented within this work comes as the result of creating a connected device that performs the activities of noninvasive biosignal monitoring, data analysis, decision-making algorithms, and actuators control for drug delivering. This research work is presented in four sections, starting with the introduction, followed by materials and methods, a description of the proposed environment for the experiments, and finally, a discussion.

2   Materials and Methods

Monitoring data from biosignals can be a difficult task in many ways. During the designing process, it is very important to select accurate sensors. This selection process must also consider noninvasive solutions due to the importance of keeping the individual life quality, besides facilitating the access and the maintenance of the device.

Regarding the biosignals to be measured, a group of vital signals were chosen together with other indicators during the requirements gathering in the beginning of the project according to the nature of the research project to which the device will be used for. The measured signals are the following: 1) Heart rate; 2) Blood oxygenation; 3) Body temperature; 4) Ambient light level; 5) Glucose; 6) Blood pH; and 7) heart electrical signals (electrocardiogram).

Therefore, the proposed device must hold the following specifications: a) Processing power for computing for mathematical and tiny machine learning models; b) Network connection; c) Serial Communication Bus for peripheral attachment; d) GPIO pins control interface for actuators control and encoders monitoring; and e) Biosignals monitoring.

For the measurement of the aforementioned biosignals, four types of input devices were selected and are described within this section: Thermometer, Lux meter, Infrared sensor, and Electrodes.

Thermometer: Body temperature During the process of finding an input device to measure the body temperature, two types of devices were considered, namely, digital temperature sensors, and thermocouple sensors. After researching the available devices' specifications, the digital temperature sensors seemed to be the best choice due to a question of accuracy. The model of the selected sensor is the TMP117, with its 0.1 °C accuracy [7].

Lux meter Measuring the ambient light level is important due to the effects of light exposure on the circadian rhythm, making the light level a critical factor to a medical device. Therefore, the BH1750 sensor was selected to measure the ambient light level. The sensor presents an accuracy ratio (Sensor reading/Actual ambient value) that varies between 0.96 and 1.44, according to sensor temperature, this ratio is very close to 1.00 for temperatures between 20 and 30 °C [19].

Infrared sensor The infrared sensor of the project consists of a few LEDs aligned with one photodiode. The idea behind this approach consists in positioning a certain part of the animal's body between the LEDs and the photodiode and measuring the differences between levels of light absorption by the photodiode. For obtaining good results in this approach, high-resolution analogical readings are vital. The ADS1115 was the Analogical-to-Digital converter selected for it due to its' 16-bit resolution and the capacity of delivering readings at rates up to 860 samples per second [8].

Electrodes There are several works dealing with the electrocardiography dilemma in rats [14]. Various types of methods have been described in the literature. Some need the animal to be sedated [17], which deviates from the purpose of the research. Other methods need the insertion of the device under the animal's skin [3]. Those methods enable measurements in freely moving rats for several weeks but are invasive, which is also a deviation from the purpose. In this research under development, there is no fully developed solution for this problem yet, but our research has tended to point to the development of electrodes attached to a Flexible Printed Circuit (FPC). This approach would work in combination with the analogical-digital converter mentioned above.

After choosing the sensors, the technical requirements of the processing central connections are established, all the sensors and converters mentioned use the I2C for communication, so having an I2C bus is the requirement for the processing central to be able to attach the selected devices.

2.1 The Processing Central

Moving forward, there are other requirements that enable a device to perform the functions of a central that can extract the data from the sensors, process this data in mathematical and tiny machine learning models, send/receive it through the network, and control actuators in function of the decisions made through the models to delivery drugs with accuracy. The device also carries some constraints, which consist of: size and weight (considering it will be attached to a rat); and GPIO pins for controlling the actuators. There are several options able to fulfill the desired functions of the prototype in specific points. We can name devices like Raspberry Pi, Arduino, BeagleBone, and ESP32, but taking into account the available models during the requirements gathering in the beginning of the project, each one of them exceeded the constraints in aspects like size and weight or was below the criteria of what was necessary for performance. Therefore, going beyond the mainstream was necessary for getting the right processing central. After research and comparison between the possible devices [2, 18], the Vocore 2 [21], which is an open hardware SBC, was selected as the best option. This SBC runs the Linux Distribution OpenWrt as its operating system [11, 15]. The device also has the following technical specifications: size: 25.6mm x 25.6mm x 3.0mm; CPU: MT7628, 580 MHz, MIPS 24K; Memory: 128MB, DDR2, 166MHz; storage: 16MB NOR on board, support SDXC up to 2TB; wireless connection: 802.11n, 2T2R, speed up to 300Mbps with onboard antenna; GPIO: 40 pins; I2C: One I2C bus; PWM: 4 pins; Temperature Operation: 0°C   40°C(full load) or −10°C   70°C; power supply: 3.6V 5.5V, 500mA; power consumption: 74mA wifi standby, 230mA with wifi at full speed, 5V input; logic level: 3.3V, current 4mA/8mA [21].

As for the programming language chosen to handle the device, we resorted to Python, since it has been increasing its performance with each update [20]. The choice of Python was due both to the growing wave of its use in embedded systems and in general IoT applications, as well as for its non-verbal structure and short learning curve, enabling easy integration between researchers in fields, not exclusive to the Computer Sciences.

This language is characterized by its high level of interpretation, with dynamic typing and binding semantics. As reported by Rossum [20], it is 5 to 10 times easier to program in Python than in C/C++, and 3 to 5 times easier than in Java. Furthermore, it is extremely embeddable in other applications, even supporting object-oriented programming, as well as its use on all types of platforms. There is also the possibility of using MicroPython in OpenWrt, which consists in a lean implementation of Python optimized to run in constrained environments [16].

2.2   Developed Libraries

Vocore 2 already provides built-in solutions on the hardware level [21]. However, when compared with well-established Arduino [2], there is still a lack of information about the usage and we identified that usability needed to be improved to be able to continue the project. Therefore, we have identified the need to develop base libraries to properly manage Vocore 2 in the context of this research project.

Seven libraries have been developed, with the following goals: 1) control the GPIO pins; 2) create a high-level stepper motor control interface through the GPIO pins; 3) allow Vocore 2 to maintain accurate monitoring of the steps taken by the motor through scroll wheel rotary encoders; and 4) interfacing with external devices using the I2C bus.

The "Vocore 2 GPIO" library was developed to make it easy to control GPIO pins in Vocore 2. Through it, it is possible to set the state of the pins as INPUT (data/electric current input) or OUTPUT (data output/electric current), and read (in case of INPUT) the state of the port (0 for low logic level being received, or 1 for high logic level being received) and write (in case of OUTPUT) signal to the port (0 for high low level being sent, or 1 for high logic level being sent) [1]. Aiming to bring a quick understanding of the library, we tried to design it with commands similar to the ones used in the Arduino framework. For the management of the states of pins and signals sent and received, the principle "Everything is a file" of Unix [15] systems was used, in which changing the state of a pin or changing the logical level of that pin occurs by simply subscribing to a file.

Thus, similarly to the Arduino framework, the configuration of a pin occurs through the process of setting the pin mode (INPUT or OUTPUT); and reading or writing digital signal on the pin (high or low). In the developed library, it is used similarly, with the addition of a command to create a file corresponding to the pin, which was omitted within the pin mode setting function.

The "Vocore 2 Stepper Motor" library was developed to control stepper motors through drivers in Legacy Mode, which consists in a simple controlling mode that enables the drivers to receive digital pulses as commands directly in the ports "STEP" and "DIRECTION" from the GPIO pins of the microcontroller. This control happens by changing the logical state of the GPIO pins using the "Vocore 2 GPIO" library. In overview, a driver performs basic control in Legacy Mode by setting the direction of rotation, via the logic state of the DIR pin, and also sending pulses in a logic high state to the STEP pin, where each pulse corresponds to a command for a rotor step. This way, the number of motor revolutions per minute can be regulated merely by changing the time between the sent pulses. There is also the option to activate the micro-stepping mode, in which, instead of the 1:1 ratio between pulses and steps, there will be a 1:2n ratio, where n, in the tested drivers, varies in the entire interval [2,8], enabling the motor to move in even smaller steps, increasing the number of steps per revolution.

This library is directly linked to the drug delivery system. By using a linear actuator connected to the stepper motor shaft and to an adapted syringe, this part of the project aims to fulfill the role of a delivering system. The Stepper Motors were chosen for this task especially because of its' capacity of high accuracy for spinning. There's still a redundancy for dealing with possible errors of stepping, which is a library for attaching incremental scroll wheel rotary encoders to the motor shaft. This library is called "Vocore 2 Scroll Wheel Rotary Encoder" and makes the Vocore 2 able to interpret the number of steps taken and the direction of those steps.

About the I2C libraries, the following libraries were developed: "Vocore 2 ADS1115", "Vocore 2 MPU6050", "Vocore 2 BH-1750", and "Vocore 2 TMP117". The first one makes up for the lack of analogical inputs in Vocore 2. Using the I2C communication protocol, data from four analogical pins present in the analog-to-digital converter are sent to the SBC with an amplitude of 16 bits. It works through the process of writing in the configuration register a tuple containing the address of the ports from which the potential difference will be calculated; waiting a certain period of time to gauge the value; and, finally, reading the measured value.

The "Vocore 2 MPU6050" is a library for interfacing the Vocore 2 with the MPU-6050 sensor, which has accelerometer and gyroscope functions, allowing the chip to be able to measure acceleration and tilt in three axes, totaling the simultaneous generation of 6 axes of data about the environment in which the device is immersed. The library acts to enable the connection and reading of sensor data. It is implemented by setting the device configuration registers as desired, according to the "Register Map and Descriptions" specification, a document provided by the device manufacturer [9]; from there, the MPU-6050 operates in continuous mode in reading acceleration and angular velocity values related to the chip itself in the environment and its positioning, simply reading the pair of registers corresponding to the required value axis and from which one is it (e.g.: X-Axis/Accelerometer, Z-Axis/Gyroscope, etc.).

When it comes to lighting, the "Vocore 2 BH-1750" was developed for interfacing with the sensor BH-1750. Similarly to the other libraries, this one also acts on the connection and acquisition of sensor readings by Vocore 2. From its connection to the power supply and the correct connection with the pins of the I2C bus, the sensor operates by sending reading requests, reading only the register corresponding to the mode (e.g.: Continuous reading mode, high-resolution 0.5lx: register 0x11) [19].

The development of a library for the TMP117 temperature sensor arose from the lack of accuracy on the use of a K-Thermocouple attached to a converter MAX6675. The thermocouple is limited to an accuracy of $\pm 5\,°C$, which is not ideal for measuring body temperature. Because of this, the TMP117 sensor, with an accuracy of $\pm 0.1\,°C$ and communication via I2C, was added to the project for replacement, as seen in the title of the library "Vocore 2 TMP117" [7]. Therefore, the reading of the temperature values of the TMP117 occurs in a simple way, similar to the BH-1750 sensor, in which it is enough to write a request with the value 0x0220 in the 0x00 register, and after a time interval, read the response from the 0x00 register.

The library "Vocore 2 ADS1115" enables the Vocore 2 to interface with the analogical-digital converter ADS1115, it works by the same principle as the other I2C libraries, where a value representing the port to get data from is written in a register and read from another. This library, besides the applications mentioned in other sections, will also enable Vocore 2 to keep a track of the battery levels.

3   Proposed Environment

In order to contain the sensors and the Vocore 2 that must be carried by each rat, a case is being designed to be 3D printed. The details of the case are not part of the scope of this paper as they are still under development to optimize the design.

The device aims to monitor biosignals and perform drug injections during long periods, so there is a great challenge in maintaining the device constantly energized. Considering that it will be monitoring a living animal, it cannot be constantly connected by cable. Given that, the solution found lies in using wireless charging, as presented in Fig. 1, to maintain the energy supply to the device.

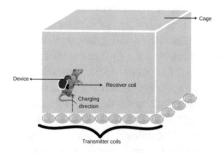

Fig. 1. Representation of the wireless charging scheme. Transmitter coils to be put under feeding areas.

The chosen technique to accomplish this was the electromagnetic induction, even though it works for very short distances, and is not as efficient as cable charging. However, since the device will be operating on a small animal in a controlled environment, this problem can be circumvented by spreading transmitter coils around the ambient floor, placed under the cage, mainly under feeding stations, to keep the device under the charging process frequently.

On the top of each cage there will be a Raspberry Pi connected to a camera with a microphone (Fig. 2), which will be in charge of processing four external signals: a) position and movement tracking, using image processing algorithms, b) tail flexing tracking, using pose estimation algorithms, c) sound recording of the cage through the microphone to monitor the sounds emitted by the animal, and d) ambient temperature, humidity, and atmospheric pressure.

The Raspberry Pi will also work as a server connected to the same Wireless Local Area Network (WLAN) [4, 23] as the device attached to the rat. By doing that, it will be able to receive data from the device through HTTP requests, sync to the aforementioned external signals, and store it.

The data of the IoMT device will be stored in a micro SD card attached to Vocore 2. The process of storing data will be given by rounds, each round consisting on:

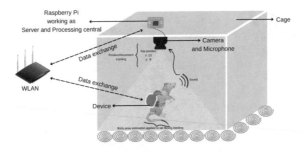

Fig. 2. Raspberry Pi functions scheme.

1. Reading the biosignals from sensors;
2. Cleaning noise and reading errors;
3. Translating raw readings to their equivalents, e.g.: turning raw infrared data into its equivalent meaning for heart rate, blood oxygenation, etc.;
4. Saving all the data readings of the round in a row of a comma-separated values (CSV) file, assuming an acceptable degree of synchronization between them.

There are three types of time interval that will be determined during the initial settings of the device:

1) between rounds of readings, which leads to a cycle of readings of biosignals from a certain instant (Fig. 3);

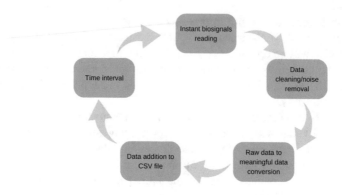

Fig. 3. Biosignals reading cycle.

2) the time interval coverage of the opening and closure of each CSV file, which leads
   to a cycle of i. CSV file creation/opening; ii. successive data additions to the file;
   and iii. CSV file closure (Fig. 4);

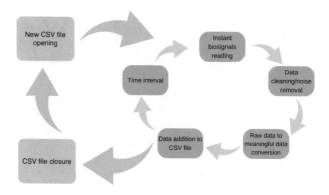

Fig. 4. CSV file creation cycle.

3) the interval between requests for uploading the files to the Raspberry Pi Server,
   which leads to a cycle of file uploading through HTTP requests every amount of
   pre-defined time (Fig. 5).

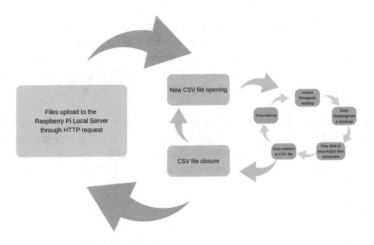

Fig. 5. File upload to Server cycle.

3.1 Data Logging Ecosystem

Given that the project is under development and the experiments have not been performed yet, in this work we present partial results regarding the usage of Vocore 2. The libraries were successfully tested, making the data logging ecosystem proposed in this work possible, as presented in the figure below:

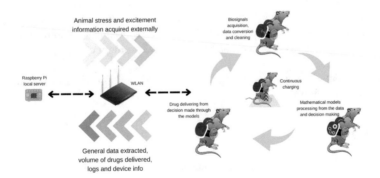

Fig. 6. Data logging ecosystem.

The ecosystem that was developed covers all the aspects of monitoring, storage, and decision-making for the data, the energy supply for the device, and the data transfer through the network. Adding to that, the wireless connections along with the Raspberry Pi will likely establish a noninvasive environment for the rats in future experiments.

Besides, the results that could be validated were satisfactory and, through what was obtained, we have the range of resources that can be implemented: (a) control reservoirs and the injection of drugs through to the stepper motor control library "Vocore 2

Stepper Motor"; (b) obtain movement/agitation data of the animal, through the "Vocore 2 MPU6050" library; (c) measure the animal's body temperature using the "Vocore 2 TMP117" library; (d) attach infrared sensors, electrodes, and measure battery levels through the analogical-digital converter ADS1115 and its respective library "Vocore 2 ADS1115"; and (d) measure the ambient luminosity, using the "Vocore 2 BH-1750" library.

Performance tests are not possible to be executed yet as the "backpack" is still under development. When mature and stable prototype designs and implementations are available, tests for data exchange, decision-making speed, and sufficiency of accuracy and delivering speed of those decisions for achieving the desired results will be executed.

4 Conclusions

We have presented a data logging architecture using single-board computers to monitor rats and inject drugs during an experiment. The proposed environment includes a "backpack" with a Vocore 2 interfacing with sensors and controlling stepper motors, to which seven libraries have been developed and made available to the community to improve the usability of the device. All the data generated in the device will be in continuous data exchange through the Local Network with a Raspberry Pi working as a server and a central for image and sound processing and also ambient data monitoring. This data will be used in mathematical models for predictive analytics and decision-making about the volume of drugs delivered during the experiment.

Finally, by developing libraries on a device such as the Vocore 2, which lacks a significant presence in the scientific research community, the present work puts into evidence its relevance. Furthermore, we consider it to be an innovative contribution to the IoMT and Computation Science fields for providing noninvasive solutions in animal experimenting, as well as propositions to surpass the problem of aging.

Acknowledgements. This project has been funded by the Center for Research and Education on Aging (CREA), UC Berkeley.

References

1. The Linux Kernel Archives. GPIO Documentation. In: The Linux Kernel (2022)
2. Arduino. Language Reference (2022). https://www.arduino.cc/reference/en/
3. Braga, V.A., Burmeister, M.A.: Applications of telemetry in small laboratory animals for studying cardiovascular diseases. In: Modern Telemetry (Oct 2011). https://doi.org/10.5772/22989
4. Crow, B.P., et al.: IEEE 802.11 wireless local area networks. IEEE Commun. Mag. **35**(9), 116–126 (1997). https://doi.org/10.1109/35.620533
5. Dwivedi, R., Mehrotra, D., Chandra, S.: Potential of Internet of Medical Things (IoMT) applications in building a smart healthcare system: A systematic review. J. Oral Biol. Craniofacial Res. **12**(2), 302–318 (2022). https://doi.org/10.1016/j.jobcr.2021.11.010
6. Ho, D., et al.: Heart rate and electrocardiography monitoring in mice. Curr. Protoc. Mouse Biol. **1**, 1–24 (2012). https://doi.org/10.1002/9780470942390.mo100159

7. Texas Instruments. TMP-117 high-accuracy, low-power, digital temperature sensor with SMBus - and I2C-compatible interface. Texas Instruments Incorporated, pp. 1–50 (September 2022)
8. Texas Instruments. Ultra-small, low-power, 16-bit analog-to-digital converter with internal reference. Texas Instruments Incorporated, pp. 1–37 (October 2009)
9. Invensense. MPU-6000 and MPU-6050 Product Specification Revision 3.4. In: Invensense, pp. 1–52 (August 2013)
10. Isikdag, U.: Internet of Things: Single-Board Computers. In: Enhanced Building Information Models, pp. 43–53 (2015)
11. Jin, T.: OpenWrt Development Guide. In: Wireless Networks Lab. CCIS, pp. 1–11 (Febuary 2012)
12. Joyia, G.J., et al.: Internet of Medical Things (IOMT): Applications, Benefits and Future Challenges in Healthcare Domain. J. Commun. 12, 240–247 (2017) https://doi.org/10.12720/jcm
13. Mitsunaga, Jr. J.K., et al.: Rat an experimental model for burns. A systematic review. Acta Cirúrgica Brasileira **27**, 417–423 (2012). https://doi.org/10.1590/S0102-86502012000600010
14. Konopelski, et al.: Electrocardiography in Rats: a Comparison to Human. Physiological research/Academia Scientiarum Bohemoslovaca 65, 717–725 (2016). https://doi.org/10.33549/physiolres.933270
15. Mauerer, W.: Professional Linux Kernel Architecture. Wiley Publishing, Inc. (2008)
16. MicroPython. MicroPython (2018). https://micropython.org/
17. Normann, S., Priest, R.E., Benditt, E.P.: Electrocardiogram in the normal rat and its alteration with experimental coronary occlusion. Circulation Res. 9, 282–287 (1961). https://doi.org/10.1161/01.RES.9.2.282
18. Raspberry Pi. Raspberry Pi Documentation (2022). https://www.raspberrypi.com/documentation/microcontrollers/rp2040.html
19. Rohm. Digital 16bit serial output type ambient light sensor IC. In: Rohm Co., pp. 1–18 (April 2010),
20. van Rossum, G.: Glue it all together with Python. In: OMG-DARPAMCC Workshop on Compositional Software Architecture (January 1998)
21. VoCore Studio. VoCore2. In: VoCore Studio (2022)
22. Al-Turjman, F., Nawaz, M.H., Ulusar, U.D.: Intelligence in the Internet of Medical Things era: A systematic review of current and future trends. Comput. Commun. **150**, 644–660 (2020). https://doi.org/10.1016/j.comcom.2019.12.030
23. Zhao, C.W., Jegatheesan, J., Loon, S.C.: Exploring IOT application using Raspberry Pi. In. J. Comput. Netw. Appli. **2**, 27–34 (2015)

Security Challenges and Recommendations in 5G-IoT Scenarios

Dalton C. G. Valadares[1,2(✉)], Newton C. Will[3], Álvaro Á. C. C. Sobrinho[4],
Anna C. D. Lima[1], Igor S. Morais[1], and Danilo F. S. Santos[1]

[1] UFCG/VIRTUS RDI Center, Campina Grande, PB, Brazil
dalton.valadares@embedded.ufcg.edu.br, anna.dantas@estudante.ufcg.edu.br,
igorsilva.morais@ee.ufcg.edu.br, danilo.santos@virtus.ufcg.edu.br
[2] Federal Institute of Pernambuco, Caruaru, Brazil
[3] Federal University of Technology - Paraná, Curitiba, Brazil
will@utfpr.edu.br
[4] Federal University of Pernambuco's Agreste, Garanhuns, Brazil
alvaro.alvares@ufape.edu.br

Abstract. The fifth-generation (5G) mobile communication systems are
already a reality. This communication technology can increase and simplify the adoption of the Internet of Things (IoT) applications (e.g.,
industrial IoT), given that it will provide the means to connect up
to one million devices in a squared kilometer. Many of these applications can generate sensitive data, which requires the adoption of security
mechanisms, and these mechanisms must consider the computational
limitations of the devices. Considering that IoT devices under security
attacks can access the 5G infrastructure, this paper presents recommendations and challenges regarding security in IoT devices. It also presents
a threat model and a few common attacks, classifying them with the
STRIDE model and CVSS and discussing how they can impact the 5G
infrastructure.

1 Introduction

The fifth-generation (5G) mobile communication systems increase communication speed and capacity, providing reliable and speedy connectivity for the Internet of Things (IoT) applications [16]. The 5G massive Machine-Type Communication (mMTC) should enable the connection of up to one million devices by
squared kilometer. The mMTC is based on the Low Power Wide Area Network
(LPWAN) standard, which provides communications of small data quantities
with low energy consumption for a long-distance range [23]. This communication type eases the deployment of many IoT applications once the IoT devices
can be connected straight to the 5G infrastructure.

The USA IoT Cybersecurity Improvement Act of 2020 [39] defines IoT devices
as devices with a transducer (sensor or actuator) to interact with the physical

L. Barolli (Ed.): AINA 2023, LNNS 661, pp. 558–573, 2023.
https://doi.org/10.1007/978-3-031-29056-5_48

world directly and, at the very least, a network interface, not considering conventional information technology devices, such as smartphones and notebooks. In addition, IoT devices can work on their own without having to function as a component of another device. This definition agrees with the one stated by the ITU Internet of Things Overview [11].

The predictions regarding the number of IoT devices in the next years always present great numbers. For instance, according to the IoT Analytics' "State of IoT - Spring 2022 report" [20], the global IoT connections grew 8% in 2021, reaching 12.2 billion active endpoints. The predictions expect this number reaches 14.4 billion in 2022 and 27 billion in 2025. The IDC predictions [12] are even greater, estimating the number of 55.7 billion IoT devices generating almost 80B zettabytes of data.

Considering these predicted huge numbers of IoT devices and the 5G mMTC's high capacity to connect such devices, security is a mandatory requirement once the devices may have minimal computational resources. This characteristic, together with the heterogeneity of hardware, standards, and protocols, makes IoT devices more vulnerable and susceptible to attacks. In the first semester of 2021, the number of attacks against IoT devices grew more than 100%, reaching 1.5 billion, with most aiming to steal data, mine cryptocurrencies, or build botnets [36]. Considering botnets, in December 2021, the most powerful Distributed Denial of Service (DDoS) attack registered had a peak of 4.4 Tb/s and involved 5,270 IP addresses [24].

Given these security concerns, many institutions worldwide are employing efforts to provide operational and technical recommendations regarding the security of IoT devices and applications. For instance, the USA government already has a law (Public Law 116–207, IoT Cybersecurity Improvement Act of 2020) [39] with recommendations that set minimum security standards for IoT devices owned or controlled by USA federal agencies. In Europe, the European Parliament recently proposed a regulation on horizontal cybersecurity requirements for products with digital elements [6], which can also include IoT devices.

In this sense, we investigated scientific papers and technical reports to gather challenges and recommendations regarding security in IoT applications, considering scenarios with 5G communication. This information can help regulatory agencies to propose recommendations and guidelines to solution providers and device manufacturers, specifying the minimum security requirements and how to validate them.

The main contributions of this paper are listed below:

- we gathered and discussed common threats to IoT devices;
- we classified the common threats with the STRIDE method and applied the Common Vulnerability Scoring System (CVSS) system to them;
- we gathered technical recommendations regarding cybersecurity in IoT devices;
- we proposed a threat model with six attack surfaces to be considered when analyzing security problems and solutions; and
- we discussed the impact in the 5G infrastructure in case adversaries perform attacks exploring the common threats.

The remainder of this paper is organized as follows: Sect. 2 presents a brief description of the main security properties as well of the STRIDE model and the CVSS system; Sect. 3 describes some common threats related to IoT devices and applications, and presents the threat model defined in this paper; Sect. 4 presents some security recommendations, with some of them based on technical reports; Sect. 5 discusses the impacts on the 5G infrastructure components if the known attacks are successful; Sect. 6 summarizes some papers related to IoT and 5G; and Sect. 7 concludes this paper, mentioning some suggestions for future work.

2 Background

2.1 Common Security Properties

The NIST FIPS 199 [9] standard introduces confidentiality, integrity, and availability as the three security objectives for information and for information systems. These three objectives are known as the *CIA triad* and are listed below.

- **Confidentiality:** It restricts the viewing of data only to those who may have the right to view it, making it difficult for those who do not have this permission to access this data. The most common technique to ensure data confidentiality is encryption, where data is encrypted using a key, and only the holder of the same key, for symmetric encryption, or a complementary key, for asymmetric encryption, can decrypt the data [14].
- **Integrity:** It regards the protection against corruption or unauthorized modification of data, ensuring that the data being read is the same as it was previously written or sent. The most common way to ensure data integrity is using the Message Authentication Code (MAC). Using MAC allows the detection of any change in the message, as different data sets will generate different hashes [10].
- **Availability:** It implies that the service is available to legitimate users at all times and without interruption, ensuring access and use of information in a reliable and timely manner. The level of availability required for a given service increases with the criticality of that service. Several attacks can cause the loss or reduction of availability of a given service, and these can be mitigated with the use of authentication systems, intrusion detection systems, and firewalls.

2.2 STRIDE Model

STRIDE is a threat classification model for security assessment that derives from an acronym of the following threat categories [34]:

- *Spoofing*: attempt to gain access to a system using a false identity to gain an illegitimate advantage;
- *Tampering*: unauthorized modification of data, whether in storage or transmission;

- **R**epudiation: the ability of users (legitimate or otherwise) to deny that they performed specific actions or transactions;
- **I**nformation Disclosure: unwanted exposure of private data to individuals who should not have access to them;
- **D**enial of Service: the process of making a system or application unavailable, or with impaired availability, to legitimate users; and
- **E**levation of Privilege: occurs when a user with limited privileges assumes the identity of a user with higher privileges to gain access to an asset.

The STRIDE model is used by developers and companies to categorize threats to systems, taking into account their effects on security. The definition of a threat model helps assess and document security risks associated with a system, allowing the definition of efficient, realistic, and meaningful security requirements. In addition, proper threat identification and selection of countermeasures help reduce attackers' ability to perform illegitimate operations.

2.3 Common Vulnerability Score System

The CVSS is a method used to classify vulnerabilities and their risks. CVSS uses a scoring system to define a vulnerability qualification level. This method is widely used in computer systems and IT environments to categorize vulnerabilities [8].

The CVSS comprises three metrics groups: Base, Temporary and Environmental. The Base Metric (BM) reflects the severity of a vulnerability according to its intrinsic and immutable characteristics, i.e., characteristics that do not change over time or in different environments. This group is subdivided into Exploitability Metrics (EM) and Impact Metrics (IM), where the first measures the technical means and the ease with which the vulnerability can be exploited, and the second involves the three security properties described in Sect. 2.1. The Temporary Metric (TM) adjusts the BM according to factors that change over time, and the Environmental Metric (EM) adjusts the BM and TM to a specific environment. Typically, only the BM is published, as it does not change over time and is common to all environments.

CVSS 3.0 defines five severity levels for a vulnerability, namely: None (score 0.0); Low (score between 0.1 and 3.9); Medium (score between 4.0 and 6.9); High (score between 7.0 and 8.9); and Critical (score between 9.0 and 10.0).

3   Threats in 5G-IoT Environments

3.1   Common Threats

In this Section we briefly describe the common threats to IoT devices and applications.

- **Hijacking:** This attack is common and quite harmful to the IoT network, as the attacker can take control of a legitimate device that is connected to the network and turn it into a bot to steal information from the system or even to flood the network traffic, resulting in a DDoS attack and causing system instability [28].

- **Tampering:** The attacker intercepts the network and manages to impersonate a legitimate user, being able to exchange information with other users/devices and thus gaining access to privileged data [17]. The attacker can also modify parameters being exchanged between users to manipulate information that pertains to the IoT scope, such as obtaining access credentials from devices or users.

- **DDoS:** This type of attack tends to affect the availability of systems, as such attacks overwhelm their targets with unwanted Internet traffic. This type of attack can intensify on a 5G network, and cause an infected packet, exchanged by several devices, to contaminate all the nodes it passed through (which could be used in a botnet). Examples of specific DDoS attacks include TCP flooding, UDP flooding, TCP SYN flooding, and TCP resynchronization [22].

- **Jamming:** Jamming attacks are used to block and disturb legitimate communications [27]. Jamming is an easy attack to apply, as all wireless communication technology is vulnerable to it. This attack can cause major impacts on the network and can be a gateway for other attacks, such as DoS.

- **Man-in-the-Middle:** The attacker places a fake node between the communication of two valid nodes [18]. Through this attack, he/she manages to convince these nodes that his/her node is valid and gets access to the information transmitted between these nodes, being able to modify it or not send it to its destination.

- **Eavesdropping:** The attacker can monitor networks, exploit security breaches and weak connections between IoT devices and the server [15]. This attack directly impacts the confidentiality of the service, as the focus of eavesdropping is snooping on information and stealing data.

- **Spoofing:** This attack compromises both wired and wireless networks [32]. The attacker modifies the address to some forged value that belongs to a valid user or device, in order to obtain illegal benefits in the network. In this way, the attacker appears to be an authorized, known, and secure device or user.

- **Relay Attacks:** This attack relays a legitimate user's authentication signal to gain network access. The content of the message that was stolen does not need to be decrypted, even if the stolen message is just resent, the attacker may still be able to succeed when no protection or additional checks are implemented [1].

- **Skimming:** In this attack, a device is used to clone the credentials of a legitimate user [21]. This device copies physical credentials, such as an ID card or chip. The attacker needs to be physically in the place where he/she will place his/her skimming device.

- **Brute Force:** Most IoT devices keep default passwords, allowing attackers, through brute force attacks, to quickly access the devices by guessing these passwords [13]. The brute force attack is just the beginning of exploiting a vulnerability that tends to become a bigger chain.

- **Sybil Attack:** This attack is characterized by the manipulation of false or stolen identities [30]. Malicious nodes in the attacker's possession impersonate the behavior of a legitimate user or node. This attack aims to steal sensitive information stored by other nodes.
- **Wormhole:** Two or more nodes, controlled by attackers, create a special path for data to pass through and record what is being transmitted [44]. The malicious node is located in a strategic place in the network, providing the shortest route for exchanging messages between the other devices. Thus, the attacker obtains all the data that is being trafficked on the network.
- **Node Replication Attacks:** The attacker captures a legitimate node, replicates it and send it back to the infrastructure [25]. This excess of nodes can end up resulting in unavailability of services, generate insecure data and modify the network configuration. The attacker can still keep the replicated node on the network, which will act as a legitimate node and collect sensitive data.

3.2   Threat Model

We defined the threat model shown in Fig. 1, considering possible attack surfaces in IoT applications with 5G infrastructure. The skull symbol represents possible attackers, whether on devices or communication channels. The defined threat model considers six attack surfaces, as follows.

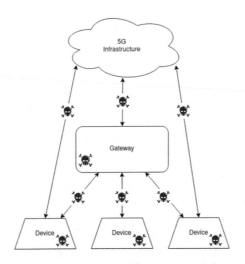

Fig. 1. The considered threat model.

- **Device:** Each device can be targeted by attackers and, when compromised, allow communication with the 5G infrastructure or gateways, expanding the attack surface to other targets.

- **Gateway:** If an attacker takes control of the gateway, all devices connected to it can be targeted to be co-opted.
- **Communication Between Device and Gateway:** The communication between devices and gateways can be targeted by eavesdroppers, who can steal sensitive data.
- **Communication Between Gateway and Infrastructure:** Attackers can exploit the communication channel between gateways and the 5G infrastructure.
- **Communication Between Device and Infrastructure:** Attackers can exploit the communication channel between the devices and the 5G infrastructure in the case of direct connection of the devices.
- **Device-to-Device Communication:** This feature opens the door to propagate attacks hop by hop to reach a vulnerable device.

4 Recommendations

4.1 ETSI 303645 - Cybersecurity for Consumer IoT

This Section lists and describes the main recommendations of the ETSI report on Cybersecurity for Consumer IoT [5].

- **Do not use default passwords:** Devices must use unique passwords, mechanisms that reduce the risk of automatic attacks (e.g., brute force), and appropriate authentication mechanisms. They must provide a way to change the authentication value used and hamper brute force attacks on devices that are not so resource-limited, making them impractical to perform.
- **Implement ways to manage vulnerability reporting:** Manufacturers must have a vulnerability disclosure policy and deal with them promptly. In addition, they must continually monitor for new vulnerabilities and mitigate them.
- **Keep software up-to-date:** All software components must be updated securely, with mechanisms for such procedures on devices with sufficient resources. The update must happen in a simple way and involve automatic mechanisms and appropriate encryption techniques in the process. The process of checking for updates can happen at device startup and periodically after that. The automatic updates must be configurable and enabled at startup, timely, and can be scheduled. The devices must verify their authenticity and integrity. Manufacturers must inform users about necessary updates and what risks are mitigated by them. They must also publish the defined period for support of updates or the reasons for not having them. In the case of devices with minimal resources, each device must clearly identify its model, either by label or physical interface.
- **Store security parameters securely (protected):** The device must secure security parameters in persistent storage. If the device encodes and stores the unique identifier, it must have ways to prevent tampering, whether by physical, electrical, or software means. Critical security parameters must

not be hard-coded in source code. Besides, any security parameters used for integrity verification and software update authenticity must be unique per device and generated in a way that reduces the risk of attacks.

- **Perform secure communication:** The device must use appropriate encryption for secure communication and evaluated implementations to provide security and networking functionality, mainly when related to encryption. The cryptographic algorithms must be updatable, and access to the device in the initial state by the network interface must be possible only after authentication on the interface. Changes to device security settings should only be possible after authentication, and the devices should encrypt critical security parameters to be secure in transit. They must protect the confidentiality of critical security parameters that are remotely accessible through network interfaces, and the manufacturers must follow standards and recommendations for the secure management of these parameters.

- **Minimize exposed attack surfaces:** All unused logical and physical interfaces should be disabled initially, and network interfaces should minimize unauthenticated disclosure of security-related information. Devices must avoid unnecessary exposure of physical interfaces that can be a target for attacks and disable the debug interfaces when physically accessible. The software must run with minimal privileges, and the device must include a hardware-level access control mechanism for memory. The manufacturers must only allow software services necessary for the device's proper functioning, with the code limited to the functionality necessary for the device's operation. They must follow secure development processes for software deployed on the device.

- **Ensure software integrity:** The device must verify its system using secure boot mechanisms. If the system detects an unauthorized change, the device must alert the user by connecting only to the networks necessary for the alert.

- **Ensure that personal data are protected:** The confidentiality of personal data transmitted between devices and services must be preserved through proper encryption, and any sensing capabilities of the device must be documented in a clear and user-accessible manner.

- **Be resilient to interruptions:** Devices must be resilient, considering power and communication interruptions. They must continue operating in case of a loss of connection to the communication network and properly recover in case of power loss. They must also consider the network infrastructure to make operational and stable connections, minimizing unnecessary overhead.

- **Examine telemetry data:** Telemetry data, when collected, must be examined to avoid security anomalies.

- **Facilitate User Data Removal:** The device and services must have functionality allowing users to remove their data easily when necessary. There must be clear instructions on how to remove personal data and clear information when the user removes such data from services and devices.

- **Facilitate installation and maintenance of devices:** Installation and maintenance of devices should involve minimal user decisions and follow security best practices regarding usability. The manufacturer should provide guidance on how to configure the device securely and verify that the device is securely configured.

- **Validate input data:** The system must validate the input data considering the type and possible ranges of values, for example.
- **Provide data protection information:** There should be information about what data are used, for what, and by whom. When necessary to process such data, users must give permission, and the processing must be only of the data necessary for the functionality. Users who have allowed the data processing must be able to cancel the permission.

4.2   ENISA - Baseline Security for IoT

The ENISA report [4] mentions the following generic issues related to the security of IoT ecosystems: huge attack surface, complex ecosystem, fragmentation of standards and regulations, limited devices resources, widespread deployment, security integration, security updates, safety aspects, lack of expertise, low cost, insecure programming, and unclear liabilities. According to the report, it is necessary to perform a risk evaluation for each IoT environment, considering the possible attack scenarios, how the threats can affect the assets, which of them are critical, and how to mitigate them if possible.

ENISA performed a study to assess the security criticality of IoT assets. This study considered interviews with experts and, according to its results, the most critical assets are the sensors, the device and network management, and communication protocols. Considering the 5G-IoT scenarios, this finding reinforces that manufacturers and providers should dedicate special attention to the IoT devices and the communication channel with the 5G infrastructure.

The study also considered the following critical attack scenarios: ransomware, DDoS using IoT botnets, stepping stones, protocol vulnerability exploitation, against the network link between controllers and actuators, against sensors, against actuators, against the administration systems, and against devices. According to the experts, the most critical attack scenarios are against the administration systems, sensors (value manipulation), and devices (botnets/commands injection).

The report defined the IoT baseline security measures according to the following three main categories: policies, organizational and process, and technical. The policy measures involve security by design, privacy by design, asset management, and risk and threat identification and assessment. The organizational and process measures consider end-of-life support, proven solutions, security vulnerabilities and incidents management, human security training and awareness, and third-party relationships. Lastly, the technical measures relate to hardware security, trust and integrity management, strong default security and privacy, data protection and compliance, system safety and reliability, secure software/firmware updates, authentication, authorization, physical and environmental access control, cryptography, secure and trusted communications, secure interfaces and network services, secure input and output handling, logging, and monitoring and auditing processes.

According to the reported study, there are some challenges in considering these defined measures. For instance, there is a fragmentation in security regulations and approaches, leading manufacturers and companies to adopt own approaches for implementing security in their IoT products and services. This fragmentation also leads to a lack of awareness and knowledge about concepts and common threats that must be considered. Regarding design and development, there is a lack of security by design and privacy by design processes, and a lack of interoperability across IoT platforms, frameworks, and devices. Besides, there is also a lack of economic incentives and product lifecycle management.

Thus, ENISA recommends that initiatives should:

- promote harmonization between security regulations and initiatives;
- raise understanding and awareness of the need for cybersecurity for IoT;
- define secure development cycle guides for IoT hardware and software;
- facilitate consensus on interoperability in IoT ecosystems;
- promote administrative and economic incentives for IoT security;
- establish lifecycle management of secure IoT applications and services;
- define responsibilities among the various stakeholders in IoT ecosystems.

Such recommendations consider the following stakeholders and their relationships: industry, providers, manufacturers, associations, academia, regulatory agencies, consumer groups, platform operators, and developers.

4.3   STRIDE and CVSS Application

We suggest using the STRIDE model to classify threats and vulnerabilities and the CVSS system to assign a criticality score to known threats and vulnerabilities. Although Ur-Rehman et al. [38] have proposed a more suitable scoring system for IoT applications ($CVSS_{IoT}$), we chose to use CVSS 3.0, as it is the most recent standard version. $CVSS_{IoT}$ adds three other metrics to evaluate, with one related to the physical damage an exploited vulnerability can cause. The classification and criticality score assists in the process of managing and prioritizing risks related to threats and vulnerabilities, helping to mitigate them.

Table 1 presents the STRIDE classification and the CVSS score for each described threat, according to the Marisetty suggestion. The colors represent the criticality levels according to the CVSS score: blue is critical (CVSS 9–10), orange is high (CVSS 7–8.9), yellow is medium (CVSS 4–6.9), and grey is not applicable.

Table 1. Threats classified with STRIDE and CVSS

	S	T	R	I	D	E
DDoS						
Tampering						
Hijacking						
Jamming						
Man-in-the-middle						
Eavesdropping						
Skimming						
Wormhole						
Bruteforce						
Spoofing						
Sybil Attack						
Node Replication						
Relay Attack						

Blue: Critical; Orange: High; Yellow: Medium; Gray: N/A.

5 Threats Impact on 5G Infrastructure

Many of the 5G components are software. For instance, the Software-Defined Network (SDN) controllers and Virtualized Network Functions (VNFs) - possible thanks to Network Function Virtualization (NFV) - are all software components. To manage such components, we need a Virtualized Infrastructure Manager (VIM), which can be another software platform such as OpenStack, a typical cloud platform used to create virtualized instances. Another 5G concept that uses software components is Network Slicing. For this reason, we can consider that a 5G infrastructure can have five main components: the physical layer, SDN, NFV, cloud, and network slices.

Thus, considering these five components and a scenario with IoT devices connected to the 5G infrastructure, we can analyze if a given threat can impact any of the components. We carried out this analysis for each threat, considering what components they can impact, and summarized the results in Table 2. Looking at the table, the "x" indicates that the threat can directly impact the component if an attack is successful. The "+" denotes that the threat can impact the component if the attacker gains more resources and evolves the attack after the successful operation.

Below, we comment the impacts considering each threat:

- **DDoS:** The attacker may intercept the communication between the device and SDN, NFV, or Cloud components, affecting these technologies by injecting malicious code into the transmitted data.
- **Tampering:** The tampering of IoT data could trick the SDN control layer making it believe that nothing is wrong with the network. Furthermore, NFV could lose the ability to discern which VMs are operating correctly. Besides, slices that are not rightly protected may also be affected.
- **Hicjacking:** Hijacking is an attack that generally tries to take control of something. In a network setting, attackers try to hijack packets by behaving

Table 2. Impacts on 5G components

	SDN	NFV	Cloud	Network Slicing	Physical Layer
DDoS	x	x	x	x	–
Tampering	x	x	–	x	–
Hijacking	+	+	+	x	x
Jamming	+	+	+	x	x
Man-in-the-middle	x	x	x	–	–
Eavesdropping	+	+	+	+	x
Skimming	–	–	–	–	x
Wormhole	–	–	–	x	x
Bruteforce	+	+	+	+	+
Spoofing	x	x	x	x	x
Sybil Attack	x	x	x	–	–
Node Replication	+	+	+	+	+
Relay Attack	+	x	+	+	x

x: Direct impact; +: Indirect impact; -: No impact.

as a secure network. Such an attack affects the physical layer and can affect network slicings since the packets will never reach their proper destination.

- **Jamming:** The jamming attack occurs at the physical layer, intending to cause network unavailability. In addition to affecting the physical layer, jamming can also affect network slicing, which directly communicates with the data received from the devices' sensors.

- **Man-in-the-middle:** The attacker may intercept the communication between the device and SDN, NFV, or Cloud technologies, affecting them by injecting malicious code into the transmitted data.

- **Eavesdropping:** The eavesdropping attack occurs at the physical layer, with the use of antennas to intercept the packets that are in transit. So this attack affects the physical layer and can also impact the SDN, NFV, Cloud, and Slicing technologies later, depending on the information obtained by the attacker.

- **Skimming:** Skimming can only affect the physical layer since the attack occurs from cloning user credentials via a physical device. The attack can also interfere with other technologies depending on the information the attacker can obtain.

- **Wormhole:** A wormhole attack creates a "tunnel" at the physical layer, with the packets being transferred to this tunnel and never reaching their proper destination. This way, the attack affects the physical layer and network slicing because packets will be lost and can affect service availability.

- **Bruteforce:** The impact of this threat depends on the device which has the authentication broken, as each device has a different role in the network infrastructure.

- **Spoofing:** In a spoofing attack, the attacker tends to send messages over the network acting as a legitimate user (spoofing the IP address). Thus, this attack can affect SDN, NFV, Cloud, Slicing, and the physical layer since, by spoofing IP addresses, the attacker can send any malicious code that damages the entire network.

- **Sybil:** Through false identities, the attacker can send fake messages to the 5G infrastructure technologies, aiming to confuse the communication and avoid the entities knowing the network's actual situation.
- **Node Replication:** The captured and replicated node's characteristics determine which technologies will be impacted by this type of threat. Thus, it can affect each of the 5G components.
- **Relay:** In a relay attack, by relaying messages to the infrastructure, the attack affects the physical layer once it is responsible for the system communication. It can also affect NFV technology, given that repeated messages will be sent back into the network. The attacker can also interfere with SDN, Cloud, and Slicing depending on the damage generated in the network from the relay attack.

As seen, the attacks on IoT devices can affect components of the 5G infrastructure. Thus, it is important to establish means of checking security mechanisms for the users and validating security requirements for the providers and manufacturers. In this direction, the NISTIR 8259 [7] provides security recommendations for IoT device manufacturers, working as a reference on the security resources the devices must offer to mitigate risks for their clients. The recommendations consider the following six basic activities, with the first four related to before the commercialization of the devices and the last two after commercialization: identify clients and define possible use cases, search the clients' needs and objectives, determine how to supply the clients' needs and objectives, plan the adequate support for the clients' needs and objectives, define approaches for communication with the clients, and decide what and how to communicate to clients (risks, vulnerabilities, support, available resources, updates, etc.).

6   Related Work

Researchers have investigated IoT, presenting challenges and recommendations for specific application scenarios [19,41]. For instance, Uddin et al. [37] discussed the risks and vulnerabilities of IoT focusing on smart homes, smart cities, wearables, and connected cars. Besides, Taimoor and Rehman [35] discussed security threats regarding the IoT architecture, focusing on personalized healthcare Services. Vangala et al. [40] analyzed the security aspects of IoT-enabled smart agriculture.

Other researchers have focused on specific problems or threats, such as the use of software clones [43] and DDoS [29]. Besides, there are studies addressing security threats for specific technologies and architectures related to 5G and IoT. Sathish and Rubavathi [31] studied the security of IoT and the use of Blockchain applications. Celik et al. [2] analyzed security aspects of 5G device-to-device communication. Ramezan et al. [27] addressed the security of multi-hop networks. Another example of related work, conducted by Chen et al [3], focused on machine-to-machine communications in ultra-dense networks.

There are some broader discussions on the security of 5G and IoT [16,26,33, 42]. However, the literature still misses more studies addressing specific security aspects of the IoT architecture for 5G application scenarios. Most discussions focus on general security aspects of the IoT architecture with few (or none) considerations of the 5G architecture.

7   Conclusion

Given that 5G enables the massive connection of IoT devices, knowing the security concerns related to the IoT ecosystem, we decided to investigate the common known threats and the recommendations proposed by technical and research institutions. Thus, we analyzed some research papers and technical documents, gathering helpful information regarding the challenges for the 5G-IoT devices and applications.

We defined a threat model with six attack surfaces that can be adopted when investigating security problems in 5G-IoT applications. We described the main attack threats and classified them properly according to the STRIDE model and the CVSS system. This information can help the risk management process to deal with security concerns. Our study can support regulatory institutions worldwide in addressing security concerns and defining mitigation strategies, mainly in those countries in the early stages of the 5G mobile communication systems deployment.

For future work, we suggest conducting a systematic literature review to get a complete list of research papers and analyze the threats and mitigations considered. We are already performing this review and planning to develop a tool that gives a critical score for an IoT application based on known vulnerabilities for each employed technology. This way, the tool will search for vulnerabilities, considering specific databases, and calculates a critical score based on the CVSS of each vulnerability.

Acknowledgements. We thank the National Telecommunications Agency (Agência Nacional de Telecomunicações - ANATEL) for supporting this research. We also thank the Virtus Research, Development and Innovation Center and Embedded and Pervasive Computing Laboratory, Federal University of Campina Grande.

References

1. Ambareen, J., Prabhakar, M., Ara, T.: LEES: a hybrid lightweight elliptic ElGamal-Schnorr-based cryptography for secure D2D communications. J. Telecommun. Inf. Technol. **2**, 24–30 (2021)
2. Celik, A., Tetzner, J., Sinha, K., Matta, J.: 5G device-to-device communication security and multipath routing solutions. Appl. Netw. Sci. **4**(1), 1–24 (2019)
3. Chen, S., Ma, R., Chen, H.H., Zhang, H., Meng, W., Liu, J.: Machine-to-machine communications in ultra-dense networks - a survey. IEEE Commun. Surv. Tutor. **19**(3), 1478–1503 (2017)
4. ENISA: Baseline security recommendations for IoT in the context of critical information infrastructures. Technical report, ENISA (2017)

5. ETSI: EN 303 645 - cyber security for consumer Internet of Things: Baseline requirements. Technical report, ETSI (2020)
6. European Commission: Proposal for a Regulation of the European Parliament and of the Council on Horizontal Cybersecurity Requirements for Products with Digital Elements and Amending Regulation (EU) 2019/1020 (2022)
7. Fagan, M., Megas, K., Scarfone, K., Smith, M.: Foundational cybersecurity activities for IoT device manufacturers. Technical report, NIST (2020)
8. Figueroa-Lorenzo, S., Añorga, J., Arrizabalaga, S.: A survey of IIoT protocols: a measure of vulnerability risk analysis based on CVSS. ACM Comput. Surv. **53**(2), 1–53 (2020)
9. Pub, F.I.P.S.: Standards for Security Categorization of Federal Information and Information Systems. Federal Information Processing Standards Publication, Gaithersburg, MD, USA (2004)
10. Hou, F., Wang, Z., Tang, Y., Liu, Z.: Protecting integrity and confidentiality for data communication. In: International Symposium on Computers and Communications. IEEE, Alexandria (2004)
11. ITU-T - Telecommunication Standardization Sector of ITU: Overview of the Internet of Things (2012)
12. Hojlo, J.: Future of industry ecosystems: shared data and insights. IDC (2021)
13. Jung, Y., Agulto, R.: Virtual IP-based secure gatekeeper system for internet of things. Sensors **21**(1), 38 (2021)
14. Katz, J., Lindell, Y.: Introduction to Modern Cryptography, 2nd edn. CRC Press, Boca Raton (2014)
15. Kwon, S., Park, S., Cho, H., Park, Y., Kim, D., Yim, K.: Towards 5G-based IoT security analysis against Vo5G eavesdropping. Computing **103**(3), 425–447 (2021)
16. Li, S., Xu, L.D., Zhao, S.: 5G Internet of things: a survey. J. Ind. Inf. Integr. **10**, 1–9 (2018)
17. Li, X., Liu, S., Wu, F., Kumari, S., Rodrigues, J.J.P.C.: Privacy preserving data aggregation scheme for mobile edge computing assisted IoT applications. IEEE Internet Things J. **6**(3), 4755–4763 (2019)
18. Lopes, A.P.G., Hilgert, L.O., Gondim, P.R., Lloret, J.: Secret sharing-based authentication and key agreement protocol for machine-type communications. Int. J. Distrib. Sens. Netw. **15**(4) (2019)
19. Mogadem, M.M., Li, Y., Meheretie, D.L.: A survey on internet of energy security: related fields, challenges, threats and emerging technologies. Cluster Comput. **25**(4), 1–37 (2021)
20. Mohammad Hasan: State of IoT 2022: Number of connected IoT devices growing 18% to 14.4 billion globally (2022). https://iot-analytics.com/number-connected-iot-devices
21. Mostefa, B., Abdelkader, G.: A survey of wireless sensor network security in the context of internet of things. In: International Conference on Information and Communication Technologies for Disaster Management, pp. 1–8. IEEE, Münster (2017)
22. Mrabet, H., Belguith, S., Alhomoud, A., Jemai, A.: A survey of IoT security based on a layered architecture of sensing and data analysis. Sensors **20**(13), 3625 (2020)
23. Ogbodo, E.U., Abu-Mahfouz, A.M., Kurien, A.M.: A survey on 5G and LPWAN-IoT for improved smart cities and remote area applications: from the aspect of architecture and security. Sensors **22**(16), 6313 (2022)
24. Oyj, N.: Nokia deepfield network intelligence report DDoS in 2021. Technical report, Nokia (2022)

25. Qadri, Y.A., Ali, R., Musaddiq, A., Al-Turjman, F., Kim, D.W., Kim, S.W.: The limitations in the state-of-the-art counter-measures against the security threats in H-IoT. Cluster Comput. **23**(3), 2047–2065 (2020)
26. Rahimi, H., Zibaeenejad, A., Rajabzadeh, P., Safavi, A.A.: On the security of the 5G-IoT architecture. In: International Conference on Smart Cities and Internet of Things. ACM, Mashhad (2018)
27. Ramezan, G., Leung, C., Wang, Z.J.: A survey of secure routing protocols in multi-hop cellular networks. IEEE Commun. Surv. Tutor. **20**(4), 3510–3541 (2018)
28. Saleem, K., Alabduljabbar, G.M., Alrowais, N., Al-Muhtadi, J., Imran, M., Rodrigues, J.J.P.C.: Bio-inspired network security for 5G-enabled IoT applications. IEEE Access **8**, 229152–229160 (2020)
29. Salim, M.M., Rathore, S., Park, J.H.: Distributed denial of service attacks and its defenses in IoT: a survey. J. Supercomput. **76**(7), 5320–5363 (2019)
30. Sankar, S.P., Subash, T.D., Vishwanath, N., Geroge, D.E.: Security improvement in block chain technique enabled peer to peer network for beyond 5G and internet of things. Peer Peer Netw. Appl. **14**(1), 392–402 (2020)
31. Sathish, C., Rubavathi, C.Y.: A survey on blockchain mechanisms (BCM) based on internet of things (IoT) applications. Multimed. Tools Appl. **81**(23), 33419–33458 (2022)
32. Sharma, P., Jain, S., Gupta, S., Chamola, V.: Role of machine learning and deep learning in securing 5G-driven industrial IoT applications. Ad Hoc Netw. **123**, 102685 (2021)
33. Sicari, S., Rizzardi, A., Coen-Porisini, A.: 5G in the internet of things era: an overview on security and privacy challenges. Comput. Netw. **179**, 107345 (2020)
34. Swiderski, F., Snyder, W.: Threat Modeling. Microsoft Press, Redmond (2004)
35. Taimoor, N., Rehman, S.: Reliable and resilient AI and IoT-based personalised healthcare services: a survey. IEEE Access **10**, 535–563 (2022)
36. Tara Seals: IoT Attacks Skyrocket, Doubling in 6 Months. Threat Post (2021)
37. Uddin, H., et al.: IoT for 5G/B5G applications in smart homes, smart cities, wearables and connected cars. In: International Workshop on Computer Aided Modeling and Design of Communication Links and Networks. IEEE, Limassol (2019)
38. Ur-Rehman, A., Gondal, I., Kamruzzuman, J., Jolfaei, A.: Vulnerability modelling for hybrid IT systems. In: International Conference on Industrial Technology. IEEE, Melbourne (2019)
39. U.S. Government Publishing Office: H.R.1668 - IoT Cybersecurity Improvement Act of 2020 (2020)
40. Vangala, A., Das, A.K., Chamola, V., Korotaev, V., Rodrigues, J.J.P.C.: Security in IoT-enabled smart agriculture: architecture, security solutions and challenges. Cluster Comput., 1–24 (2022)
41. Varga, P., et al.: 5G support for industrial IoT applications - challenges, solutions, and research gaps. Sensors **20**(3), 828 (2020)
42. Wazid, M., Das, A.K., Shetty, S., Gope, P., Rodrigues, J.J.P.C.: Security in 5G-enabled internet of things communication: issues, challenges, and future research roadmap. IEEE Access **9**, 4466–4489 (2021)
43. Zhang, H., Sakurai, K.: A survey of software clone detection from security perspective. IEEE Access **9**, 48157–48173 (2021)
44. Šarac, M., Pavlović, N., Bacanin, N., Al-Turjman, F., Adamović, S.: Increasing privacy and security by integrating a blockchain secure interface into an IoT device security gateway architecture. Energy Rep. **7**, 8075–8082 (2021)

Author Index

L. Barolli (Ed.): AINA 2023, LNNS 661, pp. 575–577, 2023.
https://doi.org/10.1007/978-3-031-29056-5

Printed in the United States
by Baker & Taylor Publisher Services